Global Development of Religious Tourism

Emilia Alaverdov
Georgian Technical University, Georgia

Muhammad Waseem Bari
Government College University, Faisalabad, Pakistan

IGI Global
PUBLISHER of TIMELY KNOWLEDGE

A volume in the Advances in Hospitality, Tourism, and the Services Industry (AHTSI) Book Series

Published in the United States of America by
IGI Global
Business Science Reference (an imprint of IGI Global)
701 E. Chocolate Avenue
Hershey PA, USA 17033
Tel: 717-533-8845
Fax: 717-533-8661
E-mail: cust@igi-global.com
Web site: http://www.igi-global.com

Library of Congress Cataloging-in-Publication Data

Names: Alaverdov, Emilia, 1978- editor. | Bari, Muhammad Waseem, 1980- editor.
Title: Global development of religious tourism / Emilia Alaverdov and Muhammad Waseem Bari, Editors.
Description: Hershey, PA : Information Science Reference, [2021] | Includes bibliographical references and index. | Summary: "This book looks at religious tourism by discussing the traditional cultures, the history of religions, national religious traditions, rituals, through the study of geography, statistics and development trends of religious tourism in the world"-- Provided by publisher.
Identifiers: LCCN 2020018673 (print) | LCCN 2020018674 (ebook) | ISBN 9781799857921 (hardcover) | ISBN 9781799857938 (paperback) | ISBN 9781799857945 (ebook)
Subjects: LCSH: Tourism--Religious aspects. | Tourism--Social aspects. | Tourism--Marketing.
Classification: LCC G156.5.R44 G58 2021 (print) | LCC G156.5.R44 (ebook) | DDC 338.4/791--dc23
LC record available at https://lccn.loc.gov/2020018673
LC ebook record available at https://lccn.loc.gov/2020018674

This book is published in the IGI Global book series Advances in Hospitality, Tourism, and the Services Industry (AHTSI) (ISSN: 2475-6547; eISSN: 2475-6555)

British Cataloguing in Publication Data
A Cataloguing in Publication record for this book is available from the British Library.

All work contributed to this book is new, previously-unpublished material. The views expressed in this book are those of the authors, but not necessarily of the publisher.

For electronic access to this publication, please contact: eresources@igi-global.com.

Advances in Hospitality, Tourism, and the Services Industry (AHTSI) Book Series

Maximiliano Korstanje
University of Palermo, Argentina

ISSN:2475-6547
EISSN:2475-6555

MISSION

Globally, the hospitality, travel, tourism, and services industries generate a significant percentage of revenue and represent a large portion of the business world. Even in tough economic times, these industries thrive as individuals continue to spend on leisure and recreation activities as well as services.

The Advances in Hospitality, Tourism, and the Services Industry (AHTSI) book series offers diverse publications relating to the management, promotion, and profitability of the leisure, recreation, and services industries. Highlighting current research pertaining to various topics within the realm of hospitality, travel, tourism, and services management, the titles found within the AHTSI book series are pertinent to the research and professional needs of managers, business practitioners, researchers, and upper-level students studying in the field.

COVERAGE

- Sustainable Tourism
- Hotel Management
- Service Management
- International Tourism
- Leisure & Business Travel
- Casino Management
- Customer Service Issues
- Service Design
- Food and Beverage Management
- Travel Agency Management

IGI Global is currently accepting manuscripts for publication within this series. To submit a proposal for a volume in this series, please contact our Acquisition Editors at Acquisitions@igi-global.com or visit: http://www.igi-global.com/publish/.

The Advances in Hospitality, Tourism, and the Services Industry (AHTSI) Book Series (ISSN 2475-6547) is published by IGI Global, 701 E. Chocolate Avenue, Hershey, PA 17033-1240, USA, www.igi-global.com. This series is composed of titles available for purchase individually; each title is edited to be contextually exclusive from any other title within the series. For pricing and ordering information please visit http://www.igi-global.com/book-series/advances-hospitality-tourism-services-industry/121014. Postmaster: Send all address changes to above address.

Titles in this Series

For a list of additional titles in this series, please visit:
http://www.igi-global.com/book-series/advances-hospitality-tourism-services-industry/121014

Handbook of Research on the Role of Tourism in Achieving Sustainable Development Goals
Filipa Brandão (GOVCOPP, University of Aveiro, Portugal) Zélia Breda (GOVCOPP, University of Aveiro, Portugal) Rui Costa (GOVCOPP, University of Aveiro, Portugal) and Carlos Costa (GOVCOPP, University of Aveiro, Portugal)
Business Science Reference • © 2021 • 385pp • H/C (ISBN: 9781799856917) • US $285.00

Role and Impact of Tourism in Peacebuilding and Conflict Transformation
Jorge Tavares da Silva (GOVCOPP, University of Aveiro, Portugal) Zélia Breda (GOVCOPP, University of Aveiro, Portugal) and Fabio Carbone (Centre for Trust, Peace, and Social Relations, Coventry University, UK & GOVCOPP, University of Aveiro, Portugal)
Business Science Reference • © 2021 • 402pp • H/C (ISBN: 9781799850533) • US $195.00

Strategies for Promoting Sustainable Hospitality and Tourism Services
Maximiliano Emanuel Korstanje (University of Palermo, Argentina) Babu George (Fort Hays State University, USA) and Alexandru-Mircea Nedelea (Stefan cel Mare University of Suceava, Romania)
Business Science Reference • © 2020 • 243pp • H/C (ISBN: 9781799843306) • US $195.00

Handbook of Research on Resident and Tourist Perspectives on Travel Destinations
Patrícia Pinto (Faculty of Economics and Research Centre for Tourism, Sustainability, and Well-Being (CinTurs), University of Algarve, Portugal) and Manuela Guerreiro (Faculty of Economics and Research Centre for Tourism, Sustainability, and Well-Being (CinTurs), University of Algarve, Portugal)
Business Science Reference • © 2020 • 529pp • H/C (ISBN: 9781799831563) • US $285.00

Global Niche Tourism and Opportunities for the Medical Tourism Industry
Manjeet Singh (Central University of Jammu, India) and Subbaraman Kumaran (Alliance University Bangalore, India)
Business Science Reference • © 2020 • 300pp • H/C (ISBN: 9781799834274) • US $215.00

Cases on Tour Guide Practices for Alternative Tourism
Gulsun Yildirim (Recep Tayyip Erdogan University, Turkey) Ozlem Ozbek (Bandirma Onyedi Eylul University, Turkey) Ceyhun Caglar Kilinc (Akdeniz University, Turkey) and Abdullah Tarinc (Akdeniz University, Turkey)
Business Science Reference • © 2020 • 351pp • H/C (ISBN: 9781799837251) • US $245.00

IGI Global
PUBLISHER of TIMELY KNOWLEDGE

701 East Chocolate Avenue, Hershey, PA 17033, USA
Tel: 717-533-8845 x100 • Fax: 717-533-8661
E-Mail: cust@igi-global.com • www.igi-global.com

Table of Contents

Preface xv

Chapter 1
Pakistan as a Destination of Religious Tourism 1
Muhammad Waseem Bari, Government College University, Faisalabad, Pakistan
Qurrahtulain Khan, University of Kotli AJK, Pakistan

Chapter 2
The Development of Islamic Knowledge and Tourism in the Modern Life of North Caucasus 11
Emilia Alaverdov, Georgian Technical University, Georgia

Chapter 3
Introducing the Cities of Ghazali in the Context of Religious Tourism 25
Aytekin Demircioğlu, Kastamonu University, Turkey

Chapter 4
Religious and Spiritual Tourism: From Its Origins to Alentejo (Portugal) 44
Bruno Barbosa Sousa, Polytechnic Institute of Cávado and Ave, Portugal & CiTUR, Portugal
Cristina Castro, Polytechnic Institute of Cávado and Ave, Portugal
Maria Emília Luís, Polytechnic Institute of Cávado and Ave, Portugal
Paulo Lopes, Polytechnic Institute of Cávado and Ave, Portugal

Chapter 5
The Technological Power of Mysticism: A New Approach to Management of Religious Destinations 65
Eda Avci, Efes Vocational School, Department of Tourism and Hotel Management, Dokuz Eylul University, Turkey
Gizem Kayar, Computer Engineering Department, Muğla Sıtkı Koçman University, Turkey

Chapter 6
Comparative Analysis of the Communication in Western and Eastern Culture: Philosophical and Religious Aspects 83
Tamara Stanislavovna Olenich, Don State Tehnical University, Russia
Andrey Aleksandrowich Mekushkin, Don State Technical University, Russia
Olesya Anatolewna Shestopalova-Todarchuk, Don State Technical University, Russia

Chapter 7
The Influence of Religion on Political Parties of the European Union 98
Kateryna Tryma, Institute of Higher Education, NAESU, Ukraine
Natalia Salnikova, Donetsk Law Institute, Ministry of Internal Affairs of Ukraine, Ukraine

Chapter 8
The Role of Religion and Culture in Modern Day Society: Religion as a Business With Geo-Political Mixtures of Faith 113
A. T. M. Abdullahel Shafi, Independent Researcher, Bangladesh

Chapter 9
Challenges of Religious Tourism in the Conflict Region: An Example of Jerusalem 135
Salome Pipia, Ivane Javakhishvili Tbilisi State University, Georgia & Georgian Technical University, Georgia
Saba Pipia, Ivane Javakhishvili Tbilisi State University, Georgia

Chapter 10
Promoting Religious Tourism Management for Creating a Soft Image of Pakistan 149
Kalsoom BeBe Sumra, COMSATS University, Pakistan
Muhammad Mehtab Alam, Riphah Institute of Public Policy, Riphah University, Pakistan

Chapter 11
Mystical Tibet Today: Growing Tourist Route for Buddhist Religion and Knowledge Seekers and Pilgrims 175
Madlein V. Kotsotsashvili, Georgian Teqnical University, Georgia

Chapter 12
The Opportunities for the Development of Religious Tourism in Georgia 192
Manana Darchashvili, Institute of Political Science, Georgian Technical University, Georgia

Chapter 13
Media and Religious Tourism: A Thematic Review 210
Rizwan Shabbir, Government College University, Faisalabad, Pakistan
Aysha Batool, Government College University, Faisalabad, Pakistan

Chapter 14
Religious Tourism in Zimbabwe: A Stakeholders` Perspective 232
Zibanai Zhou, Midlands State University, Zimbabwe

Chapter 15
Use of Social Media by Tourists at Religious Tourism Destinations in India 250
Lakhvinder Singh, Government College, Kaithal, India
Dinesh Dhankhar, Kurukshetra University, India

Chapter 16
Religious Tourism: Sanctuary Cities as the Primary Destinations 268
Luciana Aparecida Barbieri da Rosa, Universidade Federal de Santa Mari, Brazil
Maria Carolina Martins Rodrigues, CinTurs, Portugal
Waleska Yone Yamakawa Zavatti Campos, IAG Business School, Pontifical Catholic University of Rio de Janeiro, Brazil
Maria José Sousa, ISCTE, Instituto Universitário de Lisboa, Portugal
Martius Vicente Rodriguez y Rodriguez, Universidade Federal Fluminense, Brazil
Larissa Cristina Barbieri, Faculdade de Educação São Luis, Brazil

Chapter 17
Religious Tourism in Africa's Global South: Indigenous African Traditional Spirituality 283
Unathi Sonwabile Henama, Department of Tourism Management, Tshwane University of Technology, South Africa
Lwazi Apleni, Department of Recreation and Tourism, University of Zululand, South Africa

Chapter 18
Types of Religious Tourism 297
Aniqa Ijaz, University of Okara, Pakistan

Compilation of References 310

About the Contributors 356

Index 363

Detailed Table of Contents

Preface xv

Chapter 1
Pakistan as a Destination of Religious Tourism 1
Muhammad Waseem Bari, Government College University, Faisalabad, Pakistan
Qurrahtulain Khan, University of Kotli AJK, Pakistan

The objective of this chapter is to highlight the religious tourism perspective of Pakistan. This chapter sheds light on five different aspects of religious tourism in Pakistan: first, the background of tourism; second, the history of tourism; third, destinations of religious tourism in Pakistan; fourth, development of religious tourism in Pakistan; and fifth, the role of religious tourism in the socio-economic and political development of Pakistan. This chapter presents a brief history of religious tourism, famous religious places, especially religious places of Hindus, Sikhs, and Buddhists in Pakistan. In addition, this chapter highlights different historical and religious places that are sacred to Muslims such as mosques and shrines of seers and Sufi saints. The initiatives of the Pakistan government to promote religious tourism and its role in the socio-economic and political development of is also discussed in this chapter.

Chapter 2
The Development of Islamic Knowledge and Tourism in the Modern Life of North Caucasus 11
Emilia Alaverdov, Georgian Technical University, Georgia

The chapter examines the place and role of Islam in the socio-political life of the peoples in the North Caucasus, especially in the Chechen Republic, as well as relations between the state and religious organizations and society, and examines the role of Islam in the political and socio-spiritual life of the region. The work analyzes the struggles of Chechen people for independence, which were called the Chechen wars and given the cliché of Jihad, reflected in the revival of political organizations and activities. The chapter shows that having experienced two very dreadful wars it is still possible to build a strong, sovereign, safe republic with a very low criminal rate, and written a long stand strategy for the socio-economic development. It is worthy to say that the strategy has clearly defined basic priorities and development vector of the Chechen Republic until 2025 as a mechanism for their initial coordination, where the main priority in the tourism sector, since the republic has a great potential for it.

Chapter 3
Introducing the Cities of Ghazali in the Context of Religious Tourism 25
Aytekin Demircioğlu, Kastamonu University, Turkey

Ghazali is one of the most well-known scholars of the Islamic world that lived between 1058 and 1111. He was dedicated to knowledge in his short life. Ghazali made trips to take lessons and to teach. In this study, cities related to Ghazali were examined. These cities are Tus, where he was born and received his primary education; Curcan, where received Fiqh training; Nishabur, where he first studied as a student and then taught as a madrasa scholar. These three cities are located within the borders of Iran today. The fourth city is Baghdad, where he taught at the Nizamiye Madrasa. Later he was found in the cities of Damascus, Jerusalem, Al-Halil, respectively. He then went to the cities of Mecca and Medina to become a pilgrim. In this study, these nine cities, which are important for Muslims, were introduced as a religious tourism center with a descriptive method.

Chapter 4
Religious and Spiritual Tourism: From Its Origins to Alentejo (Portugal) 44
Bruno Barbosa Sousa, Polytechnic Institute of Cávado and Ave, Portugal & CiTUR, Portugal
Cristina Castro, Polytechnic Institute of Cávado and Ave, Portugal
Maria Emília Luís, Polytechnic Institute of Cávado and Ave, Portugal
Paulo Lopes, Polytechnic Institute of Cávado and Ave, Portugal

Over the years there has been an increase in competition among tourism destinations, leading to the need for a deeper understanding about the tourism realm, impact, and management. Tourism has expanded its scope, reflecting an increasing recognition in the academic community paralleled by the application of interdisciplinary concepts and methods. Indeed, research in tourism has been studying its various implications from a multitude of perspectives and with interdisciplinary insights. Areas in tourism research entail planning of tourism destinations, local development, environmental impact, territorial brand management, and tourist loyalty. With new motivations, new niches are created, and it is one of those niches that the authors address in this work, namely, spiritual tourism. Spiritual tourism derives from religious tourism because it is linked to spiritualism and is associated with wellness tourism. This manuscript aims to present the importance of spirituality in the context of tourism (Alentejo, Portugal). In the end, new lines of research will be presented for the future.

Chapter 5
The Technological Power of Mysticism: A New Approach to Management of Religious Destinations 65
Eda Avci, Efes Vocational School, Department of Tourism and Hotel Management, Dokuz Eylul University, Turkey
Gizem Kayar, Computer Engineering Department, Muğla Sıtkı Koçman University, Turkey

Technological advances have significantly influenced the tourism and the sector realities. Like other industries, in the past couple of years, the stakeholders witnessed a continuous upgrade in tourism. These technological upgrades vary from personal online travel planners to complicated immersive rooms. Tourism trend analyses, online agencies, mobile locator and interactive maps, AR/VR enhancements, virtual assistants and robots, IoT, big data, and blockchain are most popular fields that tourism industry integrate itself seamlessly. As the cities, destinations, and tourism become smarter every day, religious tourism must get its share from emerging technological benefits. Every year, millions visit various holy places. A significant amount of data can be collected from millions of pilgrims. If this data are analyzed accurately, destination management and experience enhancement can be performed much easier than today. The aim of this study is to discuss the motivation of religious tourists, and how to implement smart destination principles in religious sites.

Chapter 6
Comparative Analysis of the Communication in Western and Eastern Culture: Philosophical and Religious Aspects 83
Tamara Stanislavovna Olenich, Don State Tehnical University, Russia
Andrey Aleksandrowich Mekushkin, Don State Technical University, Russia
Olesya Anatolewna Shestopalova-Todarchuk, Don State Technical University, Russia

This chapter argues for peculiarities of intercultural communication with such a closed society as Iran. It starts from an interesting and comprehensive analysis of what is regarded to be 'intercultural communication' that is crucial important for the touristic sphere. To narrate the phenomenon on the intercultural communication within Iran, this chapter aims to reveal that though communication process within the Iranian society is based upon the tradition and religion, the country is tolerant and friendly to the foreigners. The necessity is for the people who visit the country to understand the difference between the Eastern and Western cultures and traditions. The communication in the Eastern society is based on the ancient customs and foundations, but the communication in the Western society is characterized by technological and rational knowledge of the world. Finally, this chapter attempts to propose a new understanding of communicational process in Iran, which will create possibilities for mutual understanding and cooperation with this country.

Chapter 7
The Influence of Religion on Political Parties of the European Union 98
Kateryna Tryma, Institute of Higher Education, NAESU, Ukraine
Natalia Salnikova, Donetsk Law Institute, Ministry of Internal Affairs of Ukraine, Ukraine

The chapter is aimed to carry out discourse analysis of theoretical approaches and documents that highlight the issues of transformation of party-political systems of the EU countries. The religious factor is one of the catalysts for changes in modern parties of the EU, which, being different in their ideological orientation, electoral base, and territorial character, are influenced by general development trends. An important result of the discourse analysis is the conclusion about the weakening of the traditional parties, which from the national level has become a tendency of the pan-European political process. The religious factor stimulates the formation of small parties, which performs a great role in the development of tourism in the EU. The small parties are considered to be the defenders of local cultural and religious identities. The specific position of small parties to the development of tourism is caused by the wish to follow the economic interest and by the need to defend the local identities.

Chapter 8
The Role of Religion and Culture in Modern Day Society: Religion as a Business With Geo-Political Mixtures of Faith 113
A. T. M. Abdullahel Shafi, Independent Researcher, Bangladesh

In modern day culture, religion is specifically a business, rather than a faith. After European renaissance, religion is seen as derivative of changes between effect and resistance which had forced social scientists to think about religion from different perspectives. Then with some formal mixtures of capitalism and religion, religion became a component of business, rather than just a belief. From an individual perspective to social, political, economic perspectives, it's true. Even from individual to national and international perspectives, religion is just a business concept in the geo-politics at the international, super-national, and even supra-national arena. Nothing has changed the religions; rather international geo-politics has transformed religions into their own interests sometimes as an evil and sometimes as a "divine angel," which can benefit their interests.

Chapter 9
Challenges of Religious Tourism in the Conflict Region: An Example of Jerusalem........................ 135
Salome Pipia, Ivane Javakhishvili Tbilisi State University, Georgia & Georgian Technical University, Georgia
Saba Pipia, Ivane Javakhishvili Tbilisi State University, Georgia

Pilgrimage has historically and traditionally been considered a physical journey, in order for one to know the truth and visit a place that has sacred and holy meaning. In the modern world, with the development of technology and the process of globalization, territorial boundaries are slowly shrinking, making traveling much easier, and hence, with traditional tourism, of course, religious tourism has also progressed. In the research, the authors consider the challenges that religious tourism has in the modern world, especially the case of Jerusalem. They mainly focus on two variables, legal status of Jerusalem and political instability as problems, which are associated with the development of tourism in a conflict environment. They hypothesis is that unlike conventional tourism, religious tourism has a much greater chance of overcoming various challenges and developing in a conflict-prone and volatile environment.

Chapter 10
Promoting Religious Tourism Management for Creating a Soft Image of Pakistan 149
Kalsoom BeBe Sumra, COMSATS University, Pakistan
Muhammad Mehtab Alam, Riphah Institute of Public Policy, Riphah University, Pakistan

The religious tourism management has been the subject of much attention in academic literature and has made a significant contribution in creating soft image in tourists' behavior. Pakistan has one of the best archaeological and religious historical sites in the world. Buddhist civilization at Taxila, Islamabad, Swat, and Takht Bai are of great interest for tourists from Japan, Korea, Thailand, and other countries. The aim of this chapter is to determine the promotion of religious tourism management for creating soft image in Pakistan. In this regard, initiatives of religious tourism by federal government are analyzed while incorporating input from easing immigration services, establishment of national tourism coordination board, infrastructure development, and pollution free environment, which leads to the soft image. An exploration of the nature of soft image requires examining the two key drivers: cognitive approach and the more recent cognitive-affective approach. Using quantitative method, various analysis techniques will be applied.

Chapter 11
Mystical Tibet Today: Growing Tourist Route for Buddhist Religion and Knowledge Seekers and Pilgrims 175
Madlein V. Kotsotsashvili, Georgian Teqnical University, Georgia

Tourism is one of the fastest-growing industries in the world. Currently, it provides 10% of the world's income and employs one-tenth of the total workforce. World Tourism Organisation (WTO) estimates that by 2020 1.6 billion people will travel around the globe and an estimated income of $1.5 billion will be generated from these activities. With all these financial outcomes, tourism can provide great opportunities for the local economies. The sacred geography of the Tibetan region and the sacred Mount Kailas combines the origins of the world's four religions. For several centuries Tibet was closed to foreign visitors and tourists. This is explained not only by the complex geographical location. Tibet's isolation was largely driven by both Tibetan rulers and Chinese policy.

Chapter 12
The Opportunities for the Development of Religious Tourism in Georgia 192
Manana Darchashvili, Institute of Political Science, Georgian Technical University, Georgia

In recent years, tourism in Georgia has been a priority area for the economy, and accordingly, special attention is paid to its elaboration, since it has a great impact on the stable development of society, improving the economic situation and meeting the spiritual needs. The research is devoted to a very relevant and at the same time specific type of tourism, religious tourism, pilgrimage, traveling due to religious beliefs. Georgia has very good potentials for the development of religious tourism, as it has numerous Sanctuaries, such as the robe of Christ, the cross of St. Nino, more than 12,000 churches and monasteries, and the cathedral, about 150 museums. In addition, to the Orthodox, the representatives of other religions can also visit their religious places, for instance Muslims can pray in the mosques, Jews in the synagogue, and Catholics in Old Catholic temples.

Chapter 13
Media and Religious Tourism: A Thematic Review 210
Rizwan Shabbir, Government College University, Faisalabad, Pakistan
Aysha Batool, Government College University, Faisalabad, Pakistan

This chapter aims at presenting a thematic analysis of 64 research articles on religious tourism published from 2009 to 2020 by adopting a systematic literature review method. The results indicate that prominent topics discussed were amplification of concepts, tourist motivations, and experience. The evolution of religious tourism concept through nexus of diverse scholarly terms, related domains, and religious practices needs a comprehensive literary debate to refine the subject. The socio-economic impact of religious tourism also calls for the enrichment of topics on scholarly and practical grounds. Issues such as host and tourist behavior, social interaction, visitor management, marketing components, and the role of media need to be explored for the progression of religious tourism in line with sustainability. The chapter contributes to the literature by proposing a three-dimensional model focusing on the role of media and potential research domains for further exploration.

Chapter 14
Religious Tourism in Zimbabwe: A Stakeholders` Perspective 232
Zibanai Zhou, Midlands State University, Zimbabwe

The impact of religious tourism on destination economies is clearly an important question for public policymakers wishing to develop a robust tourism industry. This chapter aimed to offer insights into the contemporary state of religious tourism in Zimbabwe. Using interpretive qualitative data obtained through observation and in-depth interviews conducted with 30 key informants purposively selected from religious tourism stakeholders, the study established that although teeming with bright prospects, as manifested through the availing of employment opportunities to communities around religious tourism sites, religious tourism in Zimbabwe remains on the back foot. Findings revealed that religious tourism is hamstrung by the absence of a proper regulatory framework, spike in stampede cases jeopardising the safety and security of pilgrims, congestion and environmental pollution. Policy recommendations were proffered.

Chapter 15
Use of Social Media by Tourists at Religious Tourism Destinations in India.................................... 250
Lakhvinder Singh, Government College, Kaithal, India
Dinesh Dhankhar, Kurukshetra University, India

The chapter assesses the social media usage by tourists during their visit to religious tourism destinations and evaluates rationale and challenges faced by religious tourism establishments and other stakeholders in using social media as tools for marketing of religious tourism. The study used secondary data sources for getting a theoretical understanding of social media and religious tourism. Primary data conveniently collected through field survey from tourists and tourism establishment at religious tourism destination to know the extent of social media usage. The chapter indicating increased use of social media by tourists in searching, planning, booking, and sharing experiences about different products and services. The study identifies a crucial role of social media in the marketing and development of religious tourism along with certain challenges related to using as a mechanism to promote destination. Finally, implication and recommendations for different stakeholders were given to ensure adequate presence on social in more effective manner.

Chapter 16
Religious Tourism: Sanctuary Cities as the Primary Destinations.. 268
Luciana Aparecida Barbieri da Rosa, Universidade Federal de Santa Mari, Brazil
Maria Carolina Martins Rodrigues, CinTurs, Portugal
Waleska Yone Yamakawa Zavatti Campos, IAG Business School, Pontifical Catholic University of Rio de Janeiro, Brazil
Maria José Sousa, ISCTE, Instituto Universitário de Lisboa, Portugal
Martius Vicente Rodriguez y Rodriguez, Universidade Federal Fluminense, Brazil
Larissa Cristina Barbieri, Faculdade de Educação São Luis, Brazil

The present study has as its general objective to analyze the differences in behavior and culture among religious tourists who visit sacred sites. The study is descriptive through research. The quantitative instrument was adapted of Irimiás and Michalkó and Irimiás, Mitev, and Michalko. The results show a heterogeneous sample that includes different age groups. Among the data it appears that the majority of respondents are female and have higher education. A large percentage were born in a religious cradle and thus are considered religious. In addition, a high number of religious tourism participants are part of a religious community. Thus, it is suggested for future research to expand the analyzed countries in order to compare results, with the aim of analyzing other contents different from those analyzed. Among other limitations of this study we have: the information presented is limited to the articles selected only with Brazilian and Portuguese tourists; the research was carried out with the snowball technique, being limited only to the sample for convenience.

Chapter 17
Religious Tourism in Africa's Global South: Indigenous African Traditional Spirituality................ 283
Unathi Sonwabile Henama, Department of Tourism Management, Tshwane University of Technology, South Africa
Lwazi Apleni, Department of Recreation and Tourism, University of Zululand, South Africa

International tourist arrivals are projected to surpass 1.8 billion by 2030 on the back of rapid growth in emerging tourism economies. Tourism has emerged as an economic messiah for a plethora of countries in Sub-Saharan Africa. It has emerged as a cost-effective means by which countries can diversify their economies, especially countries with low economies that depend on agricultural products to diversify their economies. Religious tourism can contribute to deeper economic benefit for a destination. The synthesis of literature adds to the paucity of academic gaze on religious tourism in Southern Africa. The synthesis takes the reader on a religious tourism journey that includes African spirituality, Pentecostal Christianity, and the interface between Africans spirituality and Christianity. These areas are neglected in the academic gaze and are outside the tourism beaten track, and these forms of religious tourism bring in much needed economic activities for areas on the tourism fringe.

Chapter 18
Types of Religious Tourism.. 297
Aniqa Ijaz, University of Okara, Pakistan

Religious tourism is considered as one of the interesting and oldest areas of research in the field of tourism. The process of observing and practicing the religious activity during the visit of any sacred place combined with the zeal of faith in the heart of practitioner is categorized as religious tourism. Many of the researchers have explained its different types in their own perspectives. Within the boundaries of modern scientific society, modern religious tourism can be classified into two major types. One is pilgrimage and the other one is intellectual learning excursions. The second main category has two branches. These intellectual learning excursions are based on two mottos of the tourists. One is regarding the scientific know-how, and the second one is regarding attending the religious-cultist events.

Compilation of References 310

About the Contributors 356

Index 363

Preface

One of the main aims to spread the religious tourism is that people are eager to the essence of dialogue of civilizations, thus considering the cognition process as a main course of improving society continues. The traditional values of different cultures are very vital for the development of harmonious society. Modern religious tourism is a combination of all the types of mass tourism, which is subject to commercialization and economic criteria for evaluating its effectiveness. Religious tourism contains all the main signs of tourism: planning, security, transportation, accommodation, food, cultural services, additional services, trade, escort, and others. It is important to highlight that a modern tourist choses the direction of religious destination by the motive to receive the information about the historical roots of ancient and current pilgrims.

The most ancient form of tourism is considered the pilgrimage, which has a great history and impact, since multiple religious cults and monuments have still been preserved from the ancient times and modern man is under its influence. It should be noted that in the 21st century, religious tourism is in great demand, and therefore, the theme of pilgrimage as a cultural phenomenon is becoming increasingly important. In this regards we can underline the world's two major religions and their trips to the sacred places: the trips of Muslim to Mecca and Medina and Christian to Jerusalem.

The book is a crucial reference that contains research on the current religious situation as well as the tourism industry and provides insights on their joint development. It is not possible to study any religious field without understanding the religion itself and its impact on any country's political and social system. Therefore, the work also examines the impact of religion and tourism on economic and social developments across the world. Highlighting topics that include sanctuary cities, religious tourism management, and religious tourism in regions that span Europe, Africa, the Middle East, and more, this book is targeted to managers, executives, planners, and other professionals in the tourism and hospitality industry; government officials; religious leaders; and researchers, academicians, and students working in the fields of tourism management, business management, information and communication sciences, administrative sciences and management, education, and social and political sciences.

The book covers such important topics as, the Development of Tourism in the Post Conflict Countries; it deals with the correlations of the Culture and Religion; shows example of Knowledge Sharing and Transfer, Mysticism, deals with the certain type of Mysticism, Spiritual Tourism, and gives the reader the picture Sanctuary Cities; shows the connection of Religion as Business, explains the essence of Religious Tourism Management, and the impact of social median toward the religious tourism. The book is targeted for a wide spectrum of readers: specialists in the field of tourism, religion, geography and regional studies, as well as a person interested in the mentioned topics.

ORGANIZATION OF THE BOOK

The book is organized into 18 chapters. A brief description of each of the chapters follows:

1. **Pakistan as a Destination of Religious Tourism:** The objective of this chapter is to highlight the religious tourism perspective of Pakistan and highlight five different aspects of religious tourism in Pakistan.
2. **The Development of Islamic Knowledge and Tourism in the Modern life of North Caucasus:** The chapter examines the place and role of Islam in the socio-political life of the peoples in the North Caucasus, especially in the Chechen republic. Describes the country's great potential for religious tourism.
3. **Introducing the Cities of Ghazali in the Context of Religious Tourism:** The chapter speaks about the religious cities seen with the eyes of the most well-known scholars of the Islamic world, Ghazali.
4. **Religious and Spiritual Tourism: From Its Origins to Alentejo, Portugal:** Describes the religious tourism in Alentejo, Portugal, and highlights the city's sacred places and the potential for tourism prosperity.
5. **The Technological Power of Mysticism: A New Approach to Management of Religious Destinations:** The chapter examines the technological advances that have significantly influenced the tourism sector.
6. **Comparative Analysis of the Communication in Western and Eastern Cultures: Philosophical and Religious Aspects:** This chapter argues for peculiarities of intercultural communication with such a closed society as Iran. It starts from an interesting and comprehensive analysis of what is regarded to be 'intercultural communication' that is crucially important for the touristic sphere.
7. **The Influence of Religion on Political Parties of the European Union:** The chapter is aimed to show the discourse analysis of theoretical approaches and documents that highlight the issues of transformation of party-political systems of the EU countries and their connection to the religious factor.
8. **The Role of Religion and Culture in Modern-Day Society: Religion as a Business With Geo-political Mixtures of Faith:** Describes the derivative of changes between effect and resistance which had forced social-scientists think to about religion from different perspectives after European renaissance.
9. **Challenges of Religious Tourism in the Conflict Region: An Example of Jerusalem:** Analyzes the challenges that religious tourism has in the modern world, in the case of Jerusalem. It is mainly focused on two variables: the legal status of Jerusalem and political instability as problems, which are associated with the development of tourism in a conflict environment.
10. **Promoting Religious Tourism Management for Creating Soft Image of Pakistan:** Aims to determine the promotion of religious tourism management for creating a soft image in Pakistan. In this regard, initiatives of religious tourism by the federal government are analyzed while incorporating input from easing immigration services, the establishment of national tourism coordination board, infrastructure development, and pollution-free environment which leads to the soft image.
11. **Mystical Tibet Today: Growing Tourist Route for Buddhist Religion and Knowledge Seekers and Pilgrims:** Shows the sacred geography of the Tibetan region and the sacred Mount Kailas - combines the origins of the world's four religions.

12. **The Opportunities for the Development of Religious Tourism in Georgia:** Is devoted to a very relevant and at the same time specific type of tourism - Religious tourism – Pilgrimage and describes the potential of Georgia for the development of religious tourism, as it has numerous Sanctuaries. Besides, to the Orthodoxy's, the representatives of other religions can also visit their religious places for instance.
13. **Media and Religious Tourism: A Thematic Review:** Chapter aims at presenting a thematic analysis of 64 research articles on religious tourism published from 2009 to 2020 by adopting a systematic literature review method. The chapter contributes to the literature by proposing a three-dimensional model focusing on the role of media and potential research domains for further exploration.
14. **Religious Tourism in Zimbabwe: A Stakeholder Perspective:** Aims to offer insights into the contemporary state of religious tourism in Zimbabwe. Using interpretive qualitative data obtained through observation and in-depth interviews conducted with 30 key informants purposively selected from religious tourism stakeholders.
15. **Use of Social Media by Tourists at Religious Tourism Destination in India:** Describes assesses of social media usage by tourists during their visit to religious tourism destinations and evaluates the rationale and challenges faced by religious tourism establishments and other stakeholders in using social media as tools for the marketing of religious tourism.
16. **Religious Tourism: Sanctuary Cities as the Primary Destinations:** The presented descriptive study has as its general objective to analyze the differences in behavior and culture among religious tourists who visit sacred sites.
17. **Religious Tourism in Africa's Global South Indigenous African Traditional Spirituality: Indigenous African Traditional Spirituality:** Takes the reader on a religious tourism journey that includes African spirituality, Pentecostal Christianity, and the interface between African spirituality and Christianity. These areas are neglected in the academic gaze and are outside the tourism beaten track, and these forms of religious tourism bring in much needed economic activities for areas on the tourism fringe.
18. **Types of Religious Tourism:** The chapter explains the difference between the two major types of religious tourism: Pilgrimage and Intellectual Learning Excursions.

Enjoy reading the book.

Chapter 1
Pakistan as a Destination of Religious Tourism

Muhammad Waseem Bari
https://orcid.org/0000-0003-2329-3857
Government College University, Faisalabad, Pakistan

Qurrahtulain Khan
University of Kotli AJK, Pakistan

ABSTRACT

The objective of this chapter is to highlight the religious tourism perspective of Pakistan. This chapter sheds light on five different aspects of religious tourism in Pakistan: first, the background of tourism; second, the history of tourism; third, destinations of religious tourism in Pakistan; fourth, development of religious tourism in Pakistan; and fifth, the role of religious tourism in the socio-economic and political development of Pakistan. This chapter presents a brief history of religious tourism, famous religious places, especially religious places of Hindus, Sikhs, and Buddhists in Pakistan. In addition, this chapter highlights different historical and religious places that are sacred to Muslims such as mosques and shrines of seers and Sufi saints. The initiatives of the Pakistan government to promote religious tourism and its role in the socio-economic and political development of is also discussed in this chapter.

INTRODUCTION

Religious tourism is an important part of the tourism industry. It is also called special interest tourism. Religious tourism usually related to the followers of specific faiths who visit sites that are considered holy sites. Religious sites are not only visited by the pilgrims but also visited by non-religious tourists since they have cultural, historical and religious significance. The most niche of the tourism industry is directly affected by economic instability but religious tourism is comparatively less affected by these economic issues because the reason for traveling is very robust and it plays a very important role in people's lives. So, money does not play a very vital role to stop from visiting a destination.

DOI: 10.4018/978-1-7998-5792-1.ch001

Religious tourism is the pioneering form of tourism which has begun almost with the dawn of humanity. From ancient times, religious destinations were not only a part of the cultural landscape but they also had become a vital factor in local marketing and prime parts of the economy of hosted destinations. Some of the religious places such as *Mecca in Saudi Arabia, Fatima in Portugal and Lourdes in France* attract millions of religious tourists, charity workers, missionaries and humanitarians creating a huge financial transaction in the destination. Religious tourism, in the literature of tourism, is, often described interchangeably as cultural heritage tourism, cultural tourism, cultural heritage tourism and spiritual tourism by different writers.

It is difficult to give a specific definition of religious tourism and the reason behind it is that cultural and religious travel are used synonymously. This is because the majority of cultural tourists visit religious destinations as a part of their itinerary and are thus referred to as religious tourists. The general meaning of religious tourism implies traveling with the prime motive of experiencing religious activities or the products they induce such as art, culture, traditions, and architecture. Religion and tourism are interlinked with each other. The pioneer form of tourism was in the form of visiting religious pilgrimages such as Christians traveling to Jerusalem and Muslims to Mecca. Religious pilgrimages have always been one of the dominant factors for motivating people to travel. Tracing back to the religious history it is found that major religions have globally stimulated tourism as a foundation for spreading their own religious beliefs.

Globally, Pakistan is very popular for three popular and big religions of the world, namely, Islam, Sikhism, and Buddhism. The Pakistan Tourism Development Corporation (PTDC) identified 480 locations for inbound tourism alone. Among them are 106 historical significance, 120 religions, and 26 religions and history. Additionally, it is likely to attract more religious visitors to Pakistan, including civilians and Buddhists. Recently, Pakistan opened the Central Gurdwara and the temple at Gurdwara Kartarpur Shrine in Punjab (Pakistan). Among the government's tourism and travel sector priorities, Pakistan ranks 112 out of 132 countries. However, the current government is focusing on this area by taking certain initiatives. The Khyber Pakhtunkhwa government plans to return more than 3,000 historical artifacts to attract tourists from the region. Moreover, the Khyber Pakhtunkhwa government recently launched the first mobile travel app to promote tourism in the region. The app includes information based on hotels, restaurants, service stations and roads that also will guide visitors. These initiatives can be promoted in other provinces to attract and encourage tourism. However, tourists find Mosques, Buddhist sits, churches, archeological sites of different civilizations that leave a great impact on each tourist to visit their sacred place in Pakistan.

Therefore, the objectives of this chapter are to highlight the religious tourism opportunities, destinations, and socio-economic development opportunities in Pakistan, and how Pakistan can improve its positive image through religious tourism.

History of Religious Tourism

History of humanity and traveling for religious fervor are going side by side from ancient times. Thus, religious traveling starts from the moment when people have started their journeys because of a question of faith, and belief. It starts with a religious demonstration, "the pilgrimage", where "the pilgrim" is referred to as tourist or traveler with a religious motivation. Generally, "religious tourism is any trip motivated, either exclusively or partly, by religious reasons" (Rinschede, 1992). It is broadly recognized that the most of the religious tourism is a performance of salvation and worship, except the Jews and Muslims who make their religious travel to Jerusalem and Mecca, respectively as obligatory in principle.

The majority of modern researchers agree that "Jesus" was the old historic religious figure that was born around 5 BC in the city of Nazareth, Israel. After that, there was binging of Christianity and in the 4th century, this religion becomes the leading religion in the Roman Empire. At that time, it is believed that the first pilgrimage began from the Roman Empire to places associated with Jesus. Then the history saw the rise of Islam, in the 6th and 7th century and the city of Mecca (today in Saudi Arabia) became a central point for Muslims. Muslim form all over the world travel to Mecca to make a pilgrimage (Hajj) and every year more than 15 million visits the holy city of Mecca. Hinduism is considered among the oldest world religion, including Islam and Christianity. The history of Hindus is of more than 3000 years a have the largest followers in the world. Hindus from all over the globe travel to the city "Varanasi" the holiest city in Hinduism.

Every single pilgrimage is a very distinct type of tourism: one geared towards spiritual betterment and motivated by faith. The youngest religion in the world is Sikhism, which started from the subcontinent. The founder of Sikhism was Guru Nanak and he introduced Sikhism in the 15th century in the Punjab, northern side of the Indian subcontinent. The holiest place of the Sikhism is known as Golden Temple located in Punjab (India). Nankana Sahib is another important religious site in Sikhism. Nankana Sahib is located 91 KM west of Lahore, Pakistan. Throughout the year, millions of Sikhs visit these holy places from all over the world, especially from India and western countries.

Destinations of Religious Tourism in Pakistan

Nowadays, billions of people are traveling globally and more than 50% of these trips are directed towards developing countries of the world. This global trend shows a rapid increase in tourism and Pakistan is also getting equal benefits from the same tendency at present. This increasing trend of traveling or tourism has numerous benefits and the most prominent benefit of tourism is economic growth. Religion is a significant demographic characteristic that has a foremost influence on traveler's choice and liking about a specific destination. (Raj & Morpeth, 2007) have defined religious tourism "as a distinctive type of cultural tourism rooted in religious beliefs, environments and encompasses specific actions e.g. venerating, religious sightseeing, praying, etc. by religious believers". Pakistan is full of such religious destinations with historical heritage, which not only represents the Islamic culture but also offers equally for other religions. Historical mosques, Shrines of seers and saints, temples, Gurdwaras, Buddhist monasteries, and several other monuments are located in the territory of Pakistan, which attracts the worshipers to make their sanctified journeys. Pakistan is taking serious steps to promote its religious tourism destinations because of its positive impact on the economy. Following are some famous religious destinations for which worshipers are always excited to visit:

Religions Tourism Destinations in Pakistan (Non-Muslims)

Katas Raj Temples

Katas Raj Temple/Mandir is a holy place for Hindus from the last 1500 years, located in the district Chalwal, Punjab Pakistan. There are several Mandirs/Temples at the same place that offers great allures to the Hindus. The place Katas holds the "Satgraha or set of seven old temples". In Katas Raj Temple there is also a Holy Pond for which Hindus believes that bathing in that water pool will wash away their sins (Akram, 2015; Babary & Zeeshan).

Choorio Mandir, Nagarparkar

Choorio Mandir is also a religious destination for Hindus. A famous Hindu mandir/temple "Kali Mai" is located in Choorio Village near the Pakistan-India border, Tharparkar, Sindh. Hindu worshipers from all around the world visit Choorio Mandir for their religious worships (SHAH, 2014).

Hinglaj Goddess at Hinglaj Mata Mandir

Hinglaj Mata Mandir is a Hindu temple situated in Hinglaj, Baluchistan that is constructed in the memory of a Hindu idol "Hingula Devi". Many Hindu pilgrims visit this cite to offer their ceremonials from all over the world.

Sharda Peeth

Sharada Peeth is also a reconstructed and antique Hindu temple situated in Sharda, District Neelam, Pakistan Occupied Jammu & Kashmir. In the Indian subcontinent, it was considered as the most protuberant temple university (Punjabi, 2018).

Jaulian Buddhist Monastery

Jaulian Buddhist Monastery is a reconstructed Buddhist religious place of 2nd century CE, situated in District Haripur, Khyber Pakhtunkhwa, Pakistan. Buddhists from all over the world visit this place for their spiritual calm and gratification (Khan, Arif, Butt, Pervaiz, & Arif, 2014).

Dharmarajika Stupa

The Dharmarajika Stupa "a dome-shaped shrine founded by Buddhists" is another sacred place for Buddhist people from 2nd century CE, situated near Taxila, Pakistan (Aslam & Bukhari, 2018).

Kargah Buddha

Another famous religious archaeological site for Buddhists is "Kargah Buddha" located in Gilgit, Gilgit-Baltistan, which is a 50 feet high engraved image of a Buddha. The statue is most likely belongs to the era of the 7th century (Melzer, 2014).

Buddhist Ruins of Takht-i-Bahi

The "Buddhist Ruins of Takht-i-Bahi (Throne of Origins)" is a Suren Kingdom archaeological place of a primordial Buddhist monastery complex, established in the early era of the first century. This Indo-Parthian archaeological place is located in Takht-i-Bahi District Mardan, Khyber Pakhtunkhwa province of Pakistan. This monastic complex is enormously situated on several hilltops, covering the 33ha area, characteristically famous for Buddhist holy places (Iqbal, Khan, Haq, & Hesseln, 2017; Zhang, 2018).

Gurdwara Nankana Sahib

Gurdwara Nankana Sahib is the first and foremost religious place of Sikh religion, where the founder of Sikhism "Baba Guru Nanak" was born. In the memory of their religious forefather "Baba Guru Nanak", thousands of Sikhs, yearly perform their pilgrimage. Gurdwara Nankana Sahib Is also known as "Gurdwara Janam Asthan" that is situated in district Nankana Sahib, Punjab, Pakistan (Abbasi, 2018; Singha, 2000). The Gurdwara Nankana Sahib was built in the 19th century by Maharaja Ranjit Singh.

Gurdwara Darbar Sahib Kartarpur

"Gurdwara Darbar Sahib Kartarpur", also known as "Kartarpur Sahib", is a religious worship place or gurdwara of Sikh religion in Kartarpur village (founded by Sikh Saint Baba Guru Nanak, in 1515) situated in Shakargarh, Narowal District, of Punjab Pakistan. This is a place that is very near to Pak-India international border and Pakistan has recently built the "Kartarpur Corridor" a visa-free to facilitate the followers of Sikhism (Ashraf, 2018; Kamboh, Kamal, Nisar, & Kamboh, 2018). This place is considered as the 2nd holiest place of Sikh religion, after Gurdwara Nankana Sahib, as they believe it as the death place of Baba Guru Nanak.

Religious Tourism Destinations in Pakistan (Muslims)

There are numerous religious destinations of Muslims in Pakistan, such as historical mosques and darbars/shrines of Sufi saints, and preachers of Islam. These historical and spiritual places are spread across Pakistan. Mausoleums of Shah Rukn-e-Alam, Shah Shamas Tabriz, and Hazrat Bahuddin Zakriya are the most spiritual places in Multan, Pakistan. The shrine of Baba Buley Shah is located in district Qasoor, Punjab, Pakistan. In Lahore, Pakistan, mausoleums of Hazrat Data Ganj Baksh, Hazrat Mian Mir, and Bibi Paak Daaman are the most famous religious sites. Shrines of Mangho Pir and Abdullah Shah Ghazi are located in Karachi, Pakistan. In the province of Sindh, Pakistan, Mausoleums of Shah Abdul Latif Bhitai and Hazrat Lal Shahbaz Qalandar are the places where people get their spiritual calm. In Khyber Pakhtunkhwa Province the shrine of Rehman Baba is also considered as a religious place. Moreover, the historical mosques in Pakistan such as Badshahi Mosque Lahore, Faisal Mosque, Islamabad, Abbasi Mosque, Bahawalpur, and Tooba Mosque, Karachi are famous for some state of the art construction, designs, and represent great Muslim legacy of the Indian subcontinent.

Development of Religious Tourism in Pakistan

"The British Backpacker Society," in 2018, has rated Pakistan as the "top adventure travel destination" in the world, and labeled the country as "one of the friendliest countries on earth, with mountain scenery that is beyond anyone's wildest imagination." In 2019, Pakistan is also ranked by Forbes as "one of the coolest places to visit." Pakistan has also been placed in the top 25% of world heritage locations by the "World Economic Forum's Travel & Tourism Competitiveness Report" in 2015 (Crotti & Misrahi, 2015). Pakistan is rich in historical and religious holy sites for many religions other than Islam, especially for Sikhs and Buddhists. However, hardly six thousand Indian Sikhs come to Pakistan every year that trails far behind the Golden Temple where visitors are more than 50 million. According to a survey, from 8 million Sikhs living outside India, more than 83% are keen to visit Pakistan for their religious worships.

Moreover, from 20 million Indian Sikhs, more than 79% are interested to visit their holiest places in Pakistan. But unfortunately, from these huge potential visitors, hardly 10% visit Pakistan due to several political and security reasons (Baloch & Rehman, 2015).

Every year, religious tourism has more than a 40% share in the total revenues earned by global tourism. Pakistan has great potential for religious tourism as the country has some holiest sites for Buddhists, Hindus, and Sikhs. Historical and ancient landmarks such as "Indus Valley Civilization" are also very attractive tourism destinations. To promote "religious tourism" in Pakistan, the government is taking various steps (Arshad, Iqbal, & Shahbaz, 2018). A recent joint initiative of "Pak-India Kartarpur Corridor" was an important step to facilitate Sikh religious visitors to Pakistan. To attract more and more Sikh visitors and free visa policy was launched and 5000 Sikhs were allowed to visit Gurdwara Darbar Sahib in Kartarpur. This corridor will not only contribute to durable peace, good bilateral relations and promotion of tourism but also expected to improve trade and economic relationships among Pakistan and India. Pakistan is expecting to earn more than $36.5 million per year from Kartarpur Corridor (Saher, Ahmed, Faizan, & Burney).

China-Pak Economic Corridor (CPEC) is also a significant initiative towards the development of infrastructure and attraction of both Chinese and other international tourists. According to "Pakistan Tourism Development Corporation" (PTDC), CPEC is playing an important role in tourism development in Pakistan as it helps in improving tourism services, security situation, and transportation difficulties (Arshad et al., 2018). Furthermore, CPEC will stimulate ecotourism in the north region of Pakistan, which are considered as the most beautiful areas on earth due to their wonderful meadows, beautiful mountains, green valleys, beautiful glaciers, and holy places for Buddhists that have the potential to become tourism hub in Asia (Manzoor & Wei, 2018).

Pakistan is also taking initiative to attract and encourage the Buddhists religious tourists to visit their heritage (Zhang, 2018). For this purpose, the KP government has launched a mobile app that will guide the tourists about hotels, rest-houses, service stations, roads, and restaurants. To encourage more than twenty industries related to the tourism industry, the government of Punjab province has also introduced a travel policy, and under the "Punjab travel plan," the federal government will invest $50 million to develop tourism facilities of international standards. To revive the tourism industry in Pakistan, in March 2020, the government had announced a visa-policy that allows the citizens of 175 countries to apply online visit visas to Pakistan. It is also a plan of the government to offer on arrival visas for citizens of more than 50 countries in June 2020. According to the interior ministry of Pakistan, the latest online visa policy has brought significant and satisfactory results as "44,705 applications received via Pakistan Online Visa System."

The security situation also plays an important role in bringing tourism to the country. Since 2015, there is a significant improvement in the security situation and a major decrease in terror-related threats. Many western countries have changed their travel advisory about Pakistan, and now more people in the world are willing to travel. The World Bank in its recent report on tourism has also recommended that every year Pakistan can attract thousands of tourists just by ensuring security, improving tourist infrastructure, and easing the visa-policy. As more than 83% of Sikh migrants settled in Canada, the UK, and the USA, are eager to visit their holiest religious places in Pakistan but visa application complications, security situations, and poor infrastructure are the major barriers (Muhammad et al., 2019). The World Bank is also interested in a project of $50 million for tourism development in Pakistan (Carrillo-Hidalgo & Pulido-Fernández, 2020). To revive and boost up the tourism industry, the government as decided to finish the previously mandatory condition of NOC (no objection certificate) to visit certain places in Pakistan, which will have a positive impact on tourism attraction.

Role of Religious Tourism in Socio-Economic and Political Development of Pakistan

Despite important religious destinations of different religions, Pakistan's tourism industry is still considered for behind the other regional countries like Sri Lanka, Turkey, and India (ul Haq & Ullah, 2019). There are several legal, political, and security reasons behind this deficit. Relaxed visa policy, minimizing visa restrictions, improvement in security, and development of infrastructure can attract more Sikh visitors than can lead to numerous economic benefits. For example, permitting a one-day visa/trip for Indian Sikhs to Nankana Sahib, just two-hours' drive-by road from Indian border Wagah, can increase tourists of Indian pilgrims (A. U. Khan, Zaman, & Baloch, 2019). The economic impact of this increase could be approximately increased from the current 208 million Pak rupees (Rs.) to Rs. 18 billion. This is assumed with the current spending (direct expenses on transport and hotels) of annual 44 thousand Sikh visitors (Rs. 4000 to Rs. 20000/per day per person).

However, the economic gains are not limited to these two direct expenditures only (hotel and traveling), as these generate triple economic effects through direct, induced and indirect expenditures. Induced effects include the enhancement of business activities and employment due to increased tourists. The rise in indirect expenditures occurs when purchases of hotels from their suppliers increase and also enlarge their facilities due to the increased number of visitors. All these expenditures provide a boost to the economic activities in tourist hubs. According to the estimate of the "World Travel and Tourism Council" (WTTC), every single rupee that is spent into direct expenditures is likely to result in Rs. 1.46 extra spending. Moreover, WTTC estimated that each direct job in the tourism industry can result in more 1.55 jobs in other linked industries. If we roughly calculate these multiplier effects, only Sikh visitors can contribute approximately Rs. 44 billion each year and 82 thousand jobs as well.

Likewise, Buddhist tourism is also a big potential market for the Pakistan tourism industry. The total population of Buddhists all around the world is about 500 million, and the Gandhara region of Pakistan including Taxila, Mardan, and Sawat embraces holy religious places for them (Griffin, Raj, & Yasuda, 2018). According to the survey of Gallup in 2016, the total population of Buddhists across Japan, Korea, and China is 58 million from which 5% (2.9 million) were interested to visit Pakistan. But lack of proper tourism marketing, poor infrastructure and lack of tourist facilities, and unstable security situation are presently averting this perspective from being grasped. If Pakistan can attract just 1% of interested Buddhists to visit Pakistan, that will mean 29 thousand visitors per year, with expected revenue inflow of more than $63 million. By calculating the effects of direct and indirect expenditures, Buddhist tourists can contribute more than 16 million rupees to gross domestic product (GDP) of Pakistan along with creating employment opportunities for more than 30 thousand Pakistanis.

Similarly, Pakistan also holds some special places for Hindus as well, which are nearly 1.1 billion almost 15 percent of the world's total population. If Pakistan attracts more Hindu tourists by providing facilities and proper security, this can boost the economy to implausible levels. According to the WTTC annual research report published in March 2020 about the statistics of 2019, the total contribution of tourism in GDP is 5.9% (Rs. 2,285.8 billion or $16,756.5 million). In 2019, the total tourism GDP was increased by 4.7% concerning a 3.3% rise in real economy GDP. The total share of tourism and traveling in employment is 6.2% (3,881,900 jobs). Moreover, the international visitor impact was Rs. 116.2 billion ($852.2 million) that is 3.1% of the total county exports. The research report of WTTC also reveals the inbound arrival ratios of all international tourists to Pakistan, which included 30% from the UK, 16% from the US, 8% from India, 8% from China, 5% from Canada, and 33% from rest of the world. The majority of the travelers 87% spend money for leisure purposes while 13% are business spending.

REFERENCES

Abbasi, A. S. (2018). 15 Nankana Sahib as 'A Symbol of Religious Coexistence'. *Religious Tourism in Asia: Tradition and Change Through Case Studies and Narratives*, 138.

Akram, M. (2015). *The Temples of Katasraj in Pakistan between Legend and History*. Academic Press.

Arshad, M. I., Iqbal, M. A., & Shahbaz, M. (2018). Pakistan tourism industry and challenges: A review. *Asia Pacific Journal of Tourism Research*, *23*(2), 121–132. doi:10.1080/10941665.2017.1410192

Ashraf, M. M. (2018). *Opening the Kartarpur Corridor*. Academic Press.

Aslam, N., & Bukhari, M. F. (2018). Dharmarajika Stupa: A Great Buddhist Heritage of Taxila (Pakistan). Journal of the Centre for Heritage Studies, *35*.

Babary, A., & Zeeshan, M. (n.d.). Reminiscent of Hinduism: An insight of Katas Raj Mandir. Explorer, 1(4), 121–124.

Baloch, Q. B., & Rehman, A. (2015). Regional integration of Pakistan tourism: Exploring prospects. *Abasyn University Journal of Social Sciences*, *8*(2), 405–415.

Carrillo-Hidalgo, I., & Pulido-Fernández, J. I. (2020). The Role of the World Bank in the Inclusive Financing of Tourism as an Instrument of Sustainable Development. *Sustainability*, *12*(1), 285. doi:10.3390u12010285

Crotti, R., & Misrahi, T. (2015). The travel & tourism competitiveness index 2015: T&T as a resilient contribution to national development. *The Travel & Tourism Competitiveness Report, 13*.

Griffin, K., Raj, R., & Yasuda, S. (2018). Religious tourism and sacred sites in Asia. *Religious Tourism in Asia: Tradition and Change Through Case Studies and Narratives*, 1.

Iqbal, J., Khan, Y., Haq, Z., & Hesseln, H. (2017). Estimation of Economic Value of an Archaeological Site: A Case Study of Takht-i Bahi. *Ancient Pakistan, 28*.

Kamboh, A. Z., Kamal, Y., Nisar, Q. A., & Kamboh, M. A. (2018). Kartarpur Corridor: A New Role for Political Marketing. *Journal of Indian Studies*, *4*(2), 233–238.

Khan, A. U., Zaman, G., & Baloch, Q. B. (2019). Promoting Religious Tourism in Pakistan: Empirical Evidence from Sikh Pilgrims Visiting Pakistan. *City University Research Journal, 9*(4).

Khan, M. A., Arif, S., Butt, A., Pervaiz, A., & Arif, M. (2014). Excavation at Badalpur Monastery, District Haripur (Khyber Pakhtunkhwa), Pakistan: A Preliminary Report of Season 2014. *Journal of Asian Civilizations*, *37*(1), 85.

Manzoor, F., & Wei, L. (2018). *A Review of China-Pakistan Economic Corridor and Tourism Opportunities.* Paper presented at the Third International Conference on Economic and Business Management (FEBM 2018). 10.2991/febm-18.2018.18

Melzer, G. (2014). A Palaeographic Study of a Buddhist Manuscript from the Gilgit Region. *Manuscript Cultures: Mapping the Field*, *1*, 227.

Muhammad, F., Karim, R., Qureshi, J. A., Razzaq, N., Zahra, M., & Ali, I. (2019). Environmental Degradation, Quality of Institutions and Tourism: New Evidence from Pakistan. *International Journal of Economic and Environmental Geology*, *10*(3), 125–128.

Punjabi, R. (2018). Kashmiriyat: The Mystique of an Ethnicity. *Himalayan and Central Asian Studies, 22*(4).

Raj, R., & Morpeth, N. D. (2007). *Religious tourism and pilgrimage festivals management. An international perspective*. Academic Press.

Rinschede, G. (1992). Forms of religious tourism. *Annals of Tourism Research*, *19*(1), 51–67. doi:10.1016/0160-7383(92)90106-Y

Saher, S., Ahmed, R., Faizan, M., & Burney, A. I. (n.d.). *Critical Analysis Relation between Indo-Pak: Kartarpur Corridor*. Academic Press.

Shah, D. I. (2014). The Development of Hindu Temples in Pakistan: An Historical and Archaeological Overview. Pakistan Historical Society. Journal of the Pakistan Historical Society, 62(4), 47.

Singha, H. (2000). *The Encyclopedia of Sikhism (over 1000 Entries)*. Hemkunt Press.

ul Haq, M. Z., & Ullah, R. (2019). Tourism demand function for Pakistan. Journal *of* Social Sciences & Humanities, 3(1 & 2), 1–17.

Zhang, J. (2018). To brand Gandhara, the Ancient Buddhist Centre: Pakistan's potential to develop Buddhist tourism. *Religious Tourism in Asia: Tradition and Change Through Case Studies and Narratives*, 127.

KEY TERMS AND DEFINITIONS

Pilgrim: A pilgrim refers to a traveler who is on a trip to a holy place. Traditionally, this is a physical trip to some site of special significance to the believer of a specific religious belief system. For instance, a Muslim who travel towards Makkah and Medina is considered as a pilgrim.

Religious Tourism: Religious tourism refers to the followers of specific faiths who visit sites that are considered holy sites for them. For instance, For Muslims in Makkah and Medina, for Sikhs is Golden Temple, for Christian are Jerusalem and Bethlehem.

Religious Tourism Destination: The places which are considered and visited for religious purposes are referred to as religious tourism destinations. For instance, Makkah, Medina, Golden Temple, Jerusalem, and Bethlehem, etc.

Socio-Economic Development: The process of social and economic development in a country or society refers to socio-economic development. Typically, socio-economic development is calculated with different indicators such as life expectancy, level of employment, inflation and gross domestic production.

Tourism: Tourism refers to the movement of people (both within and outside the national borders). Tourism is a multidimensional phenomenon and meanings for diverse things to different people as tourism is an abstraction of a varied range of consumption actions which require products and services from a varied range of businesses from the world.

Chapter 2
The Development of Islamic Knowledge and Tourism in the Modern Life of North Caucasus

Emilia Alaverdov
Georgian Technical University, Georgia

ABSTRACT

The chapter examines the place and role of Islam in the socio-political life of the peoples in the North Caucasus, especially in the Chechen Republic, as well as relations between the state and religious organizations and society, and examines the role of Islam in the political and socio-spiritual life of the region. The work analyzes the struggles of Chechen people for independence, which were called the Chechen wars and given the cliché of Jihad, reflected in the revival of political organizations and activities. The chapter shows that having experienced two very dreadful wars it is still possible to build a strong, sovereign, safe republic with a very low criminal rate, and written a long stand strategy for the socio-economic development. It is worthy to say that the strategy has clearly defined basic priorities and development vector of the Chechen Republic until 2025 as a mechanism for their initial coordination, where the main priority in the tourism sector, since the republic has a great potential for it.

INTRODUCTION

In recent years, in Russia, there has been a dynamic development of both domestic and foreign tourism. Many higher educational institutions of the country have begun training specialists in the field of tourism business, excursion business, etc. Manuals have been published on introducing tourism, management, marketing, advertising and insurance in tourism, organizing tourism activities, etc.

While talking about the religious tourism we have to consider: the nature and structure of religious tourism; the study of religion as a phenomenon of social life, its types and role in the life of society; study of the principles of the doctrine of the most common and famous theological schools; characteristics of the territories; description of holy places, religious centers, and pilgrimage routes; disclosure of the features of the practical organization of religious tourism.

DOI: 10.4018/978-1-7998-5792-1.ch002

In the XXI century, people have more opportunities to visit the holy places of their religion. The demand for religious tourism is due to a certain crisis of religions that have occurred in connection with the rapid changes in all aspects of human life in the past few decades.

Religious tourists and especially pilgrims embark on a journey when they need to do something more than religious actions in the usual environment of their residence. People visit holy places with different motives: to pray, solve personal problems, find a system of values, establish them on the spiritual path, and with the purpose of acquaintance with the cultural heritage of the country. On the Russian tourism market (since North Caucasus is a part of the Russian Federation), companies have already developed their practical skills in the organizations of trips excursionists for pilgrims and in the provision of services in the field of religious tourism.

Above I underlined that while talking about religion tourism we have to consider the characteristics of the territories of the spread of religions, its types, and role in the life of society, disclosure of the features of the practical organization of religious tourism. For this reason I assume that the topic "The Development of Islamic Knowledge and Tourism in the Modern life of North Caucasus" is very actual, vital and need to draw attention. Since the religion, mostly Islam is quite strictly followed and the Chechen republic situating in region has a very good potential for it.

Literary Review

The Monograph, "Twenty years of religious freedom in Russia" edited by A. Malashenko and S. Filatova analyzes various issues of the existence of religions in Russia after the collapse of the Soviet Union. Great importance is paid to the examples of Orthodoxy, Islam, and other faiths, as well as to their legislative and legal framework, and the relations between religion and the state, and to their actual practice. It examines the problems of religious "renaissance", the essence of the religion for the society, the activities of communities, and religious movements. The authors of the articles discuss the real religious freedom in the post-Soviet Russia and its constraining factors.

U. Halbach, claims that Islam is firmly rooted in the ethno-religious and cultural landscape of the region, he claims that it is an integral part of region's history, as well as the history of the Russian Federation - a multi-national and multi-confessional state. K. Khanbabaev, a famous researcher in the field of North Caucasus states that the fact that the first Muslim communities appeared in that region as early as the eighth century and quickly gained a mass of followers, merging into the life of local peoples has to be noted. Over the centuries, despite frequent conflicts, invasions, wars, Islam has not lost its unique impact on the ethnic identity of the Caucasian peoples. On the contrary, it is woven into local specifics, has absorbed part of local beliefs and traditions, and became a determining factor in distinguishing local followers of Allah.

Such researchers as A. Alikberov, V. Bobrovnikov, A. Bustanov are focused on Russian Islam in the frame of country's dimension, the formation of theological and legal schools, religious practices and social networks of Muslim communities from its historical times. The well-known researcher A. Yarlykapov explains the essence and specifics of Islamic education and highlights its challenges. The scholar highlights the challenges of historical and cultural brands formation of the regions as a resource of modern development, problems of language policy in the field of education, and the aspects of the religious situation. Another researcher, S. Markedonov presents a fundamental interdisciplinary study of the cultural complexity of Russian society its development potential and the possible risks. His context explains the historical and modern dynamics of the multi-ethnic composition of the population and the nature of interethnic relations, as well as the issues of civil and ethnic identity formation and the role of the state attitude and structure toward the existing processes.

The works of Islamic researcher V. Akaev are aimed at accounting and ensuring the rights and requests of Russian citizens in the field of ethno-cultural and religious relations. He highlights the attitude of religious and political elites in the Northern Caucasus and pays a great attention to the ideological formations, their contradictions and the correlations of religion in society.

The importance of the literature are enriched by such governmental documents as "Investor Guide to the Chechen Republic" and "Social and Economic Development Strategy Chechen Republic until 2025", where is described in detail the certain plan for the future development of the Chechen Republic in the sphere of tourism and economy.

Traditional and non-Traditional Islam in the Chechen Republic

In the numerous Islamic studies literature, a division is usually found in traditional and non-traditional Islam. In the North Caucasus, traditionally is followed Sufi Islam. The traditional Islam was widely and successfully spread in the North Caucasus until the beginning of the 20th century (Vidino, 2005). However, soon after the Bolshevik Revolution in 1917, the religious freedom came to the end and started the era of religious atheism. The modern revival of Islamic identity began in the Post Soviet territories shortly after the collapse of the Soviet Union and continues up today. Nowadays, there is a tendency towards the revival of traditional Islam, particularly, in the Chechen Republic; it is expressed in the increase in the number of mosque constructions, the spread of religious literature, Hajj to Mecca and Medina, restoration of the Islamic clerical elite, establishment of tight connections with foreign fellow believers (especially South Arabia). We have to highlight that, traditional religious (Sufi) values are used in the process of spiritual education of youth, counteraction to religious radicalism and extremism.

However, in addition to traditional Islam, alternative forms of the religion have been actively spreading in the North Caucasus since the early 1990s (Buttaeva, 2009). Numerous foreign preachers came to the region; young people went to Muslim countries to study in Islamic educational institutions. Representatives of traditional Islam - leaders of local religious administrations - found themselves in a situation of fierce competition with new Islamic movements, especially attractive for young people. These movements (usually associated with Salafism and Wahhabism) also attracted radicals and those who were opposed to the authorities.

In their study, Sitnikov, Romanov and Odaev, "Religiousness in the Chechen Republic and its Influence on Social and Political Institutions" tried to understand the extent of the traditional Islam and its knowledge in the Chechen Republic, and the number of residents following the different branches of Islam. It appeared that only 29% were followers of Tariqa, 66% of them are followers of Qadiriyya and 34% Naqshbandi (all of these movements are the branches of Sufi Islam with a certain differences). These data practically coincide with the assessment of A. Malashenko, according to which Qadiriyya is the most numerous division in Chechen Sufism, since its supporters constitute more than 60% of adherents of all believers (Sitnikov, Romanov, Odaev, 2019).

It is noteworthy that among Chechen youth the share of Tariqa followers is low, since it is more spread among middle-aged Chechens. At the same time, Chechen youth, on the one hand, are significantly more religious than older and middle aged once. Based on this, it can be assumed that the religiosity of Chechen youth is often bound not in the traditional form of Sufism but is with other movements of Islam. Many researchers share the opinion that a large part of Chechen youth is inclined towards the so-called non-traditional Islam, radical ideas find a particularly fertile ground among the full strength and energy of youth, and young people are mainly involved in so called "Pure Islam and its number is gradually increasing. (Moore & Tumelty, 2008). In this case A. Yarlykapov notes that today the communities of the North Caucasus mostly follow the "Universal Islam" based only on the Quran and the Sunnah (Yarlykapov, 2017). Ss a well-known expert in the field of religion, especially Islam V. Akaev notes that more than 1500 Russian citizens received education in Islamic universities in Saudi Arabia and it is reflected in the number of increased Sunni Islam followers in the post-communist region (Akaev, 2014).

The Chechen Republic is radically different from the majority of regions of predominantly secular Russia in terms of religiosity. At the same time, in the country, there is (at least theoretically) a unified educational space and a regulatory framework for the activities of educational institutions.

It is noteworthy that 42% support the study of Sharia and we have to say that it is a rather large share. And even those who were against compulsory Sharia study at schools often said that this should be done at home or in a religious educational institution. In general, respondents expressed unanimity about the necessity of Sharia study; there was a slight difference in the thoughts about the forms and places (Sitnikov, Romanov, Odaev T. 2019).

The confrontation between traditional and fundamentalist Islam manifested itself during the period of hostilities. One of the most chronicles of unfinished conflict is the notable clash between followers of both directions. In 1995 fundamentalists attempted to destroy the grave of St. Khedi, the mother of one of the largest Muslim authorities of the 19th century, Kunta Hadji (Kishiev). This conflict was resolved with great effort and difficulty. Even then, the question of whether to use Islam in the course of national state-building split Chechen politicians and divided them into two camps (Malashenko & Trenin, 2002). One of the factors contributing to the spread of extremist ideology in Chechnya and dividing the society in to two camps was the weakness of Orthodox Islam (Buttaeva, 2009).

The war in Chechnya had nothing to do with Islam. It had an ethno-national ideology with the aim of just to establish an independent nation-state and the entire Chechen community supported the country's independence.

Even the leaders of Chechnya themselves did not believe that Chechnya could become a completely independent country. In this regards Johar Dudaev, the president of the Chechen republic Ichkeria from 1991-1996 (until his death) often that once Chechnya became an independent republic, it would join a community of independent states and maintain close economic and political ties with Russia (Kroupenev, 2007). Before the armed conflict, the only aim of the Chechen republic was to have autonomy within the Russian Federation. The strategic goals of the Chechen leaders coincided with the goals of the Tatar elite. However, for several reasons in Chechnya (the republic is often called so), (which is still unknown to the press and the public), the situation has taken a different path, and events developed dramatically. This drastic - Moscow-Grozny conflict - military-political confrontation led Islam to become a major ideological and political vector (Akaev, 1999). Dudayev himself was the propagator of the ideology Jihad in Chechnya. Otherwise, this ideology would not be established in the republic. The fact can be proved by is J. Dudayev's famous November 1991 statement, where the phraseology and rhetoric showed that J. Dudayev was determined to use the religious factor for his political purposes (Grozny, 1992). However, the struggle for freedom in Chechen republic was a dual identity: religious and national. Since in 1998, Mufti of Chechnya Ahmed-Haji Kadirov (who was an irreconcilable enemy of Wahhabism) declared, "The Chechnya resistance movement should be considered first a religious and then a national movement" (Malashenko & Filatov, 2009). Chechen nationalism was vividly reflected in Chechen efforts to preserve and develop ethnic culture, where Islam took an undisputedly cardinal role (Kroupenev, 2007).

In 2000 Akhmad-Haji Kadyrov resigned from Mufti and was appointed at the position of head of the country's administration (Chufrin, 2001).

Many studies have been conducted about the role of Islam in the Chechen-Russia conflict. Such scholars and experts in the field of Islam as Vakhid Akayev, Alexei Kudriashenko, Vladimir Bobrunikov, Ilia Maksakov, Igor Rotaru, Aleksey Malashenko, Alexander Ignatenko and others wrote much about this topic. The papers of these scientists and experts were published exactly at the period of the armed conflict in Chechnya. From their research, it has already appeared that the Moscow- Grozny relations had no future. Even now that this large-scale conflict is over, its effects can also be assessed (International Crisis Group, 2012).

Islam has always been exactly the ideology that was opposed by the Russian domination of the North Caucasus. This is evident in the events of the XVIII-XIX centuries; when Russia tried hard to dominate in the region. At the time, Chechnya was stubbornly opposed to it and used religious-Islamic slogans, and Russia responding to Christian arguments in response.

At the same time the XVIII-XIX centuries, the idea of "Jihad" or "Holy War" spread in the region, (Souleimanov, 2017) but this idea failed to unite the Muslims, thus the establishment of the Islamic State failed. In Chechnya, the father of this ideology was Sheikh Mansur, who was destroying villages simply because the local population was not following his idea. Proponents of the idea of an Islamic state believed that only the Islamic state would be able to oppose Russian pressure.

Islam has always been one of the key factors in the formation of the Russian-Caucasian peoples and the Soviet period was no exception. In the relationship between the Communist authorities and Chechnya, Islam played a peculiar and sometimes contradictory role. Communist authorities sometimes "played" with Muslims and the Cossack and helped them to opposed the anti-Soviet movement (e.g. 1920), and even encouraging them by Islamic traditions and laws; sometimes, the communist regime destroyed mosques and declared the religion and its believer as an enemy of the Soviet system as it was happening from the 1930s to 1980s.

It is noteworthy that in the 1950s the Chechen diaspora was already indifferent to Islam. However, by the end of the Soviet regime, Islam had drawn great attention in the region, and even became a key factor in the development of the state (Savateev, 2000).

The Development of Islamic Education in the Post-Soviet Republic

Unfortunately, the Soviet period in the history of Islamic education in the North Caucasus has hardly been studied. During its course, state policy was not unchanged; periods of relief were replaced by periods of tightening the regime. In the post-war decades, only a few natives of the North Caucasus studied at the Islamic Institute named after Bukhara Mir and Arab Madrasah in the city of Tashkent. These were mainly those who stood at the leadership of the Spiritual Administration of Muslims of the North Caucasus (DUMSK). The gradual weakening of state control over religion observed since the beginning of "Perestroika" (the pre-period of Soviet Union collapse) allowed Muslims of the North Caucasus to begin to revive Islamic education in the region. It starts the period of the fastest-growing number of religious schools "mektebs" and mosques with the only goal - to raise the level of elementary Islamic education. At the very beginning of the 1990s, young mullahs who graduated from the recreated "mektebs" began actively participate in the revival of Islam in the region.

A completely new phenomenon for the region can be considered the opening of the Islamic institutes and universities. Islamic institutes were in the 1990s opened in Nalchik (Kabardino Balkaria), Grozny

(Chechnya), Ordzhonikidzevskaya (Ingushetia) and Cherkessk (Karachay Cherkessia). However, the leader of Islamic education in the region was Dagestan, where the number of local Islamic universities grew quite rapidly: in April 1998 the republic had 9 functioning universities, in May 2000 already 12 (plus another 33 branches); and in August 2002 there were 16 universities and 49 branches. In these 16 universities, 2,900 people received an education, and if you add to them those who studied in branches (2,830), you get an impressive figure of 5,730 students (Yarlykapov, 2003).

While describing Islamic education it is noteworthy to say that Islam in the North Caucasus has a long and rich history. The appearance of the first Muslims here dates back in the 7th century A.D., with the conquest of Derbent and its environs by the Arabs. Islam penetrated the region from two directions. From the south - Mesopotamia and Iran, there was the penetration of Sunni Shafi'i (madhhab) and Shiites of the Imamite (Jafarite) persuasion. Hanafi Sunnism penetrated here from the north, from Central Asia through the Golden Horde, and then from its fragments. The Hanafi sense of Sunnism was finally approved by the Crimean Khanate and the Ottoman Empire.

The modern map of the spread of Islam in the North Caucasus has changed a great deal. This map until recently was as follows: Dagestan, Chechnya, Ingushetia - the Sunni region, with a predominance of Shafiites under the strong influence of Sufism, in the south of Dagestan - a fairly significant Shiite minority, almost completely represented by Azerbaijanis; Central and North-West Caucasus - Hanafi Sunni (Vatchagaev, 2005). Today there is an increase in contacts between followers of two Sunni branches- Hanafi and Shafiite. This process is happening against the backdrop of an increase in the number of those who advocate the destruction of borders between different directions of Islam in the Islamic field of the North.

During the period of Soviet Union Islamic education as a system was almost destroyed - perhaps only the Sufi communities of Dagestan, Chechnya, and Ingushetia could manage to maintain it, albeit hopelessly backward and based on the standards of the early 20th century.

In the early 1990s a common way to revive Islam, and re-Islamisation, was to invite foreign preachers or send the youngsters to study in foreign Islamic educational institutions in the vast majority of countries in the Middle East and North Africa (Alikberov, Bobrovnikov, Bustanov, 2019). On the one hand, this led to the strengthening of the traditional forms of Islam for some regions, as it was in Adygea, where Islam was often preached by the descendants of the Adyghe Muhajirs from the Middle East and the former Yugoslavia, who consolidated the Hanafi tradition of Sunni Islam and it traditional for the republic. On the other hand, it facilitated the penetration of radical views and ideas, which intensified friction within Muslim communities. Moreover, radical ideas penetrated into regions where Islamic traditions in many places were practically not preserved, for example can be referred Kabardino-Balkaria and in those regions where Islam already had a good position for instance, Dagestan.

In the 90s of the XX century, religious education became more active and literature of such such Islamic writers and theologian as Taqī ad-Dīn Ahmad ibn Taymiyyah, Ahmad Ibn Hanbal - founder of one of the four major orthodox legal schools of Sunni Islam, Imam Hasan al-Basri, etc., began to spread in the North Caucasus and especially in Chechen republic and Dagestan. It is noteworthy to say that since the population did not know Arabic, the literature was translated from Arabic and other languages into the native and Russian. However, since some literature was spread in Arabic, young people were encouraged to learn the language. There has also been a proliferation of literature written by leaders of religious-political groups and extremists in the North Caucasus such as Bahuddin Magomedov, Magomed Tagayev, Ahmad-Kadi Akhtayev, and others (Markedonov, 2010).

As already said the revival of Islam dates back in the 1970s and 1980s, but the events of the 1990s were a pure religious renaissance: throughout Dagestan, where Islam was much more acute than in the Chechen Republic, Muslims began to unite, mosques were restored, and Muslim madrassas, institutes, and universities were opened. In January 1990, were established Muslim clerical rules. In the same year, 345 Dagestanis participated in the first USSR's Hajj, and the number of pilgrims has been increasing significantly for years (Khanbabaev, 2007).

Already in the 1990s, it became clear that the traditional picture of the spread of Islamic trends and interpretations noted above is undergoing serious changes. Internal and external migrations made serious corrections to it. The migration of people led to the migration of ideas, which complicated the main picture.

It should be noted that state policy in the North Caucasus is becoming more flexible. Efforts are being made to facilitate the process of Islamic education. A system of partnership between secular and religious universities was introduced, and among the participants in these programs, there are well-known Russian universities - Kazan Federal University, Moscow State Linguistic University, Pyatigorsk State Linguistic University. I should be emphasized that Pyatigorsk State Linguistic University has been a partner of Islamic universities in the North Caucasus, providing them with scientific and methodological assistance in the modernization of curricula, especially in particular secular and linguistic subjects, the preparation of educational and methodical literature.

The program to support the Islamic education system in the North Caucasus supports not only existing educational institutions. A project is underway to create two major Islamic educational centers - one for the predominantly Shafi's of northeast Caucasus, in Makhachkala, and the second for the predominantly Hanafi's northwestern Caucasus, in Nalchik. North Caucasus Islamic University in Makhachkala today is the most developed center of Islamic education, and the Nalchik center is still in the process of its development.

Thus, the characteristic of Islamic education in the North Caucasus has always been a part of society's life, which is characterized by a high level of dynamics. However, the culture that comes with the "modern Islam" is mainly Middle Eastern, since, during the years of Soviet modernization, the peoples of the North Caucasus lost many important cultural codes, particularly, clothing and other significant cultural elements. And yet, most likely, the cultural peculiarity of the region will continue, and Muslims will once again demonstrate the adaptive capabilities of their religion. The clear evidence is the youth movement that is eager to bring the novelty and at the same time to maintain the origins of Islam in their cultures (Yarlykapov 2017).

Modern Approach to Development of Tourism Sphere

Chechen republic is very notable for the cultural and natural views since there are located about fifty state nature monuments. Among them - ten nature reserves are created to protect certain species of animals and plants. The picturesque landscapes of the Argun historical-architectural and natural museum-reserve, the necropolis "City of the Dead", Lake Galanchozh, Kezenoy-Am Lake, waterfalls (Kharachoysky, Vashindaroysky) and Grozny dendrological garden are of significant interest to travelers and tourists. Such natural resources as mineral and mud springs have medicinal value. The Chechen Republic has everything necessary for the southern resort: there are hydropathic facilities with hydrogen sulfide and radon baths. Despite the modern architecture of Grozny, the capital of the republic managed to preserve the architectural and artistic appearance of the city center. In Grozny operate the Grozny Russian Drama Theater named after Mikhail Lermontov and the Chechen State Drama Theater named after Hanpashi

Nuradilov, as well as the State Youth Theater "Serlo" and the Chechen State Theater for Young Spectators. The main attraction of the capital is the cathedral mosque "Heart of Chechnya" named after the mufti and first president of the Chechen Republic, Akhmat-Hadji Kadyrov. The mosque is the part of a unique majestic complex, built based on the classical Ottoman style and considered the largest mosque both in Russia and in Europe. The following museums and reserves are diverse in their subjects: the National Museum of the Chechen Republic; Argun State Historical, Architectural, and Natural Museum-Reserve; State Gallery named after Akhmad-Hadji Kadyrov in the city of Grozny; Historical and Ethnographic Museum "Dondi-Yurt" (UrusMartan); Museum of Local Lore in the village of Makhkety (a branch of the National Museum of the Czech Republic); Museum of Local Lore named after Hussein Isaev; Literary and memorial museum of Arbi Mamakayev (Nadterechnoye village, Nadterechny district of the Chechen Republic); Literary and Ethnographic Museum of Lev Tolstoy; Literary Museum of Mikhail Lermontov; Memorial complex "Walk of Fame" and the museum named after Akhmat Hadji Kadyrov; Gallery of Iles Tataev (Grozny). The ethnographic museum in Urus-Martan called Dondi-Yurt, which was created on its courtyard by a resident of this city Adam Satuev, is a whole complex of buildings and premises that recreate the picture of the old village. In this museum, you can see exact copies of the architectural monuments of the mountainous Chechen republic - residential and battle towers, erected in the middle ages by ancient craftsmen using unique technology. The above mentioned ancient necropolis "City of the Dead" (Tsoi-Pede) located in the Itum-Kalinsky district of Chechnya is also unique, its exact age has not been determined, but scientists are inclined to believe that it was built no later than the 5th century AD.

Sports and recreation tourist complex "Grozny Sea" is based on the Chernorechensky reservoir of a modern complex for entertainment and recreation. It provides a separate beach area for men and women; hotel complex of a five-star category; arboretum; viewing platform; floating two-store restaurant; multifunctional entertainment complex with water attractions; color music submersible fountain; football field; sports training base. The project worth 60 billion rubles is developing on an area of 259 hectares on the coast of the Chernorechensky reservoir, created by constructing a dam more than 800 m long. The plans for the implementation of the project were announced in 2013.

The Chechen Republic has unique conditions for the development of tourism. Five climatic zones coexist in the region - from the desert to snowy mountain peaks with a height of more than 4,000 m above the sea level. The absence of harmful industrial zones that can harm the ecosystem is also one of the factors favorable for the development of tourism in the republic. The geographical position of the Chechen Republic creates favorable conditions for interaction with potential partners. The countries of the Middle East, Central Asia and the regions of the European part of Russia are within 2 600 km from the capital of the Chechen Republic, which is no more than 3.5 hours of flight.

In the tourism sector, the security factor plays a great role. According to the statistics from General Prosecutor's Office of the Russian Federation, the crime rate is certainly reduced and in recent years, the Chechen Republic has always been recognized as one of the safest regions of Russia (PricewaterhouseCoopers, 2016).

The Chechen Republic is one of the youngest and fastest-growing countries. The increase of the republic population occurs mainly due to natural growth. According to the information from chechenstat we can see that from 2006 to 2019 the population grows steadily. Since in 2006 it was 1152,3 and in 2019 it already was 1457,0 (Chechenstat, 2020).

The Republic is located in the central part of the northern slope of the Greater Caucasus Mountain Range, (height up to 4.493 m), adjacent to the Chechen Plain and the Tersko-Kum Lowland. The total area of the Republic is 16.1 thousand square kilometers (76th place in the area among the subjects of the Russian Federation). The climate is Continental; there are 265 sunny days in a year, which is twice more than the indicator in central Russia. The capital of the Chechen Republic is the city of Grozny (PricewaterhouseCoopers, 2016).

All major settlements (Argun, Gudermes, Urus-Martan, and Grozny) are in the middle part of Chechen republic, between the mountains and northern lowlands which can be considered a great convenient for tourists (International Crises Group, 2012).

We can say that Islam is the main religion not only in the Chechen republic but in all the five republics of the North Caucasus, which mostly follows the Hanafi and Shafi'i Sunni madhhabs (Muslim school of law and jurisprudence). Sufism is widespread in Chechnya, Ingushetia, and Dagestan; and Salafi religious groups are becoming more visible, especially in Dagestan. As it is known during the Soviet period, religion was prohibited, but Islam managed to survive in the North Caucasus. Official holidays of the region like for almost all Muslims are Ramadan-Bayram and Kurban-Bayram. North Caucasus republics support and do facilitate the development of Islamic education, there are Islamic higher educational institutions in every republic, and the Hajj to Mecca is attended by thousands of pilgrims from the region every year. North Caucasus leaders often emphasize their religiosity, attend religious events, and fast during Ramadan.

Despite having a dreadful experience of two devastating wars, the country has great positive changes and developments. A large reconstruction effort launched in 2006 in less than four years rebuilt the infrastructure, roads, social objects, and buildings destroyed during the two wars. Grozny acquired conspicuous new architecture: the largest mosque in Europe, a huge sports stadium, and the "Grozny City" complex of skyscrapers, unique in the North Caucasus. In winter 2012, in just one district of the city, there were simultaneously 54 streets under the reconstruction, together with the buildings of the National Library, National Museum, and a Russian drama theatre.

Power in contemporary Chechnya is concentrated in the hands of Ramzan Kadyrov, the son of the Chief Mufti of the Chechen Republic of Ichkeria, the first president of Chechen republic serving the country from 2003-2004 (till his death), Akhmad Kadyrov. Ramzan Kadyrov is the only regional leader who controls the security services on his territory and personally supervises anti-terrorist operations. The intensity of armed confrontations and casualties among the security services has significantly decreased. R. Kadyrov has developed his Islamic-nationalist state ideology and strongly supports Chechen culture, arts, and sport promotes Chechen customs and lifestyle, restored Sufi holy places (ziyarats), organizes commissions for regulating family disputes as well as campaigns against drug addiction. He actively uses Sufi Islam as a state ideology, enforces Islamic dress-code for women and even encourages polygamy. A significant part of republican TV broadcasting time is dedicated to religious issues, female TV hosts covered their heads. There are twenty madrasas, two Islamic higher education institutions, three Hafiz schools for Koran readers and over 700 mosques. Mosques, religious life, and education are under strict control, and Ramzan Kadyrov says that he will eradicate anything related to quasi religion. The Orthodox Church in Grozny has been restored as well, and the return of ethnic Russians is encouraged greatly. Before Orthodox Easter in 2012, all Christian cemeteries were cleaned, and transportation provided for former Russian residents to visit their family graves. The president enjoys much more authority than other regional leaders, partly due to close relations with Russian President Vladimir Putin. He often appears in the agenda of international politics, particularly in the Middle East, and has direct contacts with leaders of several Islamic countries. It is noteworthy to mention that the republic has many foreign famous visitors and conducts famous events, for instance, on 15 May 2012 a major contest of singers of Islamic songs in Grozny featured the famous Sheikh Mishary bin Rashid Alafasy from Kuwait. Two weeks later prominent religious leaders, Islamic scholars and officials from Kuwait, Saudi Arabia and representatives of international Islamic organizations visited the country, after attending a conference on "Islamic doctrine against radicalism" in Moscow. The republic has visits of Western celebrities such as Hollywood actors Hilary Swank and Jean-Claude Van Damme, and the violinist Vanessa Ma, etc. (International Crises Group, 2012).

Currently, the authorities and residents of the republics of the North Caucasus have focused on improving living conditions and try to establish new business projects. They are also developing long-term development strategy, the purpose of which is to make the North Caucasus an attractive place for foreign and local investors, and tourists. An important step towards the implementation of this plan was firm and certain administrative reforms. In 2010, it was decided to separate the North Caucasus District from the Southern Federal District. It included Dagestan, Chechnya, Ingushetia, North Ossetia, Kabardino-Balkaria, Karachay-Cherkessia, and the Stavropol Territory. This decision was connected, on the one hand, with the desire to increase control and ensure security in the region, and on the other, its goal was to improve the coordination of economic reforms. The strategy of the socio-economic development of the North Caucasus Federal District until 2025 assigns very serious goals for the local authorities. It involves the adoption of measures to improve the quality of life of the population. First of all, it is aimed at the reduction of unemployment and the development of the local industry, which can lead to a decrease in the flow of labor migration from the region. However, further investments in transport infrastructure and telecommunications are needed. A very interesting idea, which should contribute to the development of the entire region, is the long-term development of the tourism sector. No doubt improving the socio-economic situation in the North Caucasus can help increase its stability. The implementation of ambitious federal projects will require investments in the region of significant capital and, the attraction of foreign investors.

The modern positive changes of the republic are especially evident not only in the capital, Grozny but in the province as well. Roads, railways, pipelines, and power transmission infrastructures were modernized. The improvement of the socio-economic situation reflects on the improvement of the mood of the population.

The economic development of the region is impossible without eliminating the threat of terrorism and criminal. Only having a secure country the state market can convince investors and tourists to become interested in the North Caucasus. On the other hand, the fight against terrorists and criminals is extremely difficult without improving living conditions. To attract capital and tourists, efforts should also be made to combat widespread corruption, nepotism, and organized crime, contributing to the wasting of federal funds allocated for the reconstruction and development of the region. In these conditions, domestic and international policy of the Russian Federation in the North Caucasus is implemented reasonably, and everything indicates that in the long run, this will bring the expected results. The role of Muslims in the post-Soviet space should grow not only in the separate republics but also in the federal authorities, where currently it is not properly represented. In this regards, the followers of Allah will not have to feel like second-class citizens and will identify themselves with the state, actively forming its internal and foreign policy (Szabaciuk, 2014).

The tourism and recreation industry is one of the most important components of the economic system, the development of which characterizes both the socio-political stability of the region and its economic attractiveness for long-term investment. A feature of the tourist and recreational complex is that its development provides a balance of the various interests of society. Tourism contributes to the development of the socio-economic sphere, improvement of living conditions, preservation, and development of cultural potential, expands opportunities for self-realization of the population, and improves the overall image of the territory. The release of the tourist and recreational complex consists of the releases of the following industries: hotel and catering services; services for the organization of leisure and entertainment, culture and sports; travel agency services. The services of hotels and catering enterprises in the structure of the tourist and recreational complex of the Chechen Republic occupy the 2nd position.

The Chechen Republic, like the North Caucasus Federal District as a whole, has rather favorable conditions for the dynamic development of the tourism sector, as it has the most diverse tourism resources. As it is already described above, there are five climatic zones in the region - from the desert to snowy mountain peaks. The geographical location of the republic also contributes to the development of the tourist and recreational complex, as the region is characterized by good transport accessibility with key Russian regions, as well as countries of Europe and the Commonwealth of Independent States (almost all post soviet republics). The relatively favorable ecological situation and the absence of large industrial enterprises that can harm the ecosystem are also one of the factors determining the development of tourism in the republic.

Before the collapse of 1991, citizens from all regions of the USSR were relaxing and recovered in sanatoriums and tourist centers of the republic. Only in Grozny until the early 1990s, there functioned about 198 excursion routes. Very popular were the all-Union mountain-hiking routes Nalchik - Grozny - Shatili (Georgia), horse trails in the Shatoi region, and several other tourist destinations. Natural and climatic conditions in the Chechen Republic, landscape, and healing springs favor the development of tourism in the region. At the same time, health and ski tourism, which are one of the most popular types of tourism, has the greatest potential in the republic as well. Medical, health, and ski tourism, which are among the highest priority, will stimulate the development of some certain significant types of tourism. Health resorts can be effectively supplemented by recreational, business, cultural, religious, educational, and ecological types of tourism. In its turn, ski resorts can be most effectively supplemented by sports, extreme, events, and business tourism. The persistent interest in tourist trips to the republic today is shown by the natives of the republic living outside its borders and many other foreign visitors.

Tourism infrastructure in the Chechen Republic is also being created near holy pilgrimage sites, which is a great potential for religious tourism. The project of the Erta tourist complex being created in the region in the village of Marzoy-Mokhk, Vedeno district of the Chechen Republic is designed for pilgrims traveling to the holy place. It attracts with the beauty of landscapes, clean air, springs, it is located at an altitude of 1,500 m above the sea level. In 2008-2009 holy places of the republic were visited by 20 thousand people. Many pilgrims come from the Republics Ingushetia, Dagestan, the Kabardino-Balkarian Republic, as well as from Jordan, Syria, Turkey, Germany, China, and Kazakhstan.

In the framework of cooperation with the United Arab Emirates, agreements were reached on the construction of several tourist infrastructure facilities. So, next to the Akhmat-Arena stadium, the Arab company Troyan plans to build a five-star hotel. Thus, the tourist and recreational complex of the Chechen Republic has great potential, which in case of overcoming the main restrictions (lack of high-quality tourist infrastructure, negative image of the region, poor quality of service, insufficient transport accessibility of promising resort areas, undeveloped system ensure the safety of tourists), will be able to make a significant contribution to the economy of the region (Government of Chechen Republic, 2012).

CONCLUSION

Islam is not only a religion, but it is also a quite strong ideology and like every ideology with its certain features, which are usually expressed in societies, nationalities, and nations defines their activities and the rule of their lives, relations between the society and political force with a different ideology. Islam with its doctrinal theology is closely connected with the ideological world-view. The actuality and main point of the religion are that it deals with various aspects such as ethnical, national, economic, institutional, physiological, and as well as it is embedded in everyday life.

After the collapse of the Soviet Union, when the religious atheism came to an end, Islamic factor gained sustainable tendency, and turned into a very dynamic and local featured religion. The process of re-Islamization challenged traditional, fundamental, and modernist movements. As a result the the North Caucasus region indeed gained a tangle of political, national, religious and economic problems. However, at the present stage the situation is solved and the government led by Ramzan Kadirov has taken a firm destination to create a strong, stable, safe, criminal free place, and hospitable country with many holly places mosques, madrassas, and religious schools, (which is very important for the any kind of tourism industry), where the religion and culture are respected and followed.

The above-mentioned fact shows that the Chechen republic has great potential for many kinds of tourism such as cultural: since it has theaters, museums both, historical and ethnographic named after well-known writers, ancient buildings, etc. Recreational, health and sport tourism, in this field the republic offers sport and recreational tourist complex "Grozny Sea", with its separate beach areas for men and women, hotel complex of a five-star category; multifunctional entertainment complex with water attractions; football field; sports training base, balneological deposits of thermal hydrogen sulfide waters. Also, the republic has all the conditions for mountainous tourism including rafting, climbing, skiing, hiking, and horseback riding with the appropriate infrastructure. And naturally, being a religious, Muslim country and having many notable religious and sacred places and serving halal food the Republic of Chechen has a firm platform for the development of religious tourism.

REFERENCES

Akaev, V. (1999). *Sufism and Wahhabism in the North Caucasus. Confrontation or compromise?* The Institute of History, Archeology of Ethnography.

Akaev, V. (2014). Religious and Political Elites in the Northern Caucasus: Formation, Ideological, Ideological Contradictions, and Practical Opposition. *Religion in Society, Central Asia and the Caucasus, 15*(1).

Alikberov, A., Bobrovnikov, V., & Bustanov, A. (2019). *Russian Islam, Features of History and Cultures*. The Institute of Oriental Studies, Russian Academy of Science.

Buttaeva, A. (2009). Islamic Revival in Russia. *Bulletin of the Dagestan Scientific Center of the Russian Academy of Sciences.*

Chufrin, G. (2001). *The Security of the Caspian Sea Region*. Oxford University Press, Inc.

Collection of decrees by the President of the Chechen Republic from 1 November 1991 to 30 June 1992. (n.d.). 4.

Government of Chechen Republic. (2012). *Social and Economic Development Strategy Chechen Republic until 2025*. Ministry of Economic Development and Trade.

Halbach, U. (n.d.). *Islam in the North Caucasus*. Retrieved from http:// journals.openedition.org/assr/18403

Hanahu, R. A. (1997). *Traditional culture of the North Caucasus: the challenges of time: (Socio-philosophical analysis)*. Maikop.

International Crisis Group. (2012). *Europe Report: N°220. The North Caucasus: The Challenges of Integration (I), Ethnicity and Conflict*. Author.

Investor Guide to the Chechen Republic. (2016). PricewaterhouseCoopers.

Khanbabaev, K. (2007). *Islamic Radicalism in the North Caucasus. Ideology, Goals, Ways of Financing*. Journal Free Thought.

Kroupenev, A. (2007). *Application of Poliheuristic Methodology to a Sequence of Boris Yeltsin's Decisions During the 1995 Hostage Crisis in Budyonnovsk*. Retrieved 04 02, 2020, from https://papers.ssrn.com/sol3/papers.cfm?abstract_id=1027054

Makashenko, A., & Trenin, D. (2002). *The Time of the South: Russia in Chechnya, Chechnya in Russia*. Carnegie Endowment for International Peace.

Malashenko, A., & Filatova, S. (2009). *Twenty years of religious freedom in Russia*. Book Monograph Moscow Center Kornegi.

Markedonov, S. (2010). *Radical Islam in the North Caucasus: Evolving Threats, Challenges, and Prospects*. A Report of CSIS and Eurasia Program.

Moore, C., & Tumelty, P. (2008). Foreign Fighters and the Case of Chechnya: A Critical Assessment. *Studies in Conflict and Terrorism*, *31*(5), 412–433. doi:10.1080/10576100801993347

Moore, C., & Tumelty, P. (2008). Foreign Fighters and the Case of Chechnya: A Critical Assessment. *Journal Studies in Conflict and Terrorism*, *31*(5), 412–433. doi:10.1080/10576100801993347

Sabateev, A. (2000). *Islam and politics in the Chechen Republic*. Social Sciences and Modernity.

Sitnikov, A., Romanov, M., & Odaev, T. (2019). Religiosity in the Chechen Republic and its Influcence on Social and Political Institutions. *Monitoring of Public Opinion: Economic and Social Changes*, 157-183.

Souleimanov, E. (2017). *The North Caucasus Insurgency: Dead or Alive?* Strategic Studies Institute and U.S. Army War College Press.

Statistical Data. (n.d.). Retrieved from https://chechenstat.gks.ru/folder/38713

Szabaciuk, A. (2014). *The Russian Federation's Policy on the North Caucasus. Islam in the North Caucasus: history and challenges of our time*. Central East European Institute Publishing House.

Vatchagaev, M. (2005). *The Role of Sufism in the Chechen Resistance.* Erudition-Russian Electronic Library Report: Islam and Politics in the Chechen Republic.

Vidino, L. (n.d.). *How Chechnya Became a Breeding Ground for Terror.* Middle East Forum. Retrieved 04 02, 2020, from https://www.meforum.org/744/how-chechnya-became-a-breeding-ground-for-terror

Yarlykapov, A. (2003). Islamic education in the North Caucasus in the past and present. *Herald of Eurasia.*

Yarlykapov, A. (2017). *Islam and Tradition in the North Caucasus. Ethnic and Religious Diversity of Russia. Russian Academy of Science.* Institute of Ethnology and Anthropology.

KEY TERMS AND DEFINITIONS

Chechen Republic: A former Soviet Union republic, satiates in the North Caucasus Federal District declared its independence in 1991, however having two official/national languages: Chechen and Russian and is still a subject within the border of Russian Federation. The prevailing religion in the Chechen Republic is Islam, which has a great spiritual, cultural and political influence on many people not only in the republic but in the whole region.

Islam: Is the youngest and second-largest religion in the world, the followers of Islam believe in Allah and are called Muslims. The lifestyles of Muslims are according to the Islamic law called Sharia and a holy book called the Quran.

Islamic Education: Is one of the forms of teaching the religion, study Islam and all its aspects: rules, customs, ceremonies, etc. It implies a type of education that is mainly separate from academic ones and which as a rule considers religious beliefs as a fundamental principle and operational modality.

Jihad: Is an Arabic word and means any effort or diligence, in particular in work, study, etc. Some scholars divide Jihad into large (spiritual struggle) and small (ghazavat - armed struggle). The concept of Jihad refers to constant action for the ideals of the Islamic religion.

North Caucasus: A part of Russian Federation located in Asia, south of the European part of the Federation in the mountains and foothills of the Caucasus. It contains 7 republics (with the major faith of Islam) and 2 territories. The region has the most fertile lands, the most important resorts.

Traditional Society: A society that is governed by traditions and custom. The social structure in it is characterized by a rigid social class hierarchy, the existence of stable social communities (especially in the countries of the East), a special way of regulating the life of society, based on traditions and customs. It seeks to maintain its foundations and changes gradually - its evolution usually takes place over several generations.

Wahhabism: Formed in the 18th century, is a relatively new religious and political movement in Islam, which is opposed to all innovations. The founder of the movement was Muhammad ibn Abd al-Wahhab at-Tamimi.

Chapter 3
Introducing the Cities of Ghazali in the Context of Religious Tourism

Aytekin Demircioğlu
https://orcid.org/0000-0003-4945-5519
Kastamonu University, Turkey

ABSTRACT

Ghazali is one of the most well-known scholars of the Islamic world that lived between 1058 and 1111. He was dedicated to knowledge in his short life. Ghazali made trips to take lessons and to teach. In this study, cities related to Ghazali were examined. These cities are Tus, where he was born and received his primary education; Curcan, where received Fiqh training; Nishabur, where he first studied as a student and then taught as a madrasa scholar. These three cities are located within the borders of Iran today. The fourth city is Baghdad, where he taught at the Nizamiye Madrasa. Later he was found in the cities of Damascus, Jerusalem, Al-Halil, respectively. He then went to the cities of Mecca and Medina to become a pilgrim. In this study, these nine cities, which are important for Muslims, were introduced as a religious tourism center with a descriptive method.

INTRODUCTION

Ghazali, who was an orphan at a young age, made long trips using the money that his father saved for his education before he died. These travels were made to go to scholars in different cities and become their students. Some of his travels after becoming a teacher were in the form of going to Madrasahhs in different cities to teach and educate students. Ghazali made trips other than to take lessons and to teach. For example, he went to Damascus to get away from people and to live alone life. He also went to Mecca to increase his religious maturity and become a pilgrim.

Ghazali was born in Tus city of Horasan (Old Iran). When he was a child, he first received education from his father and then he took lessons from his city's scholars. After a while, he went to another city called Curcan to continue his fiqh education. Ghazali then went to Nishapur to continue the lessons of the

DOI: 10.4018/978-1-7998-5792-1.ch003

famous scholar al-Juvayni of the time. Ghazali did not stay in Nishapur, but when he was twenty-seven, he went to Baghdad next to the famous vizier Nizam'ul Mulk. Ghazali got out of the eye among many scholars here. When he was 33 years old in 1091, he was appointed as the chief director of the Baghdad Nizamiye Madrasah, which was the most important educational institution of that period by Nizam'ul Mulk. He especially wrote his works about fiqh, theology and philosophy in this period (Ghazali, 2020; Orman, 2013).

After a while, Ghazali began to question this active life full of fame, showiness and dignity. This interrogation caused what he did to be meaningless and Ghazali abandoned this glorious life and opted for a lengthy life. For this reason, he left Baghdad and went to Damascus where he was retired. During this retreat, Ghazali went to Mecca and visited the Kaaba and became a pilgrim. He returned to his hometown Tus in 1099.

In the meantime, from Fahru'l Mulk (Fahru'l Mulk, the son of the first protector of Ghazali, Nizamu'l Mulk, and the vizier of the Seljuk prince Sencer, who was the ruler of Horasan at the time), was in the Nizamiye Madrasahh. Insistent offers have started to arrive for lectures. When the desire to eliminate the weakness of the people's faith values combined with the insistent invitation of Fahru'l Mulk, Ghazali returned to the life of education by ending the lengthy life that he had been continuing for about 1105 or 1106 years.

Thus, at the age of 49, Ghazali became the head of Nishabur Nizamiye Madrasah. This period is his second education period. Ghazali did not stay here too long and returned to his hometown of Tus in 1109 three or four years later. However, this time his return is not for seclusion, but to continue education and education in a calm environment (Guzel, 1988).

Ghazali again received a proposal to return to the training in the Baghdad Nizamiye Madrasahh, where he was trustee by Ziyau'l Mulk (another son of Nizamu'l Mulk). However, Ghazali asked for his forgiveness from this position with a long letter. Ghazali counted three excuses in this letter. In the first of these, he stated that he currently has one hundred and fifty students in Tus, that he cannot take them with him, he does not want to leave and his hearts are broken, and it is not appropriate to leave another orphan to look after an orphan. In the second, he said that while he was in duty at Nizamiye Madrasah, his children and wife did not have, whereas he now had a family that should not be neglected. In the third, he stated that he promised him not to buy goods from a sultan and his relatives, not to greet them and not to argue with anyone. However, if he returned to Baghdad, he stated that he could not get himself from these counts, he would have to do this to manage his livelihood, that he had a small farm in Tus, but that he also made a living (Ghazali, 2020; Orman, 2013).

The main purpose of this study is to introduce the cities where roads intersect in this science-filled life of Ghazali. These cities are Tus, Curcan and Nishabur where he was born and lived, Curcan and Nişabur where he took lessons from some teachers; Baghdad where he was the chief professor; Damascus, where he went to seclusion; the cities of Jerusalem and al-Halil, where he was on pilgrimage; Mecca and Medina where he became a pilgrimage.

In this study, the mentioned cities and the relation of Ghazali with these cities are introduced through a descriptive method.

BACKGROUND

Ghazali is a scholar with a very colorful personality. He is a knowledge theorist who pursues the truth in some of his works, a mighty religious scholar in some of his works, a critic of philosophy in some of his works, a Eshari Kalam scholar according to the other Kalam scholars, a mathematician in some of his works, a sufi, sometimes a jurist according to others, and has many different personalities and characteristics.

One of the most important aspects of Ghazali is the effort and sensitivity in the formation of information, and using the suspicion about the reliability of the tools used to obtain the information, as a tool very effectively. This attitude is uncommon both in the Turkish-Islamic world and in the Western world. With this attitude, Ghazali prioritized Descartes, which put doubt on the basis of thought, centuries ago and perhaps contributed to the formation of his thoughts. Because the 'Dream' metaphor, which is used to prove the distrust of the mind in obtaining the correct information and which can not distinguish whether the mind is actually in dream or awake, is present in the works of Ghazali (Ghazali. 2020; Descartes, 2018). The use of doubt as a tool to reach the truth, and its imitation by Descartes, who lived five centuries later, led to his recognition in non-Muslim geographies.

Another dimension that makes Ghazali important is its critical personality. Suspicion and criticism are so deeply rooted in Ghazali that it gets its share from this situation. His sensitivity in questioning his life, his work has increased so much that this has prompted him to make progressive and progressive decisions about his life; one day he left all the scientific life aside and decided to live a long life, and the next day he had a lot to quit. This situation has advanced enough to affect the health of the soul and body.

The reason for choosing the cities to which Gazali is related as the subject of this study is that Gazali is an important scholar. Although he lived in the Middle Ages, his influence still continues today. For this reason, his cities have a special importance in terms of Islamic religious tourism.

LITERATURE REVIEW

In this study, more than sixty works were examined and these are stated in the bibliography. However, for a more macro level perspective in the context of religion, tourism and religious tourism, the following study can be considered: "Dallen, T. J. & Olsen, D. H. (2006). Religion and Spiritual Journeys. Routledge is an imprint of the Taylor & Francis Group" Pilgrimage is a must in Islam, Judaism and Christianity. Therefore, it would be good for the researchers to look at this article: "Kucuk, M. A. (January-June, 2013). A General Overview on the Pilgrimage Application in Judaism, Christianity and Islam in the Context of Relationship Between Religion and Tourism. Religious Studies, 16(42), 101- 128, https://dergipark.org.tr/tr/download/article-file/52135". For a discussion about whether spirituality or traveler is in the foreground in religious tourism, the following study can be considered: "Qurashi, J. (2017). Commodification of Islamic Religious Tourism: from Spiritual to Touristic Experience, International Journal of Religious Tourism and Pilgrimage: 5(1), Article 9. https://arrow.tudublin.ie/ijrtp/vol5/iss1/9".

IMPORTANT CITIES THAT GHAZALI LIVED OR VISITED

Two such as the Abbasid Caliphate (750-1258) and the Great Seljuk Sultanate (1040-1157), who led the life of Ghazali between the centers such as Tus, Curcan, Nishabur, Baghdad, Damascus, Jerusalem, El-Halil, Mecca and Medina. it was under the sovereignty of the great state (Atlas of Islamic Thoughts [AIT], 2020; Acar, 2011).

The main purpose of Ghazali to get around so many cities was to carry out scientific activities. He sometimes traveled to learn from a teacher, and sometimes went to another city to teach in a madrasa. Another purpose that directs his travels is to want to mature himself religiously. Accordingly, he was retired in Damascus for a while and went to cities such as Jerusalem, Mecca and Medina to become a pilgrim and to see holy places.

Gazali was a very productive scholar. He has been engaged in scientific activity in every city he went to. Sometimes he taught and wrote works, sometimes he became a student. The large number of books he wrote shows that he was in scientific production in every city he went to. The number of works that are agreed to belong to Gazali is approximately 105 (Bolay, 2019; Demircioglu, 2018; Ghazali, 2019b; Ghazali, 1994; Okumuş, 2000; Orman, 2000; Karlığı, 1997). Although it is not certain whether it belongs to Gazali, the number of works attributed to it is 40 (Okumuş, 2000).

The Related Cities of Ghazali

Tus

Tus city is the place where Ghazali was born and died (Ghazali, 2020). Theestablishment of Tus, an ancient city, dates back to Alexander the Great. It is located approximately 25 km northwest of the city of Mashhad in Iran. Tus is a city between the mountains and there are precious metals such as turquoise stone, copper and smut on its mountains. According to the famous geographer Ibn Batuta, the city of Tus is one of the largest settlements in Horasan in history (Kurtulus, 2012).

The manager of the city of Tus sent the letters to the governors of Kufa and Basra in 650 and offered leaving Horasan to the Muslims, provided that Horasan was left to him. Upon this, Tus, the caliph Hz. Osman's Basra governor Abdullah b. It was conquered by the chief without fighting (Kurtulus, 2012). Abbasi caliph Harun Reşit, who came to the region in 808 to resolve an internal confusion in Khorasan, felt ill and died in Tus. His grave is still here (Turkish Science, 2020).

Apart from Gazali and Harun Reşit, Tus also hosts many famous names. For example, the world-famous Iranian poet *Firdevsi* is the famous vizier of the Seljuk Empire, *Nizamu'l Mulk* and the famous Islamic scholar *Nasiruddin Tusi*(Kurtulus, 2012).

There are not many historical artifacts in Tus, which was destroyed by the Mongolian army in 1389. However, the ruins of *Tus Castle*, the *Mausoleum of Firdevsi* and *The Tus Museum*, which contains ceramics dating back to the XI. Century is worth visiting (Bozkaya, 2019). In addition, the *Tomb of Ghazali* was found in Tus in 2007 (Kamadan, 2018). This grave, which is in a neglected state, is one of the places that must be restored and brought to religious tourism.

Curcan

When Ghazali was 15 years old, he first went to Tus and started reading fiqh from one of the important scholars of the time, Muhammed er-Rezakani. Later, he went to Curcan and continued to take fiqh lessons from Abu Nasr al-Ismaili (Gazali, 2020).

Curcan, formerly known as *Astarabad*, Turkmen: Gurgen, Gurgan; Persian: ن‌اگرگ is Gorgān. The meaning of the word curcan is "Land of the Wolves". Curcan, located in the northeast of Iran, is the administrative center of Gulistan Province. Curcan, which is approximately 400 km away from Tehran, was established in the south east of the *Caspian Sea* and at the foot of the *Elbrus Mountain Range*. During the excavations in 1841 valuable items of about 3000 BC. were discovered (Kurtulus, 1993).

In Curcan, which was rich in agriculture and had a wide variety of products in the Middle Ages, silkmaking was famous. During the Mongolian invasion that took place in 1393, the city suffered great injuries and could not return to its former magnificent days. Besides the world-famous scholar Biruni, Curcan has hosted many fiqh and hadith scholars, literati and poets (Kurtulus, 1993).

Curcan is one of the important cities that should be included in religious tourism routes because of its history dating back to 3000 BC. its natural beauties and hosting many scholars together with Gazali.

Nishabur

Nishabur is a city where Gazali is both educated and taught in the madrasah and is within the borders of Iran today. It is one of the four largest cities of Horosan (Iran) with Merv, Herat and Belh in the Middle Ages. Nishabur was called Ebreshehir or Iranşehr, which means "cloudy city" in the early periods of Islam. There are statements about settlement of Nishabur related Hz. Adam's grandson Anus, Saka ruler Afrasyab and the king of Pisdadi Tahmuras. The history of the city is thought to have started in the III. BC. (AIT, 2020; Acar, 2011).

The word "Nishabur" was formed by combining the words "new, beautiful" and "Shapur" in the old Persian (Pahlavi). "Sapur" is a special name preferred by Iranian rulers (Isim Bulutu, 2020).

Nishabur is on the historical Silk Road route. The city, which had its brightest days during the Sassanids, lost its importance after Horasan was taken by the Arabs. It became an important center in the Safavids period (Masalaci, 2014). Tugrul Bey, one of the grandchildren of Selçuk Bey, who was considered the ancestor of the Seljuk Empire, declared the independence of the state in Nishabur. Thus, the Seljuk Empire was established in Nishabur. With the establishment of the empire here, Nishabur became both the capital of the Seljuk Empire and one of the favorite places of its rulers (Gezimanya, 2014). Only ruins belonging to the *Seljuk Castle* remain from the Seljuks in terms of architecture (Selcuklu Municipality, 2017).

Nishabur is in the north east of Iran. The important livelihoods of the city, located at the foot of Mount Binalud, are agriculture and trade (Turkish Information, 2020). Located 125 km from Mashhad, Nishabur is also the sister city of Konya, Turkey (Masalaci, 2014).

Nishabur, where Ghazali and his teacher *al-Cuvayni* lived, also hosted the famous poet *Omer Hayyam* and the famous scholar *Feriduddin Attar* (Ozgudenli, 2007).

The tombs, belonging to *Omer Hayyam, Feriduddin Attar, Ebubekir Taybatli* and *Sheikh Ahmet*, host many visitors (Haber 7, 2020). The city is also an ancient metropolis and was discovered through excavations by an excavation team from the American Metropolitan Museum between 1935-1940 (Gezimanya, 2020).

In the Middle Ages, Nishabur was both the administrative center and the cultural and educational center of Horasan. It is known that there are around 11-18 madrasahs in Nishabur before the establishment of *Baghdad Nizamiye Madrasa*, the most famous madrasah in the Seljuk Empire period (AIT, 2020).

The path of Gazali, which is the subject of this study, has crossed with Nishabur twice. For the first time, Ghazali came here to take fiqh lessons from the scholar named al-Cuvayni at the age of 16 and remained a student here until he was 27 years old. In the second, he came as a teacher at the age of 49 at the invitation of Seljuk ruler Fahri'l - Mulk and taught here until 53 (Demircioglu, 2018).

Since it is home to many famous names such as Ghazali and Omer Hayyam, as it is the capital of the Seljuk Empire and it contains many touristic places, Nishabur is an alternative route in terms of touristic places and religious tourism.

Baghdad

Gazali, at the age of 27, went to the Nizamiye madrasah founded by the Seljuk vizier Nizamu'l Mulk in Baghdad. Gazali, whose star shined in a short time, became the top manager of this madrasa at the age of 33. Gazali remained in this position for four years and left Baghdad at the age of 37 and went to Damascus (Demircioglu, 2018).

Baghdad was founded in 762. After the Mongol invasion in 1258, it came under the sovereignty of various states. The city, conquered by the Ottomans in 1534, came under the sovereignty of the Safavids for a short period of time between 1623 and 1638, and was once again linked to the Ottoman State (Can and Gulpınar, 2010). Baghdad, which was under the rule of the Ottoman Empire until the Lausanne Treaty signed in 1923, has been accepted as the capital of this place since 1921, when Iraq emerged as a kingdom (Halacoglu, 1991).

Baghdad, which has been the science, culture and trade center of the Islamic World for centuries, is the third largest city in the Middle East after Cairo and Tehran. Baghdad, located in the Mesopotamia region, was established on a wide alluvial plain where the Tigris River and the Euphrates River get closer to each other (Ed-Duri, 1991).

The word "Baghdad" means "gift of God" in Persian. However, this place in Hammurabi laws written between 1792 and 1750 BC. the word "Bagdasa" in Talmud and some other sources written before Christ weaken the probability that the root of the word will be Persian. Baghdad, which was built and capitalized during the Abbasid caliph Mansur period, was called *Medinetu's - Salam*, which means "city of peace" (Ed-Duri, 1991).

During the Translation Movements Period, Baghdad played a central role in translating works in various languages into Arabic and in this way made a significant contribution to the development of Islamic civilization. Thanks to this function, Baghdad has become the scientific and cultural capital of Islamic civilization. Mathematicians such as *Harezmi and Ptolemy (Ptolemy)*; Doctors such as *John, Kindi, Ebubekir er-Razi, Abbas al-Mecusi, Ali bin Rabben et-Taberi*; Astronomers such as *Yahya b. Mansur, Biruni* and *Omer Hayyam*; Sectarian scholars such as *Ahmed b. Hanbel and Imam-i Azam Abu Hanifa*; Thinkers of *Ebubekir er-Razi, Ibnu'r-Ravendi, Gazali and İhvan-ı Safa*; Sufis such as *Maruf-i Kerhi, Cuneyd-i Bağdadi, Seri es-Sakati, Ibnu's-Semmak, Haris al-Muhasibi, Ebu'l-Huseyin en-Nuri* and *Abu Said el-Harraz* are some of the celebrities who grew up in the scientific and cultural environment of Baghdad. (Ozaydin, 1991).

Baghdad is an important city that must be seen with its history of about five thousand years. Although hard times have passed after the American occupation, structures that symbolize the old view of the city such as the *El-Faw Palace*, *Umm al-Qura Mosque,sidewalk coffee shops*and *oriental markets* can still be visited (Gezimanya, 2020). Apart from these, the *Abbasid Palace*, the symbol of the city; *The Iraqi National Museum*, which hosts many important works; The golden-domed *Al-Kadhimiya Mosque* (Tenay, Nd.), Which is located in the shrines of *Imam Musa al-Kazim* and *Imam Mohammed Taqi*, which has a great place in the Shiite belief, must be visited. *TheMonument of Martyrs* built in the memory of those who lost their lives in the Iran - Iraq war, the *Bağdadi Museum*, which stands out with its wax sculptures, and the *Hatra Ruins*, which are included in the UNESCO World Heritage List, are among the must-see places of the city (Baghdad Travel Guide, 2020).

Baghdad, which is already a religious tourism center in terms of Shiites, is an important center of attraction for Islamic religious tourism due to its five thousand years of history, the fact that it has hosted many Islamic scholars along with Ghazali, and its cultural and scientific contributions to Islamic civilization.

Damascus

When Ghazali was only 33 years old, he had the highest scientific title at that time and was appointed as the top manager of the Nizamiye Madrasa in Baghdad. Ghazali left Baghdad in 1095 and went to Damascus, thinking that this authority and fame he had made him worthless in terms of religion. He was retired in Damascus for two years and started to write his most famous work, *Ihya'u Ulumi'd Din* (Gazali, 2020).

Damascus is known as the city that has been used continuously for a long time by people in the world history. In the 14th BC. century Aramaic Akkad inscriptions, the city's name is referred to as Dimašqa. The city is mentioned in the Hebrew Old Testament as Darmeśeq (קשמרד) in the sense of "well-watered place". In Arabic, its full name is Dimesk Es-Sām (ماشلا قشمد) and it means "north". In European languages, it is met with the words Damas, Damascus, Damasco (Buzpinar, 2010; Tomar, 2010).

Archaeological artifacts dating from 4.000 BC. thousand years were found in Damascus. The city, which has such an ancient history, has been occupied by many tribes in history. It came under the rule of the Ottoman Empire in 1521, was left to France in 1920 with the Sykes-Picot Treaty signed after the First World War and became the capital after Syria gained independence in 1946 (Tomar, 2010).

There were important changes in Damascus when came under Ottoman rule during the Yavuz Sultan Selim's Egypt expedition. The Ottoman Empire built many mosques madrasahs, mansions, inns and bazaars here. In addition, Damascus - Hicaz Railway project has contributed positively to the trade of the region. Besides traditional commercial products such as carpet, silk, cotton and fabric, there are also glass, textile and cement industries in Damascus. There are many libraries and museums in Damascus, which was selected as the Arab Capital of Culture in 2008 (Buzpinar, 2010).

Like almost all medieval cities, Damascus is surrounded by walls. In the Islamic period, inner castle was equipped with mosques, complexes, madrasahs, hospitals, lodges, fountains, baths, inns and residences. It is observed that the city texture of Damascus was shaped during the Turkish period, except for the Emeviye Mosque and a few buildings around it (Ulucam, 2010).

Hicaz Train Station is one of the most important works in Damascus. It was built by Sultan Abdulmecid, one of the last emperors of the Ottoman, in order to gain superiority for the French and the British in the Middle East. The railway, which was started in 1900 and will connect Damascus to the Hejaz region and Anatolia, was completed in 1911. Although the rails are being dismantled today, it is an architecturally magnificent work, although it consists of a historical building and a wagon accompanying it. *Hamidiye Bazaar* is among the historical places to visit in Damascus. The bazaar was built in

1864 by Sultan Abdulhamid. In addition, *Kasion Hill*, where the city can be watched as a whole, is one of the important stops. Due to being one of the biggest restaurants in the world, *Damascus Gate,* which is included in the Guinnes Book of Records and where 6014 people can eat at the same time, is another symbol of the city. Another important place in Damascus is the *Suleymaniye Complex*, built by Mimar Sinan in 1554 by Suleiman the Magnificent. Along with the last Ottoman sultan Vahdettin, some of the dynasty members also had graves in the complex. Right across the *Suleymaniye Complex*, there is a bazaar made of wood carving and the *Syrian National Museum*. The museum is very rich in terms of artifacts. The world's first written stone alphabet, wooden products, tiles, and silk products are exhibited in the museum. *Emeviye Mosque* is the biggest and oldest mosque in Damascus. This building, which was previously a church, was converted into a mosque in 705 after the Umayyad took the region. The grandson of Hz. Muhammad is Hussein's cut head and The tomb of Hz. Yahya is located in the garden of this mosque. Emeviye Mosque has been included in the World Heritage List by UNESCO (Yilmaz, 2010).

Apart from these, *Azem Palace* is an important historical site. Built by the Ottoman governor of Damascus, Assad, the palace is now an ethnographic museum made in the XII. century and *Nur-ad Din Hospital*, used until the XIX. century, is another important structure in which patients are treated using the sound of water. This hospital has been turned into a museum today. *The Old Town* is the most touristic part of Damascus. There is *Emeviye Mosque* in the middle. There are many touristic places such as palaces, caravanserais, mosques, cafes, market places in this region surrounded by walls. *Seyyide Rukiye Mosque*, built on behalf of Huseyin's daughter, is one of the places where Shiites are considered sacred. *The Christian Region*, where Christians lived since the Hellenistic period, is another tourist attraction. Damascus is one of the must-see places with its old houses, small shops, bars and restaurants (En Uygun, Nd.). Another important place that must be seen in Damascus is the *Selahaddin Mausoleum*. This tomb, where the tomb of Selahaddin Eyyubi, which has an important place in the history of the city, is located, is one of the places that gave Damascus its identity (Orange Smile, 2020).

Damascus is a city rich in food culture (Gezimanya, Nd.).Although it is not very easy to visit Damascus today due to the ongoing civil war in Syria, Damascus is an important city that should be included in religious tourism routes and must be visited when the opportunity arises.

Jerusalem

Jerusalem is a city where Ghazali stopped by and traveled for a while. It is known that Ghazali left Baghdad in 1095 and returned to his hometown Tus in 1099. During this period, he came to Damascus and was retired for two years, fulfilled his pilgrimage duty, wrote various works and visited scholars from various cities (Demircioglu, 2018).

When the dates and events are evaluated, it can be said that Ghazali was in Jerusalem around 1097. However, it is not easy to clearly determine how long he stayed here. In addition, it is thought that he wrote his work named *Er-Risaletü'l-Kudsiye fi Kavaidi'l-Akaid* here because it contains a reference to Jerusalem in his name.

Known as Arabic al-Kuds (القُدس) and Hebrew Jerusalem (יְרוּשָׁלַיִם), Jerusalem is one of the oldest cities in the world in the Middle East. Palestine and Israel consider Jerusalem as their capital. It is located between the Mediterranean Sea and the Dead Sea northern border. Jerusalem is considered sacred in terms of Islam, Judaism and Christianity. Since the day it was founded, the city has always been a city that someone has been trying to seize and someone has been trying to protect. For this reason, it was completely destroyed 2 times, occupied 23 times and attacked 52 times. In 1538 BC., the city was surrounded by walls during the reign of Soloman I. The city, now known as *Old Jerusalem* and surrounded by these walls, is divided into four among Muslims, Jews, Christians and Armenians. Old Jerusalem became one of the UNESCO World Heritage Sites in 1981. The city is today considered to be one of the Endangered World Heritage Sites (Harman, 2002).

Jerusalem is considered sacred in terms of all divine religions, and it takes place in different ways in the scriptures related to Islam, Judaism and Christianity. For this reason, Jerusalem has been a center of attraction that people have tried to seize in almost every period of history.

Jerusalem came under the control of many different tribes and religions in different periods. These nations have given Jerusalem various names suitable for their own culture. Therefore, Jerusalem has around 70 different known names. Jerusalem is a place of pilgrimage for Jews. Christians think that when Jesus's declaration in the *Galilee* region was not accepted, he came to Jerusalem, cleaned the temple and continued his invitation here, and his earthly life ended here. In terms of Muslims, Jerusalem, is the place where Hz. Muhammad rises to the sky to meet with God (Mirac incident). At the same time, *Masjid al-Aqsa* in Jerusalem is the first Qibla of Muslims before the Kaaba. Apart from these, Jerusalem, being a city where Hz. Abraham and many prophets came, It is an important religious center in terms of all three divine religions because of the fact that Solomon had built Beytu'l Makdis (Temple of Solomon) here (Harman, 2002).

Because Jerusalem was considered sacred by divine religions, it witnessed the most violent religious wars. The Crusader Armies came here three times to take over Jerusalem. Apart from the Crusader wars, it has been the center of many wars, large and small, in every period of history (Avci, 2002).

The walls surrounding the city and *Kubbetu's Sahara* were repaired during Ottoman period. Suleiman the Magnificent renewed the pools where water was distributed to the city repaired the water channels and built new fountains. The sultan's wife, Hurrem Sultan, had a complex built in madrasah, inn and nursing home. Here, food was distributed to madrasah students, Sufis, guests and the poor and various services were provided. These works are quite neglected today. The Ottomans established the Jerusalem telegraph system in 1865, built the Jerusalem -Yafa highway in 1868 and the Jerusalem-Jaffa railway in 1892. The science and culture life of Jerusalem was developed in the Ottoman period. Hundreds of scholars from different geographies have visited the *Masjid al-Aqsa* throughout the history of Islam and some of them have preferred to live here and have contributed to the scientific life of the city. Ghazali is one of these scholars. XVIII. As the Ottoman Empire began to weaken since the century, the scientific life in Jerusalem and the madrasah life began to weaken (al-Aseli, 2002).

There are many places in Jerusalem that are considered sacred by divine religions and must be visited (Yuksel, 2020; Avci, 2018; Gezender, 2018; Milliyet, 2018, Gezimanya, Nd.):

Kubbetu's Sahara: Kubbetu's Sahara was built in 691 by the Umayyad caliph Abdulmelik bin Marwan. The place where Kubbetu's Sahara was built is the Temple Hill, which dates back to before Christ. It has the feature of being the first domed worship place of Islam. This structure was built on the mosque built by Hz. Omer before. its current structure damaged many times in history Turkey has made with the contribution of Jordan and some Arab countries. The tiles on the walls were sent by Suleiman the Magnificent.

Masjid al-Aqsa: It is the first Qibla of Muslims. Muslims returned to Masjid Al-Aqsa for 16 months before heading to the Kaaba and performed prayers. It is located in the south east of Jerusalem. *Hacer-i Muallak Stone* in this masjid is accepted as the point where Hz. Muhammed rises to the sky to meet with God. Thousands of people can pray at this mosque, which has four minarets. Although it is not known exactly when Masjid al-Aqsa was built, it is thought that it was built forty years after the Kaaba.

Temple Mount: It is a large area where Masjid al-Aqsa and Kubbetu's Sahara are located. There are small walls surrounding it. Interesting historical fountains can also be visited here.

The Wailing Wall and the Great Temple: The Wailing Wall, considered sacred by the Jews, is believed to be the western wall of the Great Temple. The length of the wall is 488 meters. The first Great Temple built by Solomon was destroyed by Babylonians and Jews were exiled to Babylon. The temple, which was built for the second time 70 years after the exile, was destroyed by the Romans this time. The Wailing Wall is located in this destroyed part and is the place of worship of the Jews. It is rumored that Hz. Muhammad tied *Burak*, the animal he used to ascend to the sky, to this wall.

Church of the Holy Sepulcher: This place in the Old Jerusalem. It is accepted as the place where Jesus was crucified and resurrected. The building, which was built in the 4th century, is one of the pilgrimage centers for Christians. There is an accepted tomb belonging to Jesus. The path where Jesus walked on the cross is called the "path of pain" (Via Dolorosa). The last five stops of this road consisting of fourteen stops are within the Church of the Holy Sepulcher.

Hz. Omer Mosque: Jerusalem was conquered by the Islamic armies during the time of the second caliph Omar. When Omar conquered Jerusalem, he prayed somewhere. This mosque was built in the place where Omer performed the first prayer in Jerusalem.

Mount Olive (Tur): This is a hill at the height where Jerusalem can be watched, but it also has some religious meanings. It is believed that the war that will bring the end of the world *(Armagedon)* will start here. Similarly, it is believed that the *Sirat Bridge* (the bridge between heaven and hell) will be established between Harem-i Sharif in Mecca and here. In addition, It is rumored that Jesus took the prophecy here and spent the night before the crucifixion in the Garden of Getsemani here. Jews also believe that Christ, who will save them, will appear here.

Damascus Gate: This wonderful gate built by Suleiman the Magnificent is called by this name because it faces the direction of Damascus.

Halil Gate (Jaffa Gate): There is an inscription on the inside of this gate that says, "God is only one and Abraham is friend of God." This door opens to the Christian district of Jerusalem. The starting point of the city walls surrounding the city is next to this gate.

Lion Gate: The gate that takes this name is used for the transition to East Jerusalem due to the lion figures on the right and left. It is one of the most beautiful passages of the old city.

David Museum: This museum contains ruins of the First Temple, and various artifacts from the Mamluks, the Crusaders and the Ottomans.

Apart from these, there are many historical and touristic places and builds in Jerusalem such as *Virgin Mary Church, Russian Church, Arab Market, Hurva Synagogue, Betullahim Region* and *Yehuda Market*. Undoubtedly, Jerusalem is considered as sacred by all three divine religions, and it is one of the most important destinations of religious tourism.

Al-Halil

Al-Halil has a special importance for Ghazali, who went to Damascus and Jerusalem respectively after leaving Baghdad. Because Ghazali swore an oath at the grave of Abraham. In this vow, he promised himself that he would not serve the political authorities again and would not teach them in the madrasahs they established (AIT, 2020).

The city, known as El-Halil (ليلخلا) in Arabic and Hebron (ןוֹרְבֶח) in Hebrew, is located in the occupied region of Israel in the West Bank, 35 km south of Jerusalem. It is officially affiliated with

Palestine. Grape and lemon salt production is common in Hebron, which is accepted as one of the four holy cities of the Jews. Al-Halil old city center, where glass blowing art is made, is a tourism center worth seeing with its narrow streets, flat-roofed houses and old markets (Lekesiz, 2016).

The word "al-Halil" means "friend of Allah" and it is the nickname of Prophet Abraham. His origin is Sami (Amori). Samis came to this region and settled arround 1600 BC. Jewish sources says that Abraham arise when he is 25 years old in Harran, Turkey. Then he came to Al-Halil (Lekesiz, 2016).

It is believed that Abraham lived in this city 4 thousand years ago. When his wife Sare died, he buried him in an underground cave here. He was buried in the same place after his died. Above this point there is *Abraham Mosque*. This square shaped mosque has the appearance of a castle. Al-Halil is one of the oldest cities of Palestine. The market place called *The Souq*, with its different roof structures and narrow streets, is one of the religious tourism areas that must be visited (Gezimanya, Nd.).

It is believed that four prophets (Abraham, Isaac, Jacob, Yusuf) and their spouses were buried under the *Abraham Mosque*. According to the Jewish belief, it is thought that Adam and Eve were also buried here. The original structure here was built by the Romans in the 1st BC. Century. The building was transformed into a roofed mosque after 637 Muslims conquered it. The mosque seized by the Crusaders in 1100. It was named as *Abraham Castle*. When the Muslims conquered here in 1178 under the command of Selahaddin Eyyubi, the building started to be referred to as a mosque again (Gezimanya, Nd.).

There is acorn tree, 2 km west of the city, believed to be planted by Abraham. According to the belief, Abraham set up a tent at this point. Various ruins from the Roman period were found during excavations in the region. The region with this tree was later bought by the Russian Orthodox Church. Tourists were ripping off the bark of the tree trunk, as they believed it brought luck. To prevent this, the church covered the trunk of the tree with steel mesh (Turkish Cooperation and Coordination Agency (TIKA), Nd.).

Al-Rame (Memra) Harem: According to legend, Abraham met here with three angels who heralded that his infertile wife Sare's son, Isaac. Christians are also believed that is the place where Yusuf and Mary rest with her baby Jesus on his way to Egypt. One of the important bazaars of Palestine during the Roman period and the *Byzantine Church* was built here. This place has been restored as an archaeological garden by the Palestinian Ministry of Tourism and Antiquities (TIKA, Nd.).

Halhul City: It is located 5 km north of Al-Halil. Vineyards, almonds and similar trees grow here. The world-famous Al-Halil grape is produced here. According to Islamic traditions, *Nebi Yunus Masjid*, which the tomb of Yunus is built, is also located on here (TIKA, Nd.).

Susya National Heritage Site: It is an ancient Jewish village on the *Judan Mountains* in the south of the West Bank. Synagogue and mosque ruins belonging to the 5th and 8th centuries were found during the excavations. There are also strong traces of cave-dwelling life in the city. Mosaic engravings on the floor of Susya have been preserved until today (Gezimanya, Nd.a).

The city of al-Halil is one of the places of religious tourism that must be visited due to its importance in terms of divine religions and its strong historical heritage.

Mecca

Gazali showed his desire to become a pilgrim by going to Mecca while he resigned from the Chief Directorate of Nizamiye Madrasah in Baghdad. However, his real purpose was to get away from the gaudy life and glorify himself religiously. During this period when Ghazali purified himself mystically, he went to Mecca around 1097-1098 and became a pilgrim (Gazali, 2020).

Although the word "Mecca" is in Arabic (مكة), its etymological meaning has not been fully determined. The city is located in the west of the Arabian Peninsula, in the Hejaz Area and east of the Red Sea. The prophet of Islam Hz. Muhammed was born here. The Holy Book of Islam, the Qur'an, began to be downloaded here in Hira Cave. The Kaaba (House of Allah), where Muslims turn to prayer, is in this city. Mecca is referred to as the "mother of cities" in the holy book Qur'an (Qur'an, 6:92). It was administered by people who were descendants of Muhammad and called sharif. After 1925, it came under Saud rule. The Saud administration built the fourth tallest building in the world, *The Royal Clock Tower Hotel (Ebrac el-Beyt)*, right next to the Kaaba. While the Saud administration modernized the city, they damaged many historical works such as *Ecyad Castle* and the cultural texture of the city. Today, it is forbidden for non-Muslims to enter the city, which is visited by around 15 million Muslims every year (AIT, 2020).

Mecca was built on a desert valley called *Bekke*. This place is described as the "endless valley" in the Qur'an (Qur'an, 14:37). Only wild, thorny, stunted trees and shrubs with desert character grow here. The oldest document with information about Mecca is the Geography book written by Ptolemy in the second century. Ebukubeys region, where Safa and Merve hills are located in the east of this area, Sevr in the southwest, Nur (Hira) and Sebir mountains in the northeast. Arafat, Muzdelife and Mina districts, which are places of pilgrimage, are in the east of the city (Bozkurt and Kucukasci, 2003).

Since Mecca has no land suitable for agriculture, the only source of income was trade. The trade, which was carried out through the pilgrimages and fairs, was the most important economic activity of the city in history. However, some states, such as the Ottoman Empire, regularly sent aid to here and resolved the needs of those living in the city (Kucukasci, 2003).

One of the five conditions of Islam is to become a pilgrim. Muslims with no health problems and economic competence should go to Mecca and become pilgrims. For this reason, Mecca welcomes millions of visitors every year.

The places that can be visited in Mecca are as follows (Hisar Tourism, 2019; Uzun, 2001; Unal, 2001; Yasaroglu, 2001; Ets Tourism, Nd.):

Kaaba: It literally means "object in the form of a cube". It measures 10.70 x 12 meters and has a height of 15 meters. Muslims have to turn their directions to the Kaaba while performing prayers. It is also necessary to return around the Kaaba (tawaf) by reading various prayers while performing the pilgrimage. The Kaaba was damaged several times in the history due to floods and attacks. For this reason, it has been made or repaired several times. Like prophets Adam, Abraham, Ismail and Muhammad played a role in the construction or repair of the Kaaba.

The House of Prophet Muhammed was Born: This house is located in the east of Mecca. Today, the building of this house has been turned into a library. There is a tree cradle of baby Muhammad. It is a place visited by pilgrims intensely.

Mualla Cemetery: It is a historical cemetery next to the Gin Masjid, about 2 km north of the Kaaba. The tombs of Hz. Muhammad's grandfather Abdulmuttalip, his uncle Abu Talib, his wife Hz. Hatice, his sons Kasim and Abdullah are located here. This cemetery, where many famous companions also have tombs, is one of the most visited places in Mecca.

Gin Masjid: It is located next to Mualla Cemetery. Hz. Muhammed returned from Taif, three years before he migrated from Mecca o Medina performed the Morning Prayer here. After the prayer, he draw a line on the ground and asked from Abdullah b. Mesud not to cross that line. Then he started to read the Qur'an and invited the gins to Islam. According to legend, seven people from the gins who listened to the Qur'an from the prophet believed there. Later on, a masjid was built at this point and was called Gin Masjid.

Arafat: This plain, 25 km east of the city, is surrounded by mountains. It is compulsory for Muslims who make pilgrimage to visit Arafat the day before the Eid al-Adha. Cebel-i Rahme Hill, which is accepted as the place where Adam and Eve meets in the world after being expelled from heaven, is located here. Hz. Muhammed gave the Farewell Sermon on this hill.

Masjid-i Secere: It is another mosque near the Gin Masjid. It is one of the points where Muhammad prayed. According to legend, when Muhammed invited the gins to Islam, the gins asked him to bring witnesses to prove his prophecy. The Prophet then asked the nearby gum tree to testify, and when the tree witnessed, the gins accepted Islam. This masjid was built on the spot where that gum tree was located.

Sevr Cave: It is on the top of the mountain, about 4 km from the south of the Kaaba. As Hz. Muhammed fled from the oppression of the Mecca's polytheists and went to Madinah, his friend together with Hz. Ebubekir, they were hidden in this cave. According to legend, when they were inside, a spider made a net to the entrance of the cave and a pigeon nested. When the polytheists who searched for them saw the spider net and the pigeon nest on the door, they did not feel the need to look inside the cave. Such a prophet and his friend survived being caught.

Muzdelife: It is the name of the region in the Masjid Al Harem border area between Arafat and Mina. Muslims who make pilgrimage make the second visit here after making the first visit in Arafat. It literally means "to approach". Muslims are referred to by this name because they approached Allah through prayer and worship. Believers from the hill of Arafat perform the evening and night prayers here on the day before the Eid al-Adha and spend the night here.

Nur Mountain and Hira Cave: It is a rocky hill 5 km away from the Kaaba in the north east of Mecca. Before Hz. Muhammad became a prophet, he would come to this cave during Ramadan and worship here. While Hz. Muhammed was worshiping in Hira Cave in Ramadan in the year he turned 40, Gabriel came and informed him that he was a prophet. At the same time, this mountain is called *Nur Mountain*, which means "light" because the first verses of the Quran descended here and illuminated humanity.

Mina: It is the name of the region between Mecca and Müzdelife, between *Sabır Mountain* and *Murselat Mountain*, located approximately 7 km northeast of the Kaaba, and within the Harem (the limits of Hajj prohibition). Mina refers to the abundance of God's mercy as a word. Abraham threw stones at the devil here and sacrificed here in return for his son. Therefore, pilgrim candidates cut their victims here and shave their hair here. Nasr and Murselat suras descended here, Hz. Muhammed stayed here.

Cemerat: It is the place where pilgrims throw stones at the devil symbolically during the Eid al-Adha. There are three points where stones are thrown into the devil.

Erkam's House: In the first years of Islam, Muslims were performing their prayers secretly due to the pressure of the polytheists. The companion named Erkam, one of the first Muslims, opened his home for the use of Muslims. The house of Erkam was used in this way until Muslims became 40 people.

Apart from these, *Hudeybiye District*, where Hz. Muhammed met with those subject to him and signed an agreement with the polytheists; *Cebel-i Rahme Hill* where Adam and Eve meets in the world; *Masjid al-Hayf*, built in the place where Hz. Muhammed set up a tent during his farewell pilgrimage and prayed*; Mescid-i Nemire*, the place where Hz. Muhammad performed by combining the noon and afternoon prayers in Arafat; *Mescid-i Aise* built for The Prophet's wife Aise; There are important visiting points for Muslims such as *Masjid al-Cirane*, where the prophet was dressed as ihram to make umrah in the eighth years of the hijra.

Mecca is one of the holy cities of Muslims. It is obligatory for Muslims who have the opportunity to perform pilgrimage, which is one of the conditions of Islam, should go here. Therefore, Mecca is one of the indispensable centers of faith tourism.

Medina

Medina, which is affiliated to the Kingdom of Saudi Arabia, is located in the north of Mecca, west of Riyadh. The city, formerly known as *Yesrib*, was named *Medine-i Münevvere* (enlightened city) after the migration of Hz. Muhammad (Yalcin, Nd.).

Hz. Muhammed and his friends migrated to Madinah in 622, unable to withstand the pressures exerted on them in Mecca. After the Migration, Hz. Muhammed built a mosque here, which is now known as *Masjid Al-Nabawi*. Madinah came under the sovereignty of different states in history, and under the rule of the Kingdom of Saudi Arabia since 1925. Many historical monuments in the city have been destroyed due to the negative attitude of the Saudi government. Only a few buildings such as the *Amberiye Masjid* and the *Al-Muezzem Train Station* have remained from the Ottomans, which reigned for about 4 centuries, to the present day (AIT, 2020).

Medina is one of the three holy cities of Muslims, together with Mecca and Jerusalem. In addition, to the *Prophet's Mosque* and it is a city that Muslims who go to pilgrimage want to visit because of hosting Hz. Muhammed's grave.

Important places to visit in Medina are (Hisar Turizm, 2019; Bozkurt and Kucukasci, 2003; Gezimanya, Nd.):

Prophet's Mosque: After the migration to Madinah, a mosque was built in the center of the city. He himself worked in the construction of this mosque. A section for the teaching of the prophet home and a teaching room added adjacent to the masjid. His body was buried here after the Prophet died. The first two caliphs of Islam are Hz. Abu Bakr and Hz. Omar's graves are also located here. For this reason, it is a sacred place for Muslims.

Ravza-i Mutahhara (Garden of Eden): It is located in the Prophet's Mosque and it is the point where Hz. Muhammad gave sermon. It is believed that this place came to earth from heaven.

Quba Masjid: Hz. Muhammad took a break in Quba, where water wells and palm gardens, before arriving in Madina during the migration, and built a mosque here. It is the first mosque built to perform prayers collectively in the history of Islam. Hz. Muhammed worked in the construction of this mosque and stayed here for 14 days.

Uhud Mountain: Uhud War took place in this mountain, which is 5 km north of Madina. 70 companions died in this war. The tombs of these companions are located at the foot of the mountain.

Cennetu'l Baki Cemetery: In this cemetery near the Prophet's Mosque, the third caliph Hz. Osman, Ibrahim the son of Hz. Muhammed, his daughters Rukiye, Zeyneb and Fatıma, and the grandson of the prophet, Hasan and many important companions is here.

Masjid al-Kibleteyn (First Qibla Masjid): While Hz. Muhammed was praying here, the direction of Qibla was taken from Masjid Al-Aqsa and turned into Kaaba. For this reason, it is called the First Qibla Masjid.

Friday Masjid: When Hz. Muhammed left Quba during the migration, he stopped to chat with the public in the Ranuna Valley. It is the mosque built where the Prophet performed the Friday prayer here.

Mikat Masjd: While Hz. Muhammed traveled from Madina to Mecca, his approach was dressed in ihram clothes after 11 km. After this, the mosque was built at this point because it was the center of the pilgrims changing clothes.

Apart from these structures such as *Hz. Abu Bekr Masjid, Hz. Ali Masjid* and *Musalla Masjid* are places to visit in Medina.

Due to the fact that Islam is one of the three cities considered sacred and very important places it contains, Medina is one of the most important places to visit in terms of religious tourism..

CONCLUSION AND SUGGESTIONS

In this study, nine cities where Gazali, one of the most important Islamic thinkers of the Middle Ages, lived or visited for various reasons were introduced. According to the order of relations established by Gazali with these places, these cities are located within the borders of Iran today, Tus, Curcan, Nishabur; Baghdad in Iraq; Damascus in Syria; Jerusalem and Al-Halil in Palestine / Israel; Mecca and Medina are the cities within the borders of Saudi Arabia.

Among these cities, Jerusalem is considered sacred by Islam, Judaism and Christianity. Mecca and Medina are also cities that only Muslims consider sacred. Similarly, Baghdad is a city with Shiite Muslims places and places that are considered sacred by Christians in Hebron. Therefore, five of the nine cities surveyed were already cities with a religious significance in terms of religious tourism. The other three cities have the potential to be included in religious tourism routes due to their historical and cultural heritage.

This study suggests that the cities examined are included in a single tour program and included in religious tourism destinations since they are the cities in which Ghazali lives or visits, as well as their own characteristics.

Similarly, it is thought that similar travel routes can be created for the cities where scientists and artists who lived at different times and geographies lived and visited.

REFERENCES

Acar, A. (2012). Imam Ghazali and the Seljuks. In *Imam Ghazali at the 900th Year of Death, International Controversial Scientific Meeting 07 - 09 October 2011*. Istanbul: Marmara University Faculty of Theology Foundation Publications Nu: 271. Accessed from http://www.isav.org.tr/img/20150908__3924219012.pdf

Al-Aseli, K. C. (2002). Jerusalem: The Ottoman Period and After. In *Religious Foundation of Turkey [TDV] Encyclopedia of Islam*. Accessed from https://islamansiklopedisi.org.tr/kudusat 19.03.2020

Atlas of Islamic Thought (AIT). (2020). *Ghazzali*. Accessed from https://www.islamdusunceatlasi.org/timemaps/1

Avci, C. (2002). Jerusalem. In *Religious Foundation of Turkey [TDV] Encyclopedia of Islam*. Accessed from https://islamansiklopedisi.org.tr/kudus

Avci, İ. Y. (2018). *It is very easy to go to Jerusalem*. Accessed from https://www.gzt.com/mecra/kuduse-gitmek-cok-kolay-3390623

Baghdad Travel Guide. (2020). *Places to Visit in Baghdad*. Accessed from https://www.flypgs.com/sehir-rehberi/bagdat-gezi-rehberi

Bolay, S. H. (2019). *Comparison of Aristotle Metaphysics and Ghazali Metaphysics*. MEB Publications.

Bozkaya, Z. (2019). *Tus City*. Accessed from https://irangezi.com/Tûs-kenti/

Bozkurt, N., & Kucukasci, M. S. (2003a). Mekka. In *Religious Foundation of Turkey [TDV] Encyclopedia of Islam*. Accessed from https://islamansiklopedisi.org.tr/mekke at 20.03.2020

Bozkurt, N., & Kucukasci, M. S. (2003b). Medina. In *Religious Foundation of Turkey [TDV] Encyclopedia of Islam*. Accessed from https://islamansiklopedisi.org.tr/medine

Buzpinar, S. T. (2010). Damascus. In *Religious Foundation of Turkey [TDV] Encyclopedia of Islam.* Accessed from https://islamansiklopedisi.org.tr/sam--suriye#2

Can, A., & Gulpınar, S. (2010). Baghdad, 1831 - 1869. *Seljuk University Journal of Faculty of Letters, 23,* 209–217. Accessed from https://dergipark.org.tr/en/download/article-file/151651at

Demircioglu, A. (2018). *Criticism and Doubt in Ghazali* (2nd ed.). Gece Publishing.

Descartes. (2018). *Principles of Philosophy* (16th Edition). Istanbul: Say Publishing

Ed-Duri, A. (1991). Baghdad. In *Religious Foundation of Turkey [TDV] Encyclopedia of Islam.* Accessed from https://islamansiklopedisi.org.tr/bagdat#1-genel-bakisat

Gezender. (2018). *Kingdom of Heaven Jerusalem: Travel Notes*. Accessed from https://www.gezender.com/kudus-gezi-rehberi/at

Gezimanya. (2020a). *Things to Know About Baghdad*. Accessed from https://gezimanya.com/bagdatat

Gezimanya. (2020b). *Things to Know About Nishabur*. Accessed from https://gezimanya.com/iran/nisabur-hakkinda-bilinmesi-gerekenler

Gezimanya. (n.d.a). *Damascus: General Information / Guide*. Accessed from https://gezimanya.com/samat

Gezimanya. (n.d.b). *Hebron General Information / Guide.* Accessed from https://gezimanya.com/filistin/hebron

Gezimanya. (n.d.c). *Jerusalem - Israel*. Accessed from https://gezimanya.com/kudus-israil

Gezimanya. (n.d.d). *Things to Know About Medina*. Accessed from https://gezimanya.com/medineat

Ghazali. (2019). *Ihya'u Ulumi'd Din* (3rd Edition). Istanbul: Bedir Publishing.

Ghazali. (2020). *El Munkız Mine'd-Dalal (From Dalalet to Hidaday).* Istanbul: Cagdas Book.

Guzel, A. (1988). *An Essay on the Bibliography of Ghazali, Abu Hâmid Muhammed el-Gazzali (1058-1111).* Kayseri: Erciyes University Gevher Nesibe Medical History Institute Publications. Accessed from http://isamveri.org/pdfdrg/D017726/1988/1988_GUZELA2.pdf

Haber 7. (2020). *Four mausoleums belonging to famous scientists and scientists in the city of Nishabur in Khorasan*. Accessed from https://www.haber7.com/dunyayi-geziyorum/haber/2907764-horasanin-nisabur-sehrinde-unlu-ilim-ve-bilim-insanlarina-ait-4-turbe

Halacoglu, Y. (1991). Baghdad. In *Religious Foundation of Turkey [TDV] Encyclopedia of Islam*. Accessed from https://islamansiklopedisi.org.tr/bagdat#1-genel-bakis

Harman, O. F. (2002). Jerusalem. In *Religious Foundation of Turkey [TDV] Encyclopedia of Islam*. https://islamansiklopedisi.org.tr/kudus

Hisar Tourism. (2019a). *Visit Places and Mubarak Places in Makkah Al Mukarramah*. Accessed from https://hisarturizm.com.tr/mekke-i-mukerremedeki-ziyaret-yerleri-ve-mubarek-mekanlar/

Hisar Tourism. (2019b). *Visit Places and Mubarak Places in Medina-i Münevvere*. Accessed https://hisarturizm.com.tr/medine-i-munevveredeki-ziyaret-yerleri-ve-mubarek-mekanlar/

Isim Bulutu. (2020). *"Sapur" item*. Accessed from https://www.isimbulutu.com/%C5%9Fapur-isminin-anlami-nedir at 14/03/2020

Kamadan, Y. S. (2018). *Gazali's Tomb Awaits Interest*. Accessed from https://www.gzt.com/mecra/gazzlnin-mezari-ilgi-beklenen-3424638 at 16/03/2020

Karliga, B. (1997). *Gazali (Works). Religious Encyclopaedia of Islam*. Turkey Religious Foundation Publications.

Kucukasci, M. S. (2003). Mecca: The Ottoman Period. In *Religious Foundation of Turkey [TDV] Encyclopedia of Islam*. Accessed from https://islamansiklopedisi.org.tr/mekke

Kurtulus, R. (1993). Curcan. In *Religious Foundation of Turkey [TDV] Encyclopedia of Islam*. Accessed from https://islamansiklopedisi.org.tr/curcan at 17/03/2020

Kurtulus, R. (2012). Tus. In *Religious Foundation of Turkey [TDV] Encyclopedia of Islam*. Accessed from https://islamansiklopedisi.org.tr/Tûs

Lekesiz, O. (2016). *Al-Halil: The Story of the Captive City*. Accessed from http://www.gercekhayat.com.tr/yazarlar/el-halil-esir-sehrin-hikayesi/

Masalacı, M. (2014). *Luminous Horasan*. Accessed from https://www.hurriyet.com.tr/isik-sacan-horasanli-26399524 at 16/03/2020

Milliyet. (2018). *The Holy City of Faiths Jerusalem*. Accessed from https://www.milliyet.com.tr/tatil/inanclarin-kutsal-kenti-kudus-2668670 at 19/03/2020

Ghazali. (1994). Mişkatü'l Envar (Lighthouse of Lights). Bedir Publishing.

Most Affordable. (n.d.). *Damascus City Guide*. Accessed from https://www.enuygun.com/ucak-bileti/sam-sehir-rehberi/at

Okumus, M. (2000). *Understanding of the Qur'an and the Interpretation Method of Gazali* (Unpublished Doctoral Dissertation). Ankara: Ankara University Institute of Social Sciences.

Orange Smile. (2020). *Active Rest and Vacation in Damascus - Places of Interest, Entertainment and Nightlife*. Accessed from http://www.orangesmile.com/gezirehberi/sam/eglence--2541260.htm

Orman, S. (2013). *Ghazali: Biography, Truth Research, Philosophical Criticism*. Human Publications.

Ozaydin, A. (1991). Baghdad. In *Religious Foundation of Turkey [TDV] Encyclopedia of Islam*. Accessed from https://islamansiklopedisi.org.tr/bagdat#1-genel-bakis

Ozgudenli, O. G. (2007). Nishabur. In *Religious Foundation of Turkey [TDV] Encyclopedia of Islam*. Accessed from https://islamansiklopedisi.org.tr/nisabur--iran at 16/03/2020

Salman, N. (2014). *City in which Seljuk was founded: Nishabur*. Accessed from https://gezimanya.com/GeziNotlari/selcuklunun-kuruldgu-kent-nisabur

Selcuklu Municipality. (2017). *Nishabur Castle*. Accessed from http://www.selcuklumirasi.com/architecture-detail/nisabur-kalesi at 16/03/2020

Tenay, E. (n.d.). *Islamic Cities: Baghdad*. Accessed from https://www.bidunyahaber.org/islam-sehirleri-bagdat/

Tomar, C. (2010). Damascus. In *Religious Foundation of Turkey [TDV] Encyclopedia of Islam*. Accessed from https://islamansiklopedisi.org.tr/sam--suriye at 18/03/2020

Tourism, E. T. S. (n.d.). *Places to Visit in Mecca*. Accessed from https://www.etstur.com/Hac-Umre/Mekke-Ziyaret-Yerleri

Turkish Cooperation and Coordination Agency (TIKA). (n.d.). *Palestine*. Accessed from https://www.tika.gov.tr/upload/2016/Prestij%20Eserler/T%C3%BCrkce%20Filistin%20Rehberi.pdf

Turkish Knowledge. (2020). *Nishabur*. Accessed from https://www.turkcebilgi.com/

Turkish Science. (2020). *Ancient Tus City History and Oghuzes*. Accessed from http://www.turkbilimi.com/antik-Tûs-sehri-tarihi-ve-oguzlar/

Ulucam, A. (2010). Damascus. In *Religious Foundation of Turkey [TDV] Encyclopedia of Islam*. Accessed from https://islamansiklopedisi.org.tr/sam--suriye#2

Unal, S. (2001). Kaaba. In *Religious Foundation of Turkey [TDV] Encyclopedia of Islam*. Accessed from. https://islamansiklopedisi.org.tr/kabe at 20/03/2020

Uzun, M. I. (2001). Kaaba. In *Religious Foundation of Turkey [TDV] Encyclopedia of Islam*. Accessed from. https://islamansiklopedisi.org.tr/kabe at 20/03/2020

YalçınM. F. (n.d.). *Medina*. Accessed from https://www.islamdusunceatlasi.org/detail/place38 at 20/03/2020

Yasaroglu, M. K. (2001). Kaaba. In *Religious Foundation of Turkey [TDV] Encyclopedia of Islam*. Accessed from. https://islamansiklopedisi.org.tr/kabeat 20/03/2020

Yilmaz, G. (2010). *Central Damascus of Syria*. Accessed from https://gezimanya.com/GeziNotlari/suriyenin-merkezi-sam at 18/03/2020

Yuksel, G. (2020). *My Jerusalem Travel Notes*. Accessed from https://www.gorkemliyollar.com/kudus-gezi-notlarim/

ADDITIONAL READING

Caliskan, U. (2018). Faith Tourism and "Possible" Faith Tourism Routes between Turkey and Georgia. *Anemon Mus Alparslan University Social Science Journal*, *6*(3), 379–388. doi:10.18506/Anemon.337854

COMCEC Coordination Office. (2016). Muslim Friendly Tourism: Understanding the Demand and Supply Sides in the OIC Member Countries. Accessed from http://www.sbb.gov.tr/wp-content/uploads/2018/11/Muslim_Friendly_Tourism_Understanding_the_Demand_and_Supply_Sides_in_the_OIC_Member_Countries%E2%80%8B.pdf at 22/03/2020

Jamal, A., & Griffin, K. (2018). *Islamic Tourism: Management of Travel Destinations*. CABI Publications. doi:10.1079/9781786394132.0000

Kim, B., Kim, S., & King, B. (2019). Religious tourism studies: Evolution, progress, and future prospects. *Tourism Recreation Research*, 1–19. https://www.researchgate.net/publication/335943540_Religious_tourism_studies_evolution_progress_and_future_prospects/citation/download at 22/03/2020. doi:10.1080/02508281.2019.1664084

Tala, M. L., & Padurean, M. (2008). Dimensions of religious Tourism. *Amfiteatru Economic*, *10*, 242–253. https://www.researchgate.net/publication/233752241_Dimensions_of_religious_tourism at 22/03/2020

Turker, N. (2016). Religious Tourism in Turkey. In I. Egresi (Ed.), *Alternative Tourism in Turkey. Geojournal Library* (Vol. 121). Springer. doi:10.1007/978-3-319-47537-0_10

Turkey, Go. (Nd.) Religious Tourism. Accessed from https://www.goturkey.com/attractions/cat/religious-tourism at 22/03/2020

KEY TERMS AND DEFINITIONS

Ghazali: A famous Islamic scholar who lived in the Middle Ages. Apart from Islamic topics, he has many works on philosophy, logic, and Sufism. He is a thinker who has influenced the Islamic world.

Islam: The last divine religion with the Prophet Muhammad. It is one of three divine religions, along with Judaism and Christianity.

Islamic Cities: Cities that were under the rule of Muslims in the past or where Muslims currently live.

Religious Tourism: It can be defined as "Faith Tourism", which evaluates the touristic purpose trips, which are made by people in order to realize their religious beliefs and see belief centers, outside the places where they live, work and meet their usual needs.

Sacred Cities: Cities considered sacred by religious people and visited for this purpose. The cities considered sacred for Muslims in this study are Mecca, Medina and Jerusalem.

Tourism: It is a visit to a country or region to rest, travel, see or get to know.

Chapter 4
Religious and Spiritual Tourism: From Its Origins to Alentejo (Portugal)

Bruno Barbosa Sousa
https://orcid.org/0000-0002-8588-2422
Polytechnic Institute of Cávado and Ave, Portugal & CiTUR, Portugal

Cristina Castro
Polytechnic Institute of Cávado and Ave, Portugal

Maria Emília Luís
Polytechnic Institute of Cávado and Ave, Portugal

Paulo Lopes
Polytechnic Institute of Cávado and Ave, Portugal

ABSTRACT

Over the years there has been an increase in competition among tourism destinations, leading to the need for a deeper understanding about the tourism realm, impact, and management. Tourism has expanded its scope, reflecting an increasing recognition in the academic community paralleled by the application of interdisciplinary concepts and methods. Indeed, research in tourism has been studying its various implications from a multitude of perspectives and with interdisciplinary insights. Areas in tourism research entail planning of tourism destinations, local development, environmental impact, territorial brand management, and tourist loyalty. With new motivations, new niches are created, and it is one of those niches that the authors address in this work, namely, spiritual tourism. Spiritual tourism derives from religious tourism because it is linked to spiritualism and is associated with wellness tourism. This manuscript aims to present the importance of spirituality in the context of tourism (Alentejo, Portugal). In the end, new lines of research will be presented for the future.

DOI: 10.4018/978-1-7998-5792-1.ch004

INTRODUCTION

Over the years there has been an increase in competition among tourism destinations (Ferreira & Sousa, 2020), leading to the need for a deeper understanding about the tourism realm, impact and management. Tourism, as an area of study, has expanded its scope, reflecting an increasing recognition in the academic community paralleled by the application of interdisciplinary concepts and methods (Jafari & Aaser 1988). Indeed, research in tourism has been studying its various implications from a multitude of perspectives and with interdisciplinary insights (Echtner & Jamal, 1997). Areas in tourism research entail, for example, planning of tourism destinations, local development, environmental impact, territorial brand management and tourist loyalty (Getz, 1986; Embacher & Buttle, 1989; Backman & Crompton, 1991; Sousa & Rodrigues, 2019; Alves et al., 2020). The field of tourism marketing, in particular, has faced increasing challenges in capturing market dynamics, such as, market fragmentation and diversity. New habits, needs and trends in the global tourism arena create more sophisticated consumers who systematically look for different and specific experiences. Such context calls for new market approaches (Sousa & Silva, 2019).

Of particular relevance to this study is the idea of religious tourism and its experience. The history of civilizations reflects the importance of spirituality in the development of societies (Smith, 1992). Historical heritage is largely formed by religious features and religious sites are often relevant centres of attraction for visitors (Navruz-Zoda & Navruz-Zoda, 2016). Religious tourism is *"oriented around the motivations of the tourists. It usually includes religious ceremonies, festivals, sacred places and shrines, and conferences. In many cases, ceremonies and commemoration days dictate the timing of this form of tourism, which often has a seasonal character"* (Maoz & Bekerman, 2010, pp. 424). Pilgrimage centres are located in important tourist destinations, such as, Rome (Italy), Lourdes (France) or Fatima (Portugal). Thus, for some regions, religious tourism can be a platform for tourism development and, consequently, have a favourable economic impact in the local economy (Kartal, Tepeci & Atli, 2015; Nolan & Nolan, 1992; Vukonic, 2002).

The tourism system entails interactions between local communities and the tourist (e.g. Larsen, 2007; Neuhofer et al., 2014; Pinto da Silva et al., 2019) that have relevant implications for the tourist experience. Religious tourism, especially within its spiritual scope, contains unique traits considering that offer and demand are endowed with great spiritual, emotional and affective connections (Nolan & Nolan, 1992; Vukonik, 2002). Tourists who are attracted to religious sites may have different needs and expectations from pilgrims. Managers develop measures to reduce conflicts between tourists and pilgrims, for example, establishing suitable dress-code indications or behaviors (Nolan & Nolan, 1992; Olsen & Timothy, 2006; Shackley, 2006; Maoz & Bekerman, 2010). Thus, transposing the idea of experience to religious destinations raises specific aspects that are worthwhile addressing. Although there is a relevant line of research investigating experiences, the area needs further development (Ritchie & Hudson, 2009) to capture the dynamic nature of the tourism activity with experiences subject to permanent change and transformations (Neuhofer et al., 2014), as it happens in the religious context. Existing studies show limitations regarding aspects of consumer behaviour, such as, the profile, needs and motivations of pilgrims (Amaro, Antunes & Henriques, 2018).

The conception of the idea of tourism has evolved a lot in recent years. Since mankind early days human beings has traveled. In fact, in the beginning of our history, as a race, man did not stay in the same place for long. We became known as nomads, always from one side to the other, looking for the best piece of land, the best weather conditions, with the best resources. Since we started to stay in the same place, the motivations of the travels changed, the hunting trips and the commercial trips began to appear and only later, already in the classic era, the ancient Greeks began to travel for academic or sports reasons. But it is unthinkable to link that period to an idea of tourism. The very word "tourism", according to some historians, has its origin in the 19th century and derives from the term Grand Tur. The motivations of leisure, health and rest began to appear, however, at that time, only those who belonged to a higher social class had the privilege of traveling (Tosqui, 2007).

Since then, tourism has been emerging and, after the World War II, the concept of modern tourism is born. There started to be a need to get out of the comfort zone. The main motivations of the visitors were rest / leisure, sun and sea tourism and its first tourist offers began to appear. With the appearance of lowcost airlines, with the greatest literacy of the travelers and with the emergence of the Internet, tourism was segmented into multiple tourist niches. Sun and sea are no longer the tourism of choice. Cultural tourism, rural tourism and religious tourism started to gain preponderance within the sector and new motivations have been conceived by the tourism consumers (Barretto, 1995).

With new motivations, new niches were created and it is one of those niches that we will address in this article, namely, the spiritual tourism. Spiritual tourism derives from religious tourism because it is connected to spiritualism and is associated with well-being tourism; its consumers look for it as an alternative method to the cure of some psychological illnesses such as anxiety or stress. Then, there are those who seek it to achieve peace of mind In fact, it can be summarized that the search for peace of mind is the main motivation of its consumers when deciding the destination.

Based on the underlying idea of religious tourism, this chapter addresses the specific needs of the tourist market in a religious tourism context with an interdisciplinary perspective (marketing and tourism). We focus on the study of experiences in Religious and spiritual tourism. In particular, we discuss factors that may influence the attitudes and behaviours of consumers in the context of the Alentejo (Portugal). The study further allows to draw relevant managerial implications and suggestions for future research.

BACKGROUND

Spirituality

In recent years there has been a greater demand for a variety of beliefs and practices, with emphasis on Holistic practices and Yoga, for our study, it matters the Holistic practice, an experience of spiritual character. However, both promote the balance between body, spirit and mind.

The importance of spirituality, which is felt, gains prominence in a population overloaded with materialism, factors of extreme poverty, environmental and social concerns, life in constant anxiety, a population that permanently struggles to find inner peace (Willson, 2011).

A western phenomenon, where there is the movement of the tourist in order to give meaning to his life, in an attempt to better understand his inner self.

Spirituality is a 'unity' with the world, clarity of thought... just feeling the best I can be. It is any kind of connection with your inner self and with whatever is around you, alive or not. - (Sharpley & Jepson, 2011).

Taking time away from the rhythm of life, the rural community is managed at a completely different pace... people have time for each other. You have time to appreciate the things you don't have time for in the normal five-day working life - (Aitchison, MacLeod, Macleod, & Shaw, 2002)

Therefore, spiritual tourism includes the holistic, health and well-being aspect which, despite being considered different, comprise ideals that intersect in a set of characteristics that aim to provide the tourist with diverse experiences that come together. The study of social and environmental psychology was and is essential to understand the motivations of the tourist to visit a certain place.

What motivates the tourist to have a set of characteristics that go from the way he moves, where he moves, how long he moves, why he moves and, above all, what he expects to acquire from all these experiences?

It is clear that a set of economic and social factors will not only influence but also be decisive in each act or action that the tourist takes in relation to their trip; however, personality, state of mind, attitudes and values will unconsciously affect and determine each particular choice.

Dann (1981), indicates that the tourist motivation comes essentially from a state of tension that consequently promotes the travel, the same author considers that this motivation is decisive in behavior.

One of the main theories for the motivation of the tourist is the Maslow Pyramid (Maslow, 1954), that you propose a hierarchy of needs that reflect the orientation and behavior to achieve a certain goal.

Another great reference is the Push & Pull Model (Dann, 1977), which is based on two dimensions: - Psychological and social need; and - Attributes of destiny. In this specific case, push factors, are mainly factors that make the tourist want to live the experience, while pull factors have to do with the attraction that the place provides and reflect on the tourists choice, linked to the tangible aspects of the tourist destination.

Regarding spiritual tourism, Mélanie Smith & Puczkó (2009) indicates four points that will determine the tourist's choice:

1. need for relaxation;
2. healthy habits;
3. health and beauty care;
4. psychological rewards.

Studies on motivations date back to the 1930s, being the order of the day as one of the most discussed topics in consumer behavior, which reflects the tourist's need and the search for reward, with the need having a psychological character, reasons emotional and cognitive, beliefs and knowledge. The stress and daily tension are the main responsible for this behavior, in the search for a concept of spiritual nature, which makes the tourist travel long distances, in the search to satisfy their needs, combining the search for new experiences to the authenticity of the product / place.

Spiritual tourism has as its main motivation the escape from the routine, from the daily life that reveals itself to be tiring and predictable, besides escaping from the monotony of its "I", the tourist, when seeking this concept of spiritual tourism, goes in search of the divine, to connect and discover the mysteries of creation, to discover oneself in search of truth and values, in the discovery of its "I" and attempt to complete.

TOURISM: HEALTH AND WELL-BEING

Religious tourism is exclusively or strongly motivated for religious reasons. One of the oldest types of tourism and a worldwide phenomenon of religious history, it can be differentiated into several forms. Alecu (2010) and Pourtaheri et al. (2012) demarcate the specific elements and motivations of religious tourism, pilgrimage and tourism in general. There are established links of pilgrimage with belief, religious tourism with culture and tourism with entertainment (Pourtaheri et al., 2012). Figure 1 summarises this idea. In addition, religious tourism has political and social extensions (Rinschede 1992). Numerous religious sites are correspondingly national symbols. For example, Guadalupe and the Wailing Wall (Jerusalem) representing, respectively, Mexico and the Judaism. Fatima, a religious sanctuary in Portugal, constitutes another symbol of gathering for nationals and expats that come together on specific dates to celebrate the apparition of Holly Mary. Mazumdar & Mazumdar (2004) refer to the pilgrimage in Islam, as a religious motivation, that has important implications for place attachment: (i) the pilgrimage emphasizes the importance of the site, particularly in those locations associated with the Prophet; (ii) pilgrimage requires all believers to visit the place and experience its physical qualities; (iii) the identification with place becomes an integral part of Muslim identity.

The practice of activities related to health and well-being has taken place since ancient civilizations, where the medicinal demand for the cure of diseases was essentially linked to the valorization of water as a symbol of healing properties. Turkish baths, mainly practiced in greek, persian and hebrew culture, according to Melanie Smith & Kelly (2006), they were seen to have healing effects. Later the public baths in Ancient Rome which came to popularize the roman baths, turning these spaces into true social and political meetings, with the exchange of intellectual and social ideas. Today, and with the evolution of times, even though we can still consider them as a medicinal activity, contemporary society and its life in constant pressure, either by external or internal factors, assumes a greater need to seek an escape from the demands of modern times. In this sense, it adopted health and well-being activities as a concept of liberation and a new vision on how to live new experiences, a new aspect, linked to relaxation, psychological and physical well-being.

It is during the 21st century that the growing interest in motivations and expectations that lead tourists to travel to live these experiences becomes an object of study. The development of the tourism product within this concept, seeks to promote well-being and prevent illness, luxury hotels and spa, occupational therapies, aesthetic treatments, spiritual activities, massages, rehabilitation programs, among others, are then associated with this new dimension which generate an increase in demand and how to ensure a better physical, mental and spiritual condition. The result for a healthier lifestyle from a laser perspective.

According to Puczko & Smith (2008), there are five types of health and wellness tourism:

- Holistic (spiritual, yoga, meditation)
- Retreats and Festivals
- Laser and entertainment (treatment, beauty, sports, fitness)
- Medical well-being (shape recovery, rehabilitation)
- Medicinal and therapeutic (rehabilitation by disease, cure, recovery)
- Spa
- Surgical medicine (cosmetic surgery, operations)
- Hospitals and Clinics

Melanie Smith & Kelly, (2006), assume that Wellness Tourism as a Holistic lifestyle, it comprises the act of traveling to achieve personal well-being, maintenance and rejuvenation of body, mind and spirit.

SPIRITUAL TOURISM

Nowadays, the daily life of an individual is characterized by his constant movement. The lack of personal time allied to the stress caused by work, among other factors, creates negative feelings such as fear, insecurity, anger and rage. These feelings are becoming more and more cohesive in our society and, each passing day, they leave more and more marks, triggering a need to go to far and quiet places in search of peace of mind. This search has become one of the main motivations of consumers when it comes to choosing a destination.

According to Willson (2011), a growth at this level is thought of as a western phenomenon, as the eastern societies never involve the question of the spirit and don't face a series of issues that negatively influence the individual. Therefore, we can say that western society has adopted the "alternative" methods of the east in order to combat the illness of our society.

Heelas & Woodhead (2005), claims that contemporary western society has come to be defined as a "spiritual revolution". The search for peace of mind is subjective, there is a variable of methods (always with travel as its basis). Methods such as retreats, nature walks, self-discovery trips, among others, help the individual to achieve greater tranquility, well-being and, in many cases, help them to find a meaning/purpose of life.

We cannot leave aside the link between spiritual tourism and religious tourism. As already mentioned, these two concepts of tourism are associated with each other's tourist motivations. A major example of this are the pilgrimages, in which many devout travelers walk to a religious destination with faith as their main motivation and, in addition, a journey of personal development. The Caminhos de Santiago are one of the greatest examples of two tourist concepts that can be combined. In other religions, we can also highlight Buddhism with meditation and Hinduism with yoga.

SPIRITUAL TOURISM IN PORTUGAL

Despite the growing importance of spiritual tourism, this idea of tourism is not yet well established in Portugal. However, it is a country that gathers good characteristics for the development of this niche. It has a strong connection with pilgrimages (Fátima and Os Caminhos de Santiago) and regions such as Gerês, Lisbon, Azores, Minho and Algarve have excellent conditions for the organization of spiritual retreats. The website turismo.eurodicas.com.br published on June 27th 2019 the tenth best retreats in Portugal:

1. **Retiro de Silêncio**, Quinta da Costa in Famalicão and Mouquim

   Objective: To remain silent for two days, meditating to the sound of Tibetan goblets and gongs.

2. **Centro Budista**, Oporto

   Objective: To promote Buddhist knowledge and practice as a culture of peace and meditation.

3. **Centro de Retiros Karuna,** Monchique

Objective: To rest the mind throughout the silence; to relax, listen to nature and reflect on life.

4. **Center Upaya**, Lisbon

Objective: Share methodologies that can provide a more conscious experience.

5. **Vida Desperta**, Benfeita

Objective: Focus on the liberation of the spirit within personal evolution.

6. **Casa Shanti**, Mafra

Objective: To develop and disseminate methodologies that help people to deal with and combat stress, collaborating for a physical, mental and spiritual regeneration.

7. **4 Ventos**, Sobral de Abelheira

Objective: In a community and solidarity way, the participants take care of the space, tidying the bed, washing the dishes and of course, participating in various activities that strengthen the spirit of collectivity.

8. **Retiro d' Aveiras**, Aveiras de Cima

Objective: This retreat aims to show paths to inner peace, to find the light we all have and, together with Mother Nature, to discover the inner animal of each of the participants. It is one of the few spiritual retreats in Portugal that approach Shamanism.

9. **Centro Tao**, Figueira da Foz

Objective: through yoga and other types of therapies aimed at the pursuit of happiness, growth and self-knowledge, through multidisciplinary and holistic practices.

10. **Casa BMQ,** Coimbra

Objective: Dedicated to the traditional teaching of yoga and Ayurveda philosophy

Religious and spiritual tourism has been discussed in the literature as one of the growing fields of tourism, which should be capitalized in terms of marketing. The debate on the importance of, and distinction between, tourism as a ritual and as a form of spirituality is not conclusive (UNWTO, 2011). Pilgrimage is one of the oldest forms of tourism (Pourtaheri et al. 2012). The literature typically focuses on the 'tourist as pilgrim', exploring the similarities and distinctions between contemporary religious tourism and pilgrimage (Sharpley & Jepson, 2010). Nowadays, the daily life of an individual is characterized by his constant movement. The lack of personal.

THE RETREAT CENTRE OF ALENTEJO: THUBTEN PHUNTSOG GEPHEL LING RETREAT CENTRE

Spiritualism or Spiritual Tourism is seen as an escape from reality, that is, the type of tourism adopted by a group of individuals, whose reason for their displacement is the search for spiritual peace (Cunha, 2013), is linked to cultural tourism which consists of "Trips provoked by the desire to see new things, to increase knowledge, to know the particularities, and habits, (...) or even, the satisfaction of Religious Spiritual needs, Cultural Centers, (...), great Cathedral Religious Monuments, Buddhist Temples, the places where great civilizations of the world developed in the past, the great pilgrim centers, (...) are the preferences of tourists". We are facing a new spiritual social phenomenon, disconnected from traditional religions, with new forms and new experiences with oneself and with nature. Taking into account the relationship between Spiritualism, Spiritual Tourism, Holistic Tourism, Buddha, God, we focus on the one that arises from a need for adaptation, thoughts and actions, in view of new urgent challenges.

Spiritual Tourism is linked to Religious Tourism, this link is ancient, assumed as an engine in all cultures, which develop based on a religion, thus playing a fundamental role in people's culture. Religious Tourism is a universal phenomenon, (Geary, 2018) not only by visits to sacred sites, but essentially by emotional experience, spiritual living, the faith of those who believe in God, the mother of God, and all the saints (Antunes, Barroco Novais, & Dias, 2016).

The current "phenomenon" prevents people from coming together, the universe is experiencing an unprecedented economic and social change, the monuments that were visited are currently isolated from the human presence, religious retreats and spiritual retreats are closed, but each person's faith is not closed (Antunes et al., 2016). Nowadays one of the main engines of the economy are the social networks that present an alternative that allows, to put ideas in practice and at a low cost, without leaving home (Nunes, 2018), the development of spiritual tourism and religious tourism is essential today, it is the social networks that best play the role, to attract people of all nationalities (Yorgason & della Dora, 2009).

According to the Perennial Philosophy by Aldous Huxley, it quotes mystics of all religions and understands that throughout history, men and women of all religious beliefs have sought and found hidden paths that led to the same destiny, peace within. (Easwaran, 2016).

Sometimes we try to escape reality with escapism, when life is upside down. The need arises to make some adjustments, redefine priorities, rest, ask for help, say "no", delegate or even stop, when the pressure increases. Every human being goes through this, reaching that point where any little thing seems to make the person angry and frustrated. The journeys and frustrations vary from person to person, depending on what they believe in, but the God and the culture of each one leads to the same journey. In India, meditation is called The End of Suffering and Mastery of the Art of Living, the words of St. Francis have an almost universal appeal (Mackey & Sisodia, 2013).

Lord make me an instrument of your peace.
Where there is hatred, may I take love;
Where there is offense, may I take forgiveness,
Where there is doubt, let me take the faith;
Where there is despair, may I take hope;
Where there is darkness, may I take joy.
O Divine Master, let me seek more, comfort than comfort;
Understand that to be understood;

Love to be loved;
For it is in giving that one receives;
It is by forgiving that one is forgiven;
It is by dying to yourself that you are born to eternal life.
(Louro, A. Padre, S.S.P, Oliveira, R. Frei, O.P., Filipe, J. Padre, Ribeiro, 2015)

Through this prayer to the "Lord" or "God," St. Francis shows the transformation of character, guide and consciousness that the human being is capable of giving existence to something profound. He does not refer to something external, but to appeal to our deepest Self, which is the same in all human beings (Easwaran, 2016).

God is an underlying reality that is closer to each one than the body itself. In all religions there is a tradition that defends the presence of this divine flash in all creatures. It is not necessary to change religious convictions, meditation can be practiced in any religion, or even when no religion is practiced. It is a dynamic, powerful discipline, to train the mind to remain calm, free and pious in any context, it is the very essence of religion. Meditation using texts of all religions, based on excerpts from inspirational texts of the great scriptures and mystics of the whole world, thus allows a different vision (Easwaran, 2016).

AN ACCELERATED MIND IS SICK
A SLOW MIND IS HEALTHY
A STILL MIND IS DIVINE
(Eknath Easwaran, (2016, p.25)

Meditation, according to great mystics of all cultures and times, requires great effort, not relaxation, but the awakening of consciousness, an interior discipline that involves breathing. Meditation means to train the mind so that our thoughts go where we send them and remain there, until an opposite order. (Mackey & Sisodia, 2013).

The transformations in society and the changes in the lives of human beings give rise to new challenges and concerns about well-being, physical, psychological, ecological, social, cultural, emotional, ethical and spiritual (Mackey & Sisodia, 2013).

SPIRITUALITY HAS EVERYTHING TO DO
WITH THE RELIEF OF SUFFERING
BUDA

"BUDA" - was the name given to Siddartha Gautama, the prince received a formal education that should make him a warrior and future king, but contrary to what his privileged condition made him foresee, he showed a great tendency for meditation and philosophical and spiritual thinking. At the age of twenty-nine, when on a stroll through the city of Lumbini, now the territory of Nepal, he encountered for the first time the profound misery, old age, sickness and death in which the people of his country lived, he had his spiritual awakening. His birth was in Asia and is celebrated on April 8, and he died in 483 B.C., in Cadia, India, at the age of 80, gathered numerous followers who, for 45 years, followed him and recorded his thoughts. Also in Portugal, one of the largest retreat centres of Tibetan Buddhism in Europe was inaugurated in September 2014, an area of about 100 hectares that wants to put the Alentejo on the international route of Buddhist religious tourism.

THERE ARE THREE KINDS OF UNHAPPY PEOPLE:
THE ONES WHO DON'T KNOW AND DON'T ASK,
THE ONES YOU KNOW AND DON'T TEACH,
AND THOSE WHO TEACH AND DO NOT.
BUDA

Thubten Phuntsog Gephel Ling Retreat Centre, is the first of its kind in Portugal and Europe, an initiative of the Tibetan Pendê Ling Study Centre, based in Estoril, a community of practitioners with almost a decade of existence, is located in Santa Susana, in Alcácer do Sal and was born to enable "traditional and more intense" training for Buddhist students and to invite to Portugal "renowned masters in Tibetan meditative tradition", according to the Lusa o Lama Guyrme agency, one of the masters who rules the centre of Estoril. Contemporary of the Dalai Lama, Khochhen Rinpoche, who fled to India in 1959, is one of the oldest Tibetan masters still alive in the Nyingma tradition, one of the Tibetan schools.

Figure 1. Stupa Tashi Gomang Temple

There are Buddhists around the world, Tibetans, Tibetan refugees from Europe, and in the United States and Nepal, and worldwide, there are an estimated 376 million practitioners and followers of Buddhism.

In Portugal, there are an estimated 15,000 Buddhist practitioners who, according to Lama Guyrme, are leaving some marks. In the greater Lisbon area, there are 3 temples and a stupa, and they are represented by Burmese, Chinese, Japanese, Thai, Tibetan and Vietnamese traditions. The infrastructures built in Santa Susana, Alcácer do Sal so far have the capacity to accommodate 30 people. We are facing a new reality, in which living in the updated and developed world is increasingly practiced the devotion of ancient times. Buddhism is a 2,500-year-old spiritual tradition focused on personal development and the search for deep inner knowledge.

Throughout these years the need arose for people to know the benefits of Buddhist culture, the good relationship between Buddhism and science is one of the aspects that makes this religion very interesting for the modern public.

These retreats are spaces managed and inhabited by the owners just as Alcácer do Sal retreat. It will be permanently inhabited by five people and because they are small and medium sized organizations that, as a rule, are located in regions with access to the sea or in mountainous areas.

The duration of the retreats varies between one week and several months, and most of them offer pre-defined programs of one week. So, retreat visitors can reside in the space and volunteer to help keep the space clean both indoors and outdoors.

Considered a spiritual tourism, these centers have programs with high values for dealing with specific themes: health, well-being, spiritual practices, nutrition, with various activities (yoga and meditation classes, treatments and programs that promote a healthy diet).

In general, people seeking spiritualist, religious, buddhist or holistic tourism travel to destinations far from home (at his home country or abroad) looking for new ways to distance themselves from the reality in which they live (professional routine), experiencing a sense of freedom (escapism), (relaxation and inner renewal). frequenting spaces such as spas, resorts, cruises, wellness centers and retreat centers.

YOU CAN READ MANY SACRED SCRIPTURES
AND TALK A LOT ABOUT WHAT YOU READ
AND LEARN,
BUT WHAT GOOD IS IT IF YOU DON'T PUT THEM
IN PRACTICE?
BUDA

The spiritual tourism is associated with the realization of retreats in Portugal and in Europe. These retreats are also part of the forms of spiritual holistic tourism, namely in yoga and meditation, health and welfare, includes, beauty treatments, sports, fitness and pampering, also associated with these forms' leisure centers are the festivals. Spiritual tourism is a practice as the tourist decides to participate in the meditation retreat. It is also associated with the various religious currents, becoming a degree of consciousness and spiritual enlightenment accomplished through rituals, ceremonies or traditions, which may or may not be of a religious nature. From the perspective of the author Deepan, religion is not the same as spirituality, nor God it is the same thing as spirituality, therefore, science has long since discredited religion in people's thinking.

"HE WHO PROTECTS HIS MIND

OF GREED AND ANGER,
ENJOY TRUE AND LASTING PEACE"
BUDA

These retreats are characterized by being spaces managed and inhabited by the owners as in Alcácer do Sal. It will be permanently inhabited by five people, and by being small and medium sized organizations that, as a rule, are located in regions with access to the sea or in mountainous areas. The duration of the retreats varies between one week and several months, and most of them offer pre-defined programs of one week. Therefore, retreat visitors can reside in the space and volunteer to help keep the space clean both indoors and out.

The wise men say, "look inside and you will find inside yourself the true source of everything: your own conscience." Once religion has ceased to realize or prophesy, it is enough to say that God is inside each one of us, as Jesus Christ declared.

Common in all religions is God everywhere, the qualities worthy of love, passion, clemency and justice.

Spirituality, offers wisdom, freedom and fulfillment, being pointed out as a present alternative. Science and religion try to deal with the difficult issues of daily life or to withdraw passively and contemplate a religion.

In meditation we focus the mind gradually, so that when we encounter difficulties, we can eliminate everything that is not essential. We become what we pay attention to. People who dream and think about money, property, real estate, income and losses, have minds obsessed with power, pleasure, fame, and revenge.

For this reason, Buddha opens his Dhamampada with a wonderful phrase: "All that we are is the result of what we thought." Nowadays and despite all technologies and science, human beings are very insecure because they insist on wanting things that they are unable to acquire, and these things do not give them tranquility or security. Meditation becomes engraved in consciousness, words come to life within people, transforming thoughts, feelings, words and actions (Easwaran, 2016).

The recommended time to meditate in a mild climate is between five and six o'clock in the morning, always at the same time, for thirty minutes is enough, meditating for one hour can be too much, taking into account that there is loss of concentration and daily life can be affected. Starting the day early, doing physical exercise, meditating, creates a relaxed state of mind to face the day (Easwaran, 2016).

The scriptures say that the place of meditation should be calm, clean and cool and well ventilated, if possible, having a space only for meditation is ideal. If it is possible to have images that inspire you deeply, such as God, Buddha, Mother Teresa. Wearing wide and comfortable clothing, the only equipment needed is the willingness. If you can meditate with family and friends, this strengthens relationships, remembering the words of Jesus "Where two or three come together in my name, I am in the midst of them". Sitting upright with the spine, the nape of the neck and the head straight, relaxed and natural if you prefer a chair, it should have arms in the event of a fall, or on a cross-legged pillow lying on the floor. As you start to relax, the neuromuscular system improves and a drowsiness may arise, at that moment, keep your spine straight and pay more attention to what you hear, however, you may encounter physical obstacles, distractions, strong emotions, if this happens, open your eyes for a minute or two and repeat (Easwaran, 2016).

The human being needs to establish intimate relationships and feel that he is part of something in which he is accepted, although in modern times we are connected with all technology, the truth is, it is a superficial link and does not satisfy basic needs. In this way, it is necessary to be assigned a rational behavior, when one becomes without a sense of purpose and energy, it is time to take a step back and dedicate oneself. This requires resources such as the ability to concentrate in order to eliminate compulsive behavior and to be able to improve one's way of life with oneself and with others. Naturally, man faces meddling, ingratitude, arrogance, dishonesty, envy and avarice, all of this, on the part of those who do not know the difference between good and evil. But I have seen the beauty of good and the horror of evil, and I have realized that the nature of those who practice this evil is similar to mine, not of the same blood, or cradle, but with the same mind and with a portion endowed with the divine; and so none of these things can hurt me, no one can involve me in what is ugly. Neither can I be angry with my fellow man or hate him, because we were born to work together, like feet, hands and eyes, with the two rows of teeth, those above and those below. Creating difficulties with each other is against the law of nature. To be bored with someone, to turn your back on them: these are obstructions (Aurélio, 2019).

To help indelicate people, you cannot let them make their pretensions, it feeds their indelicacy, before this, the mind must remain calm and compassionate, it is important to wear it with patience. It is difficult, but important not to act in the warmth of the moment, walk if possible and repeat the mantra, meditate and then respond, the best way to change someone is with your own example, if the person shows signs of repentance, it is important to maintain friendship again and forget the wall that separated human beings (Easwaran, 2016).

The power of the mantra, is to call the thoughts where you want them to go, whenever they refer to the dark of the past, gradually with practice, the thoughts will circulate less frequently, and in time they may disappear. The mystics say that *"bad habits of the mind such as: non-stop thinking, needs, desires, fears, can make life hell, the more ingrained this pattern of thinking is, the mystics say, the more we become small islands isolated from the rest of life, with all the unhappiness that necessarily follows"*. There is, however, no instruction manual that shows the right direction, and life is a process of trial and error, which one learns as the human being advances (Hardy, 2918).

Jesus and Buddha described reality from knowledge, and after 2,000 years, we think we know more, however, and to prove the contrary ALBERT EINSTEIN said, "The further the spiritual evolution of humanity progresses, the more certain it seems to me that the genuine path of religiosity does not lie in fear of life, fear of death or blind faith, but goes through the struggle for rational knowledge.

"ENGINEERS DESIGN CHANNELS,
ARCHERS CARVE ARROWS,
THE CRAFTSMEN MODEL THE WOOD AND CLAY,
THE WISE MAN MODELS HIMSELF"
BUDA

Holistic Tourism associated with retreats has grown since 1980, with great relevance in recent years, appearing as an alternative to traditional tourism. This type of tourism is considered by research as a high-level product in the spectrum of health and well-being. In fact, Holistic Tourism is considered a thriving market in the Wellness Tourism industry, representing about 6% of domestic and international travel. It reached it peak in the year 2017, with a growth of 50%, faster than that of other markets.

As a result of the significant increase in demand for retreat centers, tourist destinations fill the gap that exists in this sector by investing in the promotion of existing spaces and also in the creation of new ones, there are multiple retreat centers in several countries that offer superior quality products such as, (meals, accommodation and activities) that allow the tourist to feel comfortable and relaxed.

AFFECTIVE AND BEHAVIOURAL INTENTIONS OF RELIGIOUS AND SPIRITUAL TOURISM

This chapter discusses the increasing phenomena of religious and spiritual tourism and the marketing perspective. Based on the literature review, we develop research propositions that connect the affective marketing dimension developed by the branding of the place (i.e., place attachment) with the consequent satisfaction with the pilgrimage experience and the behavioural intention of repeating or reinforcing the experience by taking other spiritual opportunities – (i.e., loyalty to the religious and spiritual tourism). Several studies in marketing and tourism contexts have discussed the association between affection to the place and consumer purchasing behaviour, including the study of satisfaction, loyalty or quality of service (Hwang et al., 2005). In the specific contexts of tourism, as in the cases of religious tourism and spiritual tourism, several challenges emerge (Dann, Nash & Pearce, 1988) when investigating the relationship between visitors' affection to a particular place and their behavioural intentions in the context of religious and spiritual tourism. Individuals "may have a particularly strong attachment to a particular place or structure more than others in that religion or that collective, and in some instances the attachment may be less intense or nonexistent for an individual" (Mazumdar & Mazumdar, 2004, pp. 387). For example, a shrine may be especially significant to a group because of site-related religious events. Alternatively, a non-religious person can develop a strong individual attachment to religious places.

The literature does not show convincing results with respect to the effect of the local affectivity and satisfaction (Yüksel et al., 2010). Some authors argue that satisfaction, depending on how it is defined, can positively influence local affection (Hou et al., 2005; Sousa & Rocha, 2019), whereas other authors defend that consumer satisfaction, particularly in tourism contexts, may be influenced by local affection (Halpenny, 2010; Yüksel et al., 2010). Given the above, and since the supply of religious tourism can be designed to specific markets, we suggest that place attachment in its two dimensions of place identity and place dependence, can be assumed as a relevant variable to spiritual satisfaction and consequent loyalty. According to Scannell and Gifford (2010), attachment to places is an affective, proximity-maintaining bond that can be expressed without an underlying purpose of control, as in the case of sacred places. Place attachment behaviors include pilgrimages, social support, and place restoration. Socialization through visits to religious places and pilgrimage can be among the most intense learning experiences. In spiritualism, learning through rituals, texts, and storytelling, synergistically combine so that the experiential sense is heightened. Such context enables the pilgrim to witness the place qualities described earlier, through the engagement of all the senses. In this context, acting as collectives, persons can develop strong attachment to a religious place they consider significant or a place that the collective feels is significant. The literature highlights that satisfaction tends to favour the development of behavioural loyalty. The connection between satisfaction and loyalty is particularly relevant in tourism contexts through, for example, the importance of satisfaction in revisiting the destination or the recommendation to family and friends (Kozak, 2003; Bigné et al., 2005; Prayag, 2008; Prayag & Ryan, 2012). Based on this discussion, we propose that, in a context of religious and spiritual tourism, the satisfaction facilitates the development of behavioural loyalty.

CONCLUSION

The literature widely suggests that the tourist's assessment of different destination attributes influences the overall satisfaction and, subsequently, intentions to revisit a destination (e.g. Baker and Crompton, 2000; Kozak, 2001, 2003; Alegre and Garau, 2010, Krešić, Mikulić & Miličević, 2013). Previous studies typically apply linear modelling techniques in analysing these influences, such as multiple regression analysis or structural equation modelling (Krešić, Mikulić & Miličević, 2013). Theoretical background support the propositions presented in this chapter in the context of religious and spiritual tourism. In particular, the preliminary approach allowed the perception from a marketing perspective that feelings of place identity, in the emotional dimension of the journey, and the satisfaction with the experience may increase and consequently can reinforce the loyalty to the destination (in Alentejo, Portugal contexts). These contribute to tourist satisfaction and loyalty. The research suggests that this rationale is suitable for spiritual tourism. This study is a preliminary contribution towards a greater understanding regarding the relationship between the local affection and behavioural intentions in specific contexts of religious and spiritual tourism, considering the particular case of Portugal (Alentejo). Specifically, the research suggests that the attachment to the site in the contexts of religious and spiritual tourism, through the place identity and place dependence, is related to tourist satisfaction, as well as tourist loyalty (e.g., revisiting the site, complementing the spirituality with other places, recommending the experience to others).

FUTURE RESEARCH DIRECTIONS

This study is an exploratory research based on a preliminary approach. With the insights from this study it is expected that future work can contribute to the development of empirical studies to address the developed propositions. In particular it is expected that future research explores the relationship between place attachment, satisfaction and loyalty of religious and spiritual visitors in qualitative and quantitative studies. For example, studies may look at a deeper understanding of the phenomenon in various routes of Portuguese religious and spiritual places, as well as in other locals of spirituality, in order to find differences between the place attachment to spiritual locals and the loyalty to the experience. It is important to reveal the individual destination attributes that cause satisfaction and/or dissatisfaction, and shed light on the most determinant and critical attributes in explaining the overall tourist experience, in the specific the Portuguese case. Studies could further develop a questionnaire to be applied to participants to capture a cross-sectional view of the relationships among the constructs gauging their impact.

REFERENCES

Aitchison, C., MacLeod, N. E., Macleod, N. E., & Shaw, S. J. (2002). *Leisure and Tourism Landscapes: Social and Cultural Geographies*. Psychology Press.

Alecu, I. C. (2010). Epistemological aspects of religious tourism in rural areas. International Journal of Business. *Management and Social Sciences*, *2*(3), 59–65.

Alegre, J., & Garau, J. (2010). Tourist satisfaction and dissatisfaction. *Annals of Tourism Research*, *37*(1), 52–73. doi:10.1016/j.annals.2009.07.001

Alves, G. M., Sousa, B. M., & Machado, A. (2020). The Role of Digital Marketing and Online Relationship Quality in Social Tourism: A Tourism for All Case Study. In J. Santos & Ó. Silva (Eds.), *Digital Marketing Strategies for Tourism, Hospitality, and Airline Industries* (pp. 49–70). IGI Global. doi:10.4018/978-1-5225-9783-4.ch003

Amaro, S., Antunes, A., & Henriques, C. (2018). A closer look at Santiago de Compostela's pilgrims through the lens of motivations. *Tourism Management*, *64*, 271–280. doi:10.1016/j.tourman.2017.09.007

Antunes, J., Barroco Novais, C., & Dias, H. (2016). A importância do turismo religioso no desenvolvimento das regiões: O caso do santuário da nossa senhora da lapa. *International Journal of Scientific Management and Tourism*, *2*(1), 273–285.

Backman, S., & Crompton, J. (1991). The usefulness of selected variables for predicting activity loyalty. *Leisure Sciences*, *13*(3), 205–220. doi:10.1080/01490409109513138

Baker, D., & Crompton, J. (2000). Quality, Satisfaction and Behavioral Intentions. *Annals of Tourism Research*, *27*(3), 785–804. doi:10.1016/S0160-7383(99)00108-5

Barretto, M. (1995). *Manual de Iniciação ao Estudo do Turismo*. Papirus Editora.

Bigné, J., Andreu, L., & Gnoth, J. (2005). The theme park experience: An analysis of pleasure, arousal and satisfaction. *Tourism Management*, *26*(6), 833–844. doi:10.1016/j.tourman.2004.05.006

Cunha, L. (2013). *Economia e Política do Turismo*. Lidel: edições técnicas Lda.

Dann, G. (1977). NoAnomie, ego enhancement and tourism. *Annals of Tourism Research*, 184–194.

Dann, G. (1981). Tourist motivation. An appraisal. *Annals of Tourism Research*, *8*(2), 187–219. doi:10.1016/0160-7383(81)90082-7

Dann, G., Nash, D., & Pearce, P. (1988). Methodology in Tourism Research. *Annals of Tourism Research*, *15*(1), 1–28. doi:10.1016/0160-7383(88)90068-0

Easwaran, E. (2016). *Meditação*. Centro Lusitano de Unificação Cultural.

Echtner, C., & Jamal, T. (1997). The disciplinary dilemma of tourism studies. *Annals of Tourism Research*, *24*(4), 868–883. doi:10.1016/S0160-7383(97)00060-1

Embacher, J., & Buttle, F. (1989). A repertory grid analysis of Austria's image as a summer vacation destination. *Journal of Travel Research*, *27*(3), 3–7. doi:10.1177/004728758902700302

Ferreira, J., & Sousa, B. (2020) Experiential Marketing as Leverage for Growth of Creative Tourism: A Co-creative Process. In Advances in Tourism, Technology and Smart Systems. Smart Innovation, Systems and Technologies, (vol 171. pp 567-577). Springer. doi:10.1007/978-981-15-2024-2_49

Geary, D. (2018). India's Buddhist circuit(s): A growing investment market for a "rising" Asia. *International Journal of Religious Tourism and Pilgrimage*, *6*(1), 47–57. doi:10.21427/D7PT46

Getz, D. (1986). Models in Tourism Planning: Towards Integration of Theory and Practice. *Tourism Management*, *7*(1), 21–32. doi:10.1016/0261-5177(86)90054-3

Halpenny, E. (2010). Pro-environmental behaviors and park visitors: The effect of place attachment. *Journal of Environmental Psychology, 30*(4), 409–421. doi:10.1016/j.jenvp.2010.04.006

Hardy, J. (2018). *Se eu não cuidar de mim, quem cuidará?* (1ª Edição). Pergaminho.

Heelas, P., & Woodhead, L. (2005). *The spiritual revolution: Why religion is giving way to spirituality.* Blackwell Publishing.

Hidalgo, M., & Hernandez, B. (2001). Place attachment: Conceptual and empirical questions. *Journal of Environmental Psychology, 21*(3), 273–281. doi:10.1006/jevp.2001.0221

Hou, J., Lin, C., & Morais, D. (2005). Antecedents of attachment to a cultural tourism destination: The case of Hakka and non-Hakka Taiwanese visitors to Pei-Pu, Taiwan. *Journal of Travel Research, 44*(2), 221–233. doi:10.1177/0047287505278988

Hwang, S., Lee, C., & Chen, H. (2005). The relationship among tourists' involvement, place attachment and interpretation satisfaction in Taiwan's national parks. *Tourism Management, 26*(2), 143–156. doi:10.1016/j.tourman.2003.11.006

Hwang, S., Lee, C., & Chen, H.-J. (2005). The relationship among tourists' involvement, place attachment and interpretation satisfaction in Taiwan's national parks. *Tourism Management, 26*(2), 143–156. doi:10.1016/j.tourman.2003.11.006

Jackowski, A., & Smith, V. A. (1992). Polish pilgrim-tourists. *Annals of Tourism Research, 19*(1), 92–106. doi:10.1016/0160-7383(92)90109-3

Jafari, J., & Aaser, D. (1988). Tourism as the subject of doctoral dissertations. *Annals of Tourism Research, 15*(3), 407–429. doi:10.1016/0160-7383(88)90030-8

Kartal, B., Tepeci, M., & Atli, H. (2015). Examining the religious tourism potential of Manisa, Turkey with marketing perspective. *Tourism Review, 70*(3), 214–231. doi:10.1108/TR-09-2013-0048

Kozak, M. (2001). Repeaters' behavior at two distinct destinations. *Annals of Tourism Research, 28*(3), 784–807. doi:10.1016/S0160-7383(00)00078-5

Kozak, M. (2003). Measuring Tourist Satisfaction with Multiple Destination Attributes. *Tourism Analysis, 7*(3/4), 229–240. doi:10.3727/108354203108750076

Kozak, M., & Rimmington, M. (2000). Tourist satisfaction with Mallorca, Spain, as an off-season holiday destination. *Journal of Travel Research, 38*(3), 260–269. doi:10.1177/004728750003800308

Krešić, D., Mikulić, J., & Miličević, K. (2013). The factor structure of tourist satisfaction at pilgrimage destinations: The case of Medjugorje. *International Journal of Tourism Research, 15*(5), 484–494. doi:10.1002/jtr.1891

Larsen, S. (2007). Aspects of a psychology of the tourist experience. *Journal of Hospitality and Tourism, 7*(1), 7–18. doi:10.1080/15022250701226014

Liszewski, S. (2000). Pilgrimages or religious tourism. *Peregrinus Cracoviensis, 10*, 47–51.

Lois-González, R. C., & Santos, X. M. (2015). Tourists and pilgrims on their way to Santiago. Motives, Caminos and Final Destinations. *Journal of Tourism and Cultural Change*, *13*(2), 149–164. doi:10.1080/14766825.2014.918985

Louro, A., Padre, S.S.P., Oliveira, R., Frei, O.P., Filipe, J., Ribeiro, A. C. (2015). *A Bíblia Sagrada* (9ª. Edição). Paulus Editora.

Mackey, J., & Sisodia, R. (2013). Conscious Capitalism: Liberating the Heroic Spirit of Business. Harvard Business Press.

MacLennan, J., & Moore, R. L. (2011). Conflicts Between Recreation Subworlds: The Case of Appalachian Trail Long-Distance Hikers. *The Cyber Journal of Applied Leisure and Recreation Research*, *13*(1), 1–17.

Maoz, D., & Bekerman, Z. (2010). Searching for Jewish answers in Indian resorts: The postmodern traveler. *Annals of Tourism Research*, *37*(2), 423–439. doi:10.1016/j.annals.2009.10.015

Maslow, A. H. (1954). *Motivation And Personality*. Harper & Row.

Mazumdar, S., & Mazumdar, S. (1993). Sacred space and place attachment. *Journal of Environmental Psychology*, *13*(3), 231–242. doi:10.1016/S0272-4944(05)80175-6

Mazumdar, S., & Mazumdar, S. (2004). Religion and place attachment: A study of sacred places. *Journal of Environmental Psychology*, *24*(3), 385–397. doi:10.1016/j.jenvp.2004.08.005

Millán, C. L., Garzon, D., & Navarro, S. (2016). C2C interactions creating value in the route of Santiago. *Journal of Business Research*, *69*(11), 5448–5455. doi:10.1016/j.jbusres.2016.04.153

Murray, M. (2014). The cultural Heritage of Pilgrim Itineraries: The Camino de Santiago. *Journeys*, *15*(2), 65–85. doi:10.3167/jys.2014.150204

Murray, M., & Graham, B. (1997). Exploring the dialetics of route-based tourism: The caminho de Santiago. *Tourism Management*, *18*(8), 513–524. doi:10.1016/S0261-5177(97)00075-7

Navruz-Zoda, B., & Navruz-Zoda, Z. (2016). The Destination Marketing Development of Religious Tourism in Uzbekistan. *International Journal of Tourism and Pilgrimage*, *4*(7), 9–20.

Neuhofer, B., Buhalis, D., & Ladkin, A. (2014). A typology of technology-enhanced tourism experiences. *International Journal of Tourism Research*, *16*(4), 340–350. doi:10.1002/jtr.1958

Nolan, M. L., & Nolan, S. (1992). Religious sites as tourism attractions in Europe. *Annals of Tourism Research*, *19*(1), 67–78. doi:10.1016/0160-7383(92)90107-Z

Nunes, R. I. de S. (2018). *Turismo Holístico em Portugal: Caracterização da Oferta de Retiros*. Universodade de Lisboa, Faculdade de Letras.

Oliver, R. (1980). A cognitive model of the antecedents and consequences of satisfaction decisions. *JMR, Journal of Marketing Research*, *17*(4), 46–49. doi:10.1177/002224378001700405

Oliver, R. (1999). Whence Consumer Loyalty? *Journal of Marketing*, *63*(4_suppl1), 33–44. doi:10.1177/00222429990634s105

Oliver, R. L. (1981). Measurement and evaluation of satisfaction processes in retail settings. *Journal of Retailing*, 25–48.

Olsen, D. H., & Timothy, D. J. (2006). Tourism and religious journeys. In D. J. Timothy & D. H. Olsen (Eds.), *Tourism, religion and spiritual journeys* (pp. 1–21). Routledge.

Oppermann, M. (2000). Tourism Destination Loyalty. *Journal of Travel Research*, *39*(1), 78–84. doi:10.1177/004728750003900110

Pinto da Silva, F., Brandão, F., & Sousa, B. (2019). Towards socially sustainable tourism in cities: local community perceptions and development guidelines. *Enlightening Tourism*, *9*(2), 168-198.

Pourtaheri, M., Rahmani, K., & Ahmadi, H. (2012). Impacts of religious and pilgrimage tourism in rural areas: The case of Iran. *Journal of Geography and Geology*, *4*(3), 122. doi:10.5539/jgg.v4n3p122

Prayag, G. (2008). Image, Satisfaction and Loyalty: The Case of Cape Town. *Anatolia: An International Journal of Tourism and Hospitality Research*, *19*(2), 205–224. doi:10.1080/13032917.2008.9687069

Prayag, G., & Ryan, C. (2012). Antecedents of tourists' loyalty to Mauritius: The role and influence of destination image, place attachment, personal involvement, and satisfaction. *Journal of Travel Research*, *5*(3), 342–356. doi:10.1177/0047287511410321

Puczko, L., & Smith, M. (2008). *Health and Wellness Tourism*. Routledge.

Ram, Y., Bjork, P., & Weidenfeld, A. (2016). Authenticity and place attachment of major visitor attractions. *Tourism Management*, *52*, 110–122. doi:10.1016/j.tourman.2015.06.010

Rinschede, G. (1992). Forms of religious tourism. *Annals of Tourism Research*, *19*(1), 51–67. doi:10.1016/0160-7383(92)90106-Y

Ritchie, J., & Hudson, S. (2009). Understanding and meeting the challenges of customer/tourist experience research. *International Journal of Tourism Research*, *11*(2), 111–126. doi:10.1002/jtr.721

Ross, E., & Iso-Ahola, S. (1991). Sightseeing tourists' motivation and satisfaction. *Annals of Tourism Research*, *18*(2), 226–237. doi:10.1016/0160-7383(91)90006-W

Ryan, C. (1995). Learning About Tourists from Conservations: The Over-55s in Majorca. *Tourism Management*, *16*(3), 207–216. doi:10.1016/0261-5177(95)00005-9

Santos, C. M., Ambrósio, V., Correia, A., & Peres, R. (2013). The importance of religious tourism segmentation for tourism destination management: The case of the Island of S. Miguel, Azores. *World Review of Entrepreneurship, Management and Sustainable Development*, *9*(2), 166–178. doi:10.1504/WREMSD.2013.052357

Santos, X. M. (2002). Pilgrimage and Tourism at Santiago de Compostela. *Tourism Recreation Research*, *27*(2), 41–50. doi:10.1080/02508281.2002.11081219

Scannell, L., & Gifford, R. (2010). Defining place attachment: A tripartite organizing framework. *Journal of Environmental Psychology*, *30*(1), 1–10. doi:10.1016/j.jenvp.2009.09.006

Shackley, M. (2006). Empty bottles at sacred sites. Religious retailing at Ireland's national shrine. In D. J. Timothy & D. H. Olsen (Eds.), *Tourism, religion & spiritual journeys* (pp. 94–103). Routledge.

Sharpley, R. (2002). Rural tourism and the challenge of tourism diversification: The case of Cyprus. *Tourism Management*, *23*(3), 233–244. doi:10.1016/S0261-5177(01)00078-4

Sharpley, R., & Jepson, D. (2011). Rural Tourism: A spiritual experience? *Annals of Tourism Research*, *38*(1), 52–71. doi:10.1016/j.annals.2010.05.002

Silva, J. A. (2004). Caminhos de Santiago: Uma Europa peregrina. *Theologica*, *39*(2), 331–357.

Slavin, S. (2003). Walking as spiritual practice: The pilgrimage to Santiago de Compostela. *Body & Society*, *9*(3), 1–18. doi:10.1177/1357034X030093001

Smith, M., & Puczkó, L. (2009). *Health and wellness tourism*. Elsevier.

Smith, M., & Kelly, C. (2006). Holistic tourism: Journeys of the self? *Tourism Recreation Research*, *31*(1), 15–24. doi:10.1080/02508281.2006.11081243

Smith, V. (1992). Forms of Religious Tourism. *Annals of Tourism Research*, *19*(12).

Sousa, B., & Rocha, A. T. (2019). The role of attachment in public management and place marketing contexts: A case study applied to Vila de Montalegre (Portugal). *International Journal of Public Sector Performance Management*, *5*(2), 189–205. doi:10.1504/IJPSPM.2019.099094

Sousa, B., & Rodrigues, S. (2019). The role of personal brand on consumer behaviour in tourism contexts: The case of madeira. *Enlightening Tourism. A Pathmaking Journal, 9*(1), 38-62. doi:10.33776/et.v9i1.3597

Sousa, B., & Silva, M. (2019). Creative tourism and destination marketing as a safeguard of the cultural heritage of regions: The case of sabugueiro village. *Revista Brasileira de Gestão e Desenvolvimento Regional*, *15*(5), 78-92.

Spradley, J. (1979). *The Ethnographic Interview*. Holt, Rinehart and Winston.

Steil, C. A., & Carneiro, S. S. (2008). Peregrinação, Turismo e Nova Era: Caminhos de Santiago de Compostela no Brasil. *Religião & Sociedade*, *28*(1), 105–124. doi:10.1590/S0100-85872008000100006

UNWTO. (2011). *Religious Tourism in Asia and the Pacific*. World Tourism Organization.

Uriely, N. (2005). The tourist experience: Conceptual developments. *Annals of Tourism Research*, *32*(1), 199–216. doi:10.1016/j.annals.2004.07.008

Vukonik, B. (2002) Religion, tourism and economies: a convenient symbiosis. *Tourism Recreation Research, 27*(2), 59-64.

Willson, G. (2011). The search of inner peace: Considering the spiritual movement in Tourism. *The Journal of Tourism and Peace Research*, 16–26.

Yoon, Y., & Uysal, M. (2005). An examination of the effects of motivation and satisfaction on destination loyalty: A structural model. *Tourism Management*, *26*(1), 45–56. doi:10.1016/j.tourman.2003.08.016

Yorgason, E., & della Dora, V. (2009). Geography, religion, and emerging paradigms: Problematizing the dialogue. *Social & Cultural Geography*, *10*(6), 629–637. doi:10.1080/14649360903068100

Yuksel, A., Yuksel, F., & Bilim, Y. (2010). Destination attachment: Effects on customer satisfaction, cognitive, affective and conative loyalty. *Tourism Management*, *31*(2), 274–284. doi:10.1016/j.tourman.2009.03.007

KEY TERMS AND DEFINITIONS

Place Attachment: The affectivity a tourist develops to a place.

Place Dependence: Functional connection with a particular place or context.

Place Identity: Symbolic or emotional connection for a location or context.

Spiritual Tourism: Is linked to Religious Tourism, this link is ancient, assumed as an engine in all cultures, which develop based on a religion, thus playing a fundamental role in people's culture.

Spirituality: Any kind of connection with your inner self and with whatever is around you, alive or not.

Chapter 5
The Technological Power of Mysticism: A New Approach to Management of Religious Destinations

Eda Avci
https://orcid.org/0000-0003-3834-7707
Efes Vocational School, Department of Tourism and Hotel Management, Dokuz Eylul University, Turkey

Gizem Kayar
Computer Engineering Department, Muğla Sıtkı Koçman University, Turkey

ABSTRACT

Technological advances have significantly influenced the tourism and the sector realities. Like other industries, in the past couple of years, the stakeholders witnessed a continuous upgrade in tourism. These technological upgrades vary from personal online travel planners to complicated immersive rooms. Tourism trend analyses, online agencies, mobile locator and interactive maps, AR/VR enhancements, virtual assistants and robots, IoT, big data, and blockchain are most popular fields that tourism industry integrate itself seamlessly. As the cities, destinations, and tourism become smarter every day, religious tourism must get its share from emerging technological benefits. Every year, millions visit various holy places. A significant amount of data can be collected from millions of pilgrims. If this data are analyzed accurately, destination management and experience enhancement can be performed much easier than today. The aim of this study is to discuss the motivation of religious tourists, and how to implement smart destination principles in religious sites.

DOI: 10.4018/978-1-7998-5792-1.ch005

INTRODUCTION

Religious tourism has been the pioneering form of tourism as religious destinations have become not only a part of the cultural landscape but also a vital factor in local marketing since ancient times. Nowadays, religious tourism, dating back to the history of humanity, has become one of the major parts of many hosted destinations' economies. Particularly, some religious centers that are important for all humanity, such as Mecca, Israel, Vatican create a huge financial transaction by attracting millions of religious tourists in the destination (TN Global Travel Industry News, 2014). According UNWTO 600 million national and international people visit holy places annually, also 300 million tourists visit the world's major religious sites each year (CBI, Ministry of Foreign Affairs, 2018). And more than 50,000 religious organizations are serving pilgrims by organizing religious ceremonies and pilgrim journeys (Tala & Padurean, 2008).

Annually, about 13 million Muslims visit Mecca and two million visitors go abroad to Jerusalem. The Christian pilgrimage sites like Fátima (Portugal), Our Lady of Lourdes Sanctuary (France), St Peter's Basilica, and the Vatican, the Basilica of St Francis and St Clare of Assisi (Italy) to Jasna Gora (Poland) attract millions of pilgrims and tourists (Heidari, et al., 2018). There are some important factors which are the main reasons why a large proportion of people have religious beliefs such as hope for the afterlife, superstitious and anthropomorphism nature of faiths, belief injustice. Also, children in many parts of the world legally have the same religion as their parents (Pepper, 2014).

The main motivation of religious tourists depends on their faith, as they also want to explore the historic roots, experience a unique culture, visit the world's inspiring destinations, and simply have fun on their journey, religious tourism has significantly interacted with holiday, and cultural tourism (Rinschede, 1992). The cathedrals, synagogues, and mosques located all over the world attract visitors not only for religious but also for educational and cultural purposes (Woodward, 2004). Therefore, the motivation dimensions of religious tourists can be divided into two categories such as push (internal/intrinsic) factors and pull (external/extrinsic) factors (Dann, 1981).

Although travel for a religious purpose has existed in the history of humanity from ancient times, the motivation and the characteristics of religious tourism have changed over the millennia (Raj and Griffin 2015). That's why the determination of what visitors expect from their experiences, what kind of experience visitors have (consume) at the site and what type of benefits visitors derive from their experience while leaving is becoming more essential for religious destinations (Smallman & Moore, 2010). Offering only good quality of service to consumers is not enough to manage destinations (Grung, 2016).

This study emphasizes the benefits and importance of the smart approach for the management of religious destinations by addressing two opposite issues such as technology and mysticism consists of two chapters. In the first chapter, the concept of religious tourism and the technological approach, which provides important benefits for the management of religious tourism destinations, have been examined. In the second chapter, IoT, QR&Mobile Apps and Gamification technologies and their utilization in religious tourism have been reviewed. The suggestions that significance the attributes to be considered while developing technology for the management of religious tourism destinations are presented in the conclusion part of the study.

Religious Tourism Destinations as a Smart Destination Management Approach

Finding places through internet access has become easier with an increasing number of smartphone users worldwide (Jeong & Shin, 2019). The universal digital technologies form "a hyperconnected" skin on the body of tourists (Rabari and Storper, 2015) and their relationship with the destination is articulated with mobile technology use (Wang et al., 2013; Gretzel et al., 2015; Huang et al., 2017). The smart technologies have produced a transformation for tourism not only because tourists have become more demanding, active, independent, informed and skilled but also, they have discovered new ways of searching of information, comparing, booking, interacting, sharing, complaining, reviewing and recommending (Femenia-Serra, et al., 2019). The tourist experience enhanced with the implementation of technology (Neuhofer, et al., 2013; Neuhofer, et al., 2015), such as cloud computing, sensors and GPS widespread use, virtual and augmented reality and the full adoption of social media and mobile (Xiang & Fesenmaier, 2017).

The smart information, the smart exchange and the smart processing are the principles of smart technology (Gretzel, et al., 2015a), which has significantly affected the tourism industry in various ways (Huang et al. 2017). With the impact of ICT's which have completely reshaped not only destinations management (Buhalis and Law) but also the tourist experience itself (Neuhofer et al., 2012), tourism industry has been transformed after the acception of the concept of smart technology which is the most important challenge of the tourism industry (Zhu, et al., 2014), management ways of the destinations have also been modified. Through the complexity of destinations and the difficulties of their management (Jamal & Jamrozy, 2006), smart destination management has become one of the most important concepts in tourism and also one of the most difficult to define (Pearce, 2014).

Information, access, interaction, and personalization are four core dimensions of smart technology that can be embedded in a destination (Buhalis &bAmaranggana, 2015) and for a destination to be considered smart, the investments in human social capital and traditional transport and modern ICT communication infrastructure should meet the social, cultural, economic, leisure and personal needs of visitors (Lamsfus and Alzua-Sorzabal, 2013:19). Smart destinations effort to support mobility, resource availability and allocation, sustainability and quality of life (Gretzel et al., 2015b). The main focus of smart destinations is the utilization of technologies to improve the ability of a destination to fulfill its own goals by providing quick and personalized responses to the needs of tourists, tourism firms and the government. Thus to be a smart destination is much more than a destination that applies technologies (Martin, et al., 2017).

Since its inception, the smart approach has been subsequently applied to tourism destination management which is relevant particularly_in a sector where the evolution of ICTs creates one of the most approciate factors for change (Law et al., 2014). Instead of being something related to technology adoption and used in tourism, smart tourism should be evaluated as a strategic management tool for the development and management of the destinations (Xiang et al., 2015). As a new management approach, smart tourism contains a range of smart technologies which form a smart digital ecosystem that fosters data-driven innovations such as sensors, beacons, smartphone apps, RFID, NFC, smart meters, IoT, cloud computing, relational databases (Gretzel et al., 2015b; Xiang & Fesenmaier, 2017).

One of the tourism types where this smart destination management approach can be applied is religious tourism. Religious tourism is one kind of special interest tourism, which attracts religious tourists to the holy sites. All types of niche tourism are easily affected by economic instability but although religious tourism is a kind of niche tourism it is less affected because religious sites are not only visited by pilgrims but also by religious tourists whose motivation is quite strong and this mystique experience plays a very important role in their lives (Destination World, 2010).

Because the major religions have globally stimulated tourism spreading their own religious beliefs, pilgrimages have always been one of the dominant factors for motivating people to travel (Raj & Morpeth 2007). Most of these people are interested to visit holy places for different reasons, contributing in heritage or culture, a sense of curiosity or even to enjoy the landscape of that place or site (Wong, et al., 2013).

The main focuses of the religious tourism are to visit religious sites or monuments and to explore the body-mind-spirit for inner peace (Smith, et al., 2010). Therefore, religious holidays are more isolated in which tourists seek meaning, engagement and peace through activities (Agrawal, et. al, 2008). Religious tourism as a form of spiritual tourism evaluates as an extraordinary experience (Chaline, 2002). Spiritually, religious tourism is evaluated as a self-searching journey, a moral purification or even a personal achievement (Ibănescu, et al., 2018).

The spiritual environments such as churches, synagogues, mosques, or temples have great importance in cultivating human values (Stokols, 1990), somehow related to significant moments as birth, death, critical decisions in the life of a religious personality (Ibănescu, et al., 2018). Natural beauties, traditional temples, statues and images will influence visitors' perceptions (Mazumdar & Mazumdar, 2004), and visitors select the kind of religious site that they wish to visit based on their levels of interest in religion, faith, cultural heritage, art and architecture (Bond, et al., 2015).

In sacred places, people can create a unique experience (Trauer & Ryan, 2005) by practicing rituals, engaging with symbols, and traditions (Hadden, 1995). With the effect of the cultural and natural environment, religious tourism can help to strengthen and maintain self-identity (Bond & Falk, 2013; Jafari & Scott, 2014; Eid, 2015). Hence, religious tourism should focus primarily on spiritual experiences such as social, educational or restorative experiences (Tirca & Stanciulescu, 2011). The pilgrimage experience is affected by spiritual needs (Hughes, et. al., 2013) and visitors are willing to express their feelings and thoughts to families, friends, and others who share similar interests and religious beliefs (Belhassen, Caton, & Stewart, 2008).

As it is emphasized by Rifai (2015) that religious tourism is one of the most effective tools for the sustainable development of destinations due to its benefits such as raising awareness of the common heritage of humanity, providing resources for protection, contributing to local development and creating cultural understanding. Despite the numerous negative impacts such as globalization, terrorism, racism, economic recession, climate change and epidemic threat of disease, religious tourism is one of the most important phenomena which has been spread all over the world (UNWTO, 2011). However, this situation brings some major problems such as exceeding carrying capacity, waste management, monetizing religion, lack of examination of religious trusts and air pollution (SIGA, 2012).

Therefore, smart approach is accepted as a great opportunity for tourism destination management to solve many problems when it comes to cross-sector cooperation and long-term development (Jamal & Jamrozy, 2006). The smartness improves social, economic and environmental prosperity for all partners and produces values for tourists (Bulti, et al, 2019). The smartness supplies information to organizations for the value creation to provide products and services which are the incentive for all (Bulti, et al, 2019). Smart destinations can offer realtime information, instant feedback, dynamic platform, dynamic recommender system to tourists and all stakeholders (Buhalis & Amaranggana, 2015). With the utilization of smart technologies visitors have the potential to increase the time in learning about local products and places and to enhance their engagement in the overall tourism experience (Caveda, et al., 2018). Smart destinations allow tourists to create their own experience which is named "agile experience" by making their own rhythm (Hope, 2017). Besides all its positive effects, the destinations that want to adapt themselves to this new destination management approach should manage the changes and ensure strong performance and collaboration with all stakeholders (Bornhorst, et. Al., 2010).

Enhancing Mystique Experience Through Technology

When one says tourism, it can mean one of the many types as cultural, adventure, business, culinary, medical, and so on. Some of these types are quite open for technological support, whereas some still struggle to integrate technology. However, religious tourism and mystic experiences seem quite promising for technology-based enhancements. For instance, DeAscaniis et al. (2018) discuss the usage of Information and Communication Technologies (ICT) at religious heritage sites by giving special survey results run with the pilgrims visited the Loreto sanctuary in Italy. The authors showed that even in a small-size crowd, ICTs can be of good use and people started to see the potential of the ICTs. In Ramos & Garcia (2019), the authors shared a case study on the contribution of ICTs to cultural and religious sites by giving a specific example from Valencia. The authors stated that the utilization of ICT has positive implications in the cultural and religious tourism experience, mainly when visiting a cultural point of interest and has contributed to planning a more successful trip. In this chapter, these novelties, trends and advancements will be discussed in three main sections: IoT, QR & Mobile Apps and Gamification.

IoT

Internet of Things (IoT) is one of the most popular concepts that has been frequently addressed in the last couple of years. This technology refers to the connection of devices and sensors (see Fig. 1) through internet and cloud services to gather data. This data is later analyzed and produce meaningful outcome

Figure 1. Internet of Things (iQVizyon, 2020)

about the regarding the analyzed concept. According to statista.com, the number of installed IoT devices will be around 31 billion worldwide in 2020.

The tourism industry constitutes multiple stakeholders, reaches a high level of internet usage/ interconnection and therefore, produces heavy data that need to be addressed and thoroughly analyzed. From online travel planners to museums, one can encounter with the high-level technology usage and various connected devices. This makes the industry much smarter. The word "smart" can be explained easily with examples.

For instance, transportation became smarter with travel planners, online booking systems, easier data streaming, maintenance monitoring, passenger services, safety checks and analyses over train sensors and flight sensors. Besides, the hospitality sector started to be smarter with enhanced user experiences, cost-saving strategies over IoT, Electronic Key Cards and so on. Cities started to be smarter with green energy approaches, better transportation opportunities and crowd management systems. Such cities currently take more attention and attract more tourists. Especially crowd management systems make cities more attractive and reliable as the population of cities increases every day. The host touristic cities try to find solutions for the handshaking of the citizens and tourists. Verma and Shukla explains various ways of IoT usage in the tourism industry (Verma & Shukla, 2019). In Tripathy, et al. (2018), the authors handled the issue of independent and sustainable mobility of tourists in smart cities. The authors mention that tourists generally prefer to visit known places since they concern about safety. However, this concern arises from the lack of awareness of supported IoT based applications. Many cities started to integrate IoT services that are accessible through mobile devices. However, many visitors have no knowledge about these services and therefore, those initiatives are not utilized as expected. Considering these issues, the authors introduced an IoT based independent tourist mobility framework, namely iTour (see Fig. 2) and created a smart map. This approach can be used in many cities to raise awareness and enhance the tourist experience.

At this point, mention beacon technology should be mentioned. Bluetooth beacons are small, wireless transmitters that send signals to other nearby devices. Beacons are becoming more popular every day in terms of data collection since they provide accurate data to be analyzed. Using a beacon, museums can provide interactive services to visitors and manage the crowd by taking the attention to rather empty spots. Beacons, infrareds, other wireless sensors and therefore IoT can be utilized in many touristic places including religious tourism as well. The best usage of IoT in religious areas is crowd management systems. Although this idea is rather novel and still at the beginning of improvement, companies, e.g. (Netwings Solutions Pvt Ltd), started to research this topic.

What we exactly know is, some of the most crowded touristic areas in the world is religious areas with thousands of devotee flock at the same time, e.g. St. Peter square in the Vatican as Pope calls public or Mecca in Hajj time. Devotees need to be monitored and directed accordingly via sensor systems to prevent collapse in some specific areas. Besides, such monitoring is crucial for emergency and healthcare services. However, this topic and a systematic solution is still an open research area. For instance, Netwings Solutions try to solve the safety and crowd management issues in Mecca using beacons.

As Tripathy, et al. mentioned (2018), smart maps are quite useful tools to direct the tourists. These maps can also be used for pilgrimage sites to manage crowds and redirect the pilgrims to correct spots efficiently. The authors thoroughly discussed the research and development directions of pilgrims' ICT usage channels. In terms of an engineering approach, the authors specifically stated that the below given IoT ideas/projects should be handled and elaborated for enhancements in religious visits (DeAscaniis & Cantoni,2016):

Figure 2. iTour Framework described in (Tripathy, Tripathy, Ray, & Mohanty, 2018)

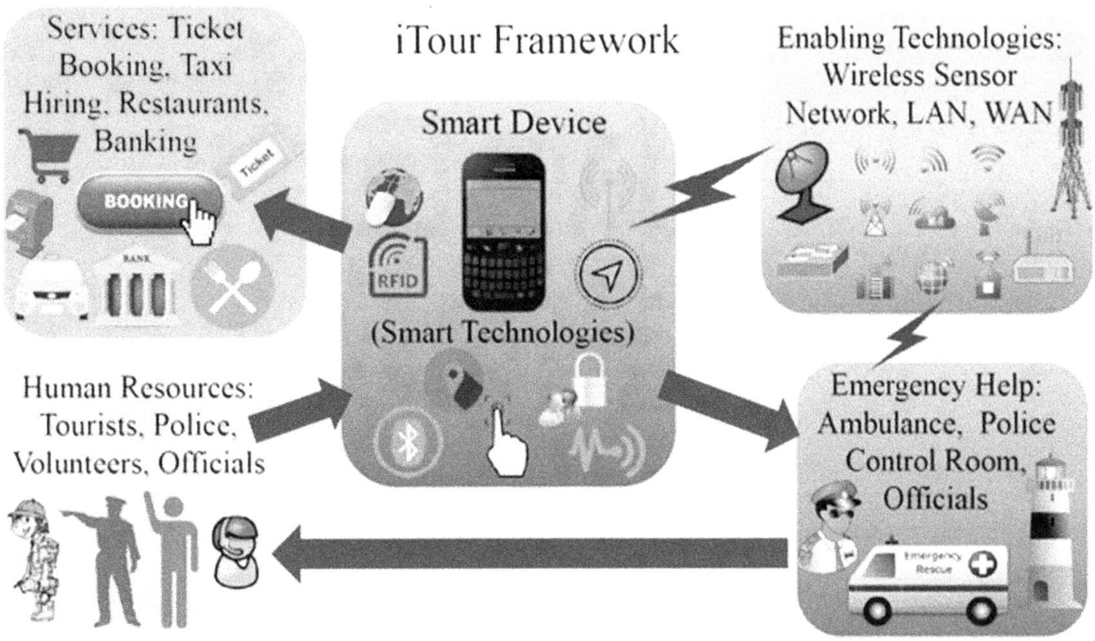

- ❖ mobile phones that cover signaling during prayers
- ❖ software to track the behavioral changes of pilgrims (e.g. digital faith companion to register daily activities)
- ❖ pop-up reminders for activities
- ❖ crowdsourcing of stories related to different experiences
- ❖ virtual P2P communities for experience sharing
- ❖ pilgrimage storybooks
- ❖ spiritual eco-tourism
- ❖ faith matcher for finding suitable travel companions

These ideas open the doors for further research in the field. More qualitative and quantitative data seem to be available soon as more researchers focus on IoT utilization in religious sites.

QR & Mobile Apps

Quick Response (QR) codes were first designed in 1994 as two-dimensional matrix barcodes (see Fig. 3) for the automotive industry in Japan. These codes are machine-readable optical labels and they contain information related to any kind of item that is attached. This information can be item identification, product tracking, time tracking, and this list grows up with much additional information. Only a mobile phone with a camera is enough to read the label and send this data to any QR-reader application for accessing the related information.

Since 1994, QR codes became very popular outside Japan and the automotive industry. As the writer stated that about 6.2% of the total mobile audience in U.S. which equals approximately 14 million users,

Figure 3. The QR code generated by the authors for the AR-based mobile application they created (Kayar, 2019)

scanned a QR code on their mobile device in 2011 (Writer, 2011). Recently, this number increased not only in U.S. but all over the world. Nowadays, QR codes are used for easy software downloads, patient identifications, many business-related document tracking, education links that vary from K12 to universities, mobile tourism and many others.

In the tourism industry, QR codes started to be used for sharing any kind of information with hotel customers and interested travelers as electronic tourist guides. This idea has also been adapted by religious tourism experts. Religious tourism is a type of tourism where visitors are interested in fulfilling their duties in a holy region rather than sightseeing. However, visiting these regions requires many steps to be taken. Each ritual should be explained carefully. Prescribing these actions to be taken in a holy place is always tedious. Besides, following these steps can be quite error-prone if you follow a simple map or a prescription paper. However, by tracking a QR code at the entrance of the region, the visitor can be directed over mobile phone prescriptions.

Alshattnawi (2012) is one of the first researchers who mentioned such utilization in a research paper. The author mentioned the ritual steps for Umrah and Hajj in the holy city Mecca and prescribed the steps of the pilgrimage. The main idea of this paper was to show the superior effect of using QR codes instead of brochures. During the pilgrimage, visitors try to find the correct brochure for each step that should be taken where there are many brochures at hand. However, by using QR codes, easy navigation is possible.

Narbona and Arasa (2016), discussed the role of instant mobile messaging of people in massive religious events and how they are affected by these messages. Later, et al. (2017) discussed the usage of mobile technologies and apps in Camino de Santiago, which is visited by many American and European pilgrims (approx. 200000 in 2016). In this research, the authors showed that smartphones, tablets, e-readers and Comino apps are frequently used by pilgrims. The authors also showed the differences between European and American pilgrims' attitudes towards technology.

Antunes and Amaro (2016) similarly discuss the mobile apps' usage in Comino de Santiago and direct the app developers for the needs of pilgrims in the region. The authors mention the strategies to develop an app for pilgrims and to encourage the pilgrims to use this app. The authors also mentioned two engineering approaches to enhance the religious visit:

- Faith status signaling – Tamagotchi type
- Church navigators for churches rich in art

These two approaches are only possible using mobile device apps. Church navigators can be easily provided using above-described QR based applications. However, Tamagotchi type faith status signaling needs more Artificial Intelligence and Machine Learning support. This is still an open research area in the religious tourism and technology integration field.

Gamification

Davis (2015) explains the myth, magic and mysticism in the age of technology. At one part of the article, the author states that "the most obvious place to track the prints of myth, magic, and mysticism through contemporary techno culture is, of course, in our fictions." This statement displays the fictious reality of us.

The place where the fiction can be best realized is the games and simulations. Everybody is an inborn gamer because of human nature shelters game instincts. Since very early ages, people learn much better through games and this continues also in older ages. Therefore, gamification is very popular nowadays in many diverse fields that range from tourism to finance, from military to health and so on.

The virtual and fictious worlds of cultural heritage sites and religious sites can be created through technology. Nowadays, there are many ways to develop games since game engines are accessible, they are easy to use and performance efficient due to current hardware and software advances. Using historical concepts in digital games is not new and still popular in the game world.

For instance, 2D Holy Bible and Quran related quiz applications are easily found in app stores. Besides, you can encounter with historical and religious figures in more complicated 3D games. As an example, ancient Greek religion is described in the game Age of Mythology, the historic figure Ahriman was portrayed in Prince of Persia or Assasin Altair Ibn-La'Ahad was characterized in the popular game Assassin's Creed.

The games can be also used for other spiritual purposes. For instance, in May 2012, the Exeter Cathedral used the game: Flower during its Sunday service. The flower is a spiritual video game (Narcisse, 2012) where the player controls the wind, blowing a flower petal through the air using the movement of the game controller. Flying close to flowers results in the player's petal being followed by other petals.

As gamification has become extensively popular, Augmented Reality and Virtual Reality Technologies have also started to be used as supporting immerse mechanisms in tourism. Augmented Reality can be described as an interactive experience where real-life background meets with virtual objects. Differently, in virtual reality, an immerse experience in provided using VR-glasses which puts the user totally into a virtual world. Both technologies can be supported with sensors, visual, audial or haptic information if necessary. AR and VR are two commonly preferred approaches for education, experience enhancement, training and similar gamifiable fields. The authors discussed the potential usage of augmented reality applications in tourist guidance education by integrating gamification modules (Avci, et al., 2019). Besides, Kayar &Cobanoglu (2019) described a way to attract the young generation to museums using games and Augmented Reality based mobile applications.

Besides, the video game industry also supports complicated Virtual reality-based 3D history games. In Historium VR, 15th century Bruges are visualized with its architecture, cobblestone walkways, and arched bridges. It is kind of a history lesson that includes landmarks such as Dampoort, St. Donatian's, Waterhalle, Crane, and Belfry. Or another example is VR Terra-Cotta Warriors (see Fig. 4). This game takes you on a trip through ancient China. Meanwhile, you can see the real-life-sized terracotta statues built for Qin Shi Huang, the first Emperor of China. This game teaches you both the history and spiritualism at the backside since it constitutes of afterlife origin.

Figure 4. VR Terra-Cotta Warriors (Viveport, 2017)

When those games are checked, they are easily seen as Technologies where religious figures and sites are integrated. Myths, mythological figures, religious sites and even the actions that need to be taken can be portrayed, visualized, gamified and simulated in games. More immersive Technologies, e.g. AR, VR, can also be utilized for further support throughout these experiences.

CONCLUSION

In this study, emerging technologies and their usege in religious tourism have been reviewed in three main sections: IoT, QR&Mobile Apps and Gamification. Besides these three fields, various sub-fields of technology can always be in use for tourism, however, the authors especially review the aforementioned techniques due to their potential in religious sites and it is indicated by the swot analysis table below.

Table 1. Swot Analysis of the Emerging Technologies (IoT, QR&Mobile Apps and Gamification) in Religious Tourism

STRENGTH	WEAKNESSES
❖ Religious tourism is a growing market ❖ The number of the people are increasing who are interested in spirituality but do not consider themselves religious ❖ Increasing interest of young generation ❖ Spirituality tends to be more important to religious tourists between 30 and 55 years than traditional religion	❖ Only few researchers study the technological advancements and their integration in religious and mysticism. This brings lack of knowledge and therefore, resources.
OPPORTUNITIES	**THREATS**
❖ Can be connect personally to a holy place ❖ Can better understand and appreciate a religion ❖ Enhancing religious tourism activities ❖ Can be organized an interesting program ❖ Health and safety measures can be increased ❖ Contributing destination image as a religious travel destination	❖ The wrong perception people have about the replacement of technology with the mystic. The actual ideas are only to bring enhancement, not the replacement. This should be clearly emphasized. ❖ The wrong perception directing people to think that the attitude toward mysticism will change when technology comes in.

As can be seen, technological advancements are becoming more accessible and prominent every day. Religious tourism has a great potential to use emerging technologies for crowd management and user experience enhancement. However, conducting further research to develop information about religious tourism features, motivations and trends is crucial. This is also significant in order to improve the policies that protect and preserve religious sites. New policies and approaches created for the development, management and promotion of tourism will allow sustainability in religious sites.

In addition, tourists need more awareness to develop a network to promote cooperation, exchange of information and good practice in tourism management in religious sites. Mobile apps should be in use for this purpose and IoT features have to be utilized for a smart management system. Fostering the participation of local religious communities and civil society in the development and management of religious tourism at the local level, as well as public-private coordination and cooperation can be automatically provided if such a system is realized.

Rather than thinking local, more global approaches should also be embraced. Therefore, the QRs and the apps are neccessary to create pilgrimage routes between countries and direct the pilgrims via these apps. Besides, encouraging the valuable contribution of pilgrimages and religious tourism to intercultural and interfaith dialogues has to be widespread and should be supported via mobile apps.

The reality of religious areas can be supported via smart business models. These models can only be created using throughly analyzed big data, and this data is accessible via IoT. Researchers should also focus on this topic in the future. Besides, entrepreneurship should be promoted in local communities by providing knowledge and skills necessary to innovate in the development and management of tourism services, activities and experiences.

Researchers should also focus more on the immerse virtual worlds to enhance the quality of visitor experience in religious places. Such the research can be easily supported using gamification, AR and VR Technologies. see What is obvious is the market needs more games and still open for further research to collect useful data from mystic experiences. Using this method, it becomes easier to create opportunities for tourists to be aware of and respect other beliefs and spiritual traditions.

Although intelligence technologies and religious tourism are one of the most researched areas in the tourism field, the interdisciplinary reviewing of the issues is not adequate. With this discussion study, it is aimed to examine the key concepts, the external influences, the challenges and the implications of the IoT, QR Mobile Apps and Gamification technologies within the scope of religious tourism. Therefore, the study has the potential to be a reference for further interdisciplinary researches.

As it is mentioned before, it is known that technology can never be replaced with mystic components of people lives but only can be used to enhance the experience they have. The only aim of this study is to show this integration.

REFERENCES

Agrawal, M., Choudhary, H., & Tripathi, G. (2010). Enhancing Buddhist tourism in India: An exploratory study. *Worldwide Hospitality and Tourism Themes*, *2*(5), 477–493. Advance online publication. doi:10.1108/17554211011090102

Alshattnawi, S. (2012). Effective use of QR codes in religious tourism. In *Proceedings - 2012 International Conference on Advanced Computer Science Applications and Technologies, ACSAT 2012*. 10.1109/ACSAT.2012.43

Anrew, P. (2014). *8 Reasons people embrace religion*. Retrieved from https://listverse.com/2013/06/13/8-reasons-people-embrace-religion/

Antunes Angela, A. S. (2016). Intentions to use a pilgrimage app: Which features really matter? A Pathway for the New Generation of Tourism Research. In F. Dias (Ed.), *Proceedings of the EATSA Conference* (pp. 525–535). Lisbon: APTUR.

Belhassen, Y., Caton, K., & Stewart, W. P. (2008). The search for authenticity in the pilgrim experience. *Annals of Tourism Research*, *35*(3), 668–689. Advance online publication. doi:10.1016/j.annals.2008.03.007

Bogdan-Constantin, I., & Alina, M. (2018). Territorial and marketing strategies related to religious tourism in Northern Moldavia, Romania. *CES Working Papers, 10*(1), 129–148.

Bond, N., & Falk, J. (2013). Tourism and identity-related motivations: Why am I here (and not there)? *International Journal of Tourism Research*, *15*(5), 430–442. Advance online publication. doi:10.1002/jtr.1886

Bond, N., Packer, J., & Ballantyne, R. (2015). Exploring Visitor Experiences, Activities and Benefits at Three Religious Tourism Sites. *International Journal of Tourism Research*, *17*(5), 471–481. Advance online publication. doi:10.1002/jtr.2014

Bornhorst, T., Brent Ritchie, J. R., & Sheehan, L. (2010). Determinants of tourism success for DMOs & destinations: An empirical examination of stakeholders' perspectives. *Tourism Management, 31*(5), 572–589. Advance online publication. doi:10.1016/j.tourman.2009.06.008

Buhalis, D., & Amaranggana, A. (2015). Smart Tourism Destinations Enhancing Tourism Experience Through Personalisation of Services. In Information and Communication Technologies in Tourism 2015. doi:10.1007/978-3-319-14343-9_28

Buhalis, D., & Law, R. (2008). Progress in information technology and tourism management: 20 years on and 10 years after the Internet-The state of eTourism research. *Tourism Management, 29*(4), 609–623. Advance online publication. doi:10.1016/j.tourman.2008.01.005

Bulti, A. G., Ray, A., & Bhuyan, P. (2019). Smart tourism system architecture design using the internet of everything (IOE) over cloud platform. *International Journal of Innovative Technology and Exploring Engineering*.

CBI Ministry of Foreign Affairs. (2018). *What are the opportunities for religious tourism from Europe?* Academic Press.

Chaline, E. (2002). *Zen and the Art of Travel*. MQ Publications.

Dann, G. M. S. (1981). Tourist motivation an appraisal. *Annals of Tourism Research, 8*(2), 187–219. Advance online publication. doi:10.1016/0160-7383(81)90082-7

Dario, C., Elahi, M., David, M., Stefano, M., Elena, N., & Francesco, R. V. A. (2018). Tangible Tourism with the Internet of Things. Information and Communication Technologies in Tourism. In Information and Communication Technologies in Tourism. Jönköping: Springer.

De Ascaniis, S., Mutangala, M. M., & Cantoni, L. (2018). *ICTs in the tourism experience at religious heritage sites: a review of the literature and an investigation of pilgrims' experiences at the sanctuary of Loreto*. Church, Communication and Culture. doi:10.1080/23753234.2018.1544835

DeAscaniis, S., & Cantoni, L. (2016). Pilgrims in the Digital Age: a research manifesto. *International Journal of Religious Tourism and Pilgrimage*. doi:10.21427/D7Z120

Eda, A., Gizem, K., & Cihan, Ç., K. E. (2019). The Potential Usage of Augmented Reality Applications in Tourist Guidance Education. In A. L. Cobanoglu Cihan (Ed.), *Global Congress On Smart Tourism* (pp. 79–81). Canakkale: Çanakkale Onsekiz Mart University.

Eid, R. (2015). Integrating Muslim Customer Perceived Value, Satisfaction, Loyalty and Retention in the Tourism Industry: An empirical study. *International Journal of Tourism Research, 17*(3), 249–260. Advance online publication. doi:10.1002/jtr.1982

Eric, D. (2015). *Myth, Magic, and Mysticism in the Age of Information*. Retrieved from https://lareviewofbooks.org/article/myth-magic-mysticism-age-information/

Femenia-Serra, F., Neuhofer, B., & Ivars-Baidal, J. A. (2019). Towards a conceptualisation of smart tourists and their role within the smart destination scenario. *Service Industries Journal, 39*(2), 109–133. Advance online publication. doi:10.1080/02642069.2018.1508458

GizemK. (2019). *PinctadaAR*. Retrieved from https://play.google.com/store/apps/details?id=com.SMK.PinctadaAR&hl=tr

Gretzel, U., Sigala, M., Xiang, Z., & Koo, C. (2015). Smart tourism: Foundations and developments. *Electronic Markets*, *25*(3), 179–188. Advance online publication. doi:10.100712525-015-0196-8

Gretzel, U., Werthner, H., Koo, C., & Lamsfus, C. (2015). Conceptual foundations for understanding smart tourism ecosystems. *Computers in Human Behavior*, *50*, 558–563. Advance online publication. doi:10.1016/j.chb.2015.03.043

Gurung, R. (2016). *Marketing religious tourism destination. Case study: the birthplace of Gautama Buddha*. Centria University.

Hadden, J. K. (1995). Religion and the quest for meaning and order: Old paradigms, new realities. *Sociological Focus*, *28*(1), 83–100. Advance online publication. doi:10.1080/00380237.1995.10571040

Heidari, A., Yazdani, H. R., Saghafi, F., & Jalilvand, M. R. (2018). The perspective of religious and spiritual tourism research: A systematic mapping study. *Journal of Islamic Marketing*, *9*(4), 747–798. Advance online publication. doi:10.1108/JIMA-02-2017-0015

Hernández-Martín, R., Rodríguez-Rodríguez, Y., & Gahr, D. (2017). Functional zoning for smart destination management. *European Journal of Tourism Research.*

Hop. (2017). *5 things you need to know about Smart Destinations*. Retrieved from https://hopu.eu/5-things-you-need-to-know-about-smart-destinations/

Huang, C. D., Goo, J., Nam, K., & Yoo, C. W. (2017). Smart tourism technologies in travel planning: The role of exploration and exploitation. *Information & Management*, *54*(6), 757–770. Advance online publication. doi:10.1016/j.im.2016.11.010

Hughes, K., Bond, N., & Ballantyne, R. (2013). Designing and managing interpretive experiences at religious sites: Visitors' perceptions of Canterbury Cathedral. *Tourism Management*, *36*, 210–220. Advance online publication. doi:10.1016/j.tourman.2012.11.022

iQvizyon. (2020). *Internet of Things*. Retrieved from http://iqvizyon.com/?page_id=223

Jafari, J., & Scott, N. (2014). Muslim world and its tourisms. *Annals of Tourism Research*, *44*, 1–19. Advance online publication. doi:10.1016/j.annals.2013.08.011

Jamal, T. (2006). Collaborative Networks and Partnerships for Integrated Destination Management. Tourism Management Dynamics. doi:10.1016/b978-0-7506-6378-6.50028-7

Jeong, M., & Shin, H. H. (2019). Tourists' Experiences with Smart Tourism Technology at Smart Destinations and Their Behavior Intentions. *Journal of Travel Research*. Advance online publication. doi:10.1177/0047287519883034

Kayar Gizem, Ç. C. (2019). How to attract young generation to museums using the idea of gamification and an augmented reality based mobile application. In *Global Conference of Business and Economics* (pp. 133–135). Istanbul: Academic Press.

Lamsfus, C., & Alzua-Sorzabal, A. (2013). Theoretical Framework for a Tourism Internet of Things: Smart Destinations. *TourGUNE Journal of Tourism and Human Mobility*.

Law, R., Buhalis, D., & Cobanoglu, C. (2014). Progress on information and communication technologies in hospitality and tourism. *International Journal of Contemporary Hospitality Management*, *26*(5), 727–750. Advance online publication. doi:10.1108/IJCHM-08-2013-0367

Li, Y., Hu, C., Huang, C., & Duan, L. (2017). The concept of smart tourism in the context of tourism information services. *Tourism Management*, *58*, 293–300. Advance online publication. doi:10.1016/j.tourman.2016.03.014

MacCannell, D. (1973). Staged Authenticity: Arrangements of Social Space in Tourist Settings. *American Journal of Sociology*, *79*(3), 589–603. Advance online publication. doi:10.1086/225585

Mazumdar, S., & Mazumdar, S. (2004). Religion and place attachment: A study of sacred places. *Journal of Environmental Psychology*, *24*(3), 385–397. Advance online publication. doi:10.1016/j.jenvp.2004.08.005

Melanie, S., & Nicola, M. R. M. H. (2010). Key Concepts in Tourist Studies. London: SAGE.

Narbona, J., & Arasa, D. (2016). The Role and Usage of Apps and Instant Messaging in Religious Mass Events The Role and Usage of Apps and Instant Messaging in Religious Mass Events. *International Journal of Religious Tourism and Pilgrimage*. doi:10.21427/D7JT6Q

Narcisse, E. (2012). *A playstation 3 game will be part of a church service this sunday*. Retrieved from https://kotaku.com/5909361/this-sunday-video-games-go-to-church

Neuhofer, B., Buhalis, D., & Ladkin, A. (2012). Conceptualising technology enhanced destination experiences. *Journal of Destination Marketing & Management*, *1*(1-2), 36–46. Advance online publication. doi:10.1016/j.jdmm.2012.08.001

Neuhofer, B., Buhalis, D., & Ladkin, A. (2015). Smart technologies for personalized experiences: A case study in the hospitality domain. *Electronic Markets*, *25*(3), 243–254. Advance online publication. doi:10.100712525-015-0182-1

News, T. G. T. I. (2014). *The importance of the religious tourism market*. Retrieved from https://www.eturbonews.com/50998/importance-religious-tourism-market

Nickerson, R. C., & Eng, J. (2017). Use of mobile technology and smartphone apps on the Camino de Santiago: A Comparison of American and European Pilgrims. In Atas da Conferencia da Associacao Portuguesa de Sistemas de Informacao. doi:10.18803/capsi.v17.260-269

Palestine, S. of. (2015). *Bethlehem declarationon religious tourism as a means of fostering socio-economic development of host communities*. Retrieved from https://www.bethlehem.edu/document.doc?id=3240

Pearce, D. G. (2014). Toward an Integrative Conceptual Framework of Destinations. *Journal of Travel Research*, *53*(2), 141–153. Advance online publication. doi:10.1177/0047287513491334

Raj. (2015). *Religious Tourism and Pilgrimage Management: An International Perspective*. Oxfort: Cabi.

Raj, R., & Morpeth, N. D. (2007). *Religious tourism and pilgrimage festivals management: An international perspective*. doi:10.1079/9781845932251.0000

Ramos, C. M. Q. G. A. M. (2019). Analysis of the Contribution of ICT to Cultural and Religious Tourism: In Communicating Religious Heritage to Visitors and Tourists. In Research on Socio-Economic Impacts of Religious Tourism and Pilgrimage (pp. 167–194). IGI Global.

Rinschede, G. (1992). Forms of religious tourism. *Annals of Tourism Research*, *19*(1), 51–67. Advance online publication. doi:10.1016/0160-7383(92)90106-Y

SIGA. (2012). *Diverse Beliefs: Tourism of Faith Religious Tourism Gains Ground*. Yes Bank.

Smallman, C., & Moore, K. (2010). Process studies of tourists' decision-making. *Annals of Tourism Research*, *37*(2), 397–422. Advance online publication. doi:10.1016/j.annals.2009.10.014

Solutions, N. (n.d.). *Use Cases*. Retrieved from https://hipla.io/crowd-management-at-religious-places/

Stokols, D. (1990). Instrumental and spiritual views of people-environment relations. *The American Psychologist*, *45*(5), 641–646. Advance online publication. doi:10.1037/0003-066X.45.5.641

Tala Madalina Lavinia, P. M. (2008). Dimensions of religious tourism. *Amfiteatru Economic*, *10*(2), 242–253.

Taleb, R. (2015). Fostering sustainable socio-economic development in host communities. In *International Conference on Religious Tourism*. Bethlehem: World Tourism Organization.

Tirca, A., & Stanciulescu, G. C. (2011). Managing the religious tourism experience in Romanian Christian Orthodoxy. *International Journal of Business and Globalisation*, *7*(1), 40. Advance online publication. doi:10.1504/IJBG.2011.040845

Trauer, B., & Ryan, C. (2005). Destination image, romance and place experience - An application of intimacy theory in tourism. *Tourism Management*, *26*(4), 481–491. Advance online publication. doi:10.1016/j.tourman.2004.02.014

Tripathy, A. K., Tripathy, P. K., Ray, N. K., & Mohanty, S. P. (2018). *ITour: The Future of Smart Tourism: An IoT Framework for the Independent Mobility of Tourists in Smart Cities*. IEEE Consumer Electronics Magazine. doi:10.1109/MCE.2018.2797758

UNWTO. (2011). *Religious Tourism in Asia and the Pacific*. Retrieved from https://www.e-unwto.org/doi/pdf/10.18111/9789284413805

Verma, A., & Shukla, V. (2019). *Analyzing the Influence of IoT in Tourism Industry*. SSRN Electronic Journal. doi:10.2139srn.3358168

Viveport. (2017). *Terra-Cotta Warriors*. Retrieved from https://www.viveport.com/apps/2220d4be-38bf-4fbc-93b9-e74b90b2dcb9/VR_Terra-Cotta_Warriors/

Wang, D., Li, X., & Li, Y. (2013). China's "smart tourism destination" initiative: A taste of the service-dominant logic. *Journal of Destination Marketing & Management*, *2*(2), 59–61. Advance online publication. doi:10.1016/j.jdmm.2013.05.004

Wong, C. U. I., Ryan, C., & McIntosh, A. (2013). The Monasteries of Putuoshan, China: Sites of Secular or Religious Tourism? *Journal of Travel & Tourism Marketing*, *30*(6), 577–594. Advance online publication. doi:10.1080/10548408.2013.810996

Woodward, S. C. (2003). Faith and tourism: accommodating visitor and worshipper in the historic city. *6th US/ICOMOS International Symposium, Managing Conflict and Conservation in Historic Cities.*

World, D. (2010). *Religious and pilgrimage tourism. In: the e-newsletter for tourism destination professionals.* Retrieved from http://www.destinationworld.info/newsletter/feature65.html

Writer, S. (2011). *14m Americans scanned QR and bar codes with their mobiles.* Retrieved from https://internetretailing.net/themes/14m-americans-scanned-qr-and-bar-codes-with-their-mobiles-in-june-2011

Xiang, Z., & Fesenmaier, D. R. (2017). Big data analytics, tourism design and smart tourism. In Analytics in Smart Tourism Design (pp. 299–307). Berlin: Springer.

Xiang, Z., & Iis, T. (2015). Special Issue on Smart Destinations: Foundations, Analytics, and Applications. *Journal of Destination Marketing & Management*, *4*(3), 143–201. doi:10.1016/j.jdmm.2015.07.001

Zhu, W., Zhang, L., & Li, N. (2014). Challenges, Function Changing of Government and Enterprises in Chinese Smart Tourism. *Enter 2014.*

ADDITIONAL READING

Alshattnawi, S. (2012). Effective use of QR codes in religious tourism. In *Proceedings - 2012 International Conference on Advanced Computer Science Applications and Technologies, ACSAT 2012.* 10.1109/ACSAT.2012.43

Antunes Angela, A. S. (2016). Intentions to use a pilgrimage app: Which features really matter? A Pathway for the New Generation of Tourism Research. In F. Dias (Ed.), *Proceedings of the EATSA Conference* (pp. 525–535). lisbon: APTUR.

Bond, N., Packer, J., & Ballantyne, R. (2015). Exploring Visitor Experiences, Activities and Benefits at Three Religious Tourism Sites. *International Journal of Tourism Research*, *17*(5), 471–481. Advance online publication. doi:10.1002/jtr.2014

Buhalis, D., & Amaranggana, A. (2015). Smart Tourism Destinations Enhancing Tourism Experience Through Personalisation of Services. In Information and Communication Technologies in Tourism 2015. doi:10.1007/978-3-319-14343-9_28

Buhalis, D., & Law, R. (2008). Progress in information technology and tourism management: 20 years on and 10 years after the Internet-The state of eTourism research. *Tourism Management*, *29*(4), 609–623. Advance online publication. doi:10.1016/j.tourman.2008.01.005

De Ascaniis, S., Mutangala, M. M., & Cantoni, L. (2018). *ICTs in the tourism experience at religious heritage sites: a review of the literature and an investigation of pilgrims' experiences at the sanctuary of Loreto.* Church, Communication and Culture., doi:10.1080/23753234.2018.1544835

DeAscaniis, S., & Cantoni, L. (2016). Pilgrims in the Digital Age: a research manifesto. *International Journal of Religious Tourism and Pilgrimage.* doi:10.21427/D7Z120

Jeong, M., & Shin, H. H. (2019). Tourists' Experiences with Smart Tourism Technology at Smart Destinations and Their Behavior Intentions. *Journal of Travel Research*. Advance online publication. doi:10.1177/0047287519883034

Law, R., Buhalis, D., & Cobanoglu, C. (2014). Progress on information and communication technologies in hospitality and tourism. *International Journal of Contemporary Hospitality Management*, *26*(5), 727–750. Advance online publication. doi:10.1108/IJCHM-08-2013-0367

KEY TERMS AND DEFINITIONS

Anthropomorphism: Transferring human qualities to another entity.

App: The shorthand of application specific mobile applications.

Gamification: Application of game elements and principles for custom, application specific problems.

IoT: Internet of Things is a term that refers to millions of connect devices and the data transfer over them.

Mysticism: Spiritual apprehension of knowledge.

Pilgrimage: Religious journeys to faith-based destinations.

QR: Quick Response code is kind of a barcode allows a user to access specific information rapidly.

Religious Tourism: A kind of niche tourism that tourists are motivated to visit faith-based places and it is also referred as faith or spiritual tourism.

Smart Destination: Describes the destinations that embrace the use of emerging technologies for data collection, data analysis and implementation of environmental, economic, and social measures for a sustainable future.

Unique Experience: Exhilaration that cannot be obtained from anything else.

Chapter 6
Comparative Analysis of the Communication in Western and Eastern Culture:
Philosophical and Religious Aspects

Tamara Stanislavovna Olenich
https://orcid.org/0000-0002-1212-9181
Don State Tehnical University, Russia

Andrey Aleksandrowich Mekushkin
Don State Technical University, Russia

Olesya Anatolewna Shestopalova-Todarchuk
Don State Technical University, Russia

ABSTRACT

This chapter argues for peculiarities of intercultural communication with such a closed society as Iran. It starts from an interesting and comprehensive analysis of what is regarded to be 'intercultural communication' that is crucial important for the touristic sphere. To narrate the phenomenon on the intercultural communication within Iran, this chapter aims to reveal that though communication process within the Iranian society is based upon the tradition and religion, the country is tolerant and friendly to the foreigners. The necessity is for the people who visit the country to understand the difference between the Eastern and Western cultures and traditions. The communication in the Eastern society is based on the ancient customs and foundations, but the communication in the Western society is characterized by technological and rational knowledge of the world. Finally, this chapter attempts to propose a new understanding of communicational process in Iran, which will create possibilities for mutual understanding and cooperation with this country.

DOI: 10.4018/978-1-7998-5792-1.ch006

INTRODUCTION

The problem of communication and communicative competence of modern society is closely connected with the sociocultural competence of the people. The sociologists study the forms of human communication existing within the various segments of the society, the role of various factors such as education, culture, language, mentality, economic and social conditions, etc. upon the communication.

Communication is a very important factor for the development of the individuals because by introducing individuals to the standards, values, and ways of thinking of their culture (Zalewska, 2019, p.66). One of the most important problems is to understand how individuals acquire new habits and behaviors, the mechanism of this acquisition and how it is socially coordinated because all abovementioned can influence the process of youth socialization and it is closely connected with the of communicative competence (Zalewska, 2019, p.66).

Intercultural communication provides social life with new frameworks and dynamics, that is, «it shapes people's ideas, habits, and expectations concerning social life and their own lives» (Szafraniec, 2017, 167). The primary purpose of intercultural communication is to increase understanding of culturally mediated communication phenomena (Mehdi, 2011). A major component of a culture is its systems of values, beliefs, and material products (Nourollah, 2012). Intercultural communication creates the context in which two or three systems of values, beliefs, and material staff which form culture are enabled to meet.

The establishment and development of contacts between people from different cultures, especially when they visit countries with different cultural and religious tradition, generates the needs of joint activities and includes the exchange of information to explore the essence, content, and structure of communication. The peculiarity of this process is the formation of a new type of communication: changes in the way people learn and socially interact (Genner and Süss, 2017, p.4). To investigate intercultural communication as a background for the development of the touristic sphere, especially as for such a closed country as Iran is an interesting academic task considered in the present paper.

Research Method

The analysis of the literature shows that communication, as well as intercultural communication as a certain type of the one, is based on skills, firstly, to correctly express and transmit information, and secondly, to listen and draw the right conclusions. The research is based on the classical works by Émile Durkheim, Maximilian Carl Emil Weber, and Marie Auguste François Xavier Comte.

The research of intercultural communication is based on a big variety of social science research methods. The first principle of intercultural communication research is a comparison of two or more cultures. There are three basic methods used by scientific researchers to study intercultural communication: questionnaire, experiment, and content analysis. It is typically to use the survey questionnaire as a research method so we had used it in the form of the interview. The second methodical foundation was a content analysis of the academic papers and informal literature about Iran and its traditions. Each form of communication research is aimed to enhance the comprehension of communication and its forms. The comparison of cultures is useful while specifying what to expect when interacting with representatives of different cultures.

Findings

Iran has an ancient history, therefore the traditions of communication and relations among Iranians have ancient historical roots and a great influence on the Iranian religious and philosophic tradition. Communication etiquette consists of a set of speech and behavioral patterns, which determine human behavior in one or another social situation and establish contacts with the others (Williams, 2016).

Gender-Based Communication in Iran

As we say of the Iranian community, first and foremost attention will be paid to communication between Iranians «male-male» and «male-female». To understand Iranian society, it is of supreme importance to see the difference between *mahram* (unmarriageable) and *namahram* (the rest). The former are close relatives: parents, grandfathers and grandmothers, sisters and brothers, their children, husband's and wife's parents, children and their spouses, grandchildren, blood uncles and aunts. The rest, whether they are relatives or strangers, is *namahram*. *Namahram* means people who are formally possible to marry, but indeed it's undesirable and inconceivable. This difference leads to various behavioral patterns (Aliev, 2010).

According to the authors' opinion and position of many researchers, specific features of religious traditions in Iranian society influence the communication process between Iranian men and women and this factor plays a crucial role in the stability of such social institute as marriage in Iranian society. The Muslims and the Muslim women in most cases completely agree with the distribution of the social roles between men and women. As a result, because the social roles are distributed and the responsibilities of husband and wife are clearly defined and recognized, one can avoid a lot of problems in his or her family. A Muslim marriage is not a romantic union, a kind of love melodramas, but it is blood and bodily kinship of two human beings. The family with the accordance to the Shari'ah definition is a group of people, which is a core of the society and a family is created through the marriage of men and women, who then have children (Sharipova, 2014). The members of the family are in constant communication with the ancestors of the couple, as well as their brothers and sisters, grandchildren, uncles, and aunts of the father and mother and their children. There is an indication of the protection, support, and the existence of a strong connection based on kinship, kinship by marriage or dairy-related in these determinations. The family order is the rules and regulations which coordinate all affairs inside a family. The relationships between husband and wife in Islam are the following:

- Sympathy and Love (vidd, mavadda).
- Mercy (Rahma).
- Leadership with responsibility entrusted to her husband (kivama).
- Council (Shura).

The first two provisions are explained in Surah Ar-Rum («The Romans»):

And among His Signs is this, that He created for you mates from among yourselves, that ye may dwell in tranquillity with them, and He has put love and mercy between your (hearts): Verily in that are Signs for those who reflect (30:21).

According to the leadership in the family, God made it clear that this is no more than the responsibility the husband should take over. This is stated in Surah «An-Nisa» («Women»). The essence of the «superiority» of men is that the Almighty Allah created the man for one task and the woman for another task. The wisdom of this rule is based on the difference in the physical and the inner spiritual structure between a man and a woman. Saying of the Allah «that male superiors of women» is interpreted by some scientists of tafsir as a distinction of life spheres between men and women, although this verse is about the complete responsibility of both a male and a female (Sharipova, 2014). However, the woman in Islam takes an important place concerning God and the society as a whole. It is regarded as a jewel that needs care and respect. There are policies designed to preserve her honor and beauty.

In general, all the Muslim community can be regarded as a big family. These include, in addition to relatives, – neighbors, friends, acquaintances, and so on. Allah ordered the Muslims to be in touch with all of them and said that these relations should be fraternal:

«Indeed, the believers – the brothers» (49:10).

This is also a principle of relations with humanity in general since Islam told people to help each other and to treat each other with warmth and affection. Thus, the Quran says:

«O, people! Verily, We have created you from male and female and made you into nations and tribes that you may know each other» (49:13).

Speaking with the Iranian families, I noticed that family members are seriously concerned about the welfare of their children. They believe that children should grow properly in the spirit of Islam, as in this case, they will be protected by their parents. According to this Surah «Al-Tahrim» («Ban») states:

O ye who believe! Save yourselves and your families from a Fire whose fuel is Men and Stones, over which are (appointed) angels stern (and) severe, who flinch not (from executing) the Commands they receive for Allah, but do (precisely) what they are commanded (66:6)

In Iran, older people are very important for them not to be left alone. And it is most important, for each member of the family to get his or her responsibilities in this connection.

While examining the role of a man in the establishment of a new family, the following results can be highlighted:

- The husband should treat his wife as an equal, and not forbid her what is allowed by Sharee'ah. There are many hadiths on this subject: «The best of you are those who are good to their wives. And I'm the best for my wives» («At-Tirmidhi»).
- The husband should patiently tolerate the insult caused by his wife and take it easy when she's angry. Indeed, it may happen that a woman will cover the anger and irritation and under the influence of them, she will be suspended from her husband.
- The husband must not under any circumstances to humiliate his wife either by word or by deed, should not humiliate her family or mock them. A true Muslim does not put his wife in an awkward position in front of other people.

- If a husband doesn't love his wife, he should not oppress her, put to the rack, to demonstrate his aversion to her or to instill a sense of insecurity and uncertainty. He must be patient. The hadith of the Prophet says: «Let no man hates believing wife, as if you do not like her behavior, find the other [good] content with her qualities» (Muslim). Islam tells to treat a woman well. The Messenger of Allah turned to the men and said, «Be merciful to the woman is a fall to bounteous souls' share». Those men are bad who treat women with no respect and insults them». Another hadith says: «Always treat women well. To tell the truth, the woman was created from a rib, and the top of it has the greatest curvature. If you try to straighten it, you will break it, and if you leave it alone, it will remain crooked. Therefore one must always treat women well» («Al-Bukhari» and «Muslim»).

This example is a kind of explanation of what is the woman, and especially of her nature. Thus, she cannot always be right as her husband wishes and he shall know that this is her inherent peculiarity so he should not try to correct her in the accordance with his opinions what is right and what is wrong. The husband must treat with indulgence to the missteps of his wife and prefers not to notice some little things, taking into account the inherent peculiarities of the sex (Sharipova, 2014). As a result, there will be peace, tranquility, and happiness in the house.

In Muslim families, the mother is traditionally occupied with childrearing. It is a great credit to a woman to raise faithful children, but it also carries an enormous responsibility to the Supreme Being. A child's first human contact is with its mother. She carries her offspring in her womb, brings it into this world, nurses, and brings it up. In Hadith Qudsi (sayings of Muhammad) says this: «Paradise lies under the feet of mothers». Abu Hurayrah narrates that a man came to the Prophet and asked him: «Who is most deserving of my close companionship?», He replied: «Your mother, your mother, your mother, then your father, then the next closest to you in kinship; then the next closest». There is no doubt that Islamic upbringing has unique features for several reasons. But as in any family, Muslim parents ought to be friends to their flesh and blood. Perhaps, there are no universal ways in which to bring up a child as a Muslim.

The Prophet Muhammad said: «A father gives his child nothing better than a good education». In consequence, millions of Muslims on the Earth bring their children up in a responsible manner. Again, parents in Iran have authority without exception and they are highly respected. Parental control is usually treated with the utmost respect. Until children reach the age of majority, parental guidance is obligatory. A child may express his or her point of view but rude disagreement, argument, or the raising of a voice to one's parents are unacceptable.

After the Islamic Revolution, an Iranian woman related quite well to her family responsibility and public duty. Iranian women with great enthusiasm played a part in the process of development of their country by establishing various organizations (Norouzi, 2018). A famous journalist and poetess, Ludmila Avdeeva, who lived a long time in Iran, with her profound knowledge of the country, suggests that the family is not just a social unit but, above all, an «intergenerational continuity» based on traditions and customs that cement a society. According to Avdeeva, «Devotion to ethical ideals, based on human values, is characteristic of every religion». In her opinion, Muslim women are the 'heart' of an Islamic family and they are «not only beautiful but are also intelligent».

Behavioral Patterns in Iranian Culture

The Iranians consider themselves as two diverse identities: «*zaher*» (public/apparent) and «*baten*» (private/hidden). Being outside they have to satisfy established behavioral patterns. Only in their houses among immediate surrounding, they are free to be themselves. Family members are part of the inner circle. The inner circle is the base of the social and business life of a person.

Before observing practical matters of communication, it is also important to realize the role of such a social factor as a reputation. Each of people want to make a good impression on the interlocutor. It makes our cultures identical – for the Russians and Iranians, the reputation plays a certain role in the business as well as in the everyday life. For the Iranians, the reputation is above all.

One of the central tasks for Iranian is to save/preserve face (Kutlaki, 2011, p.54). In this context, it is important to understand what role «face» plays in Iranian culture. Let's draw attention to two key elements of the concept «*face*»: *hakhsiyat* (syn.: aberu or abru) – personality, character, self-respect, social standing, and e*hteram* (syn.: ezzat) –respect, esteem, dignity.

Shakhsiyat, a person's identity, is the result of one's upbringing and education. *Ehteram*, in its turn, shows up when the speaker follows rules of politeness(ta'arof) – he is ready to facilitate to the interlocutor, for example, to issue an invitation. Though *shakhsiyat* is something stable and static and *ehteram* is something dynamic and mobile their manifestations are closely related. People demonstrate their *shakhsiat* through their behavior, by, for example, conforming to social manners and paying the appropriate amount of *ehteram* to an interlocutor (Kutlaki, 2011, p.55).

A notable trait of Iranian culture is so-called ritual politeness – *ta'arof*. In various cultures, ritual politeness means different things. *Ta'arof* is «ritual politeness» running through the majority of Iranian speech patterns. This practice is connected with prescribed by religious duty of hospitality and generosity. The word *ta'arof* is of Arabic origin and it means «mutual acquaintance». It underlines the significant function of ta'arof in the course of business negotiations (Aliev, 2010).

One of the main meanings «*ta'arof kardan*» («to do ta'arof») – is to offer something (as tea or fruits). The next meaning (sounds awkward,) – is «to make an offer without expecting it to be taken up». And one more meaning of ta'arof is that of «refusing something because it is an expected form». According to *ta'arof*, an offer you must be made and refused at least once.

Principles of Communication in the Iranian Culture

In general, the Iranians follow three main principles of communication – respectfulness, modesty, and benevolence. There are various forms of deference behavior. Respectful attitude to interlocutors requires the speaker to speak from the bottom upwards. These behavioral principles have roots in human characteristics, highly appreciated in Iranian customs. *Adab* (good manners, courtesy) – is a trait that means respect, in particular to teachers, older people and people of higher standing, then cordial treatment and, finally, modesty (Kutlaki, 2011, p.60).

These principles appear through the units of language. Hence, there are various ways of addressing a person. In Russian as well as in Persian two pronouns are referring to a person: unceremonious «thou» (sing.) and polite «you» (pl.) – accordingly «*thou*» и «*Shoma*». It is impolite to use the pronoun «that», it is even rude. This pronoun is appropriate to use in intimate surroundings or in the family circle. However, people who say «thou» in private would say «you» in a stranger's presence. As the usage of «*thou*» is so limited, «you» is more widespread than in other languages. Practically it is the default pronoun in the second person. «*Thou*» can be used to a child or household servants, so, when the difference in status is clear. It has also to address God. Quran teaches us that He is closer than the jugular vein. («It was We Who created man, and We know what dark suggestions his soul makes to him: for We are nearer to him than (his) jugular vein» (Aliev, 2010). In a word, the pronoun «you» is to use in almost all cases. Thus, the familiar «thou» is used to underline the difference in status or to show intentional incivility.

Communication Within the Professional Environment

In a professional environment, the Iranians refer to each other *aga/hanum* adding a surname. If female co-workers are on friendly terms they will refer to each other adding «jān» to name in private. But, in presence of others, they will use common pattern «hanum+surname». At the secondary school girls call each other by name, but broadcasting to somebody by school radio they say «*hanum* … (surname)».

As for male co-workers, working friends usually refer to each other simply by surname not adding «*aga*». Pupils in schools for boys refer to each other in the same way. Surnames are often preceded by "titles", for example, *Aga* yemohandes (Mr. engineer) + surname, or *Hanum* doctor + surname. Last refer is used for both doctors and anybody holding undergraduate education (Williams, 2016).

To refer to a person by name (even if adding aga/hanum) may only a good friend. In Iran informality is tightly connected with intimacy, that's why it's better to avoid referring to a person by name in public.

Concerning foreign names and surnames, there is a little bit different state. As for the Iranian, it can be difficult to pronounce foreign surname, he or she will probably refer to by name (far often than to nationals) (Kutlaki, 2011, p.65).

Islam strictly prohibits touches between *namahram* (strangers). Needn't even try to touch representatives of the opposite sex; don't shake his/her hand and don't respond to an invitation to handshaking. At the same time, a foreigner has to kiss and hug representatives of his/her sex more often than he/she does it at native land. Practically the same can be recommended concerning laugh and smiles. One should not smile too much if you're a man, and if you're a woman, don't smile too often in men's presence and don't laugh loudly. However, the Iranians almost always ready to put themselves in a foreigner's position and try to forgive him or her.

The Best Topics for Communication

These civilities are a kind of social lubricant helping people to begin a conversation and easing their interrelations. The fact that one interlocutor pretends to the attention of the other enquiring about spouse, children, and parents, demonstrates his good attitude to the other interlocutor, it shows that they care about the same things.

Questions about health serve to the other purpose. They permit, so to say, to reset the conversation. If the conversation is not getting on and one interlocutor wants to finish it he can start asking after fellow's family health. Sometimes they permit to channel dying conversation (Kutlaki, 2011, p.66).

The best topics are the following: Iran, its language, culture, and history. Talks about a family at large, unobtrusively. Food, especially the variety of local cuisine. Sport, especially football (soccer) – it is always a good topic. Experts like to talk about their business line/scope of activity.

Traditions of Communication in Iran and Western Philosophy

Features of the development of Iranian culture and the method of communication used in Iranian society are largely due to the tradition and religion. The role of Islam in shaping the worldview, the normative-value system, and approaches to communication was a key point. This is evident, even based on the characteristics of communication and treatment in the verbal sphere of communication. Features of communication with representatives of other cultures are also regulated by the sacred texts.

On the one hand, this regulation does not limit intercultural communication: the principle of respect and understanding is one of the key elements of communication. In the process of research, using the method of sampling, during the private visit to Tehran we interviewed 7 girls from this city, aged 18-23, who went through the process of teaching Farsi in Tehran. According to their estimates, the process of studying and living in Tehran was comfortable (Olenich & Tryma, 2013, p.67). The Iranians were friendly, prompted how to properly address them, and the norms of accepted verbal and non-verbal communication. According to students, the Iranians sought to leave the best impression of their country, then representatives of other countries even somewhat abused. The basis of communication was a respectful attitude towards girls who aspire to become graduates and subsequently work by profession. This disproves the traditional view of Iranian women as disenfranchised family members under the guidance of a spouse. Many Iranian traditions seemed similar to Christian ones: elements of clothing, ceremonial activities, and dishes of cuisine.

The best way to keep a good relationship is to visit each other. Though, a phone is the second necessity in these cases. Telephone conversation starts with greetings and questions about the state of health. This ceremony precedes the current news exchange. The distinguishing feature of Iranian phone conversations is that the calling person doesn't introduce himself unless he asked to. Such telephone conversations shouldn't be too long as their purpose is to keep up friendly relations.

The other aspect of interpersonal communication is the skill of making offers and responding to them as well as the art of giving presents and paying compliments. If the person wants his/her present to be accepted, then normally he/she needs to insist for much longer – at least twice or three times. Reassuring the person that the presenter can spare it, it's no inconvenience. If someone wants to accept a gift or service, in this case, the person should decline it, saying the following phrase: «Barayeshomazahmatmishe» («It will cause you trouble»), which conveys the desire to accept. The person will insist it is no trouble at all. After all, the taker can accept the present with: «Razi be zahmatnabudim» («I didn't want to give you this trouble») and thank the presenter with words «Kheilimamnun» or just «Merci». Declining the offer saying the person doesn't want the gift usually is not efficient: it is part of Iranian hospitality to make the guest to eat and drink. If the person, however, wants to decline an offer, be prepared to insist: use the thanking formulas again, adding «injurirâhatar-am» (I am more comfortable like this) (Koshmanova, 2018).

Special attention should be paid to Iranians' treating their guest and the status they put him up. Iranians believe the guest to bring God's blessing to the house. That's why they welcome a guest with warmth and respect. One of the popular sayings is «Mehmanhabib-e hodast» – «The guest is loved by God». Guests are always given the most comfortable seat and the best food. They are made to eat, they are offered tea without being asked whether they want any, and are generally fussed over. At the end of the visit, they are thanked for visiting and asked to come back soon. Guests also express their thanks repeatedly during the visit, ask their hosts not to go to so much trouble, assure them that they are comfortable, and at the end of the visit thank them profusely for their hospitality and ask them to repay the visit (Kutlaki, 2011, p.68).

Hosts don't set any time limits for visiting. Guests are free to arrive and leave whenever at their convenience, but it is good manners to keep in mind the hosts' comfort.

If in the Western tradition people often use phrases like «thank you» and «sorry», Iranians must be uncontested champions of ostensible apology. An Iranian apology can be anything from an expression of humility, gratitude, and thanks. Sometimes they offer a present with their apologies if the presenter feels in debt towards the recipient of a present. After the apology they can move to a closing sequence, as in: «Sorry I've taken your time».

The most common form of apology is «bebakhshid» (from the verb «bakhshidan» – «to forgive»). Such type of apology is used within the ta-arof, for example at the end of a visit. This kind of closing sequence often takes this form:

Host: *Sorry you've had a bad time [with us].*
Guest: *Sorry for the trouble we've given you.*

It is customary for host and hostess to apologise for everything possible: bad food, lack of comfort, or waste of the visitor's time. In other words, they present their hospitality as being worse than the visitors deserve to demonstrate respect to the visitors and their modesty.

The content of the compliment may express respect and kindness towards interlocutor. The fact of saying the compliment is an expression of friendliness and hospitality (even if it is officially an accolade.) In this way, there are two possible answers towards the compliment: the particular expression of gratitude and «redirection» of the compliment to the author (in this way you will show your respect towards him).

Depending on the content, compliments are often expressions of both deference and cordiality. The fact of paying a compliment is an expression of friendliness and hospitality (even though it is a formal sign of respect). And so, there are two possible responses to a compliment: special expression of gratitude and «redirection/readdressing» of the compliment to the author (in this way the person can express his/her respect) (Koshmanova, 2018).

The initiator of the compliment pays face («ehteram») to a recipient by acknowledging the recipient as superior and himself as inferior, and so in doing, enhancing his own «shakhsiat». If the compliment is about an object or an item of clothing an offer of the object in response to a compliment is often made. There may be the following conversation:

A: *This is a nice pen!*
B: *It is not worthy of you! (that is «You can have it»).*
A: *Thank you very much! (Always meaning, «No, thank you»)* (Kutlaki, 2011, p.69).

Even though a speaker humbles himself and elevates the interlocutor, the person hardly expects the latter to agree with the force of the compliment because of the humility principle. The expected response would be a «mirroring» of the speaker's behavior: the interlocutor will humble himself and elevate the speaker. A possible response to a compliment is «Ekhtiardarid, maro, sharamandemikonid» (word for word «You are free to say anything you like, but what you say embarrasses me»). One more nuance while complimenting, for example, somebody's achievements or success, or interlocutor's children is to add «masha Allah» (the literal translation is «God has will edit»). It is considered to protect from the evil eye.

Taking compliments will be considered by Iranians as utter immodesty and even arrogance. Here's an example of how to respond appropriately:

A: *That's a nice pen!*
B: *It looks nice in your eyes (that is: You're so good, you see everything as being nice).*
A: *You're very kind* (Kutlaki, 2011).

Everyone happened both to make a request and to refuse a request. The ability to refuse without worsening the relationship between two people is a true art. While in many countries this might seem like the former colleague was looking for information, in Iran his question could also be understood as an indirect request for a loan. The curious Slavic mind wonders: why did friend make the request indirectly? Because if he asked directly, and a gentleman either didn't have that amount of money or was unwilling to lend it, both of them would lose face, which might reflect adversely on their ongoing relationship. On the other hand, if a gentleman were in a position to help the former colleague out, he would make an offer of a loan, thereby showing his cordiality and strengthening their ties of friendship. The moral of this interaction is: Don't ask for something directly if you can do so indirectly.

In general, the following piece of advice may be given: if the person has to make a request, explain his/her need without focusing on how the other person can help. It is better to recount one's problem. If the other person can help, he'll make the offer. In this context if a friend confides his problems to another person (but does not ask directly), the person can offer to help or make a general comment as a default refusal: «I hope it gets sorted out eventually».

Both the starting and finishing of conversation have a significant meaning. The conversation begins with ritual questions. Such lexical expressions are necessary for interlocutors to show mutual respect. The same concerns last minutes of conversation. Near the end of the visit, guests will indicate their intentions to leave, but the hosts will usually insist that they stay longer. When they finally rise to leave and go to the exit, the hosts will escort them to the front door, often downstairs to the front gate and even to the street. This custom shows the hosts' unwillingness to part from the guests and express their deference and hospitality toward them. The usual parting expression is «*Khodahafez*» («*May God preserve you/ God be with you*») (Kutlaki, 2011, p.69).

According to the majority of scientists, the most effective method of studying the etiquette of communication is a comparative analysis of two or more societies. Comparing the differences and similarities in codes of contact in the Western culture and the Iranian tradition allows identifying and characterizing the communication models.

The intercom is ringing.
Host: «*Hello, I'm listening*»
Guest: «*I salute you. I am …*».
Host: «*Hello. Oh, what a surprise (literally: «our eyes illuminated»)! Please, sir, please, come in*».
Guest: «*No, I won't trouble you. I just was passing by and decided to greet you and to ask how you are. Just wanted to learn what you're doing. Though, I have a small request*».
Host: «*One moment, I'm coming downstairs*».
Host opens the door and after shaking hands he says: «*Well, how're the children? How's your family? Why did you come alone? Please, come in*».

Guest: «*I don't want to trouble your family. I just wanted to see you and have a little chat in such a nice day…*».
Host: «*What are you talking about, sir, come in, please. How can a cup of tea cause us a trouble? I beg you, please*».
Guest: «*No-no, I won't bother you. It's nice out here*».
Host: «*For goodness' sake! Please, come inside*».
Guest: «*Well, if you insist so much, if you wish so/as you please*».
Host: «*Please, come in!*».
Guest: «*No, please, you first*».
Host: «*I beg you, please, stop paying compliments! My humble dwelling belongs to you. The guest is loved by God*»
Guest: «*No, sir. They say courtesy precedence… I swear, it's impossible for me to go forward. Please, after you*».
Host: «*I beg your pardon, then I will go first to show the way to your excellency*».
Guest: «*Yes, please, go forward*».
The host enters the house and the guest follows him.
Guest: «*For God's sake, forgive me for making you troubles!*».
Host: «*No, please. You've brought the light into our house and you honored us with your visit. Your visit is blessed*».
The guest goes to the chairs and host says: «*No, sir, if you please. Let's go upstairs*».
Guest: «*I beg you, it's comfortable out here*».
Host: «*Please, I'm begging you. I know for sure that our house isn't worthy of you*».

The host and the guest are discussing everything and everyone, including each other families, relatives, and acquaintances for 20 minutes and, at last, they start to speak about business and an issue that led the guest to the host's house.

Now, let's consider an analogous situation between two British men.

Host calls to his guest: «…»
Guest picks up the phone: «*Hello*».
Host: «*Hey there! This is David. Good evening*».
Guest: «*Thanks, what's up?*»
Host: «*Can we meet on Saturday and discuss buying the car?*»
Guest: «*Okay. I will come to you at about four o'clock and see the car. Is this time appropriate for you?*»
Host: «*Yes, bye*».
Guest: «*See you*».
On Saturday the guest arrives and rings at the intercom.
Host: «*Who's it?*»
Guest: «*Me*»
The host opens the door, the guest comes in and sits on the chair.
Guest: «*Hello!*»
Host: «*Welcome. So, have you decided will you buy the car or not? Would you like to have a drink?*»
Guest: «*No, thank you*».

The host pours himself a coffee and discusses trade details while drinking it. Such behavior is inappropriate for Iranian. Such treatment would be considered as very dry, heartless, and rude by Iranians. It is impossible to imagine Iranians calling each other and, without introduction and forewords, proceeding to the main topic not even asking how their families are doing and not repeating constantly «Well, how are you? What else? How are your kids? What's new?», not apologizing for «causing so many troubles and not explaining the reason for the call by saying «Well, I missed you so much and decided to call you»? Certainly, the person can do it, but only after generous forewords and eloquent introductions (that is proper preparations/«soil cultivation») it is possible to say something like «If you don't mind, I would like to speak about the car».

As Iran is interested in the development and intensification of cooperation with countries of the world, it is advisable to take into consideration the elements of communication to establish a business in Iran. The distinguishing feature of Iranian culture, as was stated before, is family values. To conduct business successfully foreign businessmen, need to keep in mind that the principles of communication (deference, humility, and cordiality) apply not only to social relationships but to the business relationships too. All visitors, whether on business or for leisure, are considered guests, and such can expect to be treated according to etiquette. A lot of words and gestures shouldn't be taken into consideration literally. If the business partner behaves decently, it doesn't mean that the person underestimates his/her abilities (Kutlaki, 2011, p.65).

In keeping with the importance of collectivity, group values, and the in-group/out-group distinction it is not a surprise that Iranians prefer to conduct business with those they know and trust, or at least with those introduced by such people. Iranians are proud of their nation and its ancient civilization, and its culture. They expect to be treated as equal partners, able to offer as much as they are likely to receive.

A business needs to make a profit to remain viable, of course, but profit isn't the only measure of success. Iranians consider the moral and ethical dimensions equally important, as well as maintaining a work-life balance. This doesn't mean that there aren't any businesses purely driven by profit or businesspeople whose lives haven't been taken over by their work, but be aware that a one-sided emphasis on profits at the expense of human considerations is an attitude generally frowned upon (Kutlaki, 2011, p.66).

As in every culture, the success of a business venture depends on impression during the initial meeting. In this case, the initial meeting and its right arrangement are very important. The first appointment is an opportunity for business parties to get to know each other. Naturally, it is necessary to make an appointment beforehand. It is advisable for both telephone and e-mail confirm the meeting about a week beforehand. One or two days before a meeting, it is worthy to call and confirm the final time and place. Though, before making an appointment, it is advisable to make sure that there is no religious or worldly holiday on the day of the meeting.

After all, properties were observed and the appointment was confirmed, the individual shouldn't be late for the meeting. It would be better to come 15-20 ahead of the estimated time, showing the significance of the meeting. While a foreign businessman is waiting for the meeting (or during it) he/she will be offered tea with some cookies and, probably, fresh fruits. The custom is the following: tea waiter (every office has such an employee) puts the tea and some sweets in front of the guest. The guest must wait until the host's invitation (Communication Etiquette, 2012, p.26).

During the business meeting, the person may be asked personal questions about the trip, family, university where the person studied and personal achievements are. Iranians may ask some questions about foreigner's home country, ways of doing things, and what your city looks like. Such questions help cultivate the personal relationships, that can then establish the basis for a successful business one.

Iranians listen to stories about curious moments of their partner's life and gladly telling about their habits and customs. It would be nice to show photos of the family, hometown, and university as well as to offer some small gifts or mementos from abroad, such as keyrings, pens, or postcards.

The essential attribute of business is a visit card. On the visit card, there must be information in English and Farsi about your position and academic degree. Visit card is a guarantee for Iranian partners to remember the name of their foreign partner and will properly estimate his social position. Visit card must be given in this way: it is given on your palms stretched forward (Communication Etiquette, 2012, p.27)

The attitude of the Iranians to the deadlines is quite interesting. Don't expect that Iranians would require a strict adherence of the deadline. They understand that their partner may have unexpected guests, diseases, unforeseen family problems. This is the reason why Muslims usually say *«Insha-allah»* (*«God willing»*) in any conversation about the future. Humans make plans and arrangements, but since all affairs lie in God's hands. On the other hand, any insistence by the Iranian business partner on a deadline that the person can feel is too tight may stem from a similar expectation that it will not be punctually met and may be an attempt to factor in a delay. The person can always explain his/her position by pointing out that the deadline will be met promptly (Kutlaki, 2011).

For a successful business, not only meetings and negotiations are important, but also memories that Iranian partners can gain. If a foreign businessman is interested in long-term cooperation with an Iranian company, then he should keep up the relationship via occasional telephone calls and e-mail, as well as *Noruz* (New Year) wishes, which is celebrated in March according to Iranian calendar.

Features of the process of intercultural communication in Iran imply respectful attitude to representatives of other countries and nations that visit this country, verbally this is manifested through the respectful treatment of "*You*", the use of prefixes "*respected*", etc. Hospitality and assistance to foreigners in adapting to local conditions, as well as the absence of conflicts and intolerance, if a representative of another country does not immediately understand local customs and rules are a characteristic feature of the process of intercultural communication in Iran, which is organically interconnected with the traditions and prescriptions of Iranian culture and religion.

It must be highlighted that the western path of development is based on the destruction of old, outdated foundations but the traditions of Eastern culture and communication can be regarded as hinders of its development. The most viable and sustainable option is a combination of rationalism and initiative of the Western tradition of communication and the Eastern one.

Based on the theory and practice of cultural globalization, the process of the occurring unification of all world civilizations and cultures is taking place as we can see on the examples on the communication traditions of the Iranian, which are rather close to the European ones. Despite the differences in mentality, worldviews, traditions, and customs, the people of Western culture and the Eastern tradition have one nature and common problems.

When comparing the Iranian culture and communication approaches with cultural and philosophy of the Western civilization, one can immediately identify a number of common features: respectful attitude to the representatives of other cultures, adherence to the principle of non-interference in private life, desire to help to adapt to the local traditions and rules.

CONCLUDING REMARKS

Studies of the features of the communication in the Iranian society and intercultural communication and representatives of the different countries who communicate with the Iranian while visiting the country as tourists convincingly confirm the fact that despite the adherence to the tradition, the Iranian society tolerantly and respectfully accepts the representatives of other countries and nations. The Iranians emphasize respectful norms of treatment that manifest themselves in certain verbal language constructs, convincingly proving that the primary principles on which the communication process is built within the Iranian society, and cross-cultural communication, is respect for the interlocutor. While visiting Iran, the tourist should take into account the differences between the Eastern and Western cultures but appreciate the way the Iranian treat foreigners.

REFERENCES

Алиев, М. [Aliev, M.]. (2010). *С благословения Аллаха: основные законы, обряды, традиции и обычаи в Исламе* [With the blessing of Allah: the basic laws, rites, traditions and customs in Islam]. Academic Press.

Этикет общения в иранской культуре: обмен любезностями [Communication Etiquette in Iranian Culture: Courtesy Sharing]. (2012). *Karavan*, *11*, 60-65.

Каменева, М. [Kameneva, M.]. (2009). Россия и Иран: история и современность [Russia and Iran: history and modernity]. *East. Afro-Asian societies: history and modernity,* 157-164.

Кошманова, О. [Koshmanova, O.]. (2018). *Краткий очерк иранского коммуникативного поведения* [A brief outline of Iranian communicative behavior]. www.pags.ru/science/conferences/E-Conference/Koshmanova.doc

Кутлаки, С. [Kutlaki, S.]. (2011). Среди Иранцев. Путеводитель по нравам и обычаям самой закрытой страны мира [Among the Iranians. Guide to the mores and customs of the most closed country in the world]. Moscow: United Press.

Оленич, Т., & Трима, Е. (2013). Иран: общество традиций (заметки на полях путешественника) [Iran: Society of Traditions (traveler's notes)]. Mariupol, 112.

Шарипова, Р. [Sharipova, R.]. (2014). Иранская семья: история и современность [Iranian family: history and modernity]. *Россия и мусульманский мир* [Russia and the Muslim world], *6*(264), 124-142.

Chatri, S. (2015). Islam and socio-economic, political and cultural development of women in Iran. *Historical and Social Review*, 7-10.

Genner, S., & Süss, D. (2017). *Socialization as Media Effect. The International Encyclopedia of Media Effects*. Zurich University of Applied Sciences.

Mehdi, S. (2011). Communication, Culture, and the Essentialized Islam. *Communication Studies*, *62*(1), 113–126. doi:10.1080/10510974.2011.540975

Norouzi, M. (2018). Feminism and the modern female prose in Iran (A critical look at Feminism in the writing of Shahrnush Parsipur). *Bulletin of the Moscow State Regional University (Russian philology)*, 154-163.

Nourollah, Z. (2012). Self-Perceived Communication Competence in Iranian Culture. *Communication Research Reports*, *29*(4), 292–298. doi:10.1080/08824096.2012.723271

Szafraniec, K. (2017). The Contemporary Context of Youth Socialization: The Specificity of Post-Communist Countries. *Polish Sociological Review*, *2*(198), 167-187.

Williams, S. (2016). Iran: The Essential Guide to Customs & Culture. In Tips on Social Etiquette in Iran. Kuperard.

Zalewska, J. (2019). Practice Theory Revisited: How Flexible Meta-habit Complements Habitus. *Polish Sociological Review*, *1*(205), 65–84.

Chapter 7
The Influence of Religion on Political Parties of the European Union

Kateryna Tryma
https://orcid.org/0000-0001-5567-1387
Institute of Higher Education, NAESU, Ukraine

Natalia Salnikova
https://orcid.org/0000-0003-4931-0491
Donetsk Law Institute, Ministry of Internal Affairs of Ukraine, Ukraine

ABSTRACT

The chapter is aimed to carry out discourse analysis of theoretical approaches and documents that highlight the issues of transformation of party-political systems of the EU countries. The religious factor is one of the catalysts for changes in modern parties of the EU, which, being different in their ideological orientation, electoral base, and territorial character, are influenced by general development trends. An important result of the discourse analysis is the conclusion about the weakening of the traditional parties, which from the national level has become a tendency of the pan-European political process. The religious factor stimulates the formation of small parties, which performs a great role in the development of tourism in the EU. The small parties are considered to be the defenders of local cultural and religious identities. The specific position of small parties to the development of tourism is caused by the wish to follow the economic interest and by the need to defend the local identities.

DOI: 10.4018/978-1-7998-5792-1.ch007

INTRODUCTION

Religion is one of the most important drivers of the development of modern humanity. The most common definition of religion as a belief in the supernatural does not exhaust all its essential characteristics and the main function of religion is the realization of a person's connection with the Supreme.

In modern society, religion can be regarded as an important factor in social transformations. The sociologists study the forms of the influence the religion can make throughout within the major and minor social groups, for example upon the education, culture, language, mentality, economic and social conditions as well as upon the political systems and political parties (Vaubel, Bowen & Zubair, Weinstein).

The transformation of political systems and political parties as an element of modern political systems is objective and is determined, in addition to the religious factor, by a wide range of interaction of socio-economic, sociocultural which, in turn, are part of more global changes in the modern society (Вайнштейн, 2018). But religion itself is that particular factor that is complex and multi-level, which determines the extremely complex, contradictory, multi-dimensional, and multi-vector nature of the modernization of party-political systems at the present stage of their development.

LITERATURE REVIEW

An analysis of the scientific literature makes it possible to specify the results of the influence the religion caused in party-political system development. The religious factors also led to the conclusion about the territorial homogeneity of voters, a high degree of personalization in party systems, and the relative religious homogeneity of the countries of the European Union (Maria-Irina, 2017). A key feature of the genesis of modern parties is the formation of new types of parties, which are characterized by the absence of a membership institution, network communication channels, direct connections with voters, and focus on solving local issues (Margetts, 2006).

The growing role of the influence of religion on political processes is confirmed by the activities held by the *small parties* which are not members of the government and are not oriented towards active participation in the elections in the future, due to orientation to a limited number of local voters. The reasons which unite citizens in small parties are connected with the protection of local interests, based on local political, economic, and social peculiarities as well as on the local cultural and religious background (Margetts, 2006).

The development of tourism is also influenced by new parties. On the one hand, the small parties partly are the antagonists of the development of the tourism on local territories to protect them, but on the other hand, they are for the developing tourism in the country to increase the economic potential of the whole state and local areas (Causevic, 2013). Thus, there is a contradictory trend determined by the desire of political elites at the local, regional and national levels to respond to the needs of the local electorate: protecting national identity, on the one hand, providing jobs and developing the economy, on the other.

Therefore, this research paper aims to investigate how under the religious factor the political parties are changing their forms and activities and outline their possible contribution to tourism development according to their political agendas.

RESEARCH METHOD

With this investigation, the authors intended to demonstrate that under the influence of the religious factor the political parties are changing forms and methods on their activities. We hypothesized that religion factor is one of the strongest societal factors which caused transformation political system and particularly political parties in modern Europe. The authors believed that it is possible to find evidence that religion is highlighted in political programs of the modern political organizations of Europe and it has an influence upon the formation of new types of political organizations.

The case study method was applied with the purpose to research and describe how religion factor is engaged in the political agenda of new types of parties in some EU countries to collect the data from the proper web-sites of political organizations.

Content analysis and comparative analysis were used to conclude the collected data. During data collection and analysis, the following inadequacy was faced: while the representatives of the mass-based political party of do not mention about the role of religion in their agenda directly or do it in a very conversed form, the representatives of small parties – a new type of political organization provided more information about religious factor in their agenda.

FINDINGS

The incompleteness of the process of transformation of political organizations in Europe is influencing the whole political landscape of the European Union. The difference in the dynamics of this process and the nature of partogenesis is due to the features of the historical development of states, the formation of their political system, as well as a model of political culture, the cultural and religious types.

Political Crises, Migration, and Religious Issues

Quantitative changes of the middle class do not allow us to have a confident look into the future, most of the representatives of the middle class in Europe are convinced that the society is divided into the globalized elites and anti-globalist-minded masses. That objectively requires parties to adapt to new emerging realities, to the emergence of new social conflicts, to a change in their electoral base. The change in social structure undermined the traditional electoral base of both center-left parties based on the working class and center-right parties, which were based on the middle class. Therefore, it can be concluded that at the beginning of the XXI century the existing political consensus of right-and center-left parties, the mainstream policy, which was aimed at promoting globalization processes, was undermined by the globalization processes.

External factors that set the parameters for the process of transformation of party-political systems include the process of forming a new world order in which the EU plays a key role. The political agenda, especially systemic parties, cannot but reflect the problems associated with the development of this process. Geopolitical and geostrategic shifts associated with the struggle for spheres of influence, control over natural resources, the need to confront global threats, could not and cannot be ignored by political parties. The ongoing geopolitical shifts associated with the end of the Cold War, the collapse of the socialist camp and the collapse of the USSR had deep, not yet fully appreciated, effects on the configuration of party-political systems, they "undermined the stakes in electoral competition in the importance of the left-right ideological split lying in the basis of most Western party systems" (Katz, 2009).

Geopolitical changes and globalization processes gave impetus to the process of depoliticization and de-ideologization within the framework of party systems in the context of the "right-left-center" division, when the main basis of the conflict lay in the plane of class relations.

Thus, the transformation of the world order, which is proceeding extremely intensively, introducing new challenges to the political agenda of the Europeans, primarily related to ensuring and reliability of their security, requires a quick reaction from them by political forces, an adequate reaction of parties to the demands of voters.

The catalyst for the transformation process was the financial and economic crisis of 2008-2009 and the migration crisis of 2015-2016 that overlapped on it. The global financial crisis of 2008 not only had economic consequences, having brought down the economies of almost all EU countries and leading to an unprecedented increase in unemployment. The crisis of 2008 led to an exacerbation of imbalances and asymmetries within the European Union, the degree of which before that was quite large. So, if the GDP per capita in Bulgaria was 7056 euros, in the Czech Republic 18,000 euros, then in Luxembourg it was more than 100 thousand euros (Creutz, 2019). The growing imbalance is evidenced by the indicators of the "index of material deprivation (deprivation)", introduced as an evaluation parameter for the development of the countries of the EU member states. The growing disproportionality during and after the crisis began to be threatening. The polarization within national states kept on growing.

The debt crisis in Greece led to a deep and widespread financial crisis in the eurozone, revealing a set of deep economic problems in this group of countries, which made it possible for several experts and politicians to talk about the need to return to national currencies. The crisis of 2008 dispelled the myth that the EU is the umbrella that can protect its inhabitants from all shocks, including economic ones. The reality turned out to be different, moreover, Brussels could not quickly adapt to new conditions and find an adequate solution to the challenges that arose. Under these conditions, national governments were forced to take the necessary decisions to overcome the crisis.

The peculiarity of this crisis was that, according to Tony Barber, the economic crisis was mixed with a potentially far-reaching crisis of legitimacy of the European political system (Barber, 2012). The more the debt crisis undermines the post-war social contract of modern Europe, the less is the confidence of voters in the political system, which is considered as a deal violator. The situation in the EU countries as a result of the crisis and its consequences not only brought a new agenda to political life, but also radically affected the level of the socio-psychological climate, public consciousness, and moods of EU residents.

The destabilization of the situation in the Middle East, the Syrian crisis, led to the emergence of hundreds of thousands of refugees who wanted to find shelter in Europe. Europe was in a state of the migration crisis. In 2015, 790,000 Muslim refugees entered the EU countries, the percentage of which in Europe increased in previous years. From 2010 to 2016, the number of Muslims increased from 19.5 to 25.7 million people, and by 2018, immigrants accounted for more than 5% of the EU population, in 2015 this percentage was 3.8 (Europe's Growing Muslim Population, 2017).

The consequences of the migration crisis and its causes are rather deeply reflected in the scientific literature, the authors and experts are unanimous in their assessments of the migration crisis as one of the dominant factors determining the qualitative and quantitative parameters of the process of transformation of the party-political systems of the countries of the European Union and the whole political life of this associations, becoming the number one issue on the political agenda, the attitude towards which has become an important marker for voters making their choice.

The problem of migrants became a political problem and was actively used in the election campaign for the electorate. The migration crisis was called the main problem, the solution of which most concerns the citizens of the European Union during the opinion polls conducted in 2015-2016, and remains so today. Only the inhabitants of Portugal put it in second place in importance. Residents of all other countries gave her first place. The greatest concern is the problem of migrants in Estonia - 79%, the Czech Republic, Denmark and Germany - 70%, and in the EU as a whole - 55% (Public opinion in the European Union, 2015). The specifics of the migration crisis in Europe is that it generates a set of overt and covert threats to the future of the EU. Among them, a special danger is represented by a decrease in the level of security and protection against terrorist threats, a threat to European values, and a lack of unity within the European Union on this issue. All this objectively requires parties of political elites to adequately respond to these challenges.

The Growth of Political Parties in Europe

The experts highlight a rapid increase in the number of new political parties caused by the financial and economic (2008) and migration (2015-2016) crises, as well as the adoption in 2010 in several countries (for example, Austria, Hungary) the law on the financing of political parties, which provides for subsidies from the budget for small parties.

By the beginning of the new electoral cycle of 2015-2019, in which almost all EU countries held elections of all levels, as well as elections to the European Parliament in May 2019, the quantitative growth of parties has become one of the significant parameters to specify party systems. So, by this time in the Czech Republic there were 89 parties and 143 political movements, in Hungary - 250, of which 116 appeared in 2014-2018; in Poland – 86 parties; in Slovakia – 61; and in Italy – 50 (Parties and movements of political alternatives in modern Europe N°357, 2019).

In Spain, in the November 2019 elections, 7 new parties took part; In Bulgaria, 36 parties took part in the election campaigns of this period; in general, the number of parties in the country has a pronounced upward trend. So, in 2000 their number was 200, in 2017 - 370, at the moment - more than 390 (Parties and movements of political alternatives in modern Europe N°357, 2019).

In the Netherlands, 28 parties participated in the 2017 elections, 13 passed in parliament. In the 2012 elections, there were 12. Two new parties were created in Britain; In Estonia, 7 parties participated in the 2019 parliamentary elections, and there were 5 in the previous elections. 12 new political entities appeared in Greece from 2010 to 2015 (Parties and movements of political alternatives in modern Europe N°357, 2019).

This trend is characteristic of party systems of the EU member states. This is also confirmed by the results of the elections to the European Parliament in May 2019, which included representatives of 160 parties (in the previous ones it was 10 less). This allowed experts to argue that the party systems of nation-states have a steady tendency to further progressive development towards their fragmentation and a fundamental change in their configuration.

Within the discourse of the typology of Sartori systems, it changes from bipolar (or dominant) to "moderately pluralistic" or multi-party. Confirmation of this thesis is the analysis of party systems in the context of the Laakso-Taagepera index. So, in Slovenia, it is 5.58, in Bulgaria - 4.88, in Austria - 5.25 (Parties and movements of political alternatives in modern Europe N°357, 2019).

As a result of recent electoral cycles, bipolar party systems have been destroyed in several states. These are Sweden, the Netherlands, Germany, Austria, France. In France, following the results of the

2017 elections, a five-pole system has developed; the bipartisan system in Britain and Sweden was destroyed. In Germany, there is a process of formation of a five-party system (Вайнштейн, 2018). The party system in Spain is fundamentally changing, turning from a bipolar to a four-pole system with a tendency to transform it into a multiparty system.

The Influence of Religious Factor Upon Partogenesis

The transformation process covers all the parameters of party systems, while it has a pronounced national specificity, both in terms of its intensity and its vectors, where the fragmentation process, which is ongoing both at the national and supranational levels, is still dominant.

For the first time, an attempt to consider a new type of parties and determine their place in the process of partogenesis was carried out by H. Margetts. Using the term "cyber party" concerning them, she defined this new form of the party not as marginal and the establishment of such parties as a natural result of the development of partogenesis as part of the development of its trajectory from the dominant cadre party to the mass party and later to all-embracing and cartel, and, finally, to a new type of party, defined as a "cyber party". The researcher considers the origins of this type to be rooted in a changing form of political participation, the spread of mixed electoral systems, and the development of the Internet and network systems. According to the researcher, the key sign of these parties is the rejection of the usual formal membership and the transition to direct relations with the voters (Margetts, 2006).

Several terms to define this type of parties is used now. These are "cyber parties", "online parties", "virtual", "Internet parties" as well as "network parties". Researchers consider the absence of membership is one of the key characteristics of this type of parties, party channels of communication are network channels, internal and external - these are direct communication with voters, strong positions in political competition for the voters. These parties possess the place between civil society and the state in the modern political landscape (Margetts, 2006).

A "small party" is a party that is not a member of the government and does not lead it, and in the future will not be able to, as it has a limited number of the voters. As a rule, small parties concentrate on a relatively narrow target electoral group, which is not in the interests of large parties. At the same time, they have significant total potential and they can quickly construct network structures. They put forward their slogans, which are important for a small electoral group, whose interests the large parties do not present. The number of such parties, operating, as a rule, at regional and local levels, and sometimes not at the federal level, is constantly growing. Usually, these are parties of the same subject, interest (Party of Women), or parties of one region (Bavarian Party).

Small parties, according to several researchers, perform the functions of indicators and timely socio-political signals. They allow drawing a wide circle, including radical points of view, into the current political discourse (Parties and movements of political alternatives in modern Europe N°357, 2019). At the same time, they create a new competitive environment. Concerning small parties, the concept of tabulated parties is widely used. These are small parties that have achieved success, received a mandate in parliament, and are recognized by other actors and political parties as a possible partner for the coalition.

As a rule, these new parties are "anti-system parties", which are extremely difficult to classify within the framework of the classical "right-left center" matrix. Almost all researchers agree that the phenomenon of anti-system is extremely complex and there is a clear tendency to further complicate it because, within the framework of ideological criteria, they can be placed throughout the political spectrum: from libertarianism to ultra-right conservatism and nationalism.

One of the features of this type of party is their personalization and orientation to the cultural (traditional) and religious views of the electorate. Moreover, in our opinion, a phenomenon characteristic of the activities of religious organizations is traced - the phenomenon of a charismatic leader, who, as a technology, acts as an element of accumulation of the interests of the population. A vivid example is the Freedom Party in the Netherlands, which consists of one person - its leader G. Wilders - and is at the same time one of the most influential political forces in the country. This fully applies to the Spanish party Podemos, where 51% of its voters are not familiar with its program, but know the leader well.

Noting the diversity of these parties, the researchers conclude that all of them, as a rule, position themselves as opponents of existing parties, clearly defining their positions as anti-political associations and classify themselves as anti-system parties.

New anti-system parties are becoming increasingly popular, winning more and more votes. The 2015-2019 election cycle convincingly confirms this thesis. In Germany, Spain, Italy, Sweden, the Netherlands, Hungary, the Czech Republic, Slovakia, and many other countries, anti-system parties took leading positions in the political field of their states and strengthened their positions in the European Parliament, increasing the number of representatives.

Nationalist Parties as Defenders of Traditional Cultures and Political Systems

Today there is a clear tendency towards an increase in both the size and influence of the nationalist parties. By the beginning of the 1990s, there were 45 autonomous, nationalist, and regional parties that participated in elections and campaigns at all levels (Lane, 1996). At the beginning of the 21st century, there were already 93 such parties, 30 of which began to play a leading role in the political processes of their states and regions (Massetti, 2009).

Assessing the role of nationalist parties in the modern political life of the European Union, experts are inclined to believe that a single ideology has formed in Europe, which unites all nationalists today and requires international interaction. A common tendency for them is the protection of Christian values, the fight against immigrants, the fight against Islamism, globalization, protection life at all stages (prohibition of abortion), the confrontation of the EU, and the United States, seeking to subjugate Europe.

New anti-system parties, as a rule, have a Euro-sceptic and populist profile, opposing and opposing the ruling establishment. The common message of the new parties, the "common denominator" is the denial of the political class as a whole.

The significance of these parties is primarily determined by the fact that they managed to aggregate and represent a much more complex and diversified system of public interests than the mainstream parties did. Thus, within the discourse of the function of the party is to fulfill its main purpose.

In the discourse of the transformational process of party-political systems, which is multi-vector in nature, processes associated with the development of anti-system parties are of particular interest. The experience of their duration allows us to talk about their evolution towards institutionalization in the form of traditional political institutions. A vivid example of this is the large-scale rebranding conducted by M. Le Pen of his National Front party, as a result of which the party not only changed its name to become the National Association but radically adjusted its program for the European Union, accumulated and revised the ideas of socialists and republicans, cleared the party of the most odious radical members, formed the image of the party as more conservative national than radical nationalist. As a result, by 2017, the National Front was transformed into a party of a nationwide scale, an effective competitor to the old systemic parties, which made it possible for M. Le Pen to pass into the second round of the presidential election.

The most striking example of embedding the mainstream in the system of parties was shown by the Greek Syriza. Having come to power in 2015, as a coalition of radically alternative parties to protest against the mainstream, she, came to the same neoliberal management methods as her opponents, whom she opposed while in opposition. The result of this evolution was its crushing defeat in the 2019 elections - it received 31.56% of the vote, lost 89 seats in parliament (Greece elections: Centre-right regains power under Kyriakos Mitsotakis, 2019).

The "Five Star Movement" in Italy acted in the same discourse, becoming the ruling party. Its new leader, Di Mayo, has embarked on the transformation of the party into a systemic one, having adjusted the program settings, abandoned the demand for a referendum on leaving the eurozone, and reduced the degree of radical criticism of the EU. Moreover, during his visit to the United States, on the eve of parliamentary elections, he distanced himself from Euro-sceptical movements.

Radically adjusts its image "Swedish democracy", which became the second political force following the results of the parliamentary elections in 2018 (Sverigedemokraternas princip program, 2018). For a long time, the party was characterized as Nazi and racist. Today the party has rid itself of a number of the most odious, nationalistically oriented figures; its latest program is aimed at combining the best elements of the traditional right and left ideology, priority is given to the national factor, multiculturalism is rejected, and a rather strict anti-migration policy is being pursued. SD nationalism is logically connected with antiglobalism and opposition to supranationalism and is the most important thesis of their program.

The Danish People's Party (Dansk Folkeparti, DF) is moving in this direction, changing its strategy, which today is aimed at shifting its political platform to the center. Their rhetoric regarding EU institutions is changing, the requirement to hold a referendum on Denmark's EU membership has been removed from the agenda (Henley, 2019). The Finnish party "True Finns" also carried out rebranding, removing a number of their radical demands, which allowed them to become second in the 2019 elections (The right-wing Finns Party does well in Finland's election, 2019).

The Spanish party "Podemos" has repeatedly changed its image and program settings. From the call to follow the example of Latin America, sympathizing with the regime of Hugo Chavez, to the decline of revolutionary rhetoric on the eve of the 2014 municipal elections (Rodríguez, 2014). Then the process of radicalization went on again, rejecting the "left-right" label, calling for the necessity of transition to a new political regime, idealization of direct democracy.

The Spanish party "Podemos" has repeatedly changed its image and program settings. From the call to follow the example of Latin America, sympathizing with the regime of Hugo Chavez, to the decline of revolutionary rhetoric on the eve of the 2015 municipal elections (Хенкин, 2018). Then the process of radicalization went on again, rejecting the "left-right" label, calling for the necessity of transition to a new political regime, idealization of direct democracy. The party has a pronounced charismatic character, voters vote for its leader Iglesias, and not for the party. So, in 2014, 50% of voters knew who its leader was, and only 1% heard about the party itself (Rodríguez, 2014). According to the results of the last extraordinary elections in November 2019, the party became the fourth, having missed the far-right party Vox (the Voice). The party is constantly evolving, integrating into systemically important parties, transforming and destroying the bipolarity of the Spanish party system.

At the same time, most researchers believe that after 2008, the evolution of right-wing parties has shifted towards their radicalization. This was largely facilitated by the migration crisis, which breathed new life into the right-wing parties. The mainstream today adopts anti-migrant slogans and anti-multicultural narratives. A vivid example of such radicalization is Fides in Hungary, Law, and Justice in Poland and the Austrian Freedom Party. Some researchers refer to this process as "hadoralization", a derivative of Hader, the leader of the Austrian Freedom Party, whose experience in this regard is indicative. The process of evolution of the mainstream parties includes the search for new models of the existence of parties that would be more effective in the struggle for power and better adapted to the current situation.

The dominant vector in solving this problem, according to experts, is the process of cartelization of parties, their integration into the state, which strengthens the community of goals and identities. This process is considered as a protective mechanism for their rooting, as a barrier to the fragmentation of parties, the creation of corporate parties, in the context of which the concept of the party as a business firm is implemented, because both are aimed at making a profit, in this case, political profit (Hopkin, 1999). This party model is characterized by a high degree of professionalism of party structures, pragmatism, and a clear focus on achieving the final result — political gain for gaining, retaining, and using power. An example of such a party is Forward Italy.

According to Mair and Katz, this leads to erosion of the representative roles of parties and the preservation of procedural roles, which contributes to the strengthening of party affiliation in the state and the disappearance of party affiliation in the society (Katz R. M., 1995). The further development of this trend, according to experts, can lead to the transformation of parties only into a "professional machine" that controls election campaigns, turning into state agents and not voters' agents, which raises the question of their legitimacy and status.

The Political Discourse of the EU: Eurocentrism vs. Euroscepticism

Under these conditions, a search is underway for new ideological concepts that researchers have designated as "molecular ideologies", not focused on universal ideas or explanations of social processes, but concentrated on a rather one-sided tunnel vision of problems - from assertive feminism to "militant environmentalism". New parties are characterized by the formation, not on an ideological basis, but on an opportunistic one, that is, around those topics and problems that at that moment were most relevant for voters, and the solution of which they were most concerned about. An example is the whole series of parties created in 2014-2017, which were based on the problem of migrants (the so-called anti-migrant parties) (Гаджиев, 2016).

New ideologies that do not justify the need to protect the interests of the class and social groups, but are focused on solving specific problems and small matters, protecting the existing ways and cultural and religious values are called "molecular ideologies", replacing large, historically established, ideological narratives (Гаджиев, 2016). An example is a new philosophy, which closely links the problems of politics with the problem of information development, called "cyber libertarianism". This philosophy laid the foundation for the creation of a new type of political organization - the party of pirates, which proclaimed the transparency of the political process, freedom of information space, network economy, the secret of private life, non-commercialization of the information space.

The absence of a pronounced and clearly defined ideological concept in most parties do not prevent researchers from defining and characterizing the profile of most parties within the EU political space. This is Eurocentrism, which is understood as the new political style that should create obstacles for the representatives of the extreme right flanks of national politics to enter power - as a rule, these are tough Eurosceptics, despite the growth of their electoral support.

So, in Sweden, they went on to create a minority government in order not to include representatives of the Swedish Democrats, the party that became the second after the election. They act similarly in the Netherlands to prevent representatives of the Freedom Party, which also became the second in the election, from entering the government. In Spain, in turn, the parties abandon the coalition with Podemos and Vox.

If this characteristic relates, as a rule, to the profile of system parties, then the following concepts determine the profile of both system and non-system parties. This is Euroscepticism and populism.

Determining the profile of emerging party systems, almost all experts mark it with such markers as Euroscepticism and populism. Euroscepticism today is a trend that characterizes the development of political processes throughout the European Union, which largely determines the qualitative parameters of the transformation of party systems. Paul Taggert designed his concept back in 1998, defining it as opposition and an expression of doubt about the process of European integration (Taggart, 1998).

A confirmation of this conclusion is Brexit, which, according to experts, gave a powerful impetus to a new wave of Euroscepticism, which should lead to the "renationalization" of the Old World and become significant support for a new generation of politicians in several European countries (Grygiel, 2016).

Euroscepticism is represented today both at the level of public consciousness and at the level of politics of political elites, the so-called social and political Euroscepticism. Euroscepticism has a different level and scale of both penetration into the mass consciousness and the nature of the activities of political elites in various EU member states.

At the same time, for the EU as a whole, this phenomenon has a steady upward trend and can be seen today at several levels: this is the evolution of European public opinion to increasingly critical, and in some cases hostile, moods against Brussels, this is the support of political parties opposing the further expansion of the European integration process, criticism of EU mechanisms and institutions by traditional system parties, expansion of cooperation between Eurosceptics at the transnational level (in particular, in the frame European Parliament), the growing number of Eurosceptic groups and civil society organizations.

So, from 2007 to 2015, EU support for its citizens fell from 52% to 37%, in 2017 only 40% rated the EU positively. The largest number of Eurosceptics live in Greece, the Czech Republic, and France. In the Czech Republic, 70% of residents consider joining the EU a mistake (European Citizenship, 2017). In France, the activity of the EU in 2015 was positively assessed by 55% of citizens, in 2016 - only 38% (Wright, 2016). It is estimated that 30% of Greeks and 50% of Spaniards are Eurosceptics (Future of Europe, 2018). In Hungary, 48% of the population in 2018 did not trust the EU and did not believe in its future (Special Eurobarometer 479, 2018). In Italy, 40% of the population associate the solution to their problems with leaving the EU (Bussard, 2018).

The lack of faith in a significant part of the EU population was confirmed by studies conducted by YouGov for the European Council on Foreign Relations in April 2019, the results of which showed that 30% of EU citizens believe in the possibility of war between EU member states, and among young people, the percentage is 50. These results allow us to conclude that only 50% live within the framework of the European integration process, and the other part lives in the mentality of the expected conflict, and this, in turn, is a breeding ground for the growth of Euroscepticism.

At the same time, in several countries, citizens demonstrate a high level of trust in the EU: this is Malta, where 93% of residents positively assess their country's presence in the EU, in Lithuania this figure is 90%; in Poland - 88% (Heath, 2018). Poland demonstrates a high degree of confidence of the EU in its inhabitants, while the ruling Law and Justice party are in tight confrontation with the EU, implementing reforms in the country, in particular judicial, whose principles are completely contrary to the principles and legislation of the EU: they are directed against the principles and institutions of constitutional liberalism - the foundation on which the EU is built.

The growth of public Euro-scepticism inevitably had to find reflection in the activities of political parties, that is, go into the form of political Euro-scepticism. Euroscepticism today has become an obligatory background of their programs related to the solution of socio-economic problems. Almost all anti-system parties and a significant part of the mainstream parties position themselves as critics of Brussels, realizing a "soft", moderate or radical Euroscepticism. Most Eurosceptics of the political elite, by definition of Robert Harmen and Menno Spearing, support the basic ideas of European integration, are positive about the idea of creating the EU, but are skeptical about its current structure (Harmsen, 2005).

The Impact of a new Type of Political Parties Upon the Tourism Development

An analysis of the activities of anti-system parties allows us to highlight another characteristic tendency of partogenesis in modern Europe. Political parties are consolidating based on national and religious interests.

An example is Spain, where autonomous regions are built on a purely national principle. This is most clearly seen in Spain, where several regional nationalist parties operate. These are the Basque Nationalist Party (Basque. Euzko Alderdi Jeltzalea, Spanish Partido Nacionalista Vasco - EAJ-PNV), the New Canaries Nationalist Party (Spanish Nueva Canarias), the Left Republicans of Catalonia, the Catalan Democrats, and the Unity of the Basque Country (Euskal Herria), Party "Compromise" from Valencia. In Italy it is the Freedom of South Tyrol party, in Slovakia the Alliance for a Family, Our Slovakia, in Greece the Greek Decision party; "Swedish Democrats" in Sweden, "True Finns" in Finland. Some of them stand on the positions of separatism, which is realized in various forms from peaceful, moderate, pragmatic, to tough, requiring the creation of an independent state.

It is the tendency to protect national interests that leads small parties to consolidate on a local basis. The emergence of local, regional, national parties is also associated with the desire of political elites to respond to the specific needs of the voter in a particular region. The economic factor plays a crucial role in the activities of the abovementioned parties, but at the same time, the factor of cultural and religious traditions also affects the development of secessionism processes, of which there are about 30 in Western Europe (Vaubel, 2013).

The development of tourism as one of the areas of economic activity of the EU, on the one hand, is based on the development of the economy, the creation of new jobs, development of infrastructure (Bowen, 2017). On the other hand, the activity of small parties is a reflection of the growth of nationalist sentiments, the electorate, which is a reaction to the socio-cultural consequences of globalization, encroaching on national identity, national identity, and national culture (Maria-Irina, 2017). The development of tourism leads to the need to globalize the local culture and adapt it to the needs of the average tourist. Thus, small parties act as a defender of the interests of the local community, as guarantors of the integrity of cultural and national-religious identity.

One of the problems that also affect the development of tourism is migration processes. Under the guise of tourists, both legal and illegal migrants penetrate the EU - which directly affects the security issues of local territories and all EU states without exception (Оленич, 2016).

The activities of small parties concerning the development of tourism are aimed at simultaneously watering all the economic benefits of this process, on the other hand, reducing possible threats to "soft security". Moreover, it cannot be said that it is a religious identity that is placed at the center of the political agenda. Rather, they unite in small parties based on national religious grounds, but in practice, these parties are still oriented towards upholding, among other things, the economic interests of local territories, because it is precisely economic interests and their protection that were often the main reason for their creation.

One of the reasons why the small parties might stand against tourism is the security reason connected with Islamization processes in Europe. Professor of Religion and Politics at the University of Birmingham, Cesari, believes that Islam should not be perceived as a system of values opposing "European", Islam is "just a religion", and European-educated intellectuals are possible in it (Cesari, 2008). According to Caesari, Islam can enrich European culture, and knowledge of Islam should be included in the curriculum of European educational institutions (Cesari, 2008). Other experts claim that Islamization cannot cause serious changes in political agenda of small parties because as soon as the necessary minimum of Muslims assimilates with secular European culture, the rest of the Muslims will be able to separate religion and cultural tradition and thereby turn to the values of European culture without losing their confessional identity (Ford, 2005). Therefore, this factor also cannot be an obstacle to include the politics on tourist development into the political agenda of the small parties.

Thus. even though in modern Europe a single ideology has emerged that unites all nationalists today and requires the protection of Christian values, the fight against immigrants, opposition to Islamism, globalization, the economic factor does not allow both traditional mass parties of the EU and new types of local parties do not allow the use of national-religious identity as a factor in the development of economic processes in the region. Thus, the development of tourism as part of the economic strategy of individual regions at the moment is not limited to the political agenda of small parties, even despite the restrained relations of their representatives to this economic sphere.

CONCLUDING REMARKS

The study of the political parties has proved our initial hypothesis stated in the paper. It was verified in the course of research that religion factor is one of the societal factors which caused transformation political system and particularly political parties in modern Europe. But religion itself is not the most significant characteristic which can affect the transformation of the political systems and political parties in modern Europe. Political parties are influenced by the changing conditions of the social and economic conditions of the national states. The modern electorate is oriented upon the protection and realization of their interests which usually are local. The growth of migrants caused problems for the soft security of many European countries and created preconditions for the traditional cultural and religious homogeneity of the countries of the EU to be degrading. In its turn, this process launched the process of the formation of the political elite of a new type which is mostly oriented on the interest of the local communities rather than on the interests of united Europe.

Religion has become one of the elements on which the suitability and expediency of the new political elite in based one. Studies of the features of religion convincingly confirmed the fact that religion one of the soft security influencers and the politicians who are publicly oriented to the traditional cultural and religious values are likely to be closer to the voters even when they are against the idea of united Europe. Religion was partly the reason to found cyber parties, small parties, the renaissance of nationalist parties. Being the factor for the activities carried out by new types of parties, religion itself has not become an obstacle or ideological idea for these political organizations. But religion is important in the process of foundation of small political parties because it shows the similarity of the local political elites and the electorate as for the cultural and national identity. Moreover, some forms of the activities normally used by the religious organizations are now carried out by the political organizations, among them, the charismatic leadership technology.

Besides religion cannot be considered as a significant factor in the economic process and development of tourism as a branch of economics. The establishment of new jobs and development of local infrastructure is more important as for the national political parties as for or local political elites, therefore, none of the national or local political parties has stated in its agenda the necessity to preserve the homogeneity of the local communities by decreasing the tourists' flow.

REFERENCES

Гаджиев, К. (2016). Метаморфозы идеологий в современной мировой политике [Metamorphoses of ideologies in modern world politics]. В Party-political systems and political ideologies in Western countries at the beginning of the XXI century: factors of evolution and directions of transformation, 112-124.

Хенкин, С. (2018). Перестройка партийной системы Испании [Reorganization of the party system of Spain]. *Actual problems of Europe*, 85-108.

Оленич, Т. К. (2016). Исламский фактор в миграционной политике Запада [Islamic factor in the migration policy of the West]. *Регулирование миграционных процессов на юге России: политические, юридические и правоохранительные аспекты*, 25-29.

Вайнштейн, Г. (2018). *Трансформация западноевропейского политического ландшафта и институциализация антисистемной политики* [Transformation of the Western European Political Landscape and the Institutionalization of Antisystem Policy]. World Economy and International Relations.

Barber, T. (2012). Europe must confront crisis of legitimacy. *Financial Times*. https://www.ft.com/content/4e2c793c-8d50-11e1-8b49-00144feab49a

Bowen, D. Z., Zubair, S., & Altinay, L. (2017). Politics and Tourism Destination Development: The Evolution of Power. *Journal of Travel Research*, *56*(6), 725–743. doi:10.1177/0047287516666719

Bussard, S. (2018). En Italie, les élections de la peu. *Le Temps*. https://www.letemps.ch/opinions/italie-elections-peur

Causevic, S., & Lynch, P. (2013). Political (In)stability and its Influence on Tourism Development. *Tourism Management*, *34*, 145–157. doi:10.1016/j.tourman.2012.04.006

Cesari, J. (2008). *Muslims in Europe and the Risk of Radicalism. Jihadi Terrorism and the Radicalisation Challenge in Europe*. Asgate.

Creutz, K. I.-M. (2019). The changing global order and its implications for the EU. Finnish Institute of International Affairs, 128.

Europe's Growing Muslim Population. (2017). *Pew Research Center*. https://www.pewforum.org/2017/11/29/europes-growing-muslim-population/

European Citizenship. (2017, November). *Standard Eurobarometer*, *88*, 117.

Ford, P. (2005). Europe's rising class of believers: Muslims. *Christian Science Monitor*. https://www.csmonitor.com/2005/0224/p10s01-woeu.html

Future of Europe. (2018). *Special Eurobarometer*, *479*, 118.

Greece elections: Centre-right regains power under Kyriakos Mitsotakis. (2019). *DDC news*. https://www.bbc.com/news/world-48902766

Grygiel, J. (2016). The Return of Europe's Nation-States. The Upside to the EU's Crisis. *Foreign Affairs*, 94–101.

Harmsen, R. (2005). *Euroscepticism: Party Politics, National Identity and European Integration*. Rodopi.

Heath, R. (2018). Europeans love the EU (and populists too). *Politico*. https://www.politico.eu/article/europeans-love-the-eu-and-populists-too/

Henley, J. (2019). Denmark's centre-left set to win election with anti-immigration shift. *The Guardian*. https://www.theguardian.com/world/2019/jun/04/denmark-centre-left-predicted-win-election-social-democrats-anti-immigration-policies

Hopkin, J. P., & Paolucci, C. (1999). The business firm model of party organisation: Cases from Spain and Italy. *European Journal of Political Research*, *35*(3), 307–339. doi:10.1111/1475-6765.00451

Katz, R. (2009). The Cartel Party Thesis: A Restatement. *Perspectives on Politics*, 753-766.

Katz, R. M., & Mair, P. (1995). Changing models of party organization and party democracy. *Party Politics*, *1*(1), 5–28. doi:10.1177/1354068895001001001

Lane, J.-E. M. (1996). *Political Data Handbook: OECD Countries (Comparative European Politics)*. Oxford University Press. doi:10.1093/019828053X.001.0001

Margetts, H. (2006). Cyber parties. In Handbook of party politics (pp. 528-535). London: Sage.

Maria-Irina, A. (2017). Tourism industry in the new Europe: trends, policies and challenges. *Proceedings of the International Conference on Business Excellence*, 493-503.

Massetti, E. (2009). Explaining regionalist party positioning in a multi-dimensional ideological space: A framework for analysis. *Regional and Federal Studies,* 501-531.

Parties and movements of political alternatives in modern Europe N°357. (2019). *Reports of Institute of Europe*, 87-112.

Public opinion in the European Union. (2015). *Standard Eurobarometer*, 298.

Rodríguez, Y. (2014). Podemos preocupa a los empresarios españoles. *Bolsamania*. https://www.bolsamania.com/noticias/pulsos/podemos-preocupa-a-los-empresarios-espanoles--625313.html

Sverigedemokraternas princip program. (2018). Landsdagarna.

Taggart, R. (1998). A touchstone of dissent: Euroscepticism in contemporary Western European party systems. *European Journal of Political Research*, 363-388.

The right-wing Finns Party does well in Finland's election. (2019). *The Economist*. https://www.economist.com/europe/2019/04/17/the-right-wing-finns-party-does-well-in-finlands-election

Vaubel, R. (2013). Secession in the European Union. *Economic Affairs*, *33*(3), 288–302. doi:10.1111/ecaf.12028

Wright, O. (2016). Euroscepticism on the rise across Europe as analysis finds increasing opposition to the EU in France, Germany and Spain. *Independent*. https://www.independent.co.uk/news/world/europe/euroscepticism-on-the-rise-across-europe-as-analysis-finds-increasing-opposition-to-the-eu-in-france-a7069766.html

KEY TERMS AND DEFINITIONS

Factor: The reason, the driving force of any process, determining its nature or its features.

Islamization: A political term for the process of enhancing the influence of Islam in various areas of public policy and public life.

Political Party: A united group of people directly setting themselves the task of seizing political power in the state or taking part in it through their representatives in government bodies and local self-government.

Religion: A belief system, including a set of moral norms and types of behavior, rituals, cult actions, and the unification of people in organizations.

Small Party: A small political organization which is not members of the government or local authorities and not oriented towards active participation in the elections in the future, due to orientation to a limited number of local voters.

Tourism Development: Process of establishing new forms and type of accompanied by a growing number of tourists.

Chapter 8
The Role of Religion and Culture in Modern Day Society:
Religion as a Business With Geo-Political Mixtures of Faith

A. T. M. Abdullahel Shafi
Independent Researcher, Bangladesh

ABSTRACT

In modern day culture, religion is specifically a business, rather than a faith. After European renaissance, religion is seen as derivative of changes between effect and resistance which had forced social scientists to think about religion from different perspectives. Then with some formal mixtures of capitalism and religion, religion became a component of business, rather than just a belief. From an individual perspective to social, political, economic perspectives, it's true. Even from individual to national and international perspectives, religion is just a business concept in the geo-politics at the international, super-national, and even supra-national arena. Nothing has changed the religions; rather international geo-politics has transformed religions into their own interests sometimes as an evil and sometimes as a "divine angel," which can benefit their interests.

INTRODUCTION

Religions are unique to their own values and qualities and as a culture, it has second to none. To some people, it's bad and even worst as a devil and at the same time to some people's, religion is the core belief to live on earth. There are fundamentalist religious groups as well as fundamentalists among the nonreligious groups, atheists, common mass as per their believe systems and even among every groups of peoples of every society. There is an on-going debate among social scientists, economists and political scientists regarding the role of religion. As per our understanding, the core is economy: truly saying- "He who controls world economy controls the world". There might be difference of views and opinions regarding such saying, still we are fundamentalist, rigid and strict on this line. Due to such view, modern-day

DOI: 10.4018/978-1-7998-5792-1.ch008

powerful states have concentrated more on economic development, rather than spiritual development or concentrate less on religion and has materialized religion on the subject of business. Even due to their economic superiority, they have tried to analyze and develop newer views on everything; society, religions, cultures, development etc. Sometimes, they have showed religion as positive force and afterwards due to the economic, social and political interests, they have tried to depict religion as monster. Therefore, religion sometimes have become a force of a positive culture and faith and immediate after that common masses have become confused of their faith and other's religious faith. For example, we can talk about current trends of humiliating Muslims and Islam all over the world. A hundred years ago, the Germans specially Hitler depicted Jewism exactly like what the economic super powers are depicting Islam and Muslim today. As soon as the propaganda against Islam and Muslim (Lewis,1990; Huntington,1996) from academic, social, economic and political perspectives arises, American and European peoples forgot their own cultures and tried to show themselves as 'secularist' and at the same time, the Muslims and Islam as 'fundamentalist' and even terrorists. Now the question arises in the mind of anyone why suddenly Islam which was a peaceful religion before 1990 (Huntington,1993) and what did suddenly occur which made Islam so demonized? So, what's the political economy of such demonization of Islam and Muslim? Business, yes sir, you heard it, right: truly economic benefit of the western economic super powers. In the next few sections, we would explain in detail of this business ventures of the capitalist economies.

Religion to some peoples are only beliefs and to some, this is just a business to whatever religion it is, i.e. whether it's Islam, Hinduism, Christianity, Buddhism. Some academicians or researchers might disagree with such views, but in it's truest sense of terms, it's a business as perpetrated by the religions themselves or capitulated by the capitalist ventures. For example, we can talk about the Muslim Eid, Hindu Pooja or Christian's Christmas where the business in religiously prescribed festivity enjoyed the highest amount of business in the world. The business during the religious festivals like Muslim Ramadan, Eids, Hindu Pooja and Christian's Christmas or Buddhist's festivals are the highest. Even the Religious tourism like Muslim Hajj (Pilgrimage) and Umrah or Hindus Tirthajatra (Pilgrimage) or other religious Pilgrimage, Capitalist economy use the most of its resources to make the highest profit of it. We would discuss the major issues concerning the religious tourism in the fifth section of this chapter.

We would like to expand the pseudo contradictions of religion and so-called ideologies of capitalism, democracy, communism in the preceding sections. Rhetorically there are contradictions among the ideologies and religions, but in practice, the contradiction is only for the business of the capitalist world. To be fair, the contradiction is not among religions and capitalism, communism or democracy, the truest contradiction is: '**who would control the economy, whether it's the Religious Institutions (church, mosque, pagoda, mandirs) or the state?**' We would like to get into details of such conflicts of interest in the next section of this chapter.

Actually, international Geo-politics had undergone a lot of changes like prostitutes to change its role and shape to be accustomed to the real-world scenario. Using terms like 'Prostitutes' might be odd in the academic arena and there might be different views on using such terms, still we are more determined to use such terms as the so-called ideologies have termed themselves as most modern and most present, yet they are not at all able to define themselves with the changing adversary of human needs, both physical, mental or spiritual. Therefore, the so-called ideologies like capitalism, communism or democracy has tried to change themselves and reshape their own pattern, style and taste as per the needs of their masters not to the need of their citizenry. For examples, basic ideologies of religion are exactly the same since it's inceptions like praying to god or believing god and the form might be different in different religions, but the so-called ideologies has changed themselves as per their own needs and aspirations. For example, democracy had been reshaped with the aspects and needs of the peoples of different countries and this ideology is not still been able to face the ongoing challenges of modern world. The same is true for Capitalism, Communism or other ideologies.

Modern day capitalism has expanded its wings from economic system to social and political systems. In case of communism, it's the same. If we take into account the history of last few thousand years, we can easily perceive that so many ideologies have crippled the world and nothing has become effective for the welfare of the common citizenry. Again, this capitalist economic system has waged 'religious war' in different corners of the world to control the economic bases of their own. To be honest, an Imam of a mosque has never started war against a country or a priest of a church has never waged war against other country and even a monk of Pagoda or purohits of a mandirs has never waged wars against each other. Rather it's the political leaders for their own interests has waged war in the name of religions. Not only that, the political leaderships have sometimes propagated against religions for their own interests and at the same time, they propagated for religions for the same interests. We have termed such roles and behaviors of the systems as 'prostitutic behavior'. Now some might argue, this is for the betterment of the society, economy and politics, but as a human being and academicians, we did not see the truest changes in any arena.

Religions in Real

First of all, we would completely disagree Weber's theory of Christian theology, i. e; protestant ethics in the growth of western capitalism (Hall,1985, P: 193-223). Because never had Christian theology supported western capitalism, nor had it strengthened the power of western economic capitalism. The truth is Protestant ethics in its truest sense of terms, is, was and always will have been against the growth of capitalism in today's form. Because, the problems of western theoreticians had been stressing upon the non-basic which best fits their economic interests ignoring the basic teachings of Christianity. Basic teachings of every religions are pleasing god and more specifically in Christianity, pleasing Jesus. Pleasing Jesus never admits profit maximization. The same is true in the case of other religions like Islam, Hinduism, Buddhism, Jainism, Sikhism, Jewism and others. In Islam, the basic teaching is Pleasing Allah. Whenever One's concentration is pleasing Allah, worldly gain is not that important. Even though Business ventures during the Muslim rules had advanced the most due to its role of market friendliness. The Prophet of Islam himself had been a merchant and their religious book Quran has stressed upon business-friendly gestures of the Muslims. The basic ideology of Islamic capitalism differs from the western capitalism in exploiting workers and maximizing profits creating income inequality and exploitation for which another theory named communism evolved which shook the world during the 19th century. Even though the character of both capitalism and communism as an economic theory projected the same at least by its applications in different parts of the world. I mean what American and west European capitalism is doing today, communistic economic theory was doing the same in former Soviet Russia and other communist world during the whole time. Both the systems created bourgeoise and ultra-bourgeoise in societies and by thus exploitation became rampant and inequality became regular. The religious systems always stressed upon doing business while the major concentration would be pleasing god and welfare of human being. Apart from such ideologies, both capitalism and communism treated themselves as new 'religion', a system of new instruments of irritating alternatives as the total system was in the hands of a bunch of criminals, drug dealers and even power-hungry immoral politicians and their slave economist and theorists. By this new term of "Slave Economic Theorists and Scientists" we would mean to say immoral academicians and researchers and even scientists who create theory, showing data and figures to the world that this system of exploitation is the best economic system of the world.

Religions, ruled economy into two basics: "Economy of Truth and Honesty" and "Economy of Lies and Dishonesty" which is unchangeable.

By 'Economy of Truth and Honesty', we mean by saying the business based on morality and ethics as well as honesty where profit maximization is not rampant to make poorest of the poor in the society and as well as there won't be only 1% richest peoples having the 99% wealth of the society. Even business should not be the means of making peoples subservient to their elite masters of the society. On the other hands, you can't stock the any commodities for longer to make huge profit by creating pseudo crisis in the whole market. You can't profit by illegal sales and marketing and even you can't trade items which is dangerous to peoples, be it foods, basic commodities or drugs, cigarettes or other products which are directly hampering human lives. In the previous sections, we termed a new phrase of 'Slave Economic Theorists and Scientists', who for money even created fake research reports to show products like certain drugs which are injurious to health have some positive impacts in daily life. To a good extent and good heart, we can agree some drugs like opium could be processed to make certain medicines does not mean directly inhaling opium might be good for human being or alcohol to a certain extent is beneficial to human body does not mean consuming alcohol is good for health. But theses slave academicians and scientists, being hungry for money are doing the same for their masters.

There had been so many economic theories of capitalism, profit maximization and even demand, supply and other streams of economic theories but there had not been a single theory on economy of honesty and truth which can by thus show the real economic turn up in the world system. Whenever you are talking about such systems, there are certain amounts of academicians and researcher who would rather question, what is our measurable scale of honesty and truth. I am not an economist. From a layman's view, we can measure the economy of honesty and truth by these scales for examples for a food product like bread:

Production of seeds and crops and control mechanism:

1. Did you strictly control its production systems in the Fields?
2. Is the wheat produced originally or by fertilizers and pesticides at field?
3. Are the fertilizers and pesticides dangerous to human life?
4. In what proportion of wheat was produced in the field? Is there any illegal mechanism involved to produce more by Generic implantation of seeds or so?
5. How was the workers and farmers pay during their works?
6. If it's private property, did the farmers treat their workers properly or forced them to work more hours?
7. Did the land owner, pay the workers properly?
8. Would the farmer pay portions of his/her profit to his workers?
9. Are the farmers properly paid during the acquisition of crops by government agencies?

From, the question number 7 and 8, there might the question arises of inequality and dishonesty as if workers are not paid proportionately as per their working hours according to their profit maximization systems, there might be less inequality among the workers and owners. Question 2 and 3 answers two serious issues: are pesticide properly controlled by the authorities and government agencies and even are those having any dangerous impacts on human body. Question 1 and 4 raises issues on unnatural productions of crops and seeds which might be not healthy for consumers.

Now on the production of Flour from wheat:

1. Are crops properly controlled while getting into the machines?
2. Did the machine factory mixing up other ingredients to the wheat seeds to make flour?
3. Are those ingredients dangerous to human body?
4. Are the machines properly controlled whether it has no danger effects to workers of the factory?
5. Are the factory workers properly paid?
6. Are the factory workers working in a proper environment?
7. What is the profit ratio between the original farmer who produce crops and the factory owner?
8. What is the payment ratio among the workers of the wheat field and the workers working the factory?

Production of Flour to Bread:

1. Is the flour properly controlled?
2. Are there any controlled effects of human health in the flour?
3. Is the bread machine properly controlled?
4. Are the workers properly paid?
5. Are the working conditions in the factory OK?
6. What is the pay ratio among the owners of the bread factory and workers?
7. Are profit proportionately distributed among the workers and factory owner?
8. While suppling the bread to the stores, are the prices properly controlled?
9. How many days are bread stored in the stores which had no health effects on customers?

There might be some others issues too. If we see questions properly, every parties are involved into it, the factory owners, farmers who are directly owning fields, workers directly working in the field, machine factories and even the consumers.

On the other hands, we have termed the other as "Economy of Dishonesty and Lies" which means whenever we are having productions of wheat in the field, we would accept all the dishonest means from paying less to workers as well as forcing them to work more hours. Mixing up generic seeds as well as crops and even mixing those crops with further pesticides which might have serious health consequences as well as other illegals means. Even during sending these supplies to customers, illegal marketizing and even tagging more prices on bread and seeds and crops and illegal stock of these products and raw materials to create unnatural crisis in the market. By adopting such illegal means, the whole economy becomes an 'Economy of Lies and Dishonesty'. Not only that, at micro levels, we would stress upon the cases of drugs, illegal medicines, illegals profit accumulation, illegals foods and clothes, everything might be an example of Economy of Dishonesty and Lies.

In such cases, we would like to comprehensively quote the scriptures of modern religions. From the Christian sense of view, the Old testament showed how the Israelites suffered as a nation in the social, economic and political life due to the lack of morality and ethics and how can a human being feel the sense of morality and ethics in only through prayer and human being's desire to understand the supreme God (The Christian Science Monitor ; 1991) .The Christian scriptures never forbid profit accumulation, rather it emphasizes on charity as well which helped society to decrease income inequality in the society. The verses did not all highlight about profit maximization and capital accumulation, rather some of these verses prophesized about the spirituality about getting closer to God which is better than worldly wealth. For example, we can cite about Leviticus 25:37 which implies, god never asked for human being's earthly wealth, rather god has given human being food and silver as worldly gain for he/she would be praying and lay down to god. Proverbs 16:8 directly incurred upon little righteousness of human being, rather than income with injustice. Ezekiel 22:12-14 exerted upon the real face of modern-day capitalism which has started war economy as means to better gains whereas God warned them of illegal bloodsheds, interests and profit making and injuring neighbors which would be dreadful act and god won't speak to him or her and would act accordingly at the day of judgement.

1 Timothy 4:8 described of the next life to come and profit of the next life which is better than earthly gains. Both Mark 10:29-30, Luke 18:29-30, Matthew 19:29, Hebrews 11:35, Philippians 1:21, Philippians 1:23, Job 9:29-31 cited about eternal life, the life after death which is more beneficial than that this life and worldly achievements. Both Deuteronomy 8:18 and Deuteronomy 28:15-20 warned human being about next life to come where punishments would be severe for those who won't obey God and his commandments. 1 Timothy 6:7 expunges earthly gains and warned peoples about the next life to come after death. Micah 4:13, Habakkuk 2:9, Proverbs 28:16, Isaiah 33:15-16, Jeremiah 22:17, Exodus 18:21, 1 Samuel 8:3, Haggai 1:9 discourages all dishonest means, oppressions, bribery and perverted justice. Proverbs 14:23, Proverbs 21:5, Proverbs 31:18, Proverbs 11:24, Luke 6:38, Jeremiah 12:13, Ezekiel 33:31, Leviticus 25:37, Proverbs 11:16, Proverbs 16:8, Job 20:18-19, Job 21:15,Job 34:9,Job 35:1-3, Psalm 73:13, Malachi 3:14, Matthew 16:26, Mark 8:36, Luke 9:25, Isaiah 44:10, Philippians 3:7, spoke about poverty, richness and charity and god's warning to those who gain wealth dishonestly as God never seek wealth from human being rather expects that peoples seek refuge for their misdeeds and actions from him. Isaiah 23:17-18, Jeremiah 17:11, Proverbs 1:18-19, Ezekiel 22:12-14, 1 Timothy 3:8, Titus 1:7 warned about illegals gains and bloodshed which in return would not yield good results for them. Isaiah 56:11, Psalm 119:36, Numbers 22:16-17, 1 Samuel 2:12-16, Jeremiah 8:10, 2 Corinthians 2:17, Acts 8:18-23, 1 Timothy 6:5 asked peoples not to be greedy of world wealth, rather asked them to seek for truth which is the best means to be closest to God. Titus 1:11,1 Timothy 4:13-15,2 Timothy 3:16-17, Proverbs 3:13-14, Titus 3:8, Timothy 6:6, 1 Timothy 4:8, Hebrews 12:10, 1 Corinthians 12:7, all these verses are based on knowledge, wisdom as the truth and also asked about other scriptures for truthfulness and holiness of God. 1 Corinthians 13:1-3,1 Corinthians 14:6,Hebrews 4:2,Titus 3:9,Proverbs 10:2, Proverbs 11:4, Jeremiah 7:8,John 6:63, James 2:14, Timothy 3:16 (King James version of Bible Online, 2020) examined the understanding of human being on god's unique power and willingness to lay down to him.

Buddhism's main value is "Buddhong Shoronong Gochham, Shonghong Shoronang Gochham,Dharmang Shoronang Gochham" which means be strict in Buddhist Values, be united and be hard and fast in Buddhist teachings. This is more about the teachings and spirituality of Buddha. One of the major teaching of this Buddhist economic thought is unity, honesty, cooperation among human beings and growing harmonious life styles. It denotes the sharing prosperity, sustainability and equity on the basis of mutual respect for each other and being kind to every creatures of the nature. It exposes three interdependent wings of life: Individual, society and nature. Wherever capitalism looks for profit maximization, Buddhist economics looks for less sufferings of all living and nonliving creatures of the world. Buddhism believed that Greed, Hatred and ignorance as three poisonous effects of human life and all the cause of human sufferings. Therefore, Buddhism asked for eightfold paths to lessen the sufferings of human life and Buddhist economists viewed wisdom, morality, meditation, rights speech, right actions and right livelihood as the key to happiness of human life. By thus, Peoples are interconnected to each other and connected to nature (Schumacher, 1973/1993, P.153-170). The basic Buddhist philosophy has more on improvement of soul through Bodhisotto (Meditation) and based on Karma (works) not on the profit maximization. Even both Mahayana and Hinayana Buddhist tend to incorporate the thoughts in the daily lives which means to explore the ideas of Buddhism as Absolute truth (Paramasotto) and there might be no other means to achieve the target of Enlightenment (Brownie,2017, P.10-15) .

Hindu religion is the only religion which has divided human being into four distinct categories; namely Brahmins as priests; Kshatriyas as warriors; Vaishyas as businessman and Shudras as normal labor class (Acyuta and Suchitra, 2005, P.260) on the basis of division of labor. This classification is termed as Savarna and even there are Dalits (Untouchables) which is called on Avarna (Susan, 2001, P.189-216). Hindu economic thoughts are based on spiritual development as well as working for day to day life. These economic thoughts are all inside in the Hindu scriptures in Vedas, Upanishads, Smirits, Manu Smrutis and Epics like Ramayana and Mahabharata (Bokare, 2009, P.10).The major aims of this economic thought was the welfare of the common peoples and the saint class of the Hindu peoples prayed to god for the same. In different parts of the Mahabharata, Ramayana, Vedas, Upanishad the economic systems are discussed with the clarifications of both political and economic system. We have taken into consideration the Hindu religious scriptures and have found verses where praying to god was the major concentration and then came economy as a means to daily livelihood but charity was the best consequences of making profit and making society stable by generating equity in the society. One such example might be Ishani Upanishad which uniquely described wealth and its distribution where the ownerships never lied to human being, rather its consideration was that resources are all natural and came from God as if God is the owner (Gewali, 2013, P.2). The Rigveda, Chaturveda and Smriti and Manu Smriti specifically described the spirituality issues and closeness to god in terms of economics and business (Sanskrit Magazine, 2014). But Koutilya's Arthashastra is somehow connected to modern day capitalism which described a lot of issues like modern day war economics, power politics, national economies and even economic philosophy (Shamasastry,1915, P.1-50).Even much of this economic literature concentrates on hymns of God.

Whereas Islamic Economic thoughts assume and assure first and foremost, the deity of One God and not profit maximization. As extreme profit maximization leads to severe inequality and economic chaos in the society, Islamic economics suggested charity which is in Arabic -Zakat. Not only Zakat, but also other forms of donations and charity to the poor is a must in Islamic economics which creates economic balance inside the society. Islamic holy scripture Quran (4:29, 5:1; 16:91) clearly indicated and mentioned about business and business with good will and contracts (The Al Quran Translated into English by Yusuf Ali, 1946). Islam forbids interests and such kinds of economic activities to purify the soul and its bad effects on the society. Even Islam indicted about byat-al-mal which is equivalent to modern day Public Finance. Other than this, there are some other economic regularities like concepts of welfare society was also introduced in the Islamic era of the Caliphates as rulers. Islam suggested Profit sharing business ideas which is equivalent to modern days idea of Stock Market. Even though Islam stressed upon god's mercy and pleasures of god in every aspects of human, social, political and economic life, yet Islamic economic system has become more genuine than that of other worldly religions (Hafner, 2006, P.16-22).

We haven taken into consideration the most populist four religions of the world while unveiling the characteristics of Religions and its role in market, business and capital accumulation and profit maximization. Actually, we are supposed to describe the geopolitical effects of religions in international capitalism and socialism and how world markets behave with this essence of religious mixed capitalism and the religion's own system. We have already discussed an essence of the systems and we have a brief idea of the differences between modern capitalism and religious economic system. In short, the basic difference is, religious system never permits profit maximization for individual where social inequality is highest, rather it suggests a system where inequality among individuals should be minimized and the major concentration of every economic system should be the pleasure of God. Therefore, religions, in big scales, never permit and even in reality that is also impossible that only 1 or 2 percent peoples are having 99% wealth of any society under any religious systems.

Religions and Contradictions

The major contradictions are in the mind of human being as there are so many religions and none of the religions are effective in initializing an effective market system and ensuring social and economic equality so that the society might be more welfare based. Whatever it is, western societies with capitalism are more just and better off than that of the so-called religious societies of middle east (Muslim majority), India (Hindu Majority), Vast parts of Asia (Buddhist majority, other than Japan, China and other countries are still stagnating and societal differences among individuals are still high with huge poverty rate) and even Latin America (Christian Majority) where poverty and inequality is high. The problem with such notion is that, this is too much of simple generalizations and far from realities. Because, none of these societies are based on the perspective of religions-based economy.

Even capitalism has in all the ways been succeeded to surpass the so-called religious systems by purporting religions as evil as like as socialism. We have in our previous chapters termed Capitalism as Prostitute as it changed its faces at every conditions and circumstances whereas religions are rigid and strict to its rules and theories. What Capitalism has propagated Christianity as 'fundamentalist' during 18th century has been successful in the 19th century to term Islam as a 'fundamentalist' religion. This newest propaganda has made them, for the time being, successful and by thus creating confusion among human being of the effectiveness of any religious systems. Capitalism is good to the extent that it has not yet termed religion as opium. But capitalism while doing business, sometimes made this religion as evil then another religion as evil by way of being really Opportunistic. This has made their way more secured not failing like socialism during the late 19th century. It has still secured its position in the world scenario even though their failures has become more visible in immunizing problems of the world market and their invincible role in world politics and the bizarre mixtures of values, ideologies and even so-called myths of tradition and modernity in maximizing the problem of real life economic and political scenario.

German Sociologist Max Weber (Weber: 1905, P.50) made things easy for both capitalism and religion in the western society to make belief that capitalism is nothing but a mixture of religious belief which made the rise of capitalism easy to the whole world. Afterwards, theorists became confused about the real role of capitalism and other economic theories of business and development and the real perceptions of religion where contradictions aroused among them. Because capitalism and other economic theories had never been successful to take over the role of religion to ensure peace, harmony and suitability in the society. Even recent thinkers of economic development completely ignored the role of religion in the society and they started propagating that religion is problem creator and obsolete for the society. Walter Benjamin (2005, P.259-262) had been at the forefront to neglect the role of religion and started depicting capitalism itself as a religion. This view is also supported by other economists like Friedman (1980, P.17). Even most modern economic thinkers either denied the role of religion in economic development or completely ignored it. We have reviewed among them are Simon (2006,P.155-170); Sen (1999,P.115-133),Stiglitz (2006,P.26),Sachs (2005,P.20),Collier (2007,P.4-30). There are other theorists who tried to show religions as one of the most driving force of capitalism and even goes further to show the interconnectedness of religions and capitalism. Among them are Heslam (2015,2011,53-72), Freeman (2012,P.1-38), Selinger (2004,P.523-543), Shao (2001,P.19-30), Bernstein and Rule (2011,P.91-131),Sherman (1997,P.165),Karlan and Appel (2011,P.1-10), Narayan (2001,P.39), Moyo (2008,P.5-15),Schlemmer (2008,P.26-75)Widmer (2012,P.167-176),Freston (2013,P.3-36),Martin (2011,76-86) and Berger (2004,P.4-5)

The most interesting pats of such contradictions are the failure of capitalism and it's changing nature like amoeba. Because, capitalism and other ideologies like communism never turned up to its own faces where it's just a machine to loot up wealth from the poor's in the name of ideology and penetrating their desires to exploit world wealth. We have cited about the works of few economists or thinkers who are in the forefront to show the interconnectedness of Christianity of certain countries of Africa and Latin America are not aware of their dominant ideologies to suppress the poor of those regions in the name of conversion into Christianity. Even the furthest parts of such description and interpretation is to exploit those poor more in the name of religion and making them subservient more to capitalism.

Witham's (2010, P.34-38) analysis of capitalism is rather more interesting. He described modern capitalism as economic activity to understand religion. Even Weintraub (2012, P.40-76) described seemingly funniest feature of capitalism to accumulate itself with religion and more science by adding mathematics, number, tables and statistics in the name of science.

This is the state of capitalism and the contradictions. On the other hand, communism as a theory of economy is just opposite and baseless. Because its masters have nothing inside their brains, rather they thought that the general masses are all illiterate not to understand their pseudo ideologies. This ideology panicked peoples to accept it in Russia and other socialists' countries by their stupid rulers as that was full of forgery and hollow. Marx's (1894, P.25-60) Explanation of capital was just a myth and a pseudo mixture of reality. The theory itself had nothing new in common, rather it was adapted from the other scriptures like Muslim traditions of the 1400 years ago. Capital accumulation is the tradition of any business, whatever you do, it can't be stopped. This is the major aims of any business. But if it creates a huge inequality in the society and make the market and economy unstable, that's a huge problem for the society. Therefore Muhammad (The prophet of the Muslims) introduced Zakat and even tax paying systems to make the market stable and make less difference between rich and poor. Even every religion what we have cited everywhere in this chapter, stressed more on charity a way to minimize the gap between the rich and poor is the simplest way to ensure economic liberalism. Communism adopted its ideologies which are replica of religious identities of different religions which believe there is no difference among human beings and even less wealth is subject to less problem worldly and the world hereafter. In prayer, religious peoples can gather together in temples, mosques or in church without having any worldly economic, social and political differentiation, communism adopted same ideologies derived from the religious scriptures. It's kind of – 'Old Wine in New Bottles'.

International Geo-Politics and Religions

Today's international geo-politics is marked with confusions and a mixture of honesty and dishonesty, secularism and a mix of religions with interests and even bizarre mixtures of cultures and noncultures. With non-cultures, we mean cultures which are virtual and does not resemblance with our day to day life. For example, movie culture is not real, yet this has been well incorporated into daily life as if this is all real. Due to advent of newer technology, we have been associated with it regularly and we had accumulated to it's gist to grab. Even we have already started debating whether are we modern without adopting ourselves to this new era of development with non-cultures and assimilating ourselves with such non existing realities of cultures. Even modern day politicians are still in the 'verge of awakening' or 'sleep mode' to cope with it. By thus they are trying to belligerently accumulating themselves and their citizenry into it knowingly or unknowingly.

To some countries, religion is more a myth than reality due to their cope up strategy and mechanism and even to some extents and at certain stages they can't cope up with their new existence without a religion. Therefore, neither they are accepting it, nor are they ignoring it. For example, we can say about United States where the constitutional preamble started with 'In God, We Trust'(Article 36 of the USA Constitution,2019) and they are still coping with it. But for the most of the countries and peoples, religion is not the obligations. They are neither accepting nor ignoring it. But the conditions of geo-politics made religions a prerequisite to some countries and to some, it's just a burden. For example, we can cite about the post 9/11 George W. Bush declaration of Crusade (The Wall Street Journal,Sep.21, 2001) which was simply a 'war on terror' as they called it. Still after almost 20 years, American peoples are confused about the terrorist attacks on twin towers. The real cause was to grab the gas pipeline between a few middle Asian countries like Tajikistan, Kirghizstan and Kazakhstan and Afghanistan which Russia was blocking and the second reason behind it was to acquire huge amount of uranium, plutonium and other mineral resources from Afghanistan (Shafi,2013, P.1-15). In real, the whole mission of 9/11 was to benefit few business men of the country which was afterwards proved by the domestic scientists, academicians and other researchers. But the business severely damaged the image of the Muslims all over the world and few innocent words like Jihad, Allahu Akbar and others few became a symbol of world terrorism. On the other hands, President Barack Obama (The Reuters,May 8, 2009), after being elected as the president went to Egypt and praised Islam as one of the most peace-loving religion of the world as this image portraited by former president after 9/11 caused huge economic loss to American economy. Even they have imposed embargo on Iran and despite the facts, the few biggest companies of USA carry on business with companies of Iran and other countries and this sparked a huge criticism all over the world (Becker, Jo article in The New York Times, December 23, 2010). After the Gujrat massacre, Norendra Modi (The Guardian,22 October, 2012), the current Prime Minister of India has been omitted diplomatic visa to the USA and other European countries. The same man after being the Prime Minister of India became one of the closest friends to USA President and other Heads of the European Governments due to their business and economic interests. Even the funniest thing is capitalism's expansionist ideologies made things worse for it as USA and other European Economies wage war against Iraq, Afghanistan and other few countries which made the world a 'War Zone". This is simply described as war economy which benefitted capitalism enormously. By 'war zone' we mean the regions where war or war like situations are imposed namely the middle eastern countries like Iraq, Syria and others. We can particularly cite about the Syrian war by which Russia, USA and other European countries are supposed to be benefitted and truly this benefitted them all. But afterwards the large influx of refugees from Syria to Europe made them more flexible which sparks large protest in the European countries in the name of 'Muslim European Invasion' and some other unreal terms like "Arabization of Europe" by some European organizations like 'PEGIDA'(The Politico, 19 October, 2015) and even few main stream political parties started gaining in politics against such large influx of Refugees.

Even after 9/11, terrorists' attacks of any kinds and anywhere of the world was declared as a Muslim terrorists attacks by the main stream media and was flattered everywhere. But the same main stream media and organizations went silent whenever any organized terrorist attacks were happened to Muslims. Whenever a Christian terrorist attacked peoples and killed innocent human beings, western media either blackout or never called him a terrorist, rather sometimes a psychic patient. We have seen the same scenario with Anders Breivik (The New Yorker, May 18, 2015) who has never been called on a terrorist, rather a mass killer and even his religion Christianity has never been called on "Religion of Terror" like what in case of Islam has happened. After 9/11, against Islam several hate crimes were initiated and organized all over the world, even though Islam has nothing to do with terrorism, as claimed by USA. Osama bin Laden was happened to be a Muslim. The funniest thing is Osama bin Laden had been declared number 1 terror suspect, not the major accused.

The former Israel Prime Minister Menachem Begin (New York Time, Dec.24, 1946) had been involved into terrorists' organizations like Irgun (The New York Times, August 16, 1947) and had been declared as "Terrorists" (The New York Times, Dec.30, 1947). The same Monachem Begin afterwards became the Prime Minister (Ministry of Foreign Affairs, Government of Israel, 2019) of Israel and was awarded with Nobel Peace Prize in 1979 (Website of the Nobel Prize Committee, Oslo, 2019).

This is what we termed as 'Business of Capitalism'. It has nothing to do with religions. Whatever benefits them, they became friend with them whether it's Islam, Buddhism, Christianity, Jainism, Sikhism, Jewism and at the same time, if it goes against it, they called it enemy with several names. They need Jew and Israel state in the middle east for their security of the Arab oil resources and they have become biggest perpetrator and friend of Israelis. They learnt this lesson after 1973 World Recession (Zarnowitz and Moore,1977) whenever the Arabs nationalized the Oil resources and they have learnt the same long before during the nationalization of Suez Canal by the Egyptian president Anwar Sadat in 1956(The Guardian, UK,14 March, 2001). This is called the "Politics of Need". Whenever you need it and so long, it is your friend and as long as you don't need them, you can throw it like a used condom. Ideologies like capitalism and socialism are at the forefront of accepting such ideologies.

The Bizarre Mixture of Religion and Capitalism as a Business

Before going into details, we would show map of religiosity of the world in brief. This map showed the religiosity of the world where most European and wealthy nations are less religious and African, South Asian poorest countries showed the trends of more religiosity.

Why is so? The pervious Gallup poll showed the same trends in 2009(Gallup Poll Survey, August 21, 2010). There might be an explanation for this trend as religiosity helped poorest peoples to cope up with their sorrows and sufferings of daily life. But religion is more than tradition, rather it has become a mixture of faith in most of the countries and peoples in their daily lives. But what we wanted to see in the trends of business is that religions are a super tool of business of capitalism and communism. We would like to see how business is coping with religious festivities in countries. In USA alone (from the figure, we see USA peoples are a bit religious which is almost 54%), the national departmental stores annual sales in December was 19.4 Billion dollars, Jewelry sales rose to 18.2%, hobby, toy and game store to 12.5%, estimated electronic shopping was 67.7 Billion Dollars, Retail Trade was 20 Billion Dollars, Christmas Ornaments from china was 1.9 Billion Dollars, Imports of Christmas Tree from China was 499.4 Million Dollars, US import of Tapered Candles was 511.3 Million Dollars(Ibid, 2010).

Figure 1.
Source: https://www.pewforum.org/2018/06/13/how-religious-commitment-varies-by-country-among-people-of-all-ages/

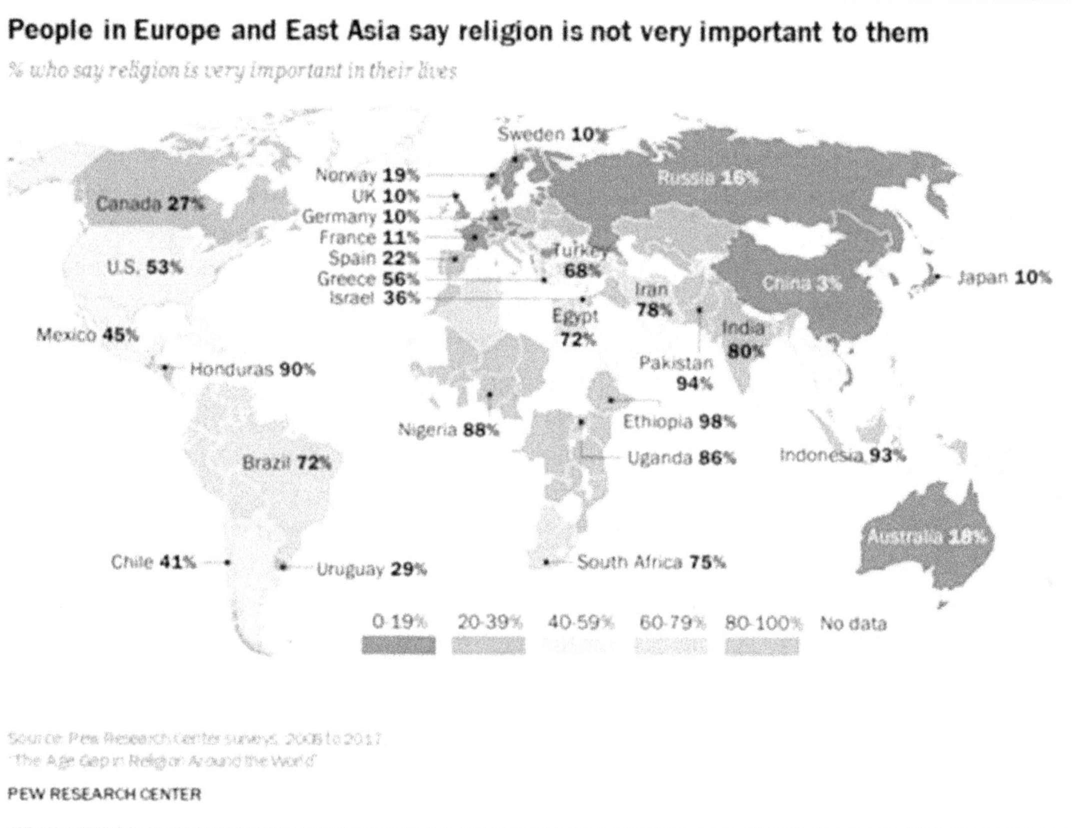

Whereas most of the European nations are not religious, still the capitalist's trends showed the biggest business trends of the European nations. In 2019, Average estimated expenditure of Christmas celebration per person was highest in UK (639 Euros), Spain (554Euros), Italy (549 Euros), Germany (488 Euros), European Averages was 460 Euros, Portugal (387 Euros), Russia (367 Euros), Poland (352 Euros) and The Netherlands (341 Euros) (United States Census Bureau, October 29,2019).

Now lets us have a look at the other markets of others parts of the world who are less religious. China is a communist country with less emphasis on religions. In 2019, China's dollar denominated export increased to 7.6% in December. China's December Trade surplus was 46.79 Billion Dollars and China's Trade surplus with USA was 23.18 Billion Dollars (Statista Website, 2020).

The above figure shows the holy trade in Indian stock market which is auspicious during Indian religious fest of 'Holi'. The Hindus termed this as 'blessings of Hindu Goddess Lakshmi.

With the Muslims, the business is extremely easy. Write the word 'Halal' in anything and Muslims would be a easy prey to that and capitalism has taken fullest chance of it. This global halal food market was 715 billion Dollars in 2018 which is estimated to grow 2043.2 billion Dollars in 2027(CNBC Website, January 13, 2020).

This is the economic trends and business in the Muslim during their religious festivals. Even during the same Muslim festival seasons of Eid and Ramadan, E-commerce increased to almost 67% and travel

Figure 2.
Source: https://www2.deloitte.com/content/dam/Deloitte/de/Documents/consumer-business/Deloitte%20Christmas%20Survey%202019%20-%20European%20results.pdf

Estimated Christmas spending (by country) (€)

	Germany	Italy	Netherlands	Poland
Gifts	217	220	122	127
Food	116	154	109	121
Socializing	51	65	32	29
Traveling	103	109	78	75
Total	488	549	341	352

	Portugal	Russia	Spain	United Kingdom
Gifts	159	123	237	336
Food	107	115	173	162
Socializing	40	31	79	71
Traveling	81	98	65	70
Total	387	367	554	639

	Gifts	Food	Socializing	Traveling	Total
European average	193	132	50	85	460

Figure 3.
Source: https://economictimes.indiatimes.com/markets/stocks/news/investors-see-ingredients-for-stocks-revival-after-diwali/articleshow/71756398.cms

Figure 4.

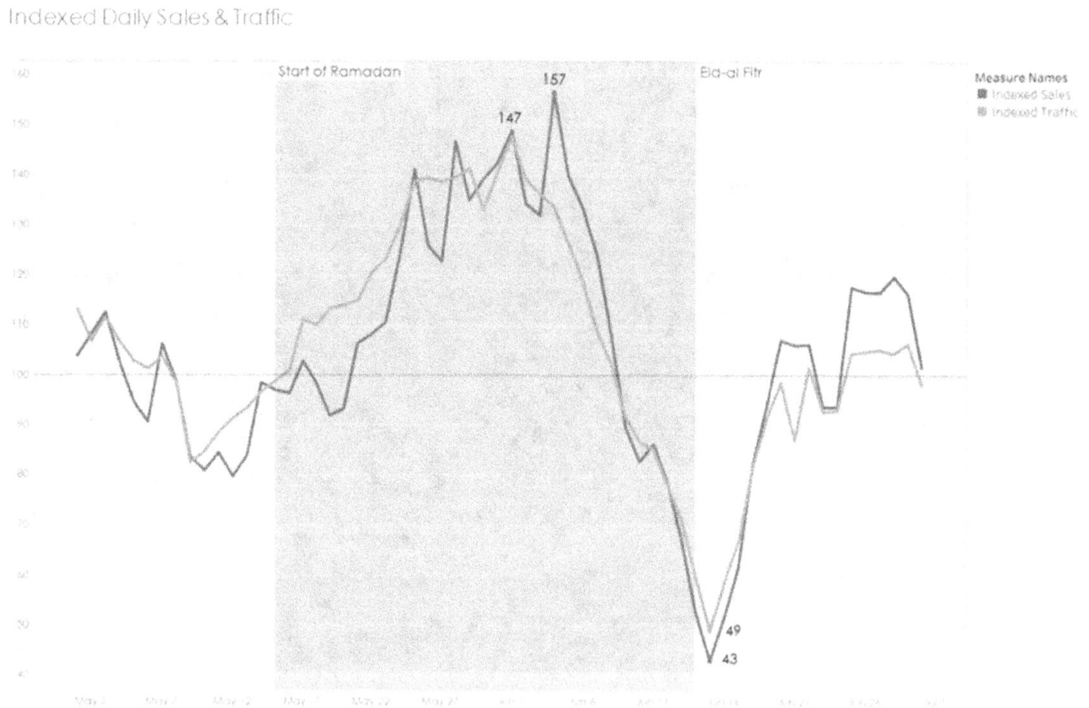

sales up to 14% as well as clothing 30%, Travel 30%, Smartphone 17%, Telecommunications 19% and consumer electronics 24% (E-commerce Guide, 2019).

This bizarre yet true relationships between business and religions still haunted the minds of the capitalism and they are taking fullest opportunity out of it.

The Real Face of Religious Tourism

Among the religious pilgrims' places, the popular destinations for each religious people are:

Muslims- Mecca, Medina and other places in Saudi Arabia.

Hindus- Shirdi, Tirupati. The Tirumala Venkateswara Temple near Tirupati is one of the holiest in India, Rameswaram, Somnath and Dwarka, Vaishno Devi, Puri, Amritsar, Mathura (Make my Travel Blog, 2019).

Buddhists- Gaya, Bodh Gaya, Rajgir, Nalanda, Sarnath, Kushinagar, Kapilavastu, Shravasti, Kaushambi, Sankisa in India (Tour my India Travel Blog, 2019)

Sikh Religion- Golden Temple, Patna Sahib, Paonta Sahib (Himachal Pradesh), Anandpur Sahib (Punjab), Baba Bakala (Punjab), Goindwal Sahib (Punjab), Taran Taran (Punjab), Sirhind (Punjab), HemKund Sahib (Uttarakhand), Nanded Sahib (Maharashtra) in India(Ibid, 2019) and, Kartarpur in Pakistan(BBC Report, Nov.9,2019)

Figure 5.
Source: https://www.criteo.com/insights/ramadan-2019-advertising-trends/

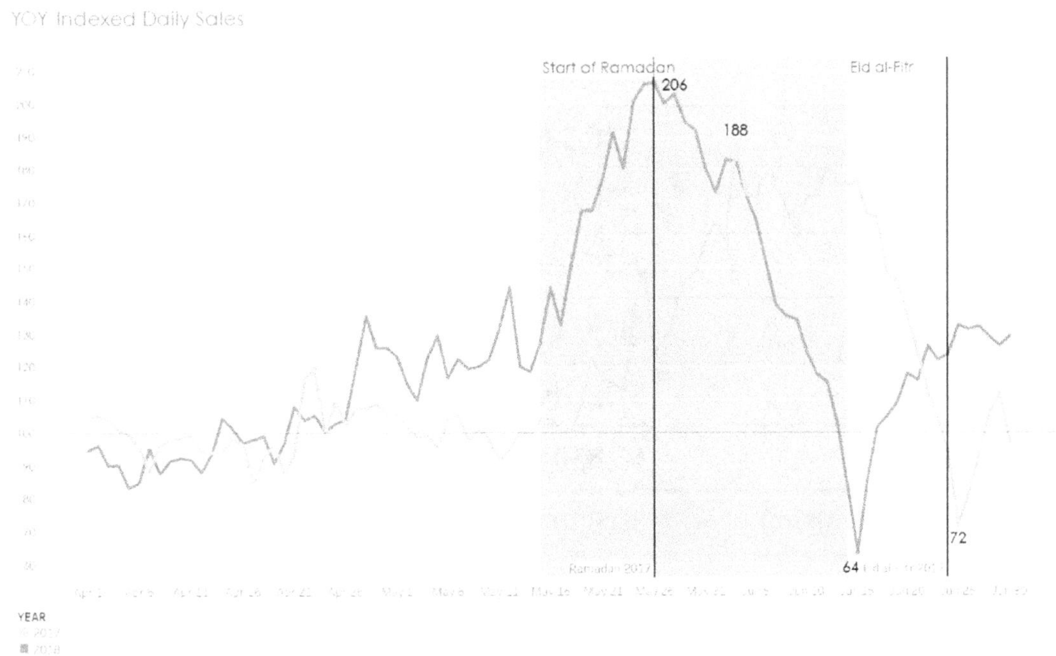

Chritian - Jerusalem, Israel, Rome, Italy, Way of Saint James, Spain, Lourdes, France, Fátima, Portugal, Seven Churches of the Revelation, Turkey, Canterbury, England, Assisi, Italy (Key Tours Blog, June 14, 2017).

Jews- Western Wall, The Western Wall Tunnels, City of David, Temple Mount, Mount Zion-Jerusalem.

Religious tourism is one of the oldest forms of tourism and in today's world, it is one of the most booming sectors. Out of the world's 600 Million tourists, 230 Million are religious tourists or in other words, Pilgrims (ORF Special Report, 97, P. 3). The biggest religious tourist business of the world is of the Muslim's hajj pilgrimage in Saudi Arabia with an estimated 16.5Billion dollars which constitutes 3% of the total GDP (Inside Arabia, August 15,2018).

The table showed the number of tourists and India's national earnings year by year at the sacred places of Buddhists religion. This is the statistics of foreign tourist visited in the sites. This statistic does not include the religious tourism of Indian Hindu tourists who visited the temples in India.

The above figure showed the tourism pattern of Europe where almost 27% are religious, VFR or health tourists. From a world perspective, this tourism is on the going and increasing day by day. For example, only a few million visited the religious shrines and temples of Europe which by now has increased to almost 20% of it's tourism industry. In Saudi Arabia, the number of umrah and Hajj pilgrimage is increasing everywhere. In 2018, the total number of umrah pilgrims was 6.3 Million which rose to 7.5 million umrah visas issued of which 7.2 million came to umrah in 2019(Umrah Weekly Update, Ministry of Religious Affairs, Government of Saudi Arabia website, 2019) which is a 12%increase.

Figure 6.
Source: Bhonsale, 2019, P.8

Destination	2015	2016	2017	2018
Gaya	2,38,617	2,88,914	3,13,817	3,26,228
Bodh Gaya	2,36,539	2,41,514	2,83,116	2,70,787
Rajgir	1,66,610	1,70,334	1,69,444	1,59,976
Nalanda	1,39,066	1,45,339	1,63,390	1,60,091
Vaishali	88,871	1,00,307	90,993	94,532
Sarnath	3,88,102	4,09,242	4,30,682	4,35,752
Kushinagar	68,830	73,514	76,221	80,147
Kapilvastu (Piprahwa)	30,820	31,213	32,064	32,791
Shravasti	1,00,070	1,06,013	1,37,184	1,49,304
Kaushambi	14,229	15,390	15,621	15,659
Sankisa	26,332	27,860	26,901	27,012
Total	1,498,086	1,609,640	1,739,433	1,752,279
No. of Foreign Tourist Visits (FTVs) to all States/UTs	23.33 million	24.71 million	26.89 million	N.A.
Percent Share of FTVs to Buddhist destinations against no. of all FTVs to all States/UTs	6.39%	6.51%	6.46 %	N.A.

Even India is known all over the world for it's dominance of Hindu culture, yet it's one of the main land for the Buddhist tourists to visit the holy temples of Buddha. Now a days, it has become a lucrative place to visit for the Buddhist pilgrims from all over the world, mainly from the south east Asian countries. Another such example would be Iran which is famous for Shiite Shrines of the world. Despite the embargo of the western world on Iran, in 2019,7.8 million (Press TV, Sep.19.2019) foreign tourists visited Iran of which almost 80% were religious tourists. The same is true for Israel also where millions of Religious tourists visited their Jews and Christian holy sites. In 2019, 1.65 Thousand Christian Tour-

Figure 7.
Source: International Tourism Highlights 2019 Edition, P.7

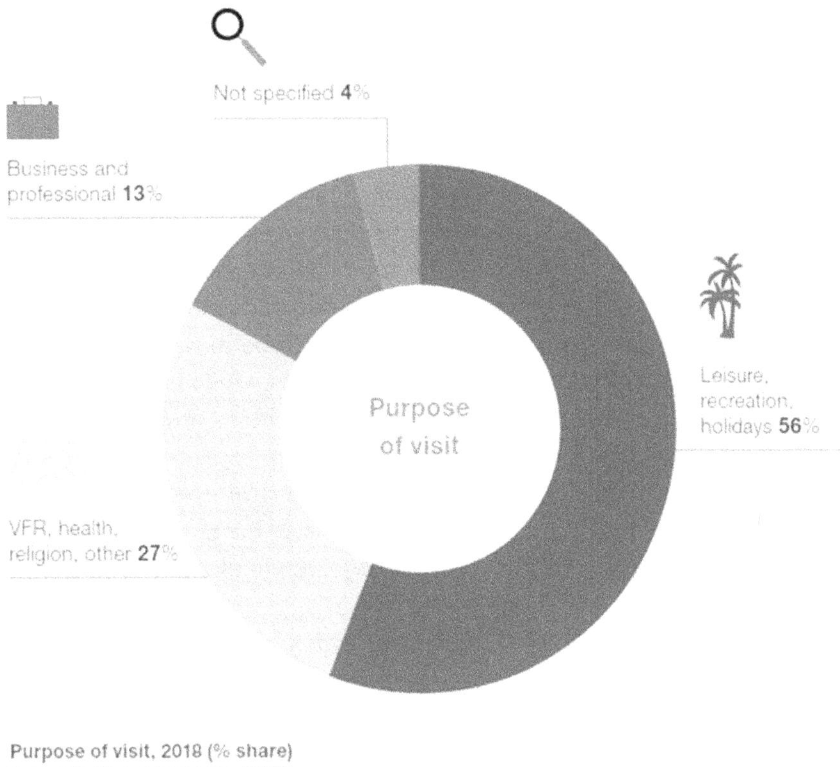

ists visited Israel to visit the Birth Place of Jesus-Bethlehem which is a 10% rise then last year. Out of 2.5 million tourists visit Israel every year, 55% are Christian (The Jerusalem Post, Dec.24, 2019).

The most interesting part of this religious tourism is, despite the embargo on some countries, peoples visited due to their pilgrims and even peoples don't care for budget. They are supposed to pay more than anything to pay their homages and tribute to the religious holy sites even sometimes that costs life risks. For example, In India, there are certain risk, stampede, flash flood and other life risk, still Hindus and Buddhists visited their temples in different non-reachable areas. Even the Muslims despite the last accident in Mecca during hajj pilgrims, visited their holy shrines even of life risks.

CONCLUSION AND REMARKS

In short, religious tourism is the best form of easy profit for any country. Religion has always played a big role of human life whether they accept it or not. For example, western secular democracies still observe the biggest festivals during the Christmas even though most of them called themselves atheist or secular. For western big companies, Christmas is still the biggest business ventures to make the highest business profit out of it. In every society religion has become the biggest culture and none can deny that. Western capitalism has sometimes tried to call itself a religion even though ideologies like capitalism or secularism or communism has no fundamental base. All these ideologies came out from the religious ideologies and even tried to have it's own look and feature which they failed to open in the face or tolerance, wellbeing and humanity.

In the first section, we have shown what actually religion is and how far that is going to benefit human kind without prioritizing profit maximization and capital accumulation, leaving hundreds of crores in destitute conditions. Even welfare economy does not solve the problems of the western world. In the next chapter, we tried to show the contradictions between the ideologies and religions and even we did not find any particular contradictions which might be rigid in case and principles. Rather in the next chapters we have shown that how ideologies like capitalism or communism tried to cope with the changing faces of religions. It's not religions changes its basic, rather all ideologies tried to change their basic and fundamental issues due to changes in the world geo-politics. In the consecutive section, we have tried to explain the bizarre mix of religions and ideologies which made it hollower. We have tried to explore how international geo-politics tried to have religion in it's grip to change their faces in mass exploitation and pseudo development of the peoples.

Then in the last section, we have produced data and empirical evidences on religious tourism by which ideologies like capitalism and communism tried to make the best benefit out of it.

This book chapter by thus created further scope of research in the following areas:

- Is there a neutral base of Capitalism or Communism which is unique and by which they can ignore religion completely?
- Why are religions so important to peoples even though it does not feed them and secure them in daily life?
- Is there any individual economic ideologies which can play it's role independently without being biased to profit maximization and capital accumulation?
- Would there be any ideology which would be at its own footsteps and can establish peace, harmony as well as moral and ethical standard in the society by establishing equal and just society?

REFERENCES

Amy, S. (1997). *The Soul of Development: Biblical Christianity and Economic Transformation in Guatemala*. Oxford University Press.

Bayly, S. (2001). *Caste, Society and Politics in India from the Eighteenth Century to the Modern Age*. Cambridge University Press.

BBC. (n.d.). https://www.bbc.com/news/world-asia-india-50342319

Becker, J. (2010). U.S. Approved Business with Blacklisted Nations. *The New York Times*. https://www.nytimes.com/2010/12/24/world/24sanctions.html

Benjamin, W. (2005). Capitalism as Religion [Fragment 74]. In Religion as Critique: The Frankfurt School's Critique of Religion. New York: Routledge.

Berger, P. (2004). *Max Weber is alive and well and living in Guatemala*. Paper prepared for the conference 'The Norms, Beliefs and Institutions of the Twenty-First Century' at Cornell University, Ithaca, NY.

Bernstein, A., & Rule, S. (2011). Flying under South Africa's Radar: The Growth and Impact of Pentecostals in a Developing Country. In P. Berger & G. Redding (Eds.), *The Hidden Form of Capital: Spiritual Influences in Societal Progress* (pp. 91–131). Anthem.

Bhonsale, M. (2019). *Religious Tourism as Soft Power: Strengthening India's Outreach to Southeast Asia, Observer Research Foundation, India Special Report No.97, Sep.2019*. https://www.orfonline.org/research/religious-tourism-as-soft-power-strengthening-indias-outreach-to-southeast-asia-55674/

Bokare, M. G. (2009). *Hindu Economics* (2nd ed.). Dr. M.G. Bokare Memorial Foundation.

Brewer, P. S. (1947, Dec. 30). Irgun Bomb Kills 11 Arabs, 2 Britons. *The New York Times*.

Brown, D. (2001). Suez and the end of empire. *The Guardian*. https://www.theguardian.com/politics/2001/mar/14/past.education1

Browne, C. (2017). *Buddhist Economics: An Enlightened Approach to the Dismal Science*. Bloomsbury Press.

Burke, J. (2012). UK government ends boycott of Narendra Modi. *The Guardian*. https://www.theguardian.com/world/2012/oct/22/uk-ends-boycott-narendra-modi

Census. (2019). https://www.census.gov/newsroom/facts-for-features/2019/holiday-season.html

Collier, P. (2007). *The Bottom Billion: Why the Poorest Countries are Failing and What Can be Done About It*. Oxford University Press.

Colvin, R. (2009). *Obama to reach out to Muslims in Egypt speech*. https://www.reuters.com/article/us-obama-muslims/obama-to-reach-out-to-muslims-in-egypt-speech-idUSTRE54754920090508?feedType=RSS&feedName=ObamaEconomy&virtualBrandChannel=10441

Crabtree, S. (2010). *Religiosity Highest in World's Poorest Nations, United States is among the rich countries that buck the trend*. https://news.gallup.com/poll/142727/religiosity-highest-world-poorest-nations.aspx

Criteo. (2019). https://www.criteo.com/insights/ramadan-2019-advertising-trends

Delcker, J. (2015). German anti-refugee movement sparks fear Officials warn a resurgent far-right has seized on the crisis to incite hatred and violence. https://www.politico.eu/article/asylum-pegida-refugees-far-right-protest-seekers-as-criminals-politicians-as-traitors/

Deloitte. (n.d.). https://www2.deloitte.com/content/dam/Deloitte/de/Documents/consumer-business/Deloitte%20Christmas%20Survey%202019%20-%20European%20results.pdf

Ecommerce Guide. (n.d.). https://ecommerceguide.com/news/effect-ramadan-ecommerce-revealed/

Economic Times. (n.d.). https://economictimes.indiatimes.com/markets/stocks/news/investors-see-ingredients-for-stocks-revival-after-diwali/articleshow/71756398.cms

Freeman, D. (2012). The Pentecostal Ethic and the Spirit of Development. In D. Freeman (Ed.), *Pentecostalism and Development: Churches, NGOs and Social Change in Africa* (pp. 1–38). Palgrave Macmillan. doi:10.1057/9781137017253_1

Freston, P. (2013). Pentecostals and Politics in Latin America. In D. E. Miller, K. H. Sargeant, & R. Flory (Eds.), *Spirit and Power: The Growth and Global Impact of Pentecostalism* (pp. 3–36). Oxford University Press. doi:10.1093/acprof:oso/9780199920570.003.0005

Friedman, M., & Friedman, R. (1980). *Free to Choose*. University of Chicago Press.

Gewali, B. R. (2013). Isha-Upanishad and Economics Consideration: An Elucidation. *Economic Literature, 11*(35-39).

Hafner, R. W. (2006). Islamic Economics and Global Capitalism, at the Symposium: Visible Hands, Religion and the Market. *Transaction Social Sciences and Modern Society, 44*(1).

Haj. (n.d.). https://www.haj.gov.sa/en/News/Details/12324

Hall, J. A. (1985). Religion and the Rise of Capitalism. *European Societies, 26*(2), 193–223. doi:10.1017/S0003975600004422

Heslam, P. S. (2015). in his article: The Rise of Religion and the Future of Capitalism. *De Ethica. A Journal of Philosophical. Theological and Applied Ethics, 2*, 3.

Huntington, S. P. (1993). The Clash of Civilizations? *Foreign Affairs Magazine*. https://www.foreignaffairs.com/articles/united-states/1993-06-01/clash-civilizations

Huntington, S. P. (1996). *The Clash of Civilizations and the Remaking of World Order*. New York, NY: Simon and Schuster.

Inside Arabia. (n.d.). https://insidearabia.com/muslim-hajj-religious-ritual-lucrative-business/

International Tourism Highlights 2019 Edition. (n.d.). https://www.e-unwto.org/doi/pdf/10.18111/9789284419470

JPost. (n.d.). https://www.jpost.com/Israel-News/Christmas-in-Israel-16500-tourists-expected-10-percent-increase-from-2018-611955

Karlan, D., & Appel, J. (2011). *More Than Good Intentions: Improving the Ways the World's Poor Borrow, Save, Farm, Learn, and Stay Healthy*. Penguin.

Keytours. (n.d.). https://blog.keytours.com/top-religious-pilgrimages-and-christian-tours-in-europe-and-mediterranean

King James version of Bible Online. (n.d.). https://www.kingjamesbibleonline.org/

Knausgaard, K. O. (2015). The Inexplicable, Inside the mind of a mass killer. *The New Yorker*. https://www.newyorker.com/magazine/2015/05/25/the-inexplicable

Lewis, B. (1990). The Roots of Muslim Rage. *Policy, 17*(4). https://www.cis.org.au/app/uploads/2015/04/images/stories/policy-magazine/2001-summer/2001-17-4-bernard-lewis.pdf

Magazine, S. (2014). *Vedic Dharma and the Principles of Economics*. https://www.sanskritimagazine.com/indian-religions/hinduism/vedic-dharma-principle-economics/

Make My Trip. (n.d.). https://www.makemytrip.com/blog/top-10-indian-pilgrimages

Martin, D. (2011). *The Future of Christianity: Reflections on Violence and Democracy, Religion and Secularization*. Ashgate.

Marx, K. (1894). [1998]: Capital: A Critique of Political Economy. In *Marx and Engels Collected Works* (Vol. 3, pp. 25–60). Progress Publishers.

MFA. (n.d.). https://mfa.gov.il/mfa/aboutisrael/state/pages/menachem%20begin.aspx

Moyo, D. (2008). Dead Aid: Destroying the Biggest Global Myth of Our Time. London: Allen Lane.

Narayan, D. (2001). Voices of the Poor. In D. Belshaw, R. Calderisi, & C. Sugden (Eds.), *Faith in Development: Partnership between the World Bank and the Churches of Africa* (p. 39). Regnum.

New York Times. (1946, Dec. 24). Zionists Condemn Palestine Terror. *The New York Times.*

New York Times. (1947, Aug. 16). Irgun's Hand Seen in Alps Rail Blast. *The New York Times.*

Nobel Prize. (n.d.). https://www.nobelprize.org/prizes/peace/1978/begin/facts/

Pew Forum. (2018). https://www.pewforum.org/2018/06/13/how-religious-commitment-varies-by-country-among-people-of-all-ages/

Press TV. (n.d.). https://www.presstv.com/Detail/2019/09/21/606743/Iran-religious-tourism-top-destinations-Mashhad-Qom

Sachs, J. (2005). *The End of Poverty: How We Can Make it Happen in Our Lifetime*. Penguin.

Schlemmer, L. (2008). *Dormant Capital: The Pentecostal Movement in South Africa and its Potential Social and Economic Role*. Centre for Development and Enterprise.

Schuhmacher, E. F. (1973/1993). *Small is Beautiful: Economics as if People Mattered*. Vintage Books.

Selinger, L. (2004). 'The Forgotten Factor: The Uneasy Relationship between Religion and Development. *Social Compass*, *51*(4), 523–543. doi:10.1177/0037768604047872

Sen, A. (1999). *Development as Freedom*. Oxford University Press.

Shafi, A. T. M. (2013, August). Critical Study of Terrorism and Counter-Terrorism and the "Business of Conflicts and Peace": Whose lose, who is Benefitted? *International Law and Policy Research Journal*, *1*(1), 1–15.

Shamasastry, R. (1915). *English Translation of Koutilya's Arthashastra*. The Government Press.

Shao, J. (2001). Alleviating Poverty in Africa. In D. Belshaw, R. Calderisi, & C. Sugden (Eds.), *Faith in Development: Partnership between the World Bank and the Churches of Africa* (pp. 19–30). Regnum.

Simon, D. (2006). *Fifty Key Thinkers on Development*. Routledge. doi:10.4324/9780203098820

Statista. (n.d.). https://www.statista.com/statistics/475509/christmas-budget-europe-by-country/

Stiglitz, J. (2006). *Making Globalization Work: The Next Steps to Global Justice*. Allen Lane.

Tan, H. (2020). *China's dollar-denominated exports and imports beat expectations in December, Jan 13, 2020*. https://www.cnbc.com/2020/01/14/china-december-2019-trade-imports-exports.html

The Al Quran Translated into English by Yusuf Ali. (1946). https://quranyusufali.com/

The Christian Science Monitor. (1991). *Honesty and Economy*. https://www.csmonitor.com/1991/0926/26171.html

"Title 36 – Patriotic and National Observances, Ceremonies, and Organizations". U.S. Government Publishing Office. Archived from the original on 12 May 2019. Retrieved 12 March, 2020. §302. National motto "In God we trust" is the national motto.

Tour My India. (n.d.). https://www.tourmyindia.com/pilgrimage/sikh-pilgrimage-tour.html

Waldman, P., & Pope, H. (2011, Sept. 21). 'Crusade' Reference Reinforces Fears War on Terrorism Is Against Muslims. *The Wall Street Journal*.

Weber, M. (1905). The Protestant Ethic and the Spirit of Capitalism. London: George Allen and Unwin.

Weintraub, E. R. (2002). *How Economics Became a Mathematical Science?* Duke University Press. doi:10.1215/9780822383802

Widmer, A. (2012). Ministering to the Pioneers of Prosperity. In M. Schlag & J. A. Mercado (Eds.), *Free Markets and the Culture of the Common Good* (pp. 167–176). Springer. doi:10.1007/978-94-007-2990-2_12

Witham, L. (2010). *Marketplace of the Gods: How Economics Explains Religion*. Oxford University Press. doi:10.1093/acprof:oso/9780195394757.001.0001

Yājñika, A., & Sheth, S. (2005). *The Shaping of Modern Gujarat: Plurality, Hindutva, and Beyond*. Penguin Books India.

Zarnowitz, V., & Moore, G. H. (1977). The Recession and Recovery of 1973-1976. Explorations in Economic Research, 4(4).

Chapter 9
Challenges of Religious Tourism in the Conflict Region: An Example of Jerusalem

Salome Pipia
Ivane Javakhishvili Tbilisi State University, Georgia & Georgian Technical University, Georgia

Saba Pipia
Ivane Javakhishvili Tbilisi State University, Georgia

ABSTRACT

Pilgrimage has historically and traditionally been considered a physical journey, in order for one to know the truth and visit a place that has sacred and holy meaning. In the modern world, with the development of technology and the process of globalization, territorial boundaries are slowly shrinking, making traveling much easier, and hence, with traditional tourism, of course, religious tourism has also progressed. In the research, the authors consider the challenges that religious tourism has in the modern world, especially the case of Jerusalem. They mainly focus on two variables, legal status of Jerusalem and political instability as problems, which are associated with the development of tourism in a conflict environment. They hypothesis is that unlike conventional tourism, religious tourism has a much greater chance of overcoming various challenges and developing in a conflict-prone and volatile environment.

INTRODUCTION

Pilgrimage has historically and traditionally been considered a physical journey, for one to know the truth and visit a place that has sacred and holy meaning. In the modern world, with the development of technology and the process of globalization, territorial boundaries are slowly shrinking, making traveling much easier and hence, with traditional tourism, of course, religious tourism has also progressed. Believers are now not only visiting the holy places for them individually or with their families, as before, but pilgrimage groups are being set up and holy places of worship are organized. There are a variety of packages that travel companies offer to believers.

DOI: 10.4018/978-1-7998-5792-1.ch009

Although the modern developed world is based on democratic principles and special attention is paid to the protection of human rights, in the 21st century, local and international conflicts are still very common. One of these sharp conflicts is the Israeli-Palestinian conflict, which has been going on for decades. When there is an armed conflict in any region, there are usually lots of political, economic, social, and other kinds of problems in such a region. In a conflict environment, tourism usually does not develop. When there is no stable and peaceful situation, people are not reluctant to visit, and investors and businesses do not show initiative to spend resources in such a region. In other words, if there is conflict, there is no place for tourism. However, there is a different situation when it comes to religious tourism. Unlike an ordinary tourist, the pilgrim is attracted to a particular location because of its sacred nature, and therefore cannot be replaced by any other location. In this article, the authors will discuss the challenges of religious tourism in the conflict region, and in particular the example of Jerusalem.

First of all, it should be noted that Jerusalem is a special case. It is a place that is considered a sacred location for the followers of the three great religions: Christians, Jews, and Muslims. Accordingly, pilgrims visit this city from many points of the world. Both Israelis and Palestinians are demanding that Jerusalem be recognized as the capital of its state. Today the sacred city is controlled by Israel. The international legal status of the city has never been definitively and definitively determined. If we follow the decisions of the international community, Jerusalem must be ruled by the United Nations today. That was the decision taken at the 1947 UN General Assembly (UN GA Res. 181(II), 1947).

Resolution No. 181 provided for the division of Palestine into a Jewish and Arab state, but according to this document, Jerusalem should not belong to any of them. The resolution meant demilitarizing the city and transferring it to international control. A referendum was to be held 10 years later, which would determine the future status of Jerusalem itself. After the resolution was adopted, clashes between Arabs and Jews increased. After the declaration of the State of Israel on May 14, 1948, the armed forces of Jordan, Egypt, Syria, Iraq, and Lebanon attacked Israel. As a result of this war, the holy city was divided into two parts. East was controlled by Jordan, West by Israel. In 1950, Israel declared Jerusalem the capital and transferred government institutions to the western part of the city. As a result of the six-day war in 1967, Israel occupied the eastern part of the city controlled by Jordan. Thereafter Israeli law is in force. UN Security Council Resolution No. 478 declared the annexation of the territory illegal and has repeatedly reaffirmed this position (UNS Res. 478, 1980). The Organization for the Liberation of Palestine declared the founding of the Palestinian state in 1988 and recognized Jerusalem as its capital. Before this decision, Jordan had refused to belong to the eastern part of the holy city. During the talks between the Palestinians and the Israelis, the sides temporarily refused to clearly define the status of Jerusalem. The 1993 peace agreement said that a decision on a holy city would be made in a new agreement in the future. It is noteworthy that according to the 2020 data, 137 out of 193 member states recognize the Palestinian state (Permanent Observer Mission of the State of Palestine to the United Nations, 2020). Alongside the ongoing political conflict over the status of Jerusalem, not one single terrorist act has been carried out by the Palestinians over the years, creating a highly volatile environment in the city. However, statistics show that many tourists visit Jerusalem every year on religious grounds. The latest survey of the world's most popular tourist cities by UK business intelligence company Euromonitor International shows Jerusalem as the world's fastest-growing destination.

Jerusalem rose six places in the rankings of the Top 100 City Destinations report in 2018 to 61st place with 3.93 overseas tourists, up 12% from 2017, and is expected to enjoy 38% growth in 2019 to 4.8 million, according to Euromonitor International. With Israel's Central Bureau of Statistics projecting 4.9 million overseas visitors to Israel in 2019, this figure might be an overestimation (Raz-Chaimovich, 2019).

In the following research, authors try to consider the challenges that religious tourism has in the modern world, in the case of Jerusalem. They mainly focus on two variables: the legal status of Jerusalem and political instability as problems, which are associated with the development of tourism in a conflict environment. The hypothesis is formulated as follows: unlike conventional tourism, religious tourism has a much greater chance of overcoming various challenges and developing in conflict and volatile environment.

THE LITERATURE REVIEW

The book "Religious Tourism and Pilgrimage Management" edited by Razaq Raj and Kevin Griffin, combines the works of various authors on the relationship between religion and tourism. The authors discuss various cases, including the Israeli case. The book supports the idea of historical linkage between travel and religion. According to this book, religion is a centuries-old phenomenon which implies extensive range of activities such as ancient, contemporary, evolving novel religious and quasi-religious movements, additionally encompassing conventional living religions of primal societies and modern secular alternatives to religion. With regard to convergence between tourism and religion, authors analyze challenges of what topics of religion intersects with tourism as a major articulation of religious tourism and pilgrimage. They not only use strong complex insights from tourism, sociology, economics, cultural theory backgrounds, but also use the strengths of their multiple educational and scientific cultural customs from both Eastern and Western philosophies to their discourses on religion, pilgrimage, tourism and events (Raj & Griffin, 2015).

In their book "Tourism, Religion, and Pilgrimage in Jerusalem" Kobi Cohen-Hattab and Noam Shoval describe developments in the tourism sphere starting from the Ottoman era Jerusalem up to present day. In their study they outline the unique religious character of the city of Jerusalem which dates back into centuries, as well as all developments and alterations which happened in the City. People who visited the City and the services provided for them, as a part of a visiting package always had a tremendous impact on the City itself. Therefore, religious tourism became one of the triggering factors for the growth of the City. This is a paradigm which other tourist destinations have also experienced over the time. Jerusalem attracted attention of spiritual visitor for ages. Given these distinctive features, Jerusalem as a travel destination and its visitors are worthy to analyze and to compare with other tourist cities. Diving into special linkage between Jerusalem and its visitors the authors focus on two major topics of the history of pilgrimage and tourism to the city. Firstly, they emphasize the integral nature of tourism and pilgrimage as crucial factors for the growth of the City. Pursuing the continuous changes in the essence of tourism, Jerusalem and its visitors also transformed their needs and consequently the city itself adapted to those changed needs. As a second topic the authors explain link between tourism and political tension. In a city where control over the territory is always challenged by claims from different actors, tourism has been used directly or indirectly as a means to exercise authority or to initiate alterations which would serve the interests of one party only. (Cohen-Hattab & Shoval, 2015).

In his academic publication "Jerusalem, Some Legal Issues" Ruth Lapidoth examines the legal status of the Holy City. The author suggests four main opinions on west Jerusalem. Firstly, she opines that Israel can be legitimate sovereign of the territory as it lawfully filled the emptiness with regard to governor of the territory when British Mandate had been terminated and Brits pulled out in 1948. Due to the fact that Israel gained title over the west Jerusalem as a result of lawful acts of self-defense it had every right to claim sovereignty of that area. The second opinion suggests that the sovereignty over Jerusalem has been put on hold until the conflict resolution deal is reached. The third option favors legality of "sovereignty claims" of Palestinian Arabs since the British Mandate was halted. Supporters of the fourth opinion stipulate that Jerusalem still falls under the status of provided in UN General Assembly resolution of 1947, which established *corpus separatum* international regime in the city. Ruth Lapidoth concludes that notwithstanding political nature of Jerusalem issue, legal aspects are also highly applicable for number of reasons. Most primarily some of the questions bear legal character. Moreover, positions of the parties tend to substantiate their positions by legal arguments. In seeking for resolution options for Jerusalem question several important factors should be taken into consideration: namely, the symbolic and religious significance of the city, its heterogeneity and its geographical location. Resolution plan should envisage complex issues such as sovereignty; delimitation; Holy Sites (their definition and their status), local governance system, urbanization and preservation, economics, public order, potential of engagement from international stakeholders, infrastructure (e.g. irrigation, rainwater drainage systems, highways and streets), disputes settlement, relations with the adjacent territories (Lapidoth, 2011).

In the book "Tourism and War" edited by Richard Butler and Wantanee Suntikul, the group of authors discusses how war and political instability can influence tourism. Amongst different types of political conflict, they also analyze religious conflict, stating that religion has long been blamed for much fighting and malcontent. Diverging assertions of rightful claims over doctrine and sacred space have sparked many instances of war and aggression, including the famous Crusades of the eleventh and twelfth centuries, which sought to free Jerusalem from the Muslims and claim it for Christians. Returning to the present day, authors claim that during the wars between Israel and Palestine religious tourism or pilgrimage demonstrates much less elasticity in demand than other forms of tourism under conditions of political instability (Butler & Suntikul,2013).

THE HISTORICAL BASIS OF THE ISRAELI-PALESTINIAN CONFLICT

Palestinians lived on Palestinian land 4,000 years before the birth of Christ. Their ancestors built Jerusalem, Jericho, Jaffa, etc. The Jews appeared on this land in 1400-1200 BC and established 80 years of control over the region. After that, these lands were first ruled by the Greeks and Romans, and later the Islamic Caliphate was established here. Jerusalem was declared the capital of Palestine. Muslim rule lasted until 1918 when the Ottoman Empire fell (Abo-Sak M & Lieutenant Colonel, 1999).

Most Jews have emigrated to Europe and North Africa since Roman rule. The Jewish Diaspora was distinguished by a high level of cultural and economic development, in all the societies in which they lived. However, their main aspiration has always been to return to the "Holy Land" (Israel Science and Technology Directory).

As for present-day Palestine, the area is inhabited by Arabs who came here in the 8th century AD, after the conquest of the caliphate.

In 1923, after the end of World War I and the fall of the Ottoman Empire, the League of Nations signed a mandate to hand over the territory of Palestine to Britain. This region stretched from present-day Jordan to the east bank of the Jordan River (Herz,2007).

Very soon the region became a destination for the so-called "Zionists". According to the ideology of Zionism, part of the Palestinian territory belonged to the Jews, and the Zionists themselves were ethnic Jews who had lived in different parts of the world for years (Cohen,2014). Although Britain was well aware of how the Zionist emigration was against the interests of the local population, it still did nothing to resist the process. Moreover, as early as 1917, British Foreign Secretary Arthur James Balform issued an official statement known as the Balfour Declaration, according to which the British government "welcomed the idea of establishing a national homeland for the Jewish people in Palestine." It is noteworthy that the Balfour Declaration did not say anything about the status of Jerusalem (Balfour Declaration,1917). From this period onwards, a conflict arose between the two societies. After World War II, the League of Nations was replaced by the United Nations, and Britain, which struggled to resolve the conflict between Arabs and Jews, imposed the whole responsibility for the Palestinian issue to the United Nations. In 1947, the UN General Assembly adopted a resolution establishing a "Palestinian Special Committee" to investigate the conflict and make recommendations. It should also be noted that during the same period, the UN composition increased to 55 members, including Egypt, Iraq, Lebanon, and Syria. Palestine was the only former mandated territory whose independence was not recognized (Hammond, 2013).

November 29, 1947, can be considered as the starting point for the conflict, when the UN General Assembly adopted a resolution approving the "Palestine Partition Plan", according to which Palestine was divided into two (Arab and Jewish) states (UN GA Res. 181(II), 1947). According to the partition plan, a special international legal regime was to be established in Jerusalem, under the control of the United Nations Supervisory and the city was referred to as a "corpus separatum" or "independent body". UN's Jerusalem partition plan clearly illustrated the city's peculiarity and its exact status of the uncertainty, the terminology, which The Assembly was also clear evidence for the resolution uses. For example, terms such as "The Holy City of Jerusalem", "Special International Legal Regime", "Independent body". The international regime established in Jerusalem included the appointment of accountable rulers before the Trusteeship Council, the introduction of special police forces to be formed outside Palestine, the election of a Legislative Council, and the full demilitarization of the city (Jones,1968). The plan was split by a Palestinian organization operating in Palestine but was rejected by the Arabs. The UN Security Council no longer took further action, so the plan to divide Palestine and establish a special status for Jerusalem had no legal effect. Although the Palestinian side did not agree with this decision and considered that the territories were being unfairly redistributed.

The conflict between the two states has been ongoing for decades, but has escalated several times, resulting in four Israeli-Palestinian wars, which are conditionally described as follows (Walzer, 2002):

1. 1948-1949 Palestinian war to destroy the state of Israel,
2. 1967 Palestinian war to gain independence from Israel and retake the occupied West Bank and Gaza Strip,
3. The 1973 Israeli war to protect security within the 1967 borders,
4. The 1982-1985 Israeli War to create a great state of Israel and to establish the occupied territories.

LEGAL STATUS OF JERUSALEM

Usually, conflict regions where the state is unable to control the situation, crime rates are high, security norms are not enforced and are dangerous areas for people. Such regions are less attractive to tourists because it is important for people to feel protected and to abide by the law while traveling. However, as mentioned above, religious tourism has a special specificity, and pilgrims are ready to pray in holy places, even in a conflict environment. The best example of this is Jerusalem. Jerusalem, a city that conjures up a sense of the holy, the historical, and the heavenly, it holds a unique attraction for people of several of the world's faiths and religions- especially Jews, Christians, and Muslims. When people dream of Jerusalem, they do not see the modern, politically controversial Jerusalem, but rather the holy biblical and religious city (Bal-El, Durst, Katz, Zislin, Strauss & Knobler, 2000). Since pilgrims in Jerusalem travel to different places at different times and establish different rituals, Jerusalem should not be regarded as one holy city, but as several holy cities-equivalent to the number of religions and denominations worshipping within. The coexistence of several holy cities side by side within the urban milieu leads to the conclusion that it is not the physical built environment which attracts pilgrims and visitors to Jerusalem; it is the images, the narratives and the dreams which each pilgrim and visitor brings, manifesting in rituals and prayers at their particular sacred space in Jerusalem (Cohen-Hattab & Shoval, 2015).

According to the Financial Times, Israel has been experiencing rise in tourism with number of visitors reaching as high as twice comparing to previous decade (around 4 million tourists as per to 2018 data), with Christian pilgrims - being the largest group among them. The figures released by Israeli ministry of tourism demonstrate that some 75, 000 Christian believers flocked to holy places of worship during the week which precedes of Easter holiday. Nazareth, Bethlehem, the Jordan River, Jerusalem and the Sea of Galilee are the top destinations for Christian visitors, however some groups in particular Faroese evangelicals, which strictly follow the script of the Holy Bible and tend to be attracted by the places where scenes from scriptures had taken place, also visit to the less "popular" destinations, situated in the areas of West Bank. Major international actors condemn Israeli settlements in the West Bank and consider them to be contrary to international law. (Financial Times, 2018).

In 1948, Israel declared independence. Egypt, Iraq, Syria, Jordan, Saudi Arabia, and Lebanon have launched an offensive against the newly formed state, despite Israel's defeat of the attackers, as a result of the war, Jerusalem was divided into two parts, the western part of which was controlled by Israel and the eastern part under the control of Jordan. In 1949, Israeli Prime Minister David Ben-Gurion declared Jerusalem the capital of the state of Israel and moved the Israeli parliament to its western part. In response, in 1950, Jordan declared the annexation of East Jerusalem.

The goal of the UN was to make Jerusalem an exemplary city, a center of spirituality, a place of learning whose influence would dispel the national and religious extremism and superstitions that had poisoned the Holy Land atmosphere over the years (MOHN, 1950). The UN Security Council has adopted the status of Jerusalem, the adoption of which took a long time, as the implementation of the "division plan" itself was in threat (U.N. Doc. A/565, I948). The status quo never came into force, as Israel and Jordan occupied the western and eastern parts of Jerusalem, prompting the United Nations to draft a second statute that would make Jerusalem no longer an "independent body" but would be integrated and fully internationalized (Remarks by the Israeli Delegation,1949). Israel and Jordan did not support the status quo, and the document was not officially ratified.

According to the UN approach, the Jerusalem sovereign has not become either party, but the city came under international control without gaining ownership over it. In other words, neither the Jews nor the Arabs have acquired ownership of Jerusalem since the British mandate. It is also noteworthy that the UN does not consider Jerusalem usually land on which sovereignty was made possible by the international law of the previous methods, but also the General Assembly was considering Jerusalem as a special area, which should be extended to the "corpus separatum" was known before that unprecedented international's legal regime.

In 1948, after Israel declared independence, Jordan occupied East Jerusalem without any warning or legal explanation, which was not protested by the Arab population living there. Jordan illegally controlled the area until 1967.

In 1967, war broke out again in the Middle East. A letter sent by Israeli Prime Minister Levi Eshkol to Jordan's King Hussein reads: "Israel will not strike any blows at Jordan if Jordan does not attack us" (Bowen, 2017).

However, Jordan still attacked Israel, which led to a battle for East Jerusalem, as a result of which Jordan lost, while Israel gained and established control over East Jerusalem as well. On November 22, 1967, the UN Security Council adopted Resolution # 242, which stated that "Israel must withdraw its military units from the territories occupied by the recent conflict." The text of the resolution was somewhat ambiguous (UN SC res 242,1967). Israel's official position on the capture of East Jerusalem and the status of the city, in general, was presented by the country's Foreign Minister, whose letter to the UN Secretary-General says: "As a result of the 1948 aggression, part of Jerusalem fell into the hands of vandals who did not allow representatives of all religions to visit the holy sites. The gross violation of freedom of religion and the right to free access to sacred places was not limited to Jews. In 1950, the Jordanian government informed the UN General Assembly's Special Ad Hoc Political Committee that it would not accept any special agreements on access to holy sites. The bombings were carried out by artillery against synagogues, the Church of the Assumption, hospitals, the Jewish University of Jerusalem, and the Museum of Israel. The holy places were used for military purposes. After June 7, 1967, peace prevailed throughout the city. Anyone can come to the holy places. The state of Israel has passed a law on the protection of holy sites, which ensures the protection of holy sites, as well as their pilgrimage and access to all religions. There has been no such legislation in Jerusalem for the last nineteen years. Since 1967, all the sacral buildings have been repaired, and the places of worship destroyed during the occupation of Jordan have been rebuilt and restored" (Israeli Letter of July 10, 1967, to the U.N. Secretary-General Concerning the Status of Jerusalem).

The occupation of Jerusalem by Jordan was temporary. Jordan has made no claim to East Jerusalem becoming its territory, so it has never claimed ownership of the land. Jerusalem retained its international status until it was occupied by Israel in 1967.

The declaration of the territory as „Terra Nullius" depends on the possession and sovereignty of this territory. As the International Court of Justice has pointed out, Terra Nullu's status is a territory in which it does not belong to anyone. Sovereignty in the territory can be obtained through its occupation, but in this case, it is necessary for this territory already to have the status of Terra Nullius at the time of occupation, in other words, at the time of occupation of this land it should not belong to anyone and its occupation should be free for everyone.

Some authors believe that the territory of East Jerusalem was Terra Nullus before its occupation by Israel. The Palestinians have never had sovereignty over Jerusalem. As Judge Shvebel said: "Because the previous owner of this territory, Jordan, illegally occupied this territory in 1948, and Israel occupied it in 1967 as a result of the exercise of its legitimate right to self-defense, so Israel has a preferential right to this territory." Under international law, Jerusalem was the territory of Terra Terra Nullus from 1948 to 1967, where sovereignty could be legally occupied. It should be noted that even the previous owner of the territory, Jordan, no longer claims to be in East Jerusalem (Ostroff, 2009).

In 1948, British experts also argued that at the time of the adoption of the UN General Assembly resolution, Jerusalem was legally the territory of Terra Nullus, and therefore the United Nations would not and could not have any territorial claims against Jerusalem. It's only interest was all holy places to be protected and that all religions be allowed to pray freely, and that's why Jerusalem needed to gain a special status (Slonim, 1998).

According to the presented opinions, we can assume that Jerusalem was indeed the territory of Terra Nullus during the transition period until Israel gained full sovereignty over it.

POLITICAL INSTABILITY AND TERRORIST ATTACKS

Political instability related with tourism can have very negative effects on tourism destinations that will result in the decline of the number of tourists who visit the destinations and also will bring negative effect to the tourism industry because tourism is a sensitive and a volatile industry and all of these effects from political instability such as armed attacks, civil wars, bombings, attacks on tourists, wars, change in government, the threat of war with another country, which will deter or obstruct the image of the destination through media reporting and government policies and even word of mouth reporting (Jafar & Badaruddin, 2010).

Hassan al-Amin, Lebanese researcher and historian notes (Al-Amin, 2002), that unlike to secular tourism, religious tourism is double-dimensional: the first serves for practicing and observance of religious rites, while the second dimension implies gaining and spreading the knowledge for wider audience. Assuming that primary objective of religious tourism is to receive Blessing from divine forces, nevertheless collateral goal is also to capturing attention of tourists. The main aim is to present a yet unexplored and unknown country to tourists based on its religious significance. Furthermore, religious tourism also boosts people –to- people relationships because if not religious importance of a particular country tourists would know nothing about people living thereto. This kind of tourism also supports rise in intelligence of people.

Although the wars between Israeli and Palestine are not necessarily an attraction for many tourists, religious tourism or pilgrimage demonstrates much less elasticity in demand than other forms of tourism under conditions of political instability. Many Christians throughout the world see Israel's struggle with its Arab neighbors as a religious conflict between the Jew's possession of their Promised Land and the centuries-old predominantly Muslim Palestinian control of these lands. This results in two mindsets among Christian pilgrims in Israel during wartimes. First, most fundamental Christian sects support the cause of Israel and the Jews, which results in some faith groups traveling the Holy Land as "solidarity tourists" in support of the state of Israel, particularly as regards the core of the conflict- Jerusalem. Most pro-Israel tourists are North American and European Jews and evangelical Christians. The second mindset is the diminished concern about war and conflict in the region among Christian pilgrims. In many travelers' view, Israel and the broader Holy Land are sacred ground, God's country, so they will be protected in their travels. Some argue that, if they were to meet their demise while traveling, few places in the world would be as sacred and honorable to die in as the Holy Land (Butler & Suntikul, 2013).

If we take a historical excursion, the Holy Land became more powerful and influential, facilitating several practices born from their original mission of protecting pilgrims. When, in 1148, the Templars began functioning as the French monarchy's bankers they soon developed into a European-wide organization not just stratified to nobles but also offering an array of services across the class divisions, including safe-deposit functions, receipts and early forms of travelers' checks. By 1233, they were responsible for ships carrying approximately 1500 pilgrims and an unknown number of merchants as well. The Templars preserved both physical and fiscal interests of Christendom in the Holy Lands and collaborated extensively with Hospitaller contemporaries.

The extent of the Hospitaller and Templar contribution to the evolution of tourism included (Butler & Suntikul, 2013):

- They policed and devoted themselves to Christian safety and well-being in the Holy Land. Without them it would have been potentially impossible for this area, perceived as the center of the world in medieval times, to be visited by Christians.
- They evolved their duties from caring for sick and homeless in Jerusalem to protecting pilgrims on the roads leading to the Holy City, to shipping them from the ports of their home nations.
- They not only ensured the safe arrival of persons but also of their money. This facilitated Christians to not only visit but also make a living in a land of extreme religious significance.
- The financial versatility of their banks made pilgrimages a possibility for some who otherwise could not have afforded them and their protection and modes of transport made it possible for those too sick or frail to attempt it by other means.

When it comes to modern-day Jerusalem, terrorism is already a major challenge for tourism. The typical scenario of political instability is the situation when the government of the country is collapsed or is divided into different self-controlling factions or primary governmental function of maintaining public order and security is shaky and frequently interrupted. Acts of terrorism, international and non-international armed conflicts, ethnic cleansing, declarations of state of emergency, armed rebellion and violence, usually pave the way towards the political instability, both inside the country and within the larger region, such as, for instance Middle East. Both, terrorism and tensions in political life have grave influence on tourism sector. Terrorist acts happen unexpectedly, are not prolonged and imminently fall into the spotlight of wider public due to extensive and largescale media coverage.

Political turbulence, which may not always be scrutinize by media, has a long lasting detrimental effect on tourism in hotspots. Luckily, tourism is increasingly discussed in light of political turmoil. It's hard to believe but tourism and terrorism have certain common features. Both of them are transnational phenomenon, engage multinational people and employ travel and communication tools. Interplay between them has previously attracted attention of some news media outlets. Accidental terrorist acts harass tourists and halt tourism flows up until public memory of such acts are vanished. However, potential for tourism business of some counties is permanently damaged due to regularity of terrorist activities. Tourism sector is threatened when terrorists injure countries image among foreign tourists and when they make the entire industry subject of their attacks. (Sönmez,1998).

The table below presents the statistics of terrorist attacks on Israeli territory for decades (Johnston R., 2019):

The number of acts of terrorism presented in the table show that dozens of innocent people die every year in the territory of Israel, including tourists who are often the victims of terrorist attacks.

The second table presents the frequency and statistics of terrorist attacks in the Israeli regions in 2018. Where Jerusalem appears to be in second place in the frequency of terrorist attacks (The Meir Amit Intelligence and Terrorism Information Center, 2018):

Because terrorist attacks often take place in the city center and crowded tourist places. Despite these statistics, which show that travel to the territory of Israel and especially Jerusalem is not safe, other statistics confirm that tourism in Israel is developing more and more. For example, according to "The

Table 1. Casualties from terrorism in Israel, 1948-2019, by year

Year	Number of Attacks	Number of Killed People	Year	Number of Attacks	Number of Killed People
1948	8	379 (223)	1984	10	9 (10)
1949	0	37 (0)	1985	22	27 (27)
1950	0	52 (0)	1986	17	14 (16)
1951	0	41 (0)	1987	16	11 (18)
1952	1	40 (1)	1988	9	16 (14)
1953	3	46 (4)	1989	10	40 (23)
1954	1	41 (11)	1990	8	33 (19)
1955	2	30 (3)	1991	12	21 (22)
1956	11	53 (24)	1992	10	34 (17)
1957	6	19 (8)	1993	26	45 (30)
1958	6	15 (11)	1994	42	65 (120)
1959	7	10 (8)	1995	17	29 (62)
1960	1	11 (1)	1996	15	56 (79)
1961	0	8 (0)	1997	10	41 (36)
1962	1	10 (0)	1998	21	16 (13)
1963	0	7 (0)	1999	6	8 (5)
1964	0	9 (0)	2000	29	44 (36)
1965	3	10 (6)	2001	118	207 (222)
1966	4	10 (5)	2002	135	452 (484)
1967	0	16 (0)	2003	67	208 (240)
1968	8	55 (21)	2004	45	117 (128)
1969	12	33 (14)	2005	28	56 (73)
1970	5	74 (18)	2006	48	30 (95)
1971	2	18 (5)	2007	10	13 (14)
1972	7	46 (41)	2008	30	36 (48)
1973	1	27 (0)	2009	14	15 (6)
1974	11	67 (72)	2010	7	11 (11)
1975	19	39 (53)	2011	11	22 (28)
1976	10	14 (9)	2012	9	10 (13)
1977	22	9 (11)	2013	11	6 (8)
1978	11	57 (70)	2014	62	25 (53)
1979	32	10 (46)	2015	108	83
1980	22	16 (26)	2016	122	90
1981	11	14 (11)	2017	34	34
1982	2	6 (2)	2018	42	26
1983	4	21 (10)	2019	16	18

Figure 1. Terrorist Attack Sites, 2018

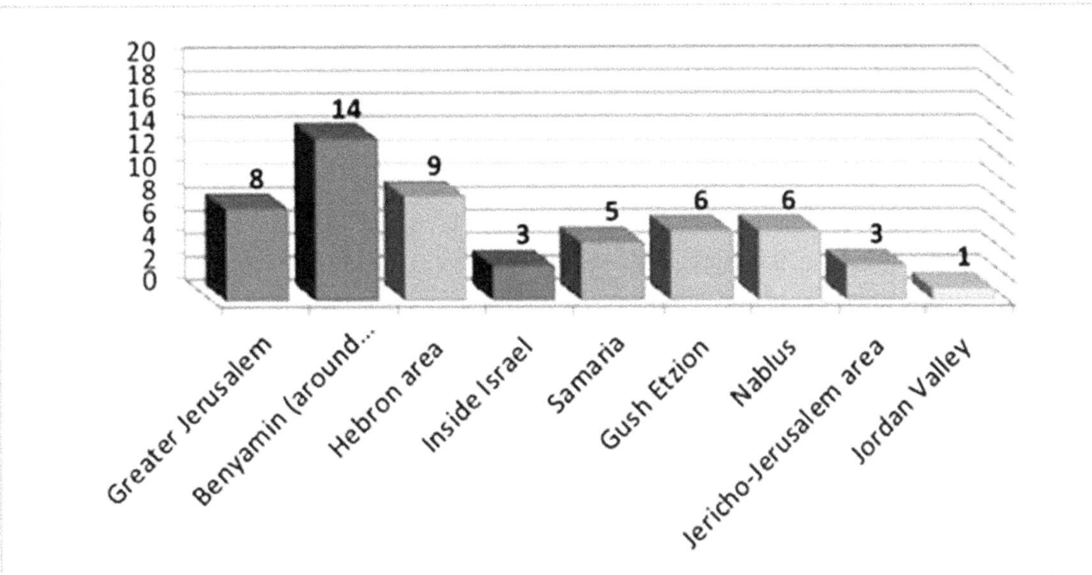

Times of Israel", in 2018 Israel hosted a record-high number of tourists 4 milllion. Data released by the ministry of tourism confirm the increase of number of tourists with 13% comparing to 2017 and with 38% comparing to 2016, while increase in income for tourism amounted to $6.3 billion. The ministry attributed such increase to a global campaign worth for $93 million of promoting Israel as tourist destination in Russia, Italy, China, Brazil, US, Germany, Ukraine, England and the Philippines. Direct investment from those countries has also raised by 13%. Another triggering factor for boosting tourism industry in Israel appeared to be the grant project with total amount of $38.5 million, which supported entrepreneurs in hospitality business and thus some 4, 000 new rooms representing annual year-on-year growth of 49%. "This historical record, of more than 4 million tourists in a year, is the realization of a decades-old dream of all those involved in the Israel tourism industry," said Tourism Minister Yariv Levin in a statement. He also noted that this success was largely caused by huge infrastructural works and due to the high number (40% of total visitors in 2018) of returned visitors. Some 61% of all tourists in Israel this year were Christian, 22% were Jews, 1.8% were Muslims, 0.6% were Buddhists, 0.5% were Hindus, and 0.1% were Bahais, the ministry indicated (Staff, 2018).

CONCLUSION

The phenomenon of globalization and the disintegration of borders have especially contributed to the development of tourism. Today, for most of the states, tourism is one of the main sources of investment and financial income and has a significant impact on the country's economic development. Political instability, high crime rates, and terrorism are the factors that directly threaten the development of tourism in the state. In their paper, authors have tried to prove that religious tourism is fundamentally different from ordinary tourism, and there are several reasons for this:

- Religious tourism is about human faith and values and not about physical relaxation or just visiting cultural monuments.
- Religious value gives certain locations the importance of a sacred place, making it impossible to replace them with any other location.
- When a pilgrim decides to visit a holy place, it is of secondary importance to him whether it is located in a politically, economically, or legally stable region.

The statistics provided by authors show that despite the high level of terrorism, religious tourism in Jerusalem is developing more and more, which gives a reason to conclude that the hypothesis presented in the introduction to the paper is true. Unlike conventional tourism, religious tourism has a much greater chance of overcoming various challenges and developing in conflict and a volatile environment and the case of Jerusalem is a perfect example of it.

REFERENCES

Abo-Sak, M. (1999). *US Involvement in Israeli-Palestinian Conflict:Solution or Complication*. Al-Hewar Center.

Al-Amin, H. (2002). Religious Tourism in Islamic Heritage. *Islamic Tourism, 2*.

Bal-El, Y., Durst, R., Katz, G., Zislin, J., Strauss, Z., & Knobler, H. (2000). Jerusalem Syndrome. *The British Journal of Psychiatry*.

Balfour Declaration. (1917). Retrieved from Israel Ministry of Foreign Affairs: https://mfa.gov.il/mfa/foreignpolicy/peace/guide/pages/the%20balfour%20declaration.aspx

Bowen, J. (n.d.). *1967 War: Six days that changed the Middle East*. Retrieved from BBC News: https://www.bbc.com/news/world-middle-east-39960461

Butler, R., & Suntikul, W. (2013). *Tourism and War*. Routledge. doi:10.4324/9780203107706

Cohen, P. (2014, October). *The Root Cause of the Never-Ending Conflict in Palestine; and How to Fix It*. Retrieved from Huffington Post: https://www.huffpost.com/entry/root-causes-in-palestine_b_5499127?guccounter=1&guce_referrer=aHR0cHM6Ly93d3cuZ29vZ2xlLmNvbS8&guce_referrer_sig=AQAAAGGfgtKWMMbxGDoS8EW3B75gVCCguodTDQ0af0TjW0q-qLOwqJYw0wNhDwwuxZBwjMxqnTQ5M6uK-tIzbmDsLA18P9beID-YSfo7ELWcQKhqD

U.N. Doc. A/565. (1948). Proceedings summarized in Annual Report of the Secretary-General on the Work of the Organization, I July I947-30 June 1948. *3 U.N. GAOR, Supp.I.*

Hammond, J. (2013, August). Washington Report on Middle East Affairs. In *The Role of the U.N in Creating the Israeli-Palestinian Conflict*. American Educational Trust. Retrieved from https://www.wrmea.org/013-august/the-role-of-the-u.n.-in-creating-the-israeli-palestinian-conflict.html

Herz, E. (2007). *Mandate of Palestine: The Legal Aspects of Jewish Rights*. Myths and Facts.

Israeli Letter of July 10, 1. t.-G. (1967, July). Brief History of Israel and the Jewish People. *Israel Science and Technology Directory, 6*, 846-850.

Jafar, S., & Badaruddin, M. (2011, July). Factors Impact on Religious Tourism Market: The Case of Palestinian Territories. *International Journal of Business and Management*, *6*, 254–260.

Johnston, R. (2018). Summary of Terrorist Attacks in Israel. *Johnston's Archive*. Retrieved from http://www.johnstonsarchive.net/terrorism/terrisraelsum.html

Jones, S. (1968). The Status of Jerusalem: Some National and International Aspects. *Law and Contemporary Problems*, *33*(1), 177. doi:10.2307/1190849

Korstanje, M., Raj, R., & Griffin, K. (2018). *Risk and Safety Challenges for Religious Tourism and Events*. CAB International. doi:10.1079/9781786392282.0000

Lapidoth, R. (2011). *Some Legal Issues*. The Jerusalem Institute for Israel Studies.

Mohn, P. (1950, october). Jerusalem and the United Nations. *International Conciliation Pamphlet*.

NationsP. O. (2020). Retrieved from https://palestineun.org/

Ostroff, M. (2009, November 24). Neocons Now Claim West Bank and East Jerusalem was 'Terra Nullius' Giving Zionists Right of Occupation. *Jerusalem Post*.

Palestinian Violence and TerrorimAgaint Israel, 2018: Data, Nature, and Trends. (2019, February 18). Retrieved from Terrorism-Info.org: https://www.terrorism-info.org.il/en/palestinian-violence-terrorism-israel-2018-data-nature-trends-2/

Raj, R., & Griffin, K. (2015). *Religious Tourism and Pilgrimage Management an International Perspective*. CAB International. doi:10.1079/9781780645230.0000

Raj, R., & Griffin, K. (2017). *Conflicts, Religion and Culture in Tourism*. CAB International. doi:10.1079/9781786390646.0000

Raz-Chaimovich, M. (2019). Jerusalem ranked the world's fastest-growing tourist destination. *Globes, Israel Business News*. Retrieved from https://en.globes.co.il/en/article-jerusalem-ranked-worlds-fastest-growing-tourist-destination-1001310945

UN GA Res. 181(II). (1947, November 29). U.N. Doc. A/RES/181(II) .

Slonim, S. (1988). *Jerusalem in America's Foreign Policy, 1947-1997*. Kluwer Law International.

Sönmez, S. (1998, April). Tourism, Terroris and Political Instability. *Annals of Tourism Research*, *25*(2), 416–456. doi:10.1016/S0160-7383(97)00093-5

Staff, T. (2018, December 29). *Israel saw record-breaking 4 Million Tourist in 2018, Says Tourism Minister*. Retrieved from Timeofisrael: https://www.timesofisrael.com/israel-saw-record-breaking-4-million-tourists-in-2018-says-tourism-ministry/

U.N. 3 CAOR, p. 2. (1949). Remarks by the Israeli Delegation. 230-237.

UN SC res 242. (1967, November 22). UNS Res. 478. (1980).

Walzer, M. (2002, Fall). The Four Wars of Israel/Palestine. *Dissent*.

KEY TERMS AND DEFINITIONS

Belligerent Occupation: Is control of part or full territory of a state by hostile army.

Conflict: Is a disagreement/problem between two or more parties. Conflicts in international relations are usually caused by states. In case of an escalation, the conflict can turn into war. Conflicts can be political, territorial, ethnic, religious, etc.

Holy Places: Are places of worship to which some believers have spiritual attachment.

Jerusalem: Is an ancient city in the Middle East, specifically in Israel, which is considered to be the holy place of three largest religions: Christianity, Judaism and Islam.

Pilgrimage: Is called the process when a believer travels to a foreign country to visit a religiously sacred place to pray and to get spiritual purification.

Political Instability: Is a crisis situation within the country that can be caused by a variety of reasons: government incompetence, economic problems, high crime rates, and so on. Political instability can also be caused by conflict or, conversely, to become the basis of the conflict itself.

Religious Tourism: Is a type of tourism when travel is based on religious motives.

Terrorism: Is armed attack against civilian population with a primary aim to spred the terror and fear among civilians. Terrorism can motivated by ideology, religion, politics, etc.

United Nations: Is an international organization founded in 1945. It is currently made up of 193 Member States. It is one of the most important guarantees for the prevention of conflicts and peace in the world. It assesses current developments in international relations and implements the principles of international law.

War: Is an armed conflict, which takes place between two or more parties and where military forces are used.

Chapter 10
Promoting Religious Tourism Management for Creating a Soft Image of Pakistan

Kalsoom BeBe Sumra
COMSATS University, Pakistan

Muhammad Mehtab Alam
Riphah Institute of Public Policy, Riphah University, Pakistan

ABSTRACT

The religious tourism management has been the subject of much attention in academic literature and has made a significant contribution in creating soft image in tourists' behavior. Pakistan has one of the best archaeological and religious historical sites in the world. Buddhist civilization at Taxila, Islamabad, Swat, and Takht Bai are of great interest for tourists from Japan, Korea, Thailand, and other countries. The aim of this chapter is to determine the promotion of religious tourism management for creating soft image in Pakistan. In this regard, initiatives of religious tourism by federal government are analyzed while incorporating input from easing immigration services, establishment of national tourism coordination board, infrastructure development, and pollution free environment, which leads to the soft image. An exploration of the nature of soft image requires examining the two key drivers: cognitive approach and the more recent cognitive-affective approach. Using quantitative method, various analysis techniques will be applied.

INTRODUCTION

Religion as an organized system of belief is typically related with trust and faith. It is defining category of getting policy results and connected with the idea of tourism (Raj & Griffin, 2017). Based on the life choices and the ideal management of holy sites the perception of visitors can be developed positively (Hoseinipor et al., 2019). Several reasons are considered as the motivating factors for people to visit sacred places. From plain curiosity to deeply religious connectivity, travelling is generally placed in getting

DOI: 10.4018/978-1-7998-5792-1.ch010

close association with supernatural ability and power (Ahmad, 2016; Korstanje et al., 2018; Wiltshier & Griffiths, 2019; Alipour et al., 2017; Tkaczynski et al., 2018). The experience of change in the form of religious tourism ascertain managing authorities to arrange facilities for the respective people in an adequate manner (Kim & King, 2019; Heidari et al., 2017).

The role of state institution to develop an interest among religious travelers through concerned institutions is undeniable. It satisfied religious actors by providing informal orientation in the form of transport, accommodation and food of their choice (Ali et al., 2019 Malodia & Singla, 2017; Garayeva & Naumov, 2016; Piva et al., 2019). The application of cognitive approaches is based on scientific assumption to support the idea that religious tourism is not only concerned with emotional attachment, but it is also relevant with the well managed religious sites (Sharma & Swamy, 2019; Durán et al., 2018; Wiltshier & Griffiths, 2016; Alipour et al., 2017). It is essential to develop a positive image because of the decades long war on terrorism which direct a negative message to the world community regarding the spiritual and religious disagreement among the stakeholders in Pakistan (He et al., 2019).

Figure 1. Nankana Sahab, Sikh Religious Destination Pakistan

The intensified situation replaces the poor law and order situation in the recent years of war against non-state actors. They are living inside the country and operated from areas around the regional political players (Almaimoni et al., 2018). This phenomenon reflect that a strong management of destinations related to the culture and history of religions that are other than Muslims be developed in accordance with international standards of the marketing (Lochrie et al., 2019). The state-of-the-art architecture and interaction of trained staff with the visitors coming from various religious background and identities project a suitable image (Ali & Maharani, 2019). In the context of perception, best management of the destinations is truly a live experience for tourists to see and realize the importance of preserved culture by the hospitable country like Pakistan. Such steps directly influence the positive image of the state, following the negative response from world community (Handriana et al., 2019; Verma & Sarangi, 2019). It is experienced that different traveling advisories are being issued in favor of Pakistan which reflect significant improvement in security situation and peaceful environment of destinations (kim & kim, 2019).

Figure 2. Chandragup Mud Volcano: Hindu Destination Pakistan
Source: Karachi Gliding Club, 2019

Context of the Study and Objectives

People with different religious backgrounds visit places of religious sites because of the motivation and exclusive interests (Gozzoli et al., 2019). Other than interest's the religious tourism is the oldest form of tourism which provides an individual to connect holy places with their emotional and spiritual attachment (Niculina, 2019; Khan et al., 2019). For that matter a well-managed destination contains health and safety measures by monitoring tour operators in checking of vehicles and instruments used for the well-being of a traveler (Almuhrzi & Alsawafi, 2017). The stable and secure environment is primary support for a safety situation in an area of destination (Korstanje & George, 2018; Leppakari & Griffin, 2016). The profile with permission of traveler may be upgraded to specific data for information and interests of other religious travelers. The rules for dormitory should be extended to single and group travelers besides the requirement and inclusion of physical activities (Kessler & Raj, 2018; Idris, 2019; Cabrera et al., 2019).

Following the are key areas for religious tourism which can develop a positive perception of the country.

Chandragup, Hingol Baluchistan

In the locality of Hingol National Park some 200 kilometers of Karachi the existence of mud volcanoes is known as holy Hindus destination where they believed that without paying tribute to Baba Chandrakup, they cannot enter Hinglaj's temple (Ahmad, 2016). Upon reaching at the top of volcano the pilgrims introduce themselves to remove their sins according to the religious instructions. The construction of roads and site facilities such as availability of public toilets, hotels and arrangements for scenic view are worthy of an off-roader's heaven (SHAH, 2016). The free movement of pilgrims with guides is available for directions enable site be used for the easy access of foreign tourists. (Web, 2020). The international traveler Mathew Paley has experienced the Baloch Muslims standing on the rim of volcano to get resell and coconut. These items are used by Hindu pilgrims who throw as a tribute against their sins and ask for forgiveness from their god. However, poor local Muslims searched for food and coconut to eat to which Mathew observe that "*it was interesting to see the collaboration of between Hinduism and Islam around the holy site.*"

Takht Bhai: An Indo-Parthian Buddhist Monastery

A well-preserved destination located at 15 kilometers form Mardan have different descriptions. Locals provided that the name (Takht I Bahi) is originated from Persian with meaning of takht as throne while Bhai shows water or spring. The archeological sites of ancient Buddhist are considered as significant remnants of Gandhara, among all Buddhism. The notable point is the preservation of site which attracts positive cognitive effects on the tourists. However, this requires more steps to prove its safety and well-being. The monastery was founded in the 1st century and used by people during 7th century and it is referred as the specific representation of religious tourism in Pakistan. In 1980 it was listed as the world heritage of UNESCO which further enhances its importance for Buddhist around the globe.

Gurudwara Pehli Patshahi, Burewala Punjab

The presence of Gurudwara in Pehli Patshahi is among those symbols of history, culture and religion that can be a source of interaction among major and minor religions of the subcontinent. Problems with this shrine is the locality which is far from the city life and inquired for the facilities which are central to attraction of foreign tourists (Durán et al., 2018). Management to this site can be part of improved image as the structure of destination allow tourists to frequently visit and enjoy the spiritual beauty of their religion. In the central courtyard there is requirement of refined clusters of stupas to be developed in accordance with the 21st century art and design. Most of Pakistani religious and non-religious destinations have scarcity of basic resources and facilities like water, quality food and the standards of living. Similarly, the availability of technological instruments followed by realization and safety of destination in all its terms and conditions is the pre-requisite for developing affective cognitive image for Pakistan.

Nankana Sahib, Recent Phenomena of Reckless Religious Destination

Nankana Sahib is the birthplace of Baba Guru Nanak who is the founder of Sikh Religion. This destination is different from the ancient and recently developed holy sites in Pakistan. Owing to the situation and location of Nankana Sahib, the management is very much serious about the provision of facilities for the religious visitors (Yousaf & Xiucheng, 2018). This is because of the geo-strategic importance connected with this destination, which has potential of attracting as many pilgrims as the people visit to Golden Temple in Amritsar city of India. Over fifty million people visits the temple with belief of their well-being in social and cultural life. Huge number of people both at external and internal front of Nankana Sahab may be managed based on the attention given to it by state own enterprises. Image building is quite relevant to the improvement in infrastructure, investment, security and restrictions of diverse site. The expected visit of eight million diaspora Sikhs can develop a very positive picture of Pakistan to the world.

In view of the enhanced interests and recently developed phenomena of religious tourism this research intends to achieve the following objectives:

1. To trace the influences of effective Religious management on cognitive and affective cognition of tourists in building soft image of Pakistan.
2. The study examines issues & impending barriers on the image of Pakistan through religious tourism facilities like;
   - 2.1.1. Darshan of Temples
   - 2.1.2. Sikh Gurdwaras accommodation & Transportation
   - 2.1.3. *Boarding and Expenditures of Buddhist Monks*

Organization of the Chapter

The chapter followed the sequence of different sections which include the section I as description and explanation of religious tourism with background and objectives. In section II the review of literature and discussion on various destinations in Pakistan. Further the Section III incorporate approaches of methodology and framework. It includes the development of survey and distribution of data according to the desired objectives. Key results and analysis are presented in section IV while section V contain the concluding directions based on the finding and discussion to be implied in future research.

GLIMPSES OF THE PREVIOUS LITERATURE

Scholars and the researchers like Almuhrzi & Alsawafi (2017); Shinde (2017) have specific views on the management of religious tourism by Pakistani authorities in terms of easing the visa services, secure movement and the simplification of processes related with tourism. The religious communities according to Badar & Chakradeo (2016) provide facilities for the standard operating procedures. For example, a receiving the application of a Sikh pilgrim, the system will send guidelines to the applicants for completion of their documents in the concerned embassy. Here Shinde (2017) reflect that Pakistan share less than one percent foreign exchange from these pilgrims compare to the Maldives 7percent, Srilanka 10percent and Indian share of 69 percent.

Further the researchers El (2016) reflect that increase study of religion is the basis for ideas of religious tourism and their management. However, the importance of data is pointed out by Enongene & Griffin (2019) as this creates problem for managing the religious sites for tourists and allow them to visit and perceive facilities accordingly. These experiences of religious tourism are also explored by Albayrak et al (2018) where the authors have provided the satisfaction of tourists and positive perception about the management. This is supported by Idrees et al (2018) with the features of knowledge sharing and role of hotel services in managing the tourists visiting to their sites. For the enhanced cognitive powers of religious tourist, the complexities require vision and leadership role as the authors Terzidou et al (2018) provide the role of intentions for motivation of human sacred interaction. This is important to understand as the authors Sharma & Swamy (2019) illustrate the prospects of progress and development of religious tourism in India.

Religious Tourism and Cognitive Approach

The authors Fujita et al (2018) discussed various approaches that are influenced by the human behavior in their interaction and visiting to destination of their interests. The cognitive approach attempts to align the difference in behavior because it is closely associated with the series of responses in respect a religious tourist who visit to the specific destination is explained by Vrij et al (2017). Key assumption in understanding and influencing the minds of tourists can be made easy by presenting the well-managed tourist site (Skavronskaya et al., 2017). Behavior of the staff related with tourism department is very important with people who are residents of specific area. Even the safe movement of a holy tourist can influence his/her mind despite the attachment and emotional belief over the super-natural power assumed by the travelers (Mikos, 2013). Each component of management contributes for reminding the visitor to remain in connection with destinations and their holy master.

Figure 3. Inter-Linking Religious Tourism & Cognitive Approach
Source: Author's Development

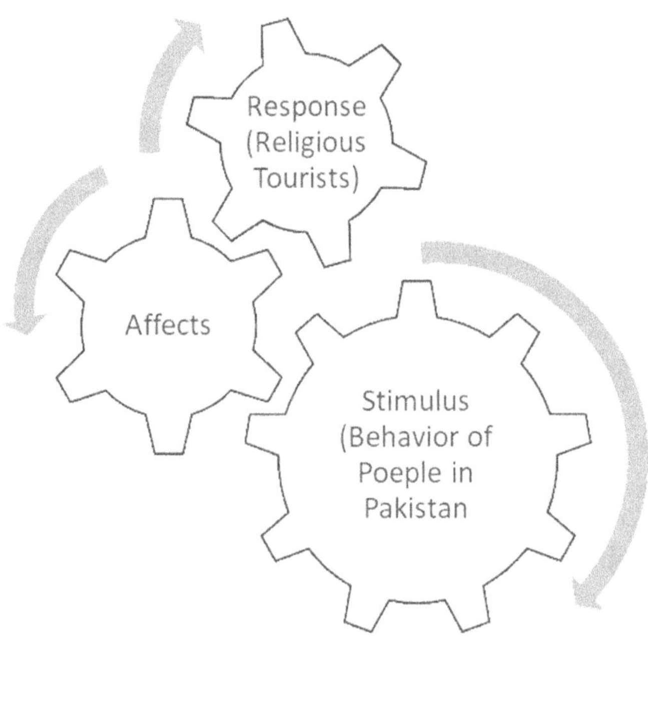

Influence of Cognitive-Affective Approach in Understanding Religious Perception

(Prayag et al., 2016). The false beliefs and the emotional are inter-related with the idea of appealing to minds of tourists but it strives for the best management of each destination which can be a source of attraction for visitors to come and easily perform their sacred rituals. As the tourists visits their respective place they are influence by the management and behavior of respective staff of the site (Michael et al., 2018). The environment provided to the person of religious nature and interests have significant impacts on future decisions and qualities for having positive or negative perception (Rivera et al., 2019). In easy terms, the reasonability of measuring the facilities is backed by intentional and unintentional situational experience of religious holiest site (Huang & Pearce, 2019). The affective variables for a religious destination incorporate the friendly behavior of local individuals followed by the activities of family-oriented living and a little fun place.

Figure 4. Affective Cognition and Perception of Management
Source: Cognitive Affective and Theory of Mind, 2013

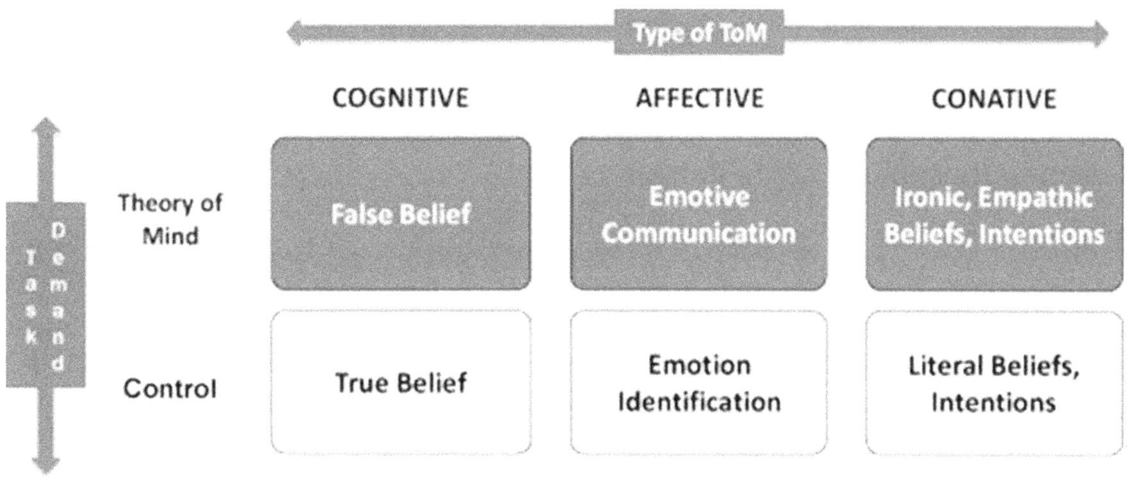

Historical View of Religious Tourism Management

The development and management of religious destinations around the globe are considered as sacred duty of the concerned people both in terms of attracting pilgrimages and providing significant facilities to travelers. A short look at the history reveals that the region of Asia-Pacific is referred as center to greater number of tourists from religious background, those gathered for religious events. As estimated, there are six million global and national voyages, among which 40% are from Europe. Same number of the groups and religious voyages are from the Asian region where it is frequently practiced that a tourist of religious nature is connected to specific destination. Further, the reflection of UNESCO provides that 60% of population are believers and practice their religion with demographic background of religion. Since the 1947, there were diverse ideals of information separated among the travelers from foreign lands. Hence, the crises conflict during 1980s and 2010 draws a negative projection which was not true to the evidences.

Projecting Pluralistic Image of Pakistan Through showcasing non-Islamic Heritage

The lapses in the management of services are because of the poor resources utilized for infrastructure development and facilities of international flair (Griffin, 2017). The identification of 480 various sites by Pakistan Tourism Development Corporation (PTDC) reflect the potential of culture and heritage. Among these identified sites there are 120 religious destinations which possess greater importance. For Buddhists and Sikh Pakistan is the center of historical reflection (Bahadori et al., 2019). Nankana Sahib, the birth-place of Baba Guru Nanak is among the holiest sites. There are 6000 Indian Sikhs visit to Pakistan and these are managed by state through the requirement of security and their political affiliation. More than eight million diaspora Sikhs with 83 percent are living outside and show their interest to visit Pakistan in respect of their holy belonging (Ghani, 2019). The estimated market of 500 million Buddhist needs to be considered as they respect their holy sites in Pakistan (Liu & Pratt, 2017). The Korean Buddhist while searching their origin trace Pakistan as the key place of religious origin where monk of Korea travelled 1300 years ago (Lücking, 2019). Various province in Pakistan are the demonstrator of historical sites as Takht-Bhai in Khyber-Pakhtunkhwa has the potential of managing 50million Mahayana Buddhist in Japan, China, and Korea. In Pakistan, Gallup Survey of 2016 show selected pool of countries provided the identification of 58million people who are likely visitors of Pakistan (Lodhi, 2017).

Managing Religious Sites in Pakistan

The management can be enhanced through the cooperative measures through mega projects like China-Pakistan Economic Corridor where this sector can be facilitated through completion of structures towards the holy sites (Rejman et al., 2016). Certain key areas are not addressed properly to attain the positive image results of people visiting to these sites. The regulatory and institutional arrangements need serious reforms to reduce the complexity and weakness of institutions concerned with religious tourism (Raj et al., 2018). Further the lack in conforming the UNESCO standards and international practice is not appropriate in relation to the management and sustainability of religious spots (Rashid, 2018).

To an extent, reformed structures like Walled City Authority are existing in managing the destination of significance for religious tourists. There is no legal authority after the 18th amendment which devolves major subjects to the provinces. This issue is explored in case of Katas Raj a Hindu Temple comprises of seven ancient temples. Here ETB exercise a complete authority over the renovation and management of Temple. However, this shows the institutional flaws where it used paint and plaster that permanently destroyed the remaining of remnants of ancient times (Suri, 2018; Tkaczynski & Arli, 2018). There lies a sacred Hindu pilgrimage site a town on the Makran Coast in Lasbela Baluchistan.

Analytical Framework and Hypotheses

The required provision for the effective management of religious tourism is possible through the support of understanding the viability of and implementation of policy and framework. For example, the contemporary influx of religious tourists to various sites in Pakistan seems increasing and requires necessary

Figure 5. Number of Visits to Cultural Sites in Pakistan 2014-2018
Source: World Bank, Data Bank, 2019

measures (Ali et al., 2019; Yousefi, 2018). The destinations and infrastructure are the primary factors that reflect religious sustainability and purity of the managers (Khan et al., 2019; Kawsar & Raj, 2017). Key factors and variables are described in the framework which reflect the importance of each component.

Rasgon et al (2019) illustrate that affective cognition is the neural correlation of affective and non-affective cognition for people and travelers. This can be done with the difference of the individuals visiting to Pakistan can be made easy through the cognitive-affective approach which is the reflection of personality in definitive situations. Further the authors (Mathew & Kuriakose, 2017), provides the responsible role of accommodation in managing the tourism destinations is significant in attracting religious tourists.

Figure 6. Analytical Framework
Source: Author's Development

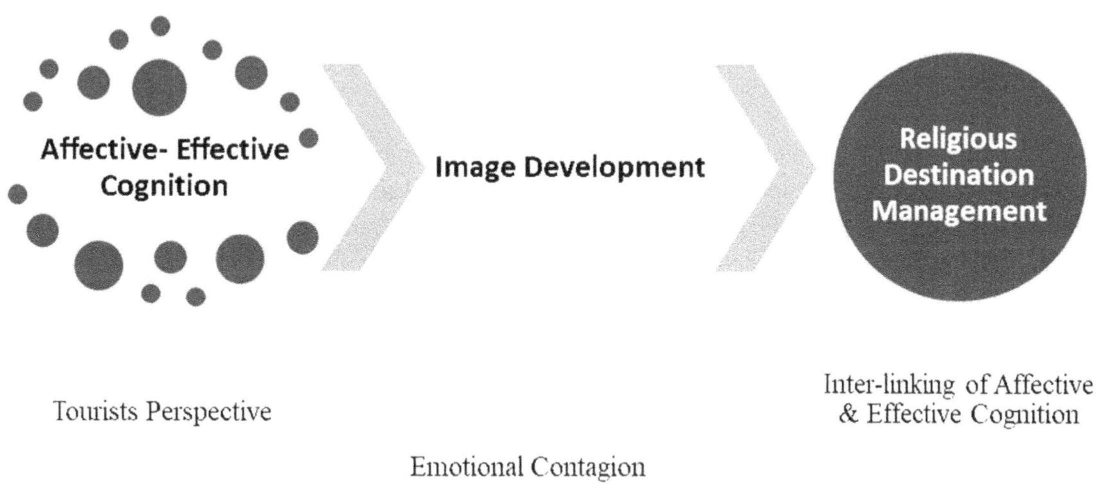

H1: There is collaborative influence of affective & effective cognitive satisfaction on the travelers visiting to religious destination in Pakistan.

The affective & effective cognitive insight is illustrated by Bohg et al (2017) in their research against the recent approaches in development and influence of image. It is argued that interaction in any environment develop positive or negative sensory signals. Same is the case with destination where travelers must support their mental capacities in favor or against the experience of tourism. Further the inter-linking of both approaches intensified the situation and compel a religious tourist to remain in search of reality through the available forces and destination.

H2: The emotional contagion of religious tourists favored the structure and development of destination in response to their requirement.

Similarly, there are certain contagious beliefs that are explained by Sugimura et al (2019) which formed the identity and structure of the ideology. These cultural and religiously sensitive beliefs are required to be managed through planned mechanism and management of destination. Non-institutional role is expected to be played for the easy access to other religions & emotional satisfaction. The thoughts and beliefs of one religious tourist transferred from one individual to other during the interaction at religious destination.

H3: The positive perspective & smooth environment provided by the management on destination changed the perception of the travelers.

For the perspective of destination management Jelacic (2017) provided certain tools and directions to develop a product. This is to enhance sharing process and interaction among the people of different ideologies and beliefs. As tourism destination is critical for every religious traveler to approach other in a way the enhanced interaction develop a process of discussion which leads to areas for identification and reducing the negative perspective. It is essential to inform the audience or the tourists visiting to respective holy site that a peaceful arrangement of interaction among people of different belonging is facilitated for kind support to understand each other.

H4: The enhanced interface among religious tourist's destination with different backgrounds reduce the negative image.

The image of tourists for destination or for any location is significant for achieving the long-term objectives and desire goals. This phenomenon is explained by Asmelash & Kumar (2019) through the indicators of validating and developing the image sustainability. Monitoring and evaluation of the work progress is one aspect of sustaining religious tourists. Further the facilities need to be reviewed after a due course of time to remain effective and efficient in satisfying the tourists. Image is well connected with experience and understanding of visitors. This is supported and explained by the authors Tomić et al (2019) destination.

Theoretical Support to the Subject of Study

Scholars have diverse view on the role of religion which exists as explanation of action and amenities found in the society. It provides satisfaction & spiritual well-being to the individual associated with the religion and sacred destinations.

Religion and Functionalism

David Emile Durkheim, a French sociologist & philosopher has established functionalism as support to religious traveling (Durkheim & Mauss, 2016). He argued that key characteristic of a religion is not a belief in supernatural force or God; instead, it is a distinction between sacred and profane. In the society sacred places are the reflection of inspired feeling & wonder humans over representation of some hidden power. However, profane are the things of mundane, with no special significance. These sacred destinations according to the author evoke powerful emotions in believers which shows that some commanding authority is associated at places of worship.

The style & nature of worship unfolded the sacred symbols & unite members of the society from different aspect to a single unity. A great source of singleness in humans are because of attachment or travelling to destination of their choice (Durkheim & Mauss, 2016). The current research is supported to address the objectives for enhanced exchange of religious tourism as every single human strive for the better. Based on their belief and trust the powerful unseen forces are same for believers. Except few, there is a great chance of streamlining people of diverse religion to see destinations and get satisfaction through harmonized values.

Functionalism- Psychological Functions of Religions

Despite the agreement with Emile Durkheim that religion promote solidarity and positive interaction, Bronislaw Malinowski the father of anthropology, approached religion as promoter of psychological actions for individual (Malinowski, 2018). In this regard the innovative technique of affective and effective cognition implied in this research are addressed by the author. Without cognitive satisfaction and stress coping the social solidarity through religious travelling would not serve the purpose.

Religious tourism or the activities minimized the social disruption which can otherwise an issue for the religious interaction (Malinowski, 2018). Further death is considered as main source of religious in majority religions & little pains like social and psychological problems are the simple reason for people to travel and remain satisfied with their life. Functionalism operates under special interest of dealing with the challenges of managing destination in smoothing mental distress of religious tourists.

RESEARCH METHODOLOGY

This study is based on the development of construct for tourism departments and attached offices which contribute for promotion of soft image through the management of religious destinations in Pakistan (Ledford & Gast, 2018). Useful insights reflect through the stimulated activities and gestures. The accurate information about the affective cognitive and effective cognition on people for promotion of tourism in the selected religious area. The role of government in executing tourism promotion initiatives is part of information collected to answer the objectives.

Development of Survey Instrument

For the achievement of objectives good survey instrument directs responses & ensure the concrete results (Leggett, 2017). It is developed in different categories and groups attached Annexure A. Group A is incorporated for Affective and effective cognition with Hindu destination in Baluchistan. The questionnaire for Group B is to ensure the responses about perspectives of tourists and it incorporates Buddhist destination of Takht Bahi. The questionnaires of Group C include the interlinking of both approaches of cognition and emotional contagion while having two destination (Nankana Sahib & Gurdwara in Burewala) of Sikh community. In brief the instrument contains number of questions with the following indicators:

- Formation of questions based on construct
- Determination of research content & reflection in the question's items
- Understandable content for the Buddhist, Hindu, & Sikh respondents
- Acceptable level of validity & reliability for key religious tourism stakeholders
- Information flow of general to specific questions with identification
- Respondent approval and revised questionnaire

Nature and Type of Survey Questionnaire

As supported by De (2011) the study used Likert scale type of question to answer among the range of option. From very unsatisfied to very unsatisfied respondent has the option of choosing one. Three main variables are included in the questionnaire (affective cognition, effective cognition & religious destination management) with three sub-variables. Most respondent are well-experienced and fall between age of 30-50 with 10-15 years of experience. The respondent selected for the survey are based on:

- Age
- Religious Education
- Experience in Religious Department
- Years of attachment with the religious destination
- Association with the subject of this research

Procedure for Data Collection and Sample

The procedure of data collection for this study is based on the nature of cognition and affective cognitive perception of religious tourists. The objectives and questions mentioned to respond to the problem of religious tourism management. The primary sources of data are considered under the survey questionnaire method (Hall and Wall, 2019). It is developed with incorporation of stakeholders like foreign religious tourists, management of the religious tourism destination, staff & workers. Further, the religious places, managers of Auqaf department /workers at hotels and the visitors in the areas of religious destinations are also included.

The stakeholders are selected as key sources for data about the tourism development and management at religious destinations. (Aykroyd et al., 2019). Questionnaires are distributed among 218 respondents where 200 returned with varied answers. 30 monks with 10 people from administrative staff in Takht Bahi are selected for data. Same are the numbers (40,40) of Sikh followers in Nankana Sahib & Gurdwara in Burewala. For Hindu temples in Baluchistan there are 45 questionnaires distributed and responded. Among 40 questionnaires there are 35 survey questions distributed between officials from foreign office (10) management (15) workers (10). The confidentiality is maintained and name or designation of staff is not mentioned. The representatives of government are also part of the survey. Four destinations have rich religious history and sites for the international travelers to fulfill their spiritual and emotional needs.

Table 1. Procedure of Data Collection

Sample	
Survey	*Workers/ Staff / Visitors* *Management of Hotels* *Officers from Foreign Office* *Representative of Religious Tourist Destination*

Statistical Analysis and Measures

For the analysis and obtaining the results following are the analytical tool used with accurate and compulsive results against the responses and objectives.

- Frequencies of Demographic Variables
- Pearson Correlation Matrix
- Regression Testing Model (Cronbach's Alpha)
- Analysis of Reliability

The procedures of analyzing problem of soft image in relation to tourism promotion are based on Primary data. After the collection of information through questionnaire where the content, narrative, discourse is analyzed through interpretation. The SPSS software is used as quantitative analytical techniques, (Ibm, 2019). Further the data is tested through the correlation, regression and descriptive analysis. Regression analysis is implied in predicting the effects of one variable on other, while correlation is to see the nature of relationship among variables (Bhatia and Bhatia, 2019).

RESULTS OF THE DATA COLLECTED AND TESTED

Respondent Characteristics anf Profile

As shown in the table of demographics there are 150 male and 50 females with percentage of 77.77 & 22.22. The age of these respondent is mostly young as 75 and 50 respondents in two categories left with

Figure 7. Responses of Survey Participants

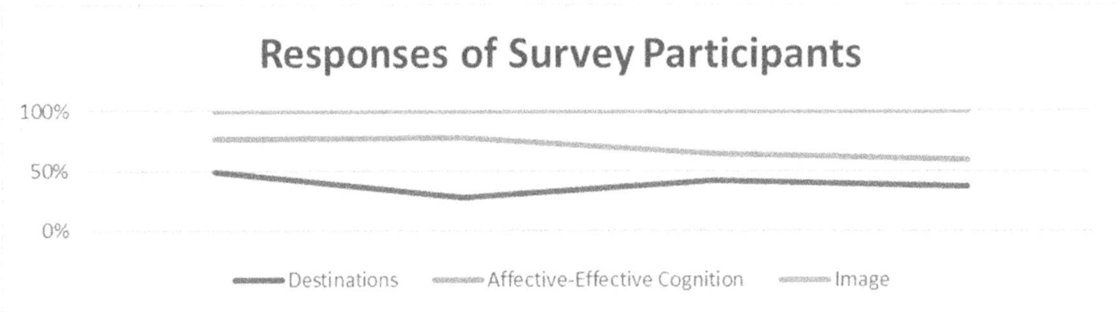

Table 2. Profile of Demography

Features	*Frequency N=200*	*Percentage%*
Male ***Female***	150 50	77.77 22.22
Age ***30-35*** ***35-40*** ***40-50***	 75 50 75	 37.5 25 37.5
Distribution ***Buddhist Monks*** ***Sikh Gurdwara*** ***Hindu Temple*** ***Officials*** ***Workers*** ***Management***	30 65 45 20 25 15	15% 32.5 22.5 10% 12.5 7.5
Total	Total	200

more than 60% of total population. This shows the related people and professionals of management relate to this research. Further the experience is intense for 3-5 years which stand for 50% of the respondents. Other numbers of experienced people are from 5-15 years having significant influence on the response rate because of their enhanced expertise. This shows that people having increase experience and age are important for managing the destination to develop soft image of Pakistan.

Table 3. Scale All Variables

Variables	*Number of Items*	*Cronbach's Alpha*
Affective-Effective Cognition	32	0.788
Emotional Contagion	35	0.810
Tourists Perspective	29	0.731
Destination Management	27	0.751

(Cronbach's Alpha) Reliability Test

For the reliability of study Cronbach's Alpha is conducted to confirm the constructs contributing this research. The reliability and validity of the research model is supported as the social scientist viewed reliability test should be more than 0.7 for considerable scale. In the table provided each variable and number of items there are values more than 0.7. This leads to 0.8 which confirmed the reliability & validity of the model.

Pearson Correlation and Descriptive Analysis

The Pearson correlation is employed to relationship among variables where it was found that affective cognition is positively correlated with all the attempts/constructs for soft image (Forster, 2018). However literal beliefs are negatively correlated because of the emotional attachment and ideological background of the religion. There is negative relation of destination management which illustrate the slow progress of development. Same is the case with difference which is negative in influencing the cognition of visitors which requires a comprehensive change. Other constructs or the variables remain positive towards development of image. H3, H5, H9 remain weak in relationship & statements while rest have strong support found in this result.

Table 4. Pearson Correlation Analysis

Description	*Mean*	*S. D*	*1*	*2*	*3*	*4*	*5*
Soft Image	4.27	1.53	1				
Affective-effective Cognition	3.31	1.64	0.0421	1			
Emotional Contagion	5.32	1.09	0.0407	0.2304	1		
Tourist Perspective	4.01	1.41	0.0792	-0.0641	0.321	1	
Destination Management	3.71	1.83	0.0393	0.0792	0.0641	0.0684	1

Analysis of Regression

The conduct of regression is to know the relationship among destination management and its influence on soft image of the country (Asmelash & Kumar, 2019). It is favorable outcome for a research to have goodness-of-fit result of the values employed. Multiple determinant is based on coefficients as it is used to enter variance of the dependent variable from independent construct. In the table the calculated equation value of $R2$, is .64 which reflect a significant variation in the destination management and soft image. Further the t-test is statistical output of a coefficient which is measured with F-test value. Here F value is 453.7 which is highly significant against the P>0.000 refer that model is strong. The hypothesis H1, H2, H4, are supported by the results while rest of the construct H4 are at negative association with soft image. The affective cognition, emotional contagion, and destination management are significantly supported by the regression analysis in contributing the soft image. However, tourist perspective is not associated with the provided results.

CONCLUSION AND INFERENCES

Table 5. Results of Regression

Description	Residual Sum of Squared	Degree of Freedom	MS	Number of Observation		200
Model	1772.37377	11	443.093442	F (4, 200)		453.70
Residual	15903.147	16284	.976611828	Prob. >F		0.0000
Total	17675.5208	16288	1.08518669	R-Squared		.64
				Adj.- R-Squared		.62
				Root MSE		.98824
Soft Image	**Coefficient**	**Stander Error**	**t-statistics**	**p-value**	**Confidence Interval [95%]**	
Affective-effective Cognition	.0299911	.0086846	3.45	0.001	.0129682	.0470139
Emotional Contagion	-.5014111	.0148529	-33.76	0.000	.5305245	.4722914
Tourist Perspective	-.0080746	.0074288	-1.09	0.277	.0226311	.0064866
Destination Management	.4994775	.0149226	26.77	0.001	.3702276	.4287260
Constant	.1997031	.0597217	3.34	0.001	.826421	.3167641

The provided literature and discussion by the scholars in their expertise and knowledge of best understanding the image development followed by models are based on their culture and context (Sugimura et al., 2019). For the research under consideration the innovative approach of cognition and affective cognitive of religious tourists is used to see the development of Pakistan's positive image. Following are the conclusion of this research that discussed below.

Main Conclusion

The finding illustrated that affective cognition is the neural correlation of affective and non-affective cognition for people and travelers. The use of cognitive-affective approach is truly purposeful and measurable in practical setting. The reflection of religious tourist personality in the definitive situations and on destination is influence by responsible role of managing the religious tourism destination. However, there is significant support for the H1, H2, H4, but H3 is not accepted. The tourist perspective is failing to satisfy the requirements of religious tourist's management. The remining constructs affective-effective cognition, emotional contagion & destination management are supportive by the data outcomes. It is argued that interaction in any environment develop positive or negative sensory signals which is true in case of destinations in Pakistan.

It is also found that perception of religious tourists favored the structure and development of temples, Sikh gurdwaras accommodation & transportation, boarding and expenditures of Buddhist Monks. It formed the identity and structure of the tourist's religious ideology. Further the cultural and religiously sensitive beliefs are need to be managed through planned mechanism policy tools. The role of strong institutions is expected to be played for the easy access to other religions. The issue of identification of knowledge for tourist needs sharing process and increase interaction among the people of different ideologies and beliefs. It is essential to inform the audience or the tourists visiting to respective holy site that a peaceful arrangement of interaction is facilitated for kind support to reduce the gap of religious knowledge. The proposed hypothesis is also tested through conclusive results of data & analysis of results.

H1: There is collaborative influence of affective & effective cognitive satisfaction on the travelers visiting to religious destination in Pakistan.

After testing the hypothesis through data analysis & implied statistical tools the development of image is varied in utilizing techniques to attract the cognitions of the religious tourists. The table 4.3 & 4.4 reflect the significant influence of cognition on religious tourists. Those visited the religious site once in their life time are new and need to increase movement with tourists that have good experience and interaction with others. Further the explanation of religious tourists having multifaceted expertise unfolds the realities of mental position influenced by destination.

The study confirmed the role of affective & effective cognition on travelers visiting to Pakistan. Hence HI is accepted.

H2: The emotional contagion of religious tourists favored the structure and development of destination in response to their requirement.

In response to the hypothesis the table 4.3 & 4.4 is analyzed where the mean & S.D values of the emotional contagion are 5.32 & 1.09. The harmonized culture is broad-mindedness of the hospitable nation as this unfolded the realities of harmonization in context of religious tourism. There are no restrictions for the travelers to visit and enjoy the holy destination in fulfillment of the desires and emotional exchange. Variety of cultural and historical resources in Pakistan are based for the development and a happy interaction among different religions.

The research recognized the emotional exchange among religious tourists along with destination development. The hypothesis H2 is accepted.

H3: The positive perspective & smooth environment provided by the management on destination changed the perception of the travelers.

For the examination of this hypothesis the study tested data in table no 4.4 where the values are inclusive at (-.0080746) & (-1.09). The negative values imply the approach that contemporary era of global inequality is pursuing the perspectives as central to perception of travelers. The option of providing smoothing environment for using tourism as tool of social integration and enhancing positive perspective. Further the values of mean & median in table no 4.3 (4.01) and (1.41) illustrate a less determination of management for building positive perspective. The sustainable growth and benefits are also included in collective and participatory actions. Same is the case with holy destination which can play their part in reducing the inter-faith conflict among religious communities. It decreases the risk of creating problems and issues related with the image of specific area. In this regard H3 is rejected.

H4: The enhanced interface among religious tourist's destination with different backgrounds reduce the negative image

The interpretation of this hypothesis provides the analysis of data mentioned in table 4.3 & 4.4 reflect positive association of image with destination management. The values (.4994775) in table 4.4 and (0.0684) in table 4.3 unfolded the that image is influenced by the difference of for religious tourist's management which requires proper facilities. This signifies the perception of religious tourists about the management of destination and increase demand of facilities at due station of holy concern. In the same way financial limitations should not be the reason for deprived state of tourists site. This hypothesis H4 is accepted.

Contributions and Policy Recommendations

The research contributed with new approach of cognitive and affective components related to initiatives of government for religious tourism and their effects upon soft image of Pakistan. The opinion, responses and observation during the interaction illustrate that institutional reforms are the primary need for the better religious tourism management (Difference on destination image and travel options, 2018). This an allow the official machinery and religious department to promote tourism through a joint venture program. It positively drives image issues and projects progress under the government. The responses stressed for the awareness among people to know the strength and potential of religious tourism industry. At holy destination no one is bothered with the specified or any difference among them. The entire approach is to purify the person as male or female for pursuing a peaceful life.

The determination for the positive or negative image of the country is based on each travel choice of the religious traveler. The image and cognitive development of tourists against the experience of travelling is like the sustainability of tourists cannot be possible. For achieving the long-term objectives & key indicators need reflection of mechanism. It must validate and develop a system of check and balance as oversight on destination. Here, the most important point is the consideration of stakeholders for fulfillment of religious facilities. Those visited the religious site once in their life need increase movement with tourists that have good experience and interaction Further the explanation of religious tourists having multifaceted expertise is fruitful for unfolding the beauty of religion. Image is well connected with experience and understanding of visitors.

It is recommended that the collaborative measures and improved security situation for the participants is an opportunity to bring foreigners in Pakistan. Through such initiatives image of the country can be changed further to achieve the ideal champion of peace. As the participants viewed progress through the removal of concern for not facilitating tourists in performing religious activities on the destinations, a revised and comprehensive analysis is required. This flaw can be approached through the initiatives like that of infrastructure development and facilities at destination can improve the image. Further, to achieve handsome results from religious tourism the state needs to be involved in person & through the inclusive approach of key stakeholders. The trained relevant staff of religious destinations is the investment by the government in realizing the worth of religious tourism management.

REFERENCES

Ahmad, A. (2016). Gwadar: A Historical Kaleidoscope. *Policy Perspectives: The Journal of the Institute of Policy Studies*, *13*(2), 149–166. doi:10.13169/polipers.13.2.0149

Ahmad, S. (2016). Tourism in Pakistan: Challenges, Prospects and Potential of Gilgit-Baltistan. *Pakistan Perspectives, 21*(2).

Albayrak, T., Herstein, R., Caber, M., Drori, N., Bideci, M., & Berger, R. (2018). Exploring religious tourist experiences in Jerusalem: The intersection of Abrahamic religions. *Tourism Management, 69*, 285–296. doi:10.1016/j.tourman.2018.06.022

Ali, S., & Maharani, L. (2019). Development of Religious Tourism in Bandar Lampung. Academic Press.

Ali, Y., Awan, M. A., Bilal, M., Khan, J., Petrillo, A., & Khan, A. A. (2019). Risk assessment of China-Pakistan fiber optic project (CPFOP) in the light of multi-criteria decision making (MCDM). *Advanced Engineering Informatics*, *40*, 36–45. doi:10.1016/j.aei.2019.03.005

Alipour, H., Olya, H. G., & Forouzan, I. (2017). Environmental Impacts of Mass Religious Tourism: From Residents' Perspectives. *Tourism Analysis*, *22*(2), 167–183. doi:10.3727/108354217X14888192562285

Almaimoni, H., Altuwaijri, N., Asiry, F., Aldossary, S., Alsmadi, M., Al-Marashdeh, I., ... Alrajhi, D. (2018). Developing and Implementing WEB-based Online Destination Information Management System for Tourism. *International Journal of Applied Engineering Research*, *13*(10), 7541–7550.

Almuhrzi, H. M., & Alsawafi, A. M. (2017). Muslim perspectives on spiritual and religious travel beyond Hajj: Toward understanding motivations for Umrah travel in Oman. *Tourism Management Perspectives*, *24*, 235–242. doi:10.1016/j.tmp.2017.07.016

Almuhrzi, H. M., & Alsawafi, A. M. (2017). Muslim perspectives on spiritual and religious travel beyond Hajj: Toward understanding motivations for Umrah travel in Oman. *Tourism Management Perspectives*, *24*, 235–242. doi:10.1016/j.tmp.2017.07.016

Asadi, M. M., Heidari, M., & Asadian, F. (2016). Forecasting religious tourism demand using artificial intelligence case study: Mashhad city. *International Journal of Humanities and Cultural Studies*, 1597-1617.

Asmelash, A. G., & Kumar, S. (2019). Assessing progress of tourism sustainability: Developing and validating sustainability indicators. *Tourism Management*, *71*, 67–83. doi:10.1016/j.tourman.2018.09.020

Badar, R., & Chakradeo, S. (2016). Sufism and tourism interface. *Space, 20*(1-2).

Bahadori, S. R., Farsani, N. T., & Shafiei, Z. (2019). Tourists' views on Islamic spiritual tours (case study: Yazd, Iran). *Journal of Islamic Marketing*, *10*(1), 323–335. doi:10.1108/JIMA-05-2017-0055

Bohg, J., Hausman, K., Sankaran, B., Brock, O., Kragic, D., Schaal, S., & Sukhatme, G. S. (2017). Interactive perception: Leveraging action in perception and perception in action. *IEEE Transactions on Robotics*, *33*(6), 1273–1291. doi:10.1109/TRO.2017.2721939

Cabrera, E. J., dela Vega, J. M., dela Vega, J. M., Dioquino Jr, A., Favor, C., & Navarro, L. (2019). Religious Tourism Assessment and Carrying Capacity: The Case of National Shrine of Saint Padre Pio in Batangas, Philippines. *Asia Pacific Journal of Multidisciplinary Research, 7*(3).

Cassidy, R., & Hawkins, B. (2017). *Interviewing key informants from the corporate sector*. Academic Press.

Difference on Destination Image and Travel Options: An Exploratory Text-Mining Study. (2018). In *2018 15th International Conference on Service Systems and Service Management (ICSSSM)* (pp. 1-5). IEEE.

Durán-Sánchez, A., Álvarez-García, J., Río-Rama, D., De la Cruz, M., & Oliveira, C. (2018). Religious tourism and pilgrimage: Bibliometric overview. *Religions*, *9*(9), 249. doi:10.3390/rel9090249

Durkheim, É., & Mauss, M. (2016). *Hobbes à lagrégation: Un cours dÉmile Durkheim suivi par Marcel Mauss*. Éditions de l'EHESS.

El-Gohary, H. (2016). Halal tourism, is it really Halal? *Tourism Management Perspectives*, *19*, 124–130. doi:10.1016/j.tmp.2015.12.013

Enongene, V., & Griffin, K. (2019). Tackling the problems of deficient data when planning for religious tourism management. *Managing Religious Tourism*, 79.

Forster, D. (2018). *Hydrodynamic fluctuations, broken symmetry, and correlation functions*. CRC Press. doi:10.1201/9780429493683

Fujita, H., Gaeta, A., Loia, V., & Orciuoli, F. (2018). Resilience analysis of critical infrastructures: A cognitive approach based on granular computing. *IEEE Transactions on Cybernetics*, *49*(5), 1835–1848. doi:10.1109/TCYB.2018.2815178 PMID:29994107

Garayeva, N., & Naumov, N. (2016). Religious sites as tourism attractions: The case of Azerbaijan. *International Hospitality and Tourism Student Journal*, *8*(1), 8–14.

Ghani, G. M. (2019). Tourist Arrivals to Muslim Countries: Is Religion Important? *Tourism, Culture & Communication*, *19*(3), 209–214. doi:10.3727/194341419X15554157596128

Gozzoli, R. B., Chen, K. T. N., & Talawanich, S. (2019). Religious Tourism and Heritage Knowledge at Phra Pathom Chedi, Thailand. *Tourism, Culture & Communication*, *19*(2), 135–152. doi:10.3727/10 9830419X15536971539416

Griffin, K. (2017). The importance of religious tourism and pilgrimage: Reflecting on definitions, motives and data. *International Journal of Religious Tourism and Pilgrimage*, *5*(3), 2.

Handriana, T., Yulianti, P., & Kurniawati, M. (2019). Exploration of pilgrimage tourism in Indonesia. *Journal of Islamic Marketing*, *11*(3), 783–795. doi:10.1108/JIMA-10-2018-0188

He, M., Li, J., Li, J., & Chen, H. (2019). A comparative study on the effect of soundscape and landscape on tourism experience. *International Journal of Tourism Research*, *21*(1), 11–22. doi:10.1002/jtr.2237

Heidari, A., Yazdani, H. R., Saghafi, F., & Jalilvand, M. R. (2017). Developing strategic relationships for religious tourism businesses: A systematic literature review. *EuroMed Journal of Management*, *2*(1), 77–98. doi:10.1504/EMJM.2017.084273

Heydari Chianeh, R., Del Chiappa, G., & Ghasemi, V. (2018). Cultural and religious tourism development in Iran: Prospects and challenges. *Anatolia*, *29*(2), 204–214. doi:10.1080/13032917.2017.1414439

Hoseinipor, S. E., Dolatabadi, H. R., & Kazemi, A. (2019). Effect of decision-making style on travel intention of religious tourism (Qom as religious tourism destination in Iran). *International Journal of Applied Management Science*, *11*(2), 113–123. doi:10.1504/IJAMS.2019.098825

Huang, K., & Pearce, P. (2019). Visitors' perceptions of religious tourism destinations. *Journal of Destination Marketing & Management*, *14*, 100371. doi:10.1016/j.jdmm.2019.100371

Idrees, I. A., Vasconcelos, A. C., & Ellis, D. (2018). Clique and elite: Inter-organizational knowledge sharing across five-star hotels in the Saudi Arabian religious tourism and hospitality industry. *Journal of Knowledge Management*, *22*(6), 1358–1378. doi:10.1108/JKM-07-2017-0276

Idris, J. (2019). Spiritual motivation for religious tourism destinations. *Spiritual and Religious Tourism: Motivations and Management*, 59.

Jawabreh, O. A. (2017). Distinction of Jordan as a destination for religious tourism. *Journal of Environmental Management and Tourism*, *8*(22), 1171–1182.

Jelacic, I. (2017). *Knowledge gap identification as a driver of learning and knowledge reuse in product development*. Academic Press.

Kawsar, M. D., & Raj, R. (2017). *Customer Relationship Marketing Tool for Growth of Religious Tourism Business: An Islamic perspective*. Academic Press.

Kessler, K., & Raj, R. (2018). 2 The Rise of Heritage and Religious Tourism to Sacred Sites in Oman. *Religious Tourism in Asia: Tradition and Change Through Case Studies and Narratives*, 10.

Khan, A. U., Zaman, G., & Baloch, Q. B. (2019). Promoting Religious Tourism in Pakistan: Empirical Evidence from Sikh Pilgrims Visiting Pakistan. *City University Research Journal, 9*(4).

Kim, B., & Kim, S. (2019). Hierarchical value map of religious tourists visiting the Vatican City/Rome. *Tourism Geographies*, *21*(3), 529–550. doi:10.1080/14616688.2018.1449237

Kim, B., Kim, S., & King, B. (2019). Religious tourism studies: Evolution, progress, and prospects. *Tourism Recreation Research*, 1–19.

Korstanje, M. E., & George, B. P. (2018). 7 Religious Tourism in the Ideological Framework of Chinese Tourism Education. *Religious Tourism in Asia: Tradition and Change Through Case Studies and Narratives*, 58.

Korstanje, M. E., Raj, R., & Griffin, K. (Eds.). (2018). *Risk and Safety Challenges for Religious Tourism and Events*. CABI. doi:10.1079/9781786392282.0000

Ledford, J. R., & Gast, D. L. (2018). *Single case research methodology: Applications in special education and behavioral sciences*. Routledge. doi:10.4324/9781315150666

Leggett, T. (2017). Survey development: Creating intended consequences. *Radiologic Technology*, *88*(5), 568–571. PMID:28500097

Leppakari, M., & Griffin, K. A. (Eds.). (2016). *Pilgrimage and Tourism to Holy Cities: Ideological and Management Perspectives*. Cabi.

Liu, A., & Pratt, S. (2017). Tourism's vulnerability and resilience to terrorism. *Tourism Management*, *60*, 404–417. doi:10.1016/j.tourman.2017.01.001

Lochrie, S., Baxter, I. W., Collinson, E., Curran, R., Gannon, M. J., Taheri, B., Thompson, J., & Yalinay, O. (2019). Self-expression and play: Can religious tourism be hedonistic? *Tourism Recreation Research, 44*(1), 2–16. doi:10.1080/02508281.2018.1545825

Lodhi, S. (2017). Terrorism and Its Impact on Economy of Pakistan. *RADS Journal of Social Sciencess & Business Management, 4*(1), 36–50.

Lücking, M. (2019). Travelling with the Idea of Taking Sides: Indonesian Pilgrimages to Jerusalem. *Bijdragen tot de taal-, land-en volkenkunde/Journal of the Humanities and Social Sciences of Southeast Asia, 175*(2-3), 196-224.

Malinowski, B. (2018). *The family among the Australian Aborigines: a sociological study*. BoD–Books on Demand.

Malodia, S., & Singla, H. (2017). Using HOLSAT to evaluate satisfaction of religious tourist at sacred destinations: The case of religious travelers visiting sacred destinations in the Himalayas, India. *International Journal of Culture, Tourism and Hospitality Research, 11*(2), 255–270. doi:10.1108/IJCTHR-10-2015-0120

Mathew, P. V., & Kuriakose, V. (2017). A Study on the Indicator Framework of Responsible Tourism in Tourism Accommodation Sector. *Asian Journal of Management, 8*(4), 989–996. doi:10.5958/2321-5763.2017.00153.6

Michael, N., James, R., & Michael, I. (2018). Australia's cognitive, affective and conative destination image: An Emirati tourist perspective. *Journal of Islamic Marketing, 9*(1), 36–59. doi:10.1108/JIMA-06-2016-0056

Mikos von Rohrscheidt, A. (2013). Religion as a subject of cognitive interest to tourists. Analysis of the phenomenon and products on offer for cultural-religious tourism. *Turystyka Kulturowa*, (5), 5-19.

Niculina, A. (2019). The economic foundation of the development strategies for the cultural religious tourism. *Sustainable Development*.

Peterson, K., Hahn, H., Lee, A. J., Madison, C. A., & Atri, A. (2016). In the Information Age, do dementia caregivers get the information they need? Semi-structured interviews to determine informal caregivers' education needs, barriers, and preferences. *BMC Geriatrics, 16*(1), 164. doi:10.118612877-016-0338-7 PMID:27662829

Piva, E., Cerutti, S., & Raj, R. (2019). Managing the Sacred: A Governance Perspective for Religious Tourism Destinations. *Managing religious tourism/Piva E. Stefania C., Razaq R*, 10-21.

Prayag, G., Chen, N. C., Hosany, S., & Odeh, K. (2016). *The Cognitive-Affective Relationship of Tourists to a Heritage Site: The Case of Petra*. Academic Press.

Raj, R., Griffin, K., & Korstanje, M. E. (2018). 1 Risk and Safety Challenges Facing Religious Tourism–An Introduction. *Risk and Safety Challenges for Religious Tourism and Events*, 1.

Raj, R., & Griffin, K. A. (Eds.). (2017). *Conflicts, Religion and Culture in Tourism*. CABI. doi:10.1079/9781786390646.0000

Rasgon, A., Lee, W. H., Leibu, E., Laird, A., Glahn, D., Goodman, W., & Frangou, S. (2017). Neural correlates of affective and non-affective cognition in obsessive compulsive disorder: A meta-analysis of functional imaging studies. *European Psychiatry*, *46*, 25–32. doi:10.1016/j.eurpsy.2017.08.001 PMID:28992533

Rashid, A. G. (2018). Religious tourism–a review of the literature. *Journal of Hospitality and Tourism Insights*.

Rejman, K., Maziarz, P., Kwiatkowski, C. A., & Haliniarz, M. (2016). Religious tourism as a tourism product. *World Scientific News*, (57), 562-575.

Rivera, D. E., Fa, M. C., & Villar, A. S. (2019). Delightful tourism experiences: A cognitive or affective matter? *Tourism Management Perspectives*, *32*, 100569. doi:10.1016/j.tmp.2019.100569

Scanlan, C. L. (2020). *Preparing for the Unanticipated: Challenges in Conducting Semi-Structured.* In-Depth Interviews. doi:10.4135/9781529719208

Scheyvens, R., & Biddulph, R. (2018). Inclusive tourism development. *Tourism Geographies*, *20*(4), 589–609. doi:10.1080/14616688.2017.1381985

Shah, S. I. A. (2016). *Economic corridor business tower Gwadar Balochistan, Pakistan* (Doctoral dissertation). Mehran University of Engineering and Technology, Jamshoro.

Sharma, A. K., & Swamy, G. A. (2019). Opportunity and threats faced on religious tourism development India. *International Journal of Scientific Research (Ahmedabad, India)*, *8*(10).

Shinde, K. A. (2017). Planning for urbanization in religious tourism destinations: Insights from Shirdi, India. *Planning Practice and Research*, *32*(2), 132–151. doi:10.1080/02697459.2016.1198197

Skavronskaya, L., Scott, N., Moyle, B., Le, D., Hadinejad, A., Zhang, R., Gardiner, S., Coghlan, A., & Shakeela, A. (2017). Cognitive psychology and tourism research: State of the art. *Tourism Review*, *72*(2), 221–237. doi:10.1108/TR-03-2017-0041

Sugimura, K., Matsushima, K., Hihara, S., Takahashi, M., & Crocetti, E. (2019). A culturally sensitive approach to the relationships between identity formation and religious beliefs in youth. *Journal of Youth and Adolescence*, *48*(4), 668–679. doi:10.100710964-018-0920-8 PMID:30187258

Suri, K. (2018). Understanding historical, cultural and religious frameworks of mountain communities and disasters in Nubra valley of Ladakh. *International Journal of Disaster Risk Reduction*, *31*, 504–513. doi:10.1016/j.ijdrr.2018.06.004

Terzidou, M., Scarles, C., & Saunders, M. N. (2018). The complexities of religious tourism motivations: Sacred places, vows and visions. *Annals of Tourism Research*, *70*, 54–65. doi:10.1016/j.annals.2018.02.011

Thomin, M. (2019). *Moving towards a harmonized heritage tourism approach: an assessment of small museums in Northwest Florida*. Academic Press.

Tkaczynski, A., & Arli, D. (2018). Religious tourism and spiritual leadership development: Christian leadership conferences. *Journal of Hospitality and Tourism Management*, *35*, 75–84. doi:10.1016/j.jhtm.2018.03.004

Tomić, S., Leković, K., & Tadić, J. (2019). Consumer behaviour: The influence of age and family structure on the choice of activities in a tourist destination. *Economic research-. Ekonomska Istrazivanja, 32*(1), 755–771. doi:10.1080/1331677X.2019.1579663

Verma, M., & Sarangi, P. (2019, August). Modeling attributes of religious tourism: A study of Kumbh Mela, India. *Journal of Convention & Event Tourism, 20*(4), 296–324. doi:10.1080/15470148.2019.1652124

Vrij, A., Fisher, R. P., & Blank, H. (2017). A cognitive approach to lie detection: A meta-analysis. *Legal and Criminological Psychology, 22*(1), 1–21. doi:10.1111/lcrp.12088

Wang, R., & Hao, J. X. (2018, July). Difference on Destination Image and Travel Options: An Exploratory Text-Mining Study. In *2018 15th International Conference on Service Systems and Service Management (ICSSSM)* (pp. 1-5). IEEE.

Web, S. A. C. (2020). *Cult Pakistan: Forgotten mysteries, bygone strangeness & odd folk| Nadeem F.* Paracha–South Asia Citizens Web.

Wiltshier, P., & Griffiths, M. (2016). Management Practices for the Development of Religious Tourism Sacred Sites: Managing expectations through sacred and secular aims in site development; report, store and access. *International Journal of Religious Tourism and Pilgrimage, 4*(7), 1–8.

Wiltshier, P., & Griffiths, M. (2019). *Managing Religious Tourism*. CABI.

Yousaf, S., & Xiucheng, F. (2018). Humanizing stigmatized places: Inter-group contact and attitude change toward Pakistan and Iran in the 'Humans of New York' Facebook space. *Journal of Business Research.*

Yousefi, J. (2018). *Importance-performance Analysis of Religious Tourism Services (The Case of the City of Birjand).* Academic Press.

APPENDIX

Survey Questionnaire

Table 7. Effective Cognition

Questions	Strongly Disagree	Disagree	Neutral	Agree	Strongly Agree
Do you consider religious tourism department are promoting step in improving departmental coordination?					
Did you expect the positive results of cooperation among federal and provincial tourism sectors on religious destination management?					
Tourist are satisfied in performing their religious activities					
Is there any Impact of improved coordination on online visa?					

Table 8. Affective-Cognitive & Tourist Contagion

Questions	Strongly Disagree	Disagree	Neutral	Agree	Strongly Agree
Clean and green religious destination is the reflection of positive image on cognition of tourists					
Do you agree that effective communication can improve security of the tourists?					
The affective-cognition of religious tourists is influence by facilities provided by religious and tourism department					
The standardized food and transport services can attract religious visitors to visit Pakistan					

Table 9. Tourist Perspective & Destination Management

Questions	Strongly Disagree	Disagree	Neutral	Agree	Strongly Agree
Is the implementation of online visa contributed for positive cognition of religious tourists?					
Transport and communication can be facilitated by support of security organizations					
Food standards on religious destinations is relevant with various religious tourists					
Pakistan can promote soft image of the country through affective cognitive approach about destination management					
The use of social media can be effective in presenting Pakistani religious destinations					

Chapter 11
Mystical Tibet Today:
Growing Tourist Route for Buddhist Religion and Knowledge Seekers and Pilgrims

Madlein V. Kotsotsashvili
Georgian Teqnical University, Georgia

ABSTRACT

Tourism is one of the fastest-growing industries in the world. Currently, it provides 10% of the world's income and employs one-tenth of the total workforce. World Tourism Organisation (WTO) estimates that by 2020 1.6 billion people will travel around the globe and an estimated income of $1.5 billion will be generated from these activities. With all these financial outcomes, tourism can provide great opportunities for the local economies. The sacred geography of the Tibetan region and the sacred Mount Kailas combines the origins of the world's four religions. For several centuries Tibet was closed to foreign visitors and tourists. This is explained not only by the complex geographical location. Tibet's isolation was largely driven by both Tibetan rulers and Chinese policy.

INTRODUCTION

Religious pilgrimage is not only an innate characteristic sociocultural phenomenon of world religions but also one of the most important integrative, communicative, and economic factors of our modernity. This is due to the specificity of inter-civilizational contacts at the global level, as well as its active influence on the nature and orientation of intercultural interaction between peoples.

Tourism is one of the fastest-growing industries in the world. Currently, it provides10% of the world's income and employs one-tenth of the total workforce. WorldTourism Organisation (WTO) estimates that by 2020 1.6 billion people will travel around the globe and an estimated income of 1.5 billion $ will be generated from these activities (Bozkurt 2014). With all these financial outcomes, tourism can provide great opportunities for the local economies. According to the latest forecasts of the World Tourism

DOI: 10.4018/978-1-7998-5792-1.ch011

Organization, such areas as Tibet, Bhutan, China, the Philippines, Indonesia will be actively developed in Asia (UNWTO.org)

The sacred geography of the Tibetan region and the sacred Mount Kailas - unite the four major religions of the world. For several centuries, Tibet was closed to foreign visitors and tourists. This is explained not only by the difficult geographical location. The isolation of Tibet was largely driven by the policies of both the Tibetan rulers and the Chinese. The Chinese occupation of Tibet in 1949 was followed by an uprising that was severely suppressed by the Chinese army and special measures were introduced in the region. Tibet has been turned into a closed colony for the international community by China, which unfortunately has so-called He carried out genocide by the name of the Cultural Revolution. And Tibetan spiritual leader Dalai Lama is still in exile and is actively trying to restore Tibet's independence and legitimate government.

It was only in 1984 that the Tibetan Gate opened to tourists and a real opportunity to travel to this unique country emerged. Nevertheless, only a few can travel there ... Of course, getting to know all the sights and even the most elementary ones requires a lot of time. Over the years, several major routes have been developed that you can best use if you are traveling for one to two weeks.

The situation in Tibet has improved for more than two decades. China has rebuilt and developed socio-economic infrastructure, communications, transport, and is contributing in every possible way to the development of tourism. According to China's Department of Tourism Development, Tibet's tourism budget increased by $ 390 million in the spring of 2019 to 41.1% (http://zwgk.mct.gov.cn/)The number of tourists visiting the region amounted to two and a half million, which is 84.2% more than in the same period last year. To attract more tourists during the winter season, the government has introduced several measures. Tibet is a special region of China that attracts tourists not only to monasteries and theological schools but also to the top of Mount Everest.

The article discusses the reasons why it was Indian Buddhism that became the spiritual foundation of Tibetan civilization, and Tibet - the "central country" of Buddhism. Based on data from Tibetan historical sources. The reasons for this were not only the historical and socio-political circumstances of Tibet and neighboring countries. Much more important in the historical choice and civilization strategy of Tibet were the spiritual and moral advantages of Buddhism and those qualities of the Indian tradition practics of Buddhism.the state, trends, and prospects of the development of religious tourism at the regional level are considered; classification of religious tourism is proposed; the motivation of religious tourists and pilgrims is investigated, their current position in the system of the tourism industry and recreation is analyzed. Because the developt of religious tourism in Tibet has a great future.

By examining the invariant and specific characteristics of religious tourism and pilgrimage, we can identify the general patterns of the spread of different forms of individual pilgrimage trends that are characteristic of many peoples and civilizations, including the world of Tibetan Buddhism.

Literary Review

For a long time, the topic of religious tourism had been controversial. Only at the end of the 20th century have been published works on the causes and circumstances that make tourists and pilgrims similar or different. Literature is increasingly uniting tourists and pilgrims in ethnographic research. Anthropological studies of tourism have also contributed to this effort, and these papers also connect tourists with the world of pilgrims. Smith 1989 publishes anthropological studies that discuss tourism and the tourist movement. Smith defined tourism as a form of spiritual travel, this contributed to the launch of studies comparing tourism and pilgrimage, it presents tourism as a form of pure travel.

Badone and Roseman 2004 and Morinis contributed to the anthropological analysis of pilgrims and tourism in 1992, studying thematic studies and theories to illustrate the different meanings of religious tourism. These efforts have also been made by sociologists who have added an understanding of the economic and socio-cultural aspects that influence the movement of tourists. Sociological studies have used a variety of approaches: McCann in 1999 and Charles in 2008 studied theoretical issues related to tourism, while Cohen in 2004 wrote on thematic research. These works contextually influence religiously motivated travel theories, the reasons why people make travel decisions, their meanings and values, and their role in different societies. Works such as Smith and Al. 2010 and Timothy 2011 helps the sociological perspective by considering cultural research. Both of them view tourism in terms of heritage, which helps us understand how religious tourism relates to broader cultural concepts. The International Journal of Religious Tourism and Pilgrims, founded in 2013, is dedicated to interdisciplinary and international research in religious tourism and is an important resource for scientists. These sources contribute to the study of Buddhism and religious tourism by proposing ways to conceptualize their motivated travel, while anthropological and sociological theories regarding the motivation of tourists create the initial framework for studying Buddhism and tourism.

Ashild Kolas' s book, "Tourism and Tibetan culture in transition: A place called Shangrila", 2008 explores the relationship between tourism, culture and ethnic identity in Shangrilla, a Tibetan region in southwest China, to show how local 'Tibetan culture' is reconstructed as a marketable commodity for tourists. It analyses the socio-economic effects of Shangrila tourism, investigating who benefits economically, while also considering its political implications and how tourism might be linked to the reassertion of ethnic identity.

The Chapter, Influence of Religions in the Development of Hospitality Culture in Asia—The Case of Tibet" In book: Tourism Product Development in China, Asian and European Countries 2020, by Xueying Lu, Duojie Suolang is discussed the influence of religion on the field of tourism. According to the authors, Religion, as one of the fundamental elements in culture, is believed to exert some influence on the traveler's behavior and the interaction between the host and the guest.

In another chapter, Religious tourism and cultural pilgrimage: A Chinese perspective" 2007 whose authors are W. Jian-Hong L. Ji Discuss the perspective of religious tourism in terms of service development. Religion, which is deeply rooted in human nature and is widely shared among all walks of people around the world, is among these factors. According to United Nations Educational, Scientific, and Cultural Organization (UNESCO) (cited in Mu et al., 2007), about 60% of the world's population is religious, and the four most popular religions are Christianity, Islam, Hinduism, and Buddhism. These religious groups, together with non-religious travelers who are interested in learning about different cultures by visiting religious sites, attractions, and activities, form a large customer base for religious tourism. ...

Tibet- The chapter In the Book: Wiley Blackwell Encyclopedia on Race, Ethnicity, and Nationalism by Isabelle Henrion-Dursi, Analyzes in detail the life of modern Tibet and the struggle for independence.

The importance of the literature is enriched by such governmental documents as a report of Foreign Affairs, Defense, and Trade Division (U.S) - Tibet: Problems, Prospects, and U.S. Policy which prepared by Kerry Dumbaugh, the Expert in Asian Affair. And "White Paper", which was published in 2019 and was developed by the People's Republic of China and which reflects the official policy of totalitarian communist China about Religious Freedom in Tibet Protected from,.Democratic Reform".

Tibet - The Central Country of Buddhism

The popularity of Tibetan Buddhism has been well established in the West for some time. Less recognized, perhaps, is the growing interest in Tibetan Buddhism not only in China but also throughout Southeast Asia in areas with substantial Chinese populations. Indeed, Tibetan Buddhist ideas have been in circulation in China for some time, at least since the Yuan dynasty(Tsomu, 2018).

Why did Buddhism, and not Bon, which had has deep roots in the Tibetan environment by the time of the emergence of the Tibetan state, become the spiritual basis for the integration of Tibet? Why did it turn out to be the Indian tradition of Buddhism? The explanation for this lies not only in the historical circumstances of the era but also in the properties of Buddhism, thanks to which it became a world religion, which by that time were preferred by many peoples of Asia, as well as in the characteristics of its Indian tradition as the "mother" (Hofmann. 2018). Besides, some more reasons related to the peculiarities of the historical consciousness of Tibetans and the spiritual strategy of Tibetan kings (Tsenpo) should also be taken into account. Historical circumstances were such that the Tibetan state that arose at the beginning of the 7th century quickly became the most influential force in Central Asia. It waged successful conquest wars, and even the great neighbor, China, was forced to reckon with Tibetan military power.

First of all, it is necessary to take into account the international status of Tibet at the time of its political construction and civil ambitions, as well as the fact of the widespread of Buddhism throughout Asia, that the Tibetan kings could not help but attract attention to the universal spiritual and moral potential in it. They understood that the civil power of Buddhism is so great that it was able to "subdue" even such an old civilization, resistant to foreign influences like China. In light of this, it seems quite logical that Tibetan leaders drew attention to Buddhism as the main civilizational factor in the process of building a Tibetan state.

Rather, in the interests of public administration, by that time Buddhism turned out to be the most effective civilizing factor, for the Buddhist countries, as proved in the fifth chapter of the historical study of the famous sinologist E .Zurher contributed to the peacefulness in the population and lowering the level of social aggression, which facilitated the overall management(Zurher. 2014). To the extent that the Bon is loyal to Buddhism, it is still preserved in the Tibetan culture as the traditional worldview of Tibetans, although it is not certain that it is an autochthonous Tibetan religion (Tucci. 2001. Stein. 2010) Part of the Bonnians who remained in positions hostile to Buddhism was forced to go underground. Bon was given a canonical form similar to Buddhism. Buddhism established itself in Tibet, going through lengthy trials and struggles.

Buddhism in Tibet became a continuation of the Indo-Buddhist civilization, and the Tibetan civilization based on it in the main - spiritually and philosophically - is typologically identical to the Buddhist civilization of India. As you know, this civilization in addition to the Himalayas also covers the Mongolian world. Tibetans deliberately chose the form of world Buddhism as the state religion, which researchers consider to be "the most difficult" since it is "the North Indian syncretism of Hinayana, Mahayana, and Vajrayana". This was the form of Buddhism that found expression not only in the secular and monastic culture of early medieval India but also in its public policy. And it also became the spiritual foundation of the entire state system of Tibet and the private life of people. As His Holiness the Dalai Lama often emphasizes during his public teachings, Tibetan Buddhism fully embraced the entire Buddhist heritage of India and is an authentic continuation of the Indian tradition and Tibet Become a "Central Country" of Buddhism. (Safronova-Urbanaeva. 2013).

For a thousand years, Tibet was the stronghold of Buddhism. Tibet embraced Indian culture, which gave him writing, literary language, and the foundations of civilization. The word "Buddhism" was coined by Europeans in the 19th century. Buddhists themselves call their religion Dharma (law, teaching) or Buddharma (Buddha teaching). It is traditionally believed that Tibet is the country of Avalokiteshvara, sympathetic to Bodhisattva C and the Tibetan people are its descendants. The Tibetans traced their origins to Avalokiteshvara, who was once embodied in the form of a huge monkey, and the goddess Tara, who took the form of a terrible demon, the union. According to legend, from them arose people who later settled in the Yarlung Valley.

China and Tibet - Historical and Political Aspects of the Controversy

Before the invasion of China in 1950, Tibet for centuries preserved a unique culture, religion, and language. Today, this culture is threatened by mass immigration of Chinese and strict control over all manifestations of Tibetan culture and national identity. China boasts huge investments in Tibet, but its economic development is primarily designed to strengthen its position in Tibet and increase its ability to exploit the natural resources of Tibet. Economic development has improved the conditions for some Tibetans but overwhelmingly favors Chinese migrants, continuing to exert economic influence on Tibetans. Along with other countries of the world, Tibet has changed a lot over the past 70 years. However, Tibetans continue to work to preserve their culture and daily oppose repressive policies. Tibet has a long and rich history as a country existing side by side with China, while political power in Asia moved between empires and kingdoms. In 1913, the 13th Dalai Lama issued a proclamation confirming the independence of Tibet, and the country retained its national flag, currency, stamps, passports, and army.

After the Chinese Communist Revolution in 1948, it invaded Tibet in 1950. Suppressed, Tibet was forced to abandon its independence. After the failed uprising against Chinese rule in 1959, the 14th Dalai Lama, the political and spiritual leader of Tibet at that time, fled into exile in India, and then tens of thousands of Tibetans. Since 1959, the Chinese government has exercised full political control over Tibet, using all the means of repression to deter and punish Tibetan resistance.

There are Important a few dramatic fragments from the history of theocratic Tibet in the twentieth century. The background for violent resistance is all that China has done to the Tibetan population and the religion since their invasion in 1950. The Chinese terrorized monasteries and villages, monks and gurus were tortured, many were burnt alive, others sent to work-camps. Children were forced to shoot to their parents. Monks were forced to kneels in front of Chinese soldiers. Villages were destroyed (Donnet. 1993).

As for the Tibetan people 'they have not only been shot, but also beaten to death, crucified, burned alive, drowned, vivisected, starved, strangled, hanged, scalded, buried alive, disemboweled, and beheaded' (Shakya. 2000). A large statue of the protector of Tibet, the bodhisattva of compassion, Avalokiteshvara/ Chenrezig in the Jokhang temple was shattered to pieces (Donnet. 1993). It would be easy to interpret this as symbolically related to the exile of the Dalai Lama, the country's religious and political leader, some years earlier. Dalai Lama was the pivot of Tibetan society. He was the incarnation of Avalokiteshvara, the Buddha of compassion and the patron deity of Tibet … The decline in the Dalai Lama's authority was more complex than a mere loss of political power in the Western sense; it was equated with the devolution of Buddhist Tibet. The Chinese, therefore, were seen not as political foes but as "enemies of the faith" (Shakya. 2000).

This legitimizes the use of violence for some Buddhists. (Gillberg. 2005). Tibetan history is noteworthy in two specific aspects. The influence of Buddhism in all aspects of everyday life. Significant at one time the male population of Tibet was monks and llamas, and ultimately this the spiritual group became the temporary rulers of Tibet, as well as its spiritual leaders. The Dalai Lama is considered of reincarnation of the patron deity of Tibet, is the highest and most revered among this ruling monastic theocracy. The second notable aspect of Tibetan history is the ambiguity and disagreements over Tibet's long-term political relationship with China. Generally, Tibetans consider Tibet as a historically independent nation, a relationship with the succession of Chinese empires a series of Chinese governments, on the other hand, argues that Tibet was the political and geographical part of China for many centuries.

Competing statements about the past of Tibet are one of the obvious obstacles in progress in the Sino-Tibetan negotiations. Although the Dalai Lama has stated many times that he does not seek Tibetan independence, he claimed that Tibet was once independent. One of the requirements of China is the recognition of the Dalai Lama's historical Chinese sovereignty over Tibet. Tibet as a geographical entity. The second argument concerns The borders of Tibet are the very definition of what Tibet consists of. The members The Tibetan community constantly speaks of "Tibet" in its broader ethnographic and historical context, including not only the entire current area mentioned in Chinese maps like "Xizang", or Tibet Autonomous Region (TAR), but also part The Chinese provinces of Qinghai, Sichuan, Gansu and Yunnan claimed by the Tibetans like the former Tibetan provinces of Kham and Amdo. Beijing defines Tibet as soon as TAP, and at the mention of "Tibet" does not include those Tibetan autonomous prefectures and countries in other provinces where high concentrations exist ethnic Tibetans. In the Dalai Lama, five points of the 1987 peace plan have proposed the transformation of Tibet into a "zone of peace", including these historical areas currently in other provinces of the PRC.(Dembaug. 2007)

On September 9, 1951, thousands of Chinese troops entered Lhasa. The violent occupation of Tibet was marked by the systematic destruction of monasteries, the suppression of religion, the denial of political freedom, mass arrests and imprisonments and killings of innocent men, women, and children

In 1956, at the invitation of the Mahabodhi Dalai Lama Indian Society went to India to attend an important Buddhist festival. According to him During the visit, the Dalai Lama met with Prime Minister Nehru and said that that he believed that he should stay in India and not return to Tibet. Both prime ministers Nehru and PRC Prime Minister Zhou Enlai, who also visited India, advised Dalai Lama to return to Tibet. He did so. Three years later, in March 1959, the PRC troops acting to forcibly suppress demonstrations in Tibet against Chinese rule, began shelling the Dalai Lama's summer palace in Lhasa, Norbulingka, ultimately destroy it. But before these military operations, China March 17 In 1959, at the age of 24, the Dalai Lama secretly left his palace and fled to India with a group of his followers. There he asked for political asylum for himself and his assistants. Indian Prime Minister Nehru granted the request for refuge and tens of thousands of Tibetans have since joined the Dalai Lama in exile.

Most of China's stay in Tibet was problematic, especially during the rapid cultural revolution (1966-1976) when most monasteries, palaces, and other aspects of Tibetan Buddhism and culture were either damaged or destroyed. The international campaign for Tibet claims that more than 1 million Tibetans died during the first 30 years of the reign of China. Beijing refutes this, arguing that material life and the health of Tibetans in Tibet has improved significantly under the control of the PRC.

March 10, 2008, began a series of demonstrations in Lhasa and other Tibetan regions of China will mark the 49th anniversary of the failed Tibetan rebellion against Chinese rule in 1959. Demonstrations began peacefully with small groups, which were then held by security forces. Both protests and The response of the Chinese authorities intensified in the following days, spreading from Tibet Autonomous

Region (TAR) on the part of Sichuan, Gansu, and Qinghai Tibetan provinces. By March 14, 2008, crowds of evil people were burning and looting establishments in the center of Lhasa. People's authorities In response, the Republic of China (PRC) closed Tibet and moved to security forces. Beijing defended its actions, as appropriate and necessary to restore civil order and prevent further violence. However, China's response led to new calls for a boycott of the opening ceremony of the Beijing Olympics on August 8, 2008, and for China to negotiate with the Dalai Lama.

China believes it has provided extensive economic assistance to Tibet and development using money from the treasury of the central government and Chinese officials often seem perplexed by the anger that many Tibetans nevertheless retain against them. Despite economic development, Tibetans claim that China interferes with Tibetan culture and religion. They cite as examples: Beijing interference in 1995 in choosing the Panchen Lama, Tibet's second-largest character; adoption in 2007 of a "reincarnation law" requiring Buddhism monks who want to reincarnate to receive prior approval from Beijing; and in China "patriotic education" campaign policies, as well as efforts to promote atheism among the Tibetan religious community. China defends campaigns as a tool to help monks become loyal, law-abiding citizens of China. The debate about the role of the Dalai Lama and the impact of PRC control on Tibet's language, culture, and religion prompted Congress to take regular action in support of Tibetan traditions - actions regularly condemned by Beijing.

Nationalism is expressed differently by Tibetans in exile and Tibetans in Tibet. Tibetans in South Asia and the West, supported by supporters of the West, are participating in the form of diaspora nationalism embodied by the Dalai Lama and the Free Tibet movement, which consists of a network of loosely connected organizations conducting street and online activities around the world. They directly oppose the Chinese government on issues of historical legality, cultural and religious restrictions, and human rights. Tibetans in Tibet, on the other hand, participate in ethnic and cultural nationalism, mainly in the form of passive resistance to the rhetoric and policies of the Chinese state and through the assertion of ethnic and cultural identity in the choice of everyday life. There are also local and inconsistent eruptions of public disagreements, such as mass protests on the eve of the Beijing Olympics and, more recently, self-immolation. Religion is highly politicized, like faith in the Dalai Lama. (Henrion-Dourcy 2016)

Tibetan culture and identity are inextricably linked with Tibetan Buddhism. Religious practice and Buddhist principles are part of the daily life of most Tibetans. Monks and nuns play a key role in their communities, providing guidance and education. They often very actively protect and promote the environment, language, and culture of Tibet. Almost all Tibetans are deeply devoted to the Dalai Lama, and his expulsion and conversion from the Chinese government are sources of grief and anger. The Tibetan commitment to the Dalai Lama and Tibetan Buddhism is seen as a danger to the occupying Chinese state, and as a result, all aspects of religious practice are carefully monitored and controlled. Simply possessing the image of the Dalai Lama can lead to arrest and torture. Monks and nuns are often the victims of security restrictions, and they make up a significant proportion of political prisoners in Tibet.

China has controlled Tibet with an iron fist since 1951 when its troops entered the region. The implementation of democratic reforms put an end to theocracy in Tibet by separating the government from religion. This is stated in a White Paper published by the State Council of China. "Freedom of religion of all ethnic groups is protected by the Constitution and laws, and all religions and religious movements equally respect and protect it, which means achieving true religious harmony,"- said the White Paper - Democratic Reform in Tibet after sixty years (www.eng.mod.gov.cn). The reincarnation of the living Buddha, which is a unique hereditary system of Tibetan Buddhism, is respected by the state and governments at various levels of the autonomous region. Since the 1980s, the central government has allocated more than $ 1.4 billion ($ 208.53 million) to rebuild Tibetan cultural relics and major monasteries and with full coverage of accident insurance - says the White Paper(Tak-ling. Woo 2019).

There are currently 1,787 buildings in Tibet for religious events, more than 46,000 monks and nuns, and 358 living nests. Tibet. China plans to replace the current Dalai Lama with its puppet when the time comes.

The UN High Commissioner for Human Rights has repeatedly said that the population of Tibet is a victim of repression. Usually, China denies the allegations, saying its government has put an end to the dictatorship and brought prosperity to this backward region, and that it fully respects the rights of the Tibetan people. China has historically considered Tibet to be part of its territory since the mid-13th century. However, Tibetans say the region was virtually independent for most of its history. Beijing considers the Dalai Lama a dangerous separatist who is trying to divide a quarter of China's territory. Nobel Peace Prize laureate Tibet's spiritual leader says he is only seeking greater rights for Tibetans, including those on religious freedom and autonomy. For Dalai Lama's supporters, China's strategy is to strengthen its control and restrict the personal freedom of the Tibetan people. "

Communist China has lifted most of its citizens out of abject poverty and promises to continue until all its people are fed properly, clothed well, gainfully working, and given an education. But it has failed to understand the depth of cultural and religious feelings of some of its minorities, most notably the Uyghurs and Tibetans. It has assumed that what has worked for the majority Han population would work for all other ethnic groups. This misunderstanding, it seems to me, requires education and not recrimination. On one hand, scholars, activists, and journalists in the West might more seriously examine the differences in how the relationship between religion and politics has developed and is understood in West Asian (Tak-ling. Woo 2019).

China's strategy is especially often focused on the anti-religious repression and violence in the Tibetan Autonomous and Xinjiang Uyghur Autonomous Regions on the western border of the country.

Chinese authorities called 7,000 Tibetan pilgrims home from their pilgrimage to the annual Kalachakra teachings held by the Dalai Lama in India. China intervened in religious teaching, which was attended by 100,000 people from all over the world in previous years, as it accuses the event of an extremist political instrument. The Chinese authorities began to pick up passports from Tibetans in October and November 2017 to prevent people from making a pilgrimage in January. The remaining monks and civilian Tibetans, who had the necessary documents and headed to the Dalai Lama or had already arrived in Bodh Gaya, India, were now forced to return with China, threatening them with losing their jobs, pensions, and monasteries, as well as threatening the relatives of the pilgrims (unpo.org/article/20589)

Tensions and total control of the region are hampering tourism and complicating travel conditions. (Juan,Yujie, Zhenbin. LeiLei. 2015)

Religious Tourism and Pilgrimage in Tibet

Modern global tourism studies, religious pilgrimage is often seen not only in the discourse of socio-economic activity but also in the context of cross-cultural communication (Cole, 2006; Timothy.2011). In the modern globalized world, pilgrimage is protected, as soon as a universal ancient religious tradition, as one of the most common and actively developing social and cultural phenomenon.

Pilgrimage is a complex and diverse spiritual and cultural tradition in the lives of many people. Pilgrimage, as a form of religious activity, dates back to ancient times and is associated with the emergence of developed forms of religion and the high stage of sociocultural relations in society. Pilgrimage is characteristic of all world monotheistic religions(Badone, Roseman.2004). It is motivated by different types of travelers because the pilgrim wants to have a "sense of belonging to religious or spiritual heritage," and the tourist travels for general purposes such as curiosity or entertainment (Collins-Kreiner, 2009). At the same time, they differ in terms of the tourism industry (Turner and Turner, 2011; Smith, 2012; Scaperlanda and Scaperlanda, 2004). For example, pilgrims to Tibet probably want to explore the palace of Potala, admiring the beauty of Imedrok Lake, except for particularly sacred places, such as the Jokhang Temple or the Trinity Monastery, the birthplace of Tibetan Buddhism. They can also do the same shopping as regular tourists on busy shopping streets. Both organizations need the same infrastructure and services. Pilgrims have become tourists, but as a specific segment of the pilgrimage market segment (Fleischer, Felsenstein. 2000). They say that "a tourist is half a pilgrim if a pilgrim is half a tourist" (Turner & Turner, 2011).

To put it more clearly, this trend, which is associated with profound changes in the mass consciousness and spiritual life of society (primarily the liberation from excessive political indoctrination and increased interest in the history of national culture), was revealed at the turn of the millennium, especially in those countries. The era of "state atheism" and with special difficulty try to overcome the consequences of the influence of atheistic ideologies. Moreover, the "cultural reconstruction" process, which can not be reduced ideologization up, but that means the mass consciousness manipulation of the most odious forms of rejection, culturally highly debatable because the education system is not sufficiently developed in the humanitarian segment of the existence or lack of education (religious and soul -zneobrivis included) creates a "doctrinal eclecticism" effect. An example of this is the so-called "Shrine tourism", which involves the open commercialization of the activities of pilgrims, which deprives it of sacred significance and insults the feelings of believers. For this reason, the study of pilgrimage as a sociocultural phenomenon has been almost completely forgotten in communist China, and now seems to be relevant both scientifically (pilgrimage was a taboo subject, religious education in general), and in practical terms, in particular, cultural policy guidelines and priorities.

In the second half of the nineteenth century - early twentieth century, the flow of pilgrims to the eastern shrines increased dramatically, attracting the attention of representatives of religious philosophy, began to study the history of pilgrimage in foreign scientific centers. China was part of historical and geographical research because, in the atheist communist state, it was impossible to study the ancient tradition of pilgrimage openly. Pilgrimage is also less studied as a tourist destination - a growing segment of religious tourism.

An important socio-economic feature of our era has become a social network. The results cluster analysis of social media shows that when traveling to Tibet, foreigners concentrate on nine categories of hot categories: Tibetan Buddhism and Tibetan culture; Snowy plateau; Tourist services and process; Climate care; And access levels in Tibet. This shows that foreigners are interested not only in the traditional tourist resources of Tibet but also in the fact that they are interested in the process of travel and that they consider this process to be a complete experience. Analysis of high-frequency precinct names has shown that foreign tourists are focusing on imbalances in each region of Tibet: high in eastern and northeastern Tibet, and low in western and southern Tibet. (Juan .Yujie; Zhenbin. LeiLei. 2015)

This shows that East and North-East Tibet has more international and domestic interactions than West and South Tibet. Concerning foreign relations, foreign blogs show that Tibet has strong ties only with Beijing and its surrounding provinces or countries such as Qinghai, Sichuan, and Nepal, and this suggests that Tibet does not have relations with much of China: this may limit some of Tibet's internal affairs. .(Juan .Yujie; Zhenbin. LeiLei. 2015).

Development of tourism. Besides, Yunnan, Xinjiang, and Shanghai were at a high frequency at the bottom of the list of words: this indicates a diversification of the route to Tibet. In general, the identified hotspots are relatively consistent with the location of high-quality tourist resources. When studying the perception of the tourist image of Tibet, studying the hotspots of the blog can give another idea of the real thoughts and feelings of foreign tourists. Currently, cluster analyzes of social networks are mainly used in the fields of education and research. This method and conclusion can be useful for the development of tourism in Tibet and to stimulate further research.

Pilgrims in Tibet have always been an important part of a culture that is so intertwined with religion that it is impossible to say where it ends and the second begins. Although the idea and practice of ancient pilgrimage are not always popular in urban areas, far from the remote provincial districts of Tibet, far from Western influence and modernization of Tibet, pilgrimage is still considered the most important journey of life. For many people, pilgrimage can get them halfway to the Tibetan Plateau, seeking their well-being through enlightenment and rescuing them from the constant revival of reincarnation.

Buddhism is the oldest of the three world religions. The Buddhist world includes many countries in South, Southeast, and East Asia, as well as many regions of Russia. There are many Buddhist temples in Western Europe. Some scientists believe that there are more than 325 million followers of Buddhism in the world. This figure does not take into account those believers who are followers of Buddhism and other religions at the same time. According to other statistics, there are about 500 million Buddhists in the modern world. About 320 million live in Asia, about 1.5 million in the United States, 1.6 million in Europe, and about 38,000 in Africa.: 72 million in Japan, 52 million in Thailand, 37 million in Myanmar, 35 million in Vietnam, 34 million in China, 12 million in Sri Lanka, and 12 million in Korea. In Cambodia - 7 million, in India - 82 million, in Laos - 2.4 million, in Nepal - 1.3 million, in Malaysia - 3 million. (https://www.pewresearch.org/)

Today, a pilgrimage around the world is regaining its place, and religious tourism is one of the most developing areas of tourism in many countries. A fundamentally important aspect is the clear separation of the concepts of "religious tourism" and "religious pilgrimage": religious tourism is part of the socio-economic activity, and pilgrimage is related to the field of religious activity.

What are the goals of the pilgrim and the tourist during their journey? The main purpose of pilgrims is to worship religious shrines and shrines, ie to recite prayers and participate in sacred services in holy places, although pilgrims may have other less important purposes, such as visiting friends, relaxing, buying medicines or practicing religious practices and experiences. Representatives of religions respect tourists who visit holy places and consider religious tourism as an important means of the spiritual upbringing of modern people.

The main purpose of religious tourism is to get acquainted with the history of holy places, the life of saints, the architecture, and the art of the temples. The tourist is informed of all this during the tour, which is the most important element of the trip for him in terms of obtaining information, besides, personal impressions are important. It should be noted here that the excursion may also be part of the pilgrimage, but not the main and not the obligatory, but only the ancillary component.

The term "Pilgrimage tourism" is now often used, but it is not correct in terms of respect for either scientific or religious traditions, as the pilgrimage is a religious tradition. There is also a difference in the tourist infrastructure. For example, Muslims fly to Hajj by plane, and they usually have a special plane for that. And Buddhists prefer separate, individual travel and do not attach much importance to comfort.

No one denies that there is an educational and cultural component to elite religious tourism and its development is needed. It is important who speaks on the topics of religion and at what level. In particular, who prepares guides and excursion-drivers? This is a problem in the reality of Tibet. If there is a talented monk in the monastery, he will make an interesting excursion. But then he will be given another obedience, and he will not do it again. There are very few professional guides and guide-translators who can guide tourists along religious routes and, unfortunately, there are virtually no pieces of training on the subject. This is a problem not only for Tibet but for many other states.

Today there is no conflict of interest between pilgrimage and religious tourism. The field of religious travel is so vast that the interests of pilgrims and tourists have long been harmoniously developing. It is practically difficult to distinguish between tourists and pilgrims, as tourists often attend religious services or celebrations, and pilgrims also visit recreational and cultural-cognitive facilities. Pilgrimage service monasteries that work to receive guests are interested in getting as many pilgrims and tourists as possible to come with them.

Interest in religious heritage was very high in all eras. Tibet has always attracted great attention with its religion and culture, and this interest is growing every year around the world in terms of religious heritage: the Holy Land ‘s and the places of origin of the Eastern religions - Nepal, India, and China.

Usually, pilgrims make much fewer demands on the level of service, food, accommodation. They stay in monastic cells at night or in the surrounding area, in tents (for example, during Hajj), they can spend the night on the floor in temples, or even on the street. The food choices are usually small and quite modest. In Christian countries, it is fasting, when it is a fasting period, while in Buddhism it is vegetarian.

There are four main traditions of Buddhist holidays and ceremonies: Tibetan, Japanese-Chinese, Terevada, and Western Buddhist society. Buddhist holidays are different in different countries and regions. Buddhist holidays are mainly about remembering one’s duty and confirming one’s intentions. Vesak commemoration holiday (celebrated in late May-early June). This day, the birthday of the Enlightenment and the day of the Buddha’s death is celebrated, but in Japan, all three of these events are celebrated separately. The Asala holiday is celebrated in October. The examples of Tibetan culture include also wedding and burial ceremonies, cooking, and table manners. There are five ways to bury a person in Tibet: burial in a temple, cremation, water burial tradition, body burial, and heavenly burial. Among them, the heavenly burial is the most mysterious for touristsThe places of worship of Tibetan Buddhism are also found in the form of natural phenomena, such as lakes and mountains located on the Tibetan Plateau. Mount Kailash is considered to be the holiest mountain on the planet and is one of the most important shrines in Tibet. This unusual Black Mountain, located in West Tibet, in the prefecture of Far Ngari, is a place of pilgrimage for many pilgrims, and the 52-kilometer shrine route is one of the most famous in the world. Kailah is a sacred place of pilgrimage for about a billion pilgrims (!) From four world religions (Hinduism, Buddhism, Jainism, and Bonn). Pilgrimage to Kailash is the highest in the world and is held at an altitude of 4600 to 5830 m. The configuration of the valleys surrounding Kailash allows you to travel around Kailash in one day (about 50 kilometers). There is no other mountain on our planet that is sacred to this huge number of people who are members of different religions.

Acquisition of Religious Knowledge and Experience

Why are Western European and American people interested in Buddhism in their search for spirituality?,, I think this is completely natural. Christianity is what is familiar to them, so they are eager to find a new religion. I think Tibetan Buddhism is something completely new for them. In the past, people considered Tibet to be a mysterious country, so the religion of this country - Buddhism - is something new and mysterious. There is a lot to be learned in Tibetan Buddhism. That’s why some people here are more likely to find answers to their questions. In general, Buddhism pays more attention to the mind than to faith. It attracts Westerners, - This is how the Dalai Lama answered the question in one of the interviews. Many Buddhist temples in Tibet are the true center of Buddhist knowledge and spiritual practice.

The first Samey monastery, called the School of Magic, was built in the 8th century. The exact date of Samye's founding is not known and has long been the subject of discussion. Most scientists are inclined to believe that this event took place in the years 760-780. It was at this monastery that thousands of years ago, at a meeting, ended an argument over the development of Buddhism in Tibet. Probably no region of Tibet is as famous for its legends and mysteries as those around the former School of Magic. On one of the nearest mountains, there are still caves where the great Padmasambava (Indian master who founded the Tibetan Buddhism and the School of Tantric Buddhism in the 8th century) hid in the world. Ningimas School respects him as the second Buddha). According to legend, Samye miraculously came close to the settlement where the king was born and where the famous magician defeated the witches. By equipping their monastery, the monks fought fiercely against the demons, who destroyed what was being built every day during the day. Padmasambava not only completed the evil machinations but also created the action of alcoholic beverages. They resigned and completed the construction of the Samei in just a few days. Probably, this story was not invented because the priests of Bonn could act as devils, as you know, they did not benefit from the new faith. One way or another, during the reign of the dean of Trizong, the city was already standing in the sand. Only available to the elite, the wonderful and the imaginary man, he was exactly what the magician was supposed to be. The architectural ensemble Samee was arranged according to Buddhist cosmology. Undoubtedly, one of the reasons for the emergence of pilgrims in Tibet was the monasteries in which the fundamental principles of Tibetan medicine were established over the centuries. The best Tibetan doctors come from the walls of Shakpor - the monks who studied in ancient times called this monastery a temple and a school of medicine. Since its inception, Shakpori has had a very strong reputation. Not everyone could enter it, stay in it, and start training even more. Only the children of aristocrats received training in Shakor. However, the rules here were much stricter than in other monasteries in Tibet. Extremely high demands were explained by the fact that the Temple of Medicine trained not only doctors but also future leaders of the country. The basic principles of Tibetan medicine include Buddhist philosophy, understanding that all phenomena, including diseases, are related to the mind (mind). Thus, a person's spiritual, magical, and physical health is determined by the harmony of his or her spiritual, magical, and physical realms. All of these parties are closely related to each other, and according to Tibetan tradition, man is seen as an integral system. Tibetan doctors do not fight diseases, but treat people, avoid the cause of the disease, try to prevent their origin in the past. The dry air of Tibet has kept medicinal plants for many y cxgears. From time immemorial, practicing lamas-physicians have been guided by the four-volume manuscript of Judas, which has been the main source of local medical science for more than 1,000 years. Legends link the book's oral tradition to the Buddha's revelations. The canonical treatise on the dwarf summarizes the knowledge accumulated throughout the existence of Tibetan medical science. Gradually, the popularity of Tibetan medicine spread throughout the world, which served as a stimulus for the development of pilgrimage tourism in Tibet, among other factors described above. Here we should mention Buddhist practice, which also attracts pilgrims to Tibet.

Tibet Between Tradition and Modernization

For more than two decades, the situation in Tibet has improved. China has restored and developed socio-economic infrastructure, communications, transport, and has all contributed to the development of tourism. According to China's Tourism Development Department, Tibet's budget for tourism in the spring of 2019 increased by 41.1% to $ 390 million. The number of tourists arriving in the region amounted to two and a half million, which is 84.2% more than the same period of the previous year. To attract more tourists during the winter season, the government introduced several activities, canceled entrance tickets to some protected areas, and discounted airline tickets. Tibet is a special region of China that attracts tourists not only to monasteries and theological schools but also to the Peak of Mount Everest.

Since the opening of the region to tourists, tourism has become one of the most important transformational effects of the ongoing socio-economic progress in Tibet. With improved access, especially when the Qinghai-Tibet Railway was launched in 2006 and the project of constantly expanding the construction of infrastructure and service facilities, Tibet gradually established itself as one of the most interesting, attractive places in the global tourism industry. Chinese (Cui, Xu,Yang, 2014). Tibetan tourism has begun to boom. At the same time, studies on Tibetan tourism have accumulated (Chen and Huang, 2010), (Wu and Pearce. 2012). Dominant studies have been conducted by various scientists on the following topics: tourism development and strategic management, tourism resources and development, alternative products and experience, tourism education, tourism public relations, and tourism marketing and management. The research topics focused on the regional development of Tibetan tourism at the macro level. In terms of global coverage of Tibetan tourism topics, widespread concern has been fueled by Western fantasies and imaginations about Tibet in the cultural dimension. Little attention has been paid to the study of tourist behavior or experience on a personal level (Chen & Huang, 2010). In general, the economic policies of the Chinese government in Tibet have helped to raise living standards, but some researchers believe that this is costly to Tibet's own historical, religious, and cultural traditions. To break with the traditional Tibetan way of life and to feel the tension between the Tibetans (Dumbaugh, 2008), to neutralize the reaction, Beijing spent significant sums of money to rebuild the Buddhist temples, most of which were destroyed by Chinese troops. During the Cultural Revolution. At a major conference in Tibet in 1994, China approved a plan to increase economic activity in Tibet by 10 percent a year. To this day, it continues to provide significant subsidies to help less developed Tibetans. Numerous infrastructure projects have been implemented to develop tourism. However, economic development will benefit not only the Tibetans but also the Han Chinese, which undermines Tibet's cultural identity, in part by encouraging large-scale migration of Han Chinese to the region - who work as technical staff on infrastructure projects and seek new economic opportunities. (www.China.org.cn)

Tibetan monasteries have a very positive attitude towards tourists; They believe that they have to bring some spirituality to the materialistic world. The main obstacle to this type of relationship is the language problem. All monasteries have monks who speak Chinese well, but knowledge of another foreign language is rare. In some monasteries, some monks study English, but such initiatives are not common. The assets of the monasteries are their control over many cultural events, some of which are related to Buddhism, in particular religious dance festivals.

However, it should also be noted that the number of tourists from the West who travel to Tibet only because of Tibetan Buddhism or to greet the shrines is limited. As for the Han and Huei Chinese, on the contrary, their number is increasing in the direction of religious tourism. In the West, this direction of tourism is somewhat marginalized, but other assets will attract a large army of tourists. It is primarily the Himalayan Mountains and the extraordinary landscape, which in addition to extreme tourism, ecotourism, or wild tourism and is a major driving force for adventure lovers, will be a topical and real treasure for intellectuals and tourists interested in sacred geography and mystical knowledge. But there is another circumstance to consider here - the geographical landscape, due to its complexity, requires a moderately good physical condition for walking.

The Han Chinese believe that only Tibetan Buddhism allows them to find the spiritual roots of their civilization. Many Chinese travelers travel to Tibet not only to pray in Tibetan monasteries but also to see the sights and to explore. It is travel to Buddhist temples and religious tourism that unites Han Chinese to gain proper knowledge of the great traditions of China.

Tourism has become an important sector of the Tibetan economy. However, so far, it has only the following directions:

- Cultural and religious tourism related to the Tibetan monastic system
- High level walking in the mountains
- Excursion in nature and visit to historical and cultural monuments (essentially Chinese tourists)

The Tibetan landscape and culture can be seen as separate areas of tourism because of its enormous potential, but it has not yet been affected by economic activity: there are mountains in many regions of Tibet where Tibetans love to go out and relax like Europeans in the Alps, Scandinavia and elsewhere. The US and Australia also enjoy their mountains and try to please tourists. Bringing these types of tourists to the Tibetan districts will not only contribute to the economic development of the region but can also improve intercultural dialogue, mutual understanding. (Kolbi. 2012) The experience of mountain resorts in Western Europe would help the Tibetan areas in the development of tourism. The out-of-town socio-economic context of Tibet has changed dramatically in recent years: rising rural economy and food prices have led to capital accumulation and the emergence of a class of local businessmen operating in trade and tourism. Who have little experience in the field of tourism. As a result, this sector is extremely fragmented, with no connection to the national and international distribution system. (Kolbi. 2012) The language barrier is a big problem in dealing with both local Chinese and foreign tourists. The weak point of Tibetan tourism is the issue of nutrition. As far as tourists are concerned, they don't like Tibetan food. Many Tibetans themselves do not like Tibetan cuisine and prefer Chinese Sichuan cuisine. Pilgrims in Tibet have always been an important part of a culture that is so intertwined with religion that it is impossible to say where it ends and the second begins. Although the idea and practice of ancient pilgrimage are not always popular in urban areas, far from the remote provincial districts of Tibet, far from Western influence and modernization of Tibet, pilgrimage is still considered the most important journey of life. For many, pilgrimage can take them halfway to the Tibetan Plateau, seeking their well-being through enlightenment and rescuing them from the constant revival of reincarnation.

CONCLUSION

The English word "pilgrims" was used in the translation of the Tibetan Nekri, or rather, it means to go or visit the appropriate place, respectively. When talking about Tibet, we mean only the modern autonomous region of Tibet, not the ethnographic Tibet, which includes the three provinces of Ustang, Kkham, and Amdo, as well as areas outside China's modern political borders, Ladakh, Zangskar, Bhutan, Dolpo, and

Mustang. In these regions, people have common writing and language, largely similar to social institutions and values, and a common sense of historical connection. Although they are less well known in the West than the Tibetan doctrinal and meditative traditions, pilgrimage has always been a major element of people's religious life in the cultural regions of Tibet. Although teaching and meditation were the responsibility of the elite monastic scientific minority, the pilgrimage was much more widespread and practically practiced for both charitable as well as theological, secular, and serological purposes. While religious elites or even ordinary Tibetans may characterize pilgrims from a sophisticated Buddhist doctrinal point of view, what they do is often rooted in local and purely authoritative local Tibetan ideas, as in Buddhism (Donnelly. 2010)

The concept of the holy place preceded the introduction of Buddhism in Tibet, while the holy places were important to the followers of both religions Bonnie and Buddhism. Pilgrimage to sacred mountains, lakes, caves, and "hidden lands" has been and remains a practice in Bonnian religion. Tibetan Buddhists visited and revered these powerful places, either copying pre-Buddhist views in a Buddhist sense or allowing autochthonous authorities to honor them along with Buddhist practice. After the conquest of Tibet by China, pilgrimage became difficult for many Tibetans. In Tibet, this continued until the liberalization of China in the mid-1980s. This change allowed the Tibetans to continue the practice of pilgrimage and opened Tibet to Western science interested in this practice. After the mid-1990s, it became possible for Tibetan and other pilgrims to embark on a religious journey and receive Buddhist teachings in Tibet. However, the problem of Tibet's political status, forced assimilation, forced internal migration to change the demographic picture at the expense of the Chinese, as well as repressions from China, total control and so-called The fight against nationalism and separatism - hinders and negatively affects the development of international tourism in Tibet, including religious tourism.

Another negative factor for the development of religious tourism is the exile of the Dalai Lama, the spiritual leader of Tibet, and the persecution of his fellow believers. The peaceful return of IDPs to Tibet has dramatically increased the number of visitors interested in Buddhist beliefs and teachings.

Better coordination between system elements is needed to improve religious tourism and foster an ecosystem conducive to tourism. A multifaceted approach that focuses on deepening the tourism segment, improving communication, and promoting and marketing spiritual culture that will help achieve an effective strategy for religious tourism based on the above topics and directions.

REFERENCES

Badone, E., & Roseman, S. R. (2004). *Approaches to the Anthropology of Pilgrimage and Tourism*. University of Illinois Press.

Bozkurt, A. (2014). Branding Adana as a Touristic Destination in the Eastern. In International Interdisciplinary Business-Economics Advancement Conference. University of South Florida M3 Center Publishing.

Chen, G.-h., & Huang, Y.-s. (2008). A study on the degree of international impact of China's tourism research: Comparative analysis and upgrading approach. *Luyou Xuekan*, *23*(5), 91–96.

CihinaS. C. (2019). Retrieved from http://eng.mod.gov.cn/Database/WhitePapers/: http://eng.mod.gov.cn/

Cole, S. (2005). Cultural tourism, community participation and empowerment. In M. Robinson & S. E. Phipps (Eds.), *Cultural Tourism in a Changing World* (Vol. 2). Channel View Publications.

Collins Kreiner, N. (2009). Researching Pilgrimage: Continuity and Transformations. *Annals of Tourism Research, 37*(2), 440–456. doi:10.1016/j.annals.2009.10.016

Cui, Q., Xu, H., & Yang, Y. (2014). Secular Pilgrimage; Tourist Experience in Tibet. *Luyou Xuekan, 29*(2), 110–117.

Donnelly, M. (2009). *Building Customer Loyalty: A customer experience based approach in a tourism context.* Retrieved from http://repository.wit.ie/id/eprint/1396

Donnet, P. (1994). *Antoine.* Survival in Question. Oxford University Press.

Dumbaugh, K. (2008). *China-U.S. Relations: Current Issues and. Foreign Affairs, Defense, and Trade Division, Asian Affairs.* Congressional Research Service.

Fleischer, A., & Felsenstein, D. (2000). *Support for rural tourism: Does it make a difference?* Academic Press.

Gillberg, C. (2006). *Warriors of Buddhism: Buddhism and violence as seen from a Vajrayana Tibetan Buddhist perspective.* Retrieved from https://www.researchgate.net/: https://www.researchgate.net/publication/326888242_Warriors_of_Buddhism_Buddhism_and_violence_as_seen_from_a_Vajrayana_Tibetan_Buddhist_perspective

Henrion-Dourcy, I. (2016, January). Tibet. In *The Wiley-Blackwell Encyclopedia of Race, Ethnicity and Nationalism.* Retrieved from https://www.researchgate.net

Hoffmann, H. (2018). *Religions of Tibet* (Vol. 4). Routledge. doi:10.4324/9780429441486

Kolbi, O. (2012). *Developing a high value added tourism in. University of Lausanne, Switzerland.* University of Lausanne.

Safronova-Urbanaeva, I. (2013). *How and why Tibet became the central country of Buddhism.* Humanitarian Vector.

Śākya, M. (2000). Sacred Art of Nepal: Nepalese Paubha Paintings: Past and Present (N. H. Saṅgha, Ed.). Handicraft Association of Nepal.

Scaperlanda, M., & Scaperlanda, M. R. (2004). *The Journey: A Guide for the Modern Pilgrim.* Layiola Press.

Smith, V. L. (2012). *Hosts and guests: The anthropology of tourism.* University of Pennsylvania Press.

Stein, R. A. (2010). *Rolf Stein's Tibetica Antiqua* (A.P. McKeown, Ed.). Boston: Brill NV Leiden.

Tibet: Problems, Prospects, and U.S. Policy. Foreign Affairs, Defense, and Trade Division, Asian Affairs. (2007). Washington, DC: Congressional Research Service.

Li, Chu, Zhao, & Wang. (2015). Tibetian tourism hotspots:co-word cluster analysis of English blogs. *Luyou Xuekan, 30,* 35–43.

Timothy, D. J. (2011). *Cultural Heritage and Tourism: An Introduction* (Vol. 4). Channel View Publications.

Tsomu, Y., & Roche, G. (2018). *Tibet's Invisible Languages and China's Language Endangerment*. Academic Press.

Tucci, G. (2001). *The Theory and Practice of the Mandala*. Dover Publications Inc.

Turner, V., & Turner, E. (2011). *Image and Pilgrimage in Christian Culture*. Columbia University Press.

UNWTO. (2019). Retrieved from UNWTO.org: https://www.e-unwto.org/doi/pdf/10.18111/9789284421152

Woo, T.-T. (2019). Religion and Politics in the People's Republic of China: An Appraisal of Continuing Mistrust and Misunderstanding. *Religions*, 31.

Wu, M.-Y., & Pearce, P. L. (2014). Chinese recreational vehicle users in Australia: A netnographic study of tourist motivation. *Tourism Management*, *43*, 22–35. doi:10.1016/j.tourman.2014.01.010

Zurher, E. (2013). Buddhism in China. In J. A. Silk (Ed.). *Sinica Leidensia*. Academic Press.

KEY TERMS AND DEFINITIONS

Bon: The old religion of Tibet.

Buddhism: It is a world religion that has no concept of a Creator God. It is called "inner religion" or "religion of the inner world."

Buddhist Knowledge: Study of Buddhist canonical texts and religious practice.

Buddhist Practice: Carrying out religious rituals, practicing meditation, as well as sharing experiences of applying other knowledge in practice.

Phenomenon: A remarkable person or thing.

Pilgrimage: To travel to pray.

Sacred: Blessed, the object of worship.

Tibet: Western Region of China, Autonomous Region in Himalaya.

Chapter 12
The Opportunities for the Development of Religious Tourism in Georgia

Manana Darchashvili
Institute of Political Science, Georgian Technical University, Georgia

ABSTRACT

In recent years, tourism in Georgia has been a priority area for the economy, and accordingly, special attention is paid to its elaboration, since it has a great impact on the stable development of society, improving the economic situation and meeting the spiritual needs. The research is devoted to a very relevant and at the same time specific type of tourism, religious tourism, pilgrimage, traveling due to religious beliefs. Georgia has very good potentials for the development of religious tourism, as it has numerous Sanctuaries, such as the robe of Christ, the cross of St. Nino, more than 12,000 churches and monasteries, and the cathedral, about 150 museums. In addition, to the Orthodox, the representatives of other religions can also visit their religious places, for instance Muslims can pray in the mosques, Jews in the synagogue, and Catholics in Old Catholic temples.

INTRODUCTION

The modern world is well aware that tourism is a field which brings together the peoples, cultures, and religions and contributes to the sustainable, harmonious development of humanity. According to the available data, tourism is developing widely in the world and is one of the leading sectors of the economy. Tourism development strategy in Georgia is based on the world-recognized principles of sustainable tourism development; on the potential of tourism which increases the country's income, and the importance of promoting the development of other sectors of the economy.

Nowadays in the globalized world, one of the main indicators of the quality of democracy is the issue of religious freedom. Accordingly, in the Georgian reality, the issue of religious freedom is a very relevant topic and is always on the political agenda. According to the thought of the field experts, Georgia has a great potential of development of the religious tourism due to the presence of numerous

DOI: 10.4018/978-1-7998-5792-1.ch012

religious shrines, and the authorities have to make more effort to use this opportunity and elaborate its further development.

Since, Georgia, as a multiethnic and multi-confessional state with centuries-old tolerant traditions and culture, religious diversity and the high religious self-identification of society are important factors of political and social life. Thus, according to the interests of international cooperation, tourism is acknowledged as one of the country's priority spheres. However, the country has to develop it in a more interesting segment such as religious tourism, which is less known in Georgian society. According to the country's potential, there is a chance of having more interesting results in the nearest future than we can observe today, and gaining the desired effect depends on the proper utilization of the existing possibilities.

The aim of the paper: is to study religious tourism in Georgia, present the problems of its elaboration, and to finds ways to solve them. As well as to elaborate a plan for its future development based on the country's existing situation– the state policy in the field of religious freedom and tourism development, following the certain experience and the recommendations in the field of religious tourism.

Paper is based on a review of the literature, certain empirical materials, various researched-studied documents, scientific papers, analysis of data processing, and observation.

The study shows the country's great potential for the development of religious tourism. Identifies the hindering factors of the elaboration of religious freedom and tourism in the country - the state politics on the development of religious policy, and seek the ways for future cultivation based on the conclusions and recommendations to overcome the existing imperfections.

The hypothesis of paper deals with the possibilities of religious tourism development in Georgia with the active involvement of the government and the private sector, and international organizations.

The conclusion/recommendations developed by the analysis made based on empirical material and documents from the relevant agencies, studied in the paper represents significant educational-practical contribution for the interested persons.

Literature Review

Even though the issue of religious tourism in Georgia emerged recently, there been written many outstanding and valuable works in the field of tourism, which describes the development of various types of tourism in the country. Among the very valuable works, the chapter uses and refers to such interesting books which will add precious value to it. Particularly the work of Avtandil Okrotsvaridze, Magda Vadachkoria, and Lali Okrotsvaridze, "Tourism and Host Management", which states the importance and priority of tourism development in Georgia dased on the data and facts of local and foreign countries. The book expresses professional aspects required for tour operators, guides and other employees in the field of tourism; it discusses the main directions of tourism; management system and structure; tourism industry, hotel classification; protected area management; legal basis of tourism and state regulations, regional management, advertising, and vocational training, preparation of advertising and professional staff, the modern state of international tourism, tourist hospitality, and local traditions. The work generally discusses the main types, forms, and directions of tourism such as social, ecological, agricultural, cultural, religious, health, cognitive, professional-business, scientific, sports, shopping tours, adventure, food, national traditions, etc. As well as it focuses on the types of ecotourism and agritourism and presents the most important tourist regions of the world and Georgia; are described the potential of Georgia's tourism resources and outlined the prospects for tourism development in the country (Okrotsvaridze, Vadachkoria, Okrotsvaridze, 2011).

Nodar Grdzelishvili in his work "Religious Tourism" discusses religious tourism and its evolution, historical aspects of pilgrimage, the geography of religious tourism, socio-economic significance, the legal aspect of religious tourism, religious tourism and culture, religious tourism in different religions (Christianity, Islam, Buddhism, Judaism), Orthodox fundamentals of religious tourism in Georgia, Georgian Apostolic Orthodox Church, Patriarchs and Dioceses, Multiethnic and Multicultural Georgia, Monuments of Georgian Culture and Religion Abroad, Influence of Religious Tourism on Regional Development - unique religious tourism opportunities in Georgia (Grdzelishvili, 2018).

Eka Devidze in her book - "Cultural Heritage and Imereti Tourism Development Problems" discusses the development of modern tourism in a very significant direction, the strategy of tourism development in Georgia, following world-recognized principles and the use of the country's potential for revenue growth. Describes - Imereti region and the ways of tourism development there, as well as the relationship of cultural heritage and tourism. Indicates existing problems and gives recommendations for solving them (Devidze, 2012).

Sergo Vardosanidze, an outstanding historian of the Georgian Churches, in his book **"Georgian Orthodox Apostolic Church in the XIX-XX centuries"** describes the situation of the Georgian Church and the work of His Holiness and Beatitude Ilia II, Catholicos-Patriarch of All Georgia. Naturally, religious tolerance plays a special role in the development of religious tourism, and in this respect, a certain part of the book highlights the years of the Soviet regime (particularly in 1938-1939), and the friendship of Patriarch with the persecuted Muslims and Christians. He recalls the details of his supports in providing the shelters to the exiled Muslims and Christians' families and remembers their praying in his spare room according to their own rules, and then gathering together for dinner. After these facts, it is not surprising that the leader of the Georgian Orthodox Church, throughout his tenure, has a tolerant attitude towards other religions (Vardosanidze, 2017).

It is noteworthy to mentions the reference book "Religions in Georgia" published by the Center for Tolerance at the Public Defender's Office, which presents information about the established religious diversity in the country. The book consists of short religious stories and presents objective information about religions, in the country's past and present life. The paper aims to promote interreligious dialogue, to create a sense of mutual respect, and to influence the impact of peaceful cooperation on the present stage by recalling the history of religious tolerance in Georgia (Kiknadze, 2008). According to the Administration of the President of Georgia, the mentioned book was given as a gift to the presented Donald Trump by the President Salome Zurabishvili, during her participation in the 74th session of the UN General Assembly, hosted by the US President and First Lady.

To understand the materials presented in the chapter about the Catholic Church several papers have been reviewed based on the study-discussion of sources in this regard; however, the book refers to a few of them it is important to focus on the Ekaterine Pavlenishvili's book "Catholicism in Georgia (in the XIX century)" where the author refers to the issue of the activities of Catholics in Georgia in the nineteenth century and the existing difficulties and challenges. It discusses the outstanding work of distinguished Georgian Catholics within the country and abroad; the relations between Orthodox and Catholic Georgians (Pavlenishvili, 2019).

Diplomatic-political or cultural-educational relations of Georgia with the Vatican have been studied by professor M. Papashvili in his book "Georgian-Roman Relations". Here, the author mentions Georgia's long-standing relationship with the Vatican and sees the roots of their friendship long before the arrival of missionaries in Georgia - in the VI century (Papashvili, 1995). Another valuable resource is considered the book of the same author's "Georgia's Orientation towards Western Europe and Sulkhan-

Saba Orbeliani" (Papashvili, 1998). The book shows the diplomatic efforts of Sulkhan-Saba (Catholic) and Georgia's interests in France's Eastern policy. Papashvili M. believes that Georgian diplomat and writer, Sulkhan-Saba Orbeliani was the first Georgian far-sighted figure, who tried to connect the idea of Georgia's freedom with the religious and political interests of Rome and France. There has been a growing interest in Georgia over the past 40 years concerning such issues as the connection between architecture and religious, and cultural or ethnic identities. In this regards another valuable work, "Catholic Churches in Georgia: History and Architecture" by researcher N. Natsvlishvili, studies the architecture of the Catholic churches of the XVIII-XIX centuries and the early XX century in the historical context of Georgia, and the artistic-architectural nature of these churches. It shows the reflections of different cultural traditions (Georgian, Armenian, Western European) on the architectural forms of Catholic churches and the revealed general artistic-stylistic tendencies, also it highlights the patronage and donations to the process of the buildings of churches and as well as the determined there tastes and traditions. Here are examined almost all Catholic churches in Georgia and the history of their construction, as well as the most important - Georgian Catholic center outside the country - the Ferikoy Monastery in Istanbul (Natsvlishvili, 2019), (this paper clearly shows the perspective for the Catholic world to travel to Georgia).

Due to its geopolitical location, Georgia is an important place for those who try to understand the interfaith context, since the country has many shrine places for Muslims. Also, several studies have examined the state of the Islamic community in Georgia and among them is one of the significant resources published by the Center for Cultural Relations operating in the Caucasian House in Georgia, "Islam in Georgia: Politics and Integration". The paper notes that despite the historical experience of the coexistence of different religious denominations, there remains a problem of the involvement of religious minorities in the country's social and political life. Particularly, the works show not only the religious but also the civil aspects of Muslim integration and the country's modern policy. It aims to raise the awareness of the Muslim community mostly of Azerbaijani origin and to develop appropriate recommendations for involvement in the process actors (Sikharulidze, Urushadze, Khmaladze, Chichinadze, 2016).

Tina Ivelashvili provides information about Judaism in Georgia and her book - "Religious Minorities in Samtskhe-Javakheti", is devoted to the arrival and settlement of Jews in Georgia, their ways of life, holidays, shrines in Georgia, particularly in Samtskhe-Javakheti. In the same paper, the author refers to the rules and ways of establishing Islam in Georgia and its correlations with other religions. Numerous primary sources discussed in the book are the indicators of religious tolerance in Georgia (Ivelashvili, 2018).

The above-mentioned literature is just a shortlist from the extensive literature of religious diversity in Georgia, which describes the history of long religious tolerance and coexistence, and the potential of a country's religious tourism.

The Main Challenge of Religious Tourism in Georgia

Peaceful coexistence is an important guarantee of religious tourism.

One of the main goals of the Georgian government and society is to strengthen and develop democracy in a multi-ethnic, multi-religious country, which can be achieved only by ensuring their equality.

The Constitution of Georgia defines the main principles of the attitude towards religion, as well as the freedom of conscience, religion, and belief. The state shall declare complete freedom of belief and religion, as well as shall recognize the special role of the Apostle Autocephalous Orthodox Church of

Georgia in the history of Georgia and its independence from the state. The relations between the state of Georgia and the Apostle Autocephalous Orthodox Church of Georgia shall be determined by the Constitutional Agreement. The Constitutional Agreement shall correspond completely to universally recognized principles and norms of international law, in particular, in the field of human rights and fundamental freedoms. (change is added by the Constitutional Law of Georgia of 30 March 2001) The fundamental principle of the Constitution highlights that all people are free by birth and equal before the law (The Constitution of Georgia, 1995).

There is a Georgian Constitutional Agreement with the Georgian Orthodox Church signed between the State of Georgia and the Apostolic Autocephalous Orthodox Church of Georgia (2002), which also recognizes the special role of the Apostolic Autocephaly Orthodox Church of Georgia in the history of Georgia and its interdependence from the State.

At the same time, along with the main ethnic group – Georgians, the representatives of other ethnic-religious groups played an important role in the development of the country, at the different stages of its history. The percentage of the major religious groups which form religious diversity in Georgia is as follows: Orthodox - (83.9%), Muslims - (9.9%), member of the Armenian Apostolic Church - (3.9), Catholics - (0.8%), Jewish - (0.1%), others - (0.8%), none - (0.6%) (The assessment of the Strategy for the Development of Religious Policy of the State of Georgia, 2015). Protecting the rights of religious and ethnic minorities and promoting diverse society and pluralism significantly contributes to the quality and development of democracy in the country. The presented major religious diversity, with their culture and experience of peaceful coexistence, creates an interesting driving source for the development of religious tourism.

In the modern world, religious tourism - Pilgrimage, excursion tours with a religious purpose, special tours, which unite excursionists and pilgrims is a very popular destination and is considered as one of the main components of the tourism industry. Churches, mosques, religious centers, and cult museums are among the tourist attractions of religious tourism.

World tourist destinations include the Vatican, Israel, several European cities, Mecca and Medina (for Muslims), and worshipers of other religions in India and other Asian countries.

Many places in Georgia are famous for their religious sanctity. In this respect, the capital itself is a very important destination. Since for centuries, there have been Catholic, Armenian community shrines, as well as mosques and synagogues, along with Sioni, Anchiskhati, and Metekhi churches in ancient Tbilisi. Each of the shrines has an extensive and interesting history due to the special peculiarities of their coexistence, which draw great touristic interest.

The Most Important Religious Shrines and Holy Places Preserved in Georgia

Georgia is a Christian, Orthodox country, where the distinctively tolerant local population has always had a good relationship with people of different nationalities, cultures, and religions. Peaceful relations have been the solid ground for the establishment of different faiths in Georgia since the ancient times, and therefore, in addition to Orthodox religious shrines in Georgia, religious-chapels and monuments of the other religions are abundant in the country, which is a good opportunity for the development of religious tourism.

Christian, Orthodox shrines, and holy places: Georgia is an ancient Christian-Orthodox country, therefore the sanctuaries and the holy places are abundant in the country, and the chapter deals a great deal to describe them.

Georgia, as one of the most important representatives of the Orthodox world, is distinguished by the liturgy of ecclesiastical relics, however, some of them are less known. The chapter will describe some of them.

The Lord's Robe has been kept in the Svetitskhoveli Cathedral in the ancient, former capital of Georgia, Mtskheta since the beginning of the first century and is considered as one of the main shrines of the Orthodox world. However, there are many versions of the location of the Lord's robe, however, the fact that it is located in the Svetitskhoveli Cathedral in Mtskheta is confirmed by archival materials and historical sources.

Christ's Crucifixion - a nail with which Christ was crucified, and according to various sources, including Leonti Mroveli's "Life of Kartli", (which is the main written source of the history of Georgia, a collection of historical works and chronicles, and is considered one of Georgia's main written source of Georgian IV century history) the emperor of the Byzantine Constantine, sent the nail to a newly Christianized King of Kartli, Mirian. In Georgia, are kept Christ's two nails, according to the certain sources one was taken to Russia and another is still preserved in Georgia, in the Sioni Cathedral (Alasania, Chkhartishvili, 2014).

All-Holy Mother of God - numerous Georgian historical sources references her as a special protector of Georgia. According to the divine traditions, Mother of God preached Christianity in Georgia but later she gave her "Wooden Icon" to Andrew the First-Called and sent him to Georgia. Therefore, there is a great love for the Mother of God among believers in Georgia. As for the arrival of the All-Holy Virgin's robe in Georgia, there are different opinions. Following the first version, it was brought from Jerusalem in the early 15th century and kept in a monastery in Khobi (according to historical sources, the robe of the Virgin Mary is associated with many sacraments). Later in 1923 (after the establishment of the Soviet government), it was moved to the repository of the Dadiani Palace Historical-Architectural Museum in Zugdidi, it is worth to say that the belt of the Virgin Mary, which was originally kept in the Bedia Monastery in Abkhazia, is also preserved there and the icon of the Mother of God of Vlacherna, which was recognized as a miracle, all of and was considered to be the first relics of Samegrelo. It is worthy to say that every year on July 15, a service is held at the Vlacherna Cathedral of the Mother of God in Zugdidi, a robe is taken from the Dadiani Palace Historical-Architectural Museum. On the occasion of the holiday, the representatives of Orthodox Churches from all over the world arrive there the robe.

Svimon Kananeli (he preached Christianity with Andrew the First-Called and Matata on the shores of the Black Sea) at the site of his martyrdom, near Mount Athos, in Abkhazia, there is a church on the mountains named after the Saint. **Matata's** grave is also in the Gonio Museum-Reserve in Georgia. It should be highlighted that he was a disciple of Christ and made a great contribution to the spread of Christianity.

Sait Nino - in 326, Christianity was declared the state religion of Georgia by the preachings of St. Nino. According to the legend, before St. Nino left from Jerusalem for Georgia the Virgin Mary gave her the vine cross to protect her and Georgian Orthodox Church (although there are other versions that Nino herself made this cross), many miracles are associated with the name of St. Nino in Georgia. She is buried in Sighnaghi (Bodbe), and the vine cross is kept in the Sioni Cathedral in Tbilisi (Bubulashvili, 2011). The Georgian Church celebrates the day of St. Nino twice a year: on January 27 (the day devoted to her death) and May 1 (the day of her arrival in Georgia).

Saint George is especially revered in Georgia, commemorating him twice a year on May 6 and November 23. However, in the Orthodox world, November St. George's Day (the day of his torture) is celebrated only in Georgia. His worship began in the early 4th century. According to historical sources,

St. Nino, the preacher of Georgians, was a cousin of St. George, and exactly she was the one who introduced Georgians the martyrdom of St. George (particle of St. George's knee and three small sacred particles are kept in the Holy Trinity Cathedral). It is worthy to mention that there are legends in Georgia about the visible penetration of St. George's battles.

It is necessary to describe the magnificent masterpieces of Georgian architecture - Ishkhan (VII c.), Khandzta (IX c.), Oshki and Otkhta (X c.), etc., presently preserved at the territory of Turkey. Initially, this territory, Tao-Klarjeti, belonged to Georgia, however, since the last, territorial redistribution, in 1921 (after the annexation of Georgia by Russia) is the territory belongs to Turkey. which are very significant in cultural and religious terms, For example, in Ishkhan, in 973, an engraved silver-gold cross was created, which is now housed in the Georgian Museum of Art. The figure of the crucified Christ is executed sculpturally, conveyed with great artistic power and dynamics. In addition to the religious shrines mentioned in the chapter, in Georgia, there are other types of shrines, for instance, in the Zugdidi Museum, in addition to the relics, parts of various saints (St. Marine's arm, St. Kvirike's hand cluster, part of St. George's thigh bone) are kept in the silver-plated Luskuma. There are various sacred parts in other museums as well. Unique cultural monuments of ancient architecture, ancient manuscripts, icons, etc., which are also preserved in state art museums and temples, and are very precious and vital from a scientific point of view.

When emphasizing Georgian Christian-Orthodox sanctities, it is important to note that this is an important challenge for the development of religious tourism. It should be noted that recently the speaker of the Parliament of Georgia during his visit to Bulgaria met the Bulgarian Patriarch Neophit. At the meeting, Patriarch Neofit praised Georgian Patriarch Ilia II and described him as one of the most distinguished figures in the Orthodox world, and stated that the Bulgarian church will add Georgia to a list of pilgrimage destinations, including countries the local church advises its parish to visit. The step will encourage religious tourism in Georgia and increase interest in Georgian holy sites" (The parliament's press office, 2019).

Catholic Churches in Georgia

One of the most important components of religions in Georgia is the Catholic Church, which has been existing in Georgia for nine centuries. Since the early 13th, the dispatched missioners by the popes of Rome had close relations with the Georgian Orthodox Church and Georgian Catholics propagated either within the country (mainly in Samtskhe-Javakheti) or abroad. With centuries of coexistence, Catholicism was represented in the multinational country by ethnic diversity: Georgians, Armenians, Russians, Czechs, Germans, Poles, and French were among the parishioners of the Catholic Church.

The construction activities of Catholics in Georgia dates back to the 13th century and was initially limited within the scope of the country's capital. The first Catholic church in Georgia was built in Tbilisi in 1243, nowadays, there are Catholic churches in Tbilisi, Kutaisi, Batumi, Gori, Akhaltsikhe, Samtskhe-Javakheti, Imereti, Guria, and Kakheti.

Jewish Shrine Synagogues in Georgia

The first research source of the Jewish arrival in Georgia dates back to 586 BC and is named "Life of Kartli", which connects the arrival of Jews on the land of Kartli with the fact of the Nebuchadnezzar conquest of Jerusalem and devastating it (Ninikashvili, 2017).

At the current stage, there are two synagogues in Guramishvili Street close to Rabati Fortress – they are upper and lower synagogues. The upper synagogue of Akhaltsikhe was constructed in 1863 and in 2011 it received the status of the cultural heritage. It is the oldest among the current synagogues in Georgia – it is 153 years old. In 2012, when the Rabat Fortress was reconstructed, the upper synagogue was also renovated when wooden elements were cleaned from oil paints, and damaged parts of paintings were restored. Jews in the capital of Georgia, Tbilisi have two synagogues: the Great Synagogue of Akhaltsikhe, which is one of the most important tourist attractions in the city, especially for tourists from Israel, and the Ashkenazi Synagogue (Ministry of Culture and Monument Protections, 2014), the third so-called Dome Synagogue, which is not operating now, and there has been set up David Baazov Historical and Ethnographic Museum of Georgian Jews (Meparidze, Chachanidze, 2019).

The restoration works of Sinagoga located in Batumi (built in 1904) are important from the point of tourism; the project was prepared in 2015 and funded by private expenses. The Oni Synagogue, which opened in 1895, was rehabilitated in 2012-2014. Oni still retains a small number of Jewish families - remnants of the one powerful and large historic Jewish community.

In Georgia, where the Georgian Jewish community was located, there was a synagogue in every commune. In some towns and villages, there were even several synagogues.

The rules and forms of prayer of the Georgian Jews, the observance of religious traditions, are considered to be one of the most unchanging and correct continuations of the Jewish traditions of ancient Palestine. It should be noted that there are synagogues of different sizes in large and small cities of different parts of Georgia. Most have been granted cultural heritage status by the state.

Mosques in Georgia

Islam is one of the most followed religions among the ethnic minorities, especially, followed by Georgian ethnical Azerbaijanis. Muslim culture is diverse here, with different ethnic groups and Islamic trends. Islam began to spread in Georgia in the eighth century, and since then several Muslim settlements have significantly emerged in the country. Thus Muslim culture is abundant in various parts of Georgia. According to certain sources, mosques in Tbilisi have been operating since the 8th century. There was the Shiite mosque erected on the left bank of the river Mtkvari, which dates back to the 16th century. However, it was destroyed during the construction of the new Metekhi bridge by the communists. At the foot of Narikala fortress exactly in the heart of old Tbilisi, there is a Sunni mosque, built in the first half of the 18th century, in which both Shiites and Sunnis pray together.

In 2012, by order of the Minister of Culture and Monument Protection, the status of a real monument of cultural heritage was additionally granted to 24 mosques in different regions of Georgia (Saneblidze, 2016).

Armenian Holy Orthodox Apostolic Church

Armenians living in the capital of Georgia have a long history, the sources link their establishment here to the 12th-13th centuries. Accordingly, there are several churches in old Tbilisi, the two the main Armenian churches are - Surb Gevorg (restoration completed in 2015) and Surb Echmiadzin (restored in 2012). It should also be noted that several churches have become the matter of dispute between the Armenian Apostolic Church and the Georgian Orthodox Church. Many Armenian temples are restored, one of which is Norashen - the Church of the Nativity.

The most important religious shrines preserved in Georgia are enough to give a general impression about the presents of wide range sanctuaries for the development of religious excursions and pilgrimage (on holidays and in general) and religious-cultural excursions. Such kinds of tours should be offered by travel agencies with high-tech and highly professional guide services, in a safe and positive environment.

The Strategy for the Development of Religious Policy of the State of Georgia

Georgia has chosen the Western European path as its political goal since the restoration of state independence. Due to the taken commitments under the European Framework Convention for the "Protection of National Minorities" since 2005, Georgia's State Strategy for Civil Consent and Integration aims to achieve civic consensus and ensures human equality (Georgian Government Decree, 2015). Georgia joined the Council of Europe in 1999 and since that pursues democratic values following various international agreements, such as the Association Agreement between Georgia and the European Union in 2014 - democratic reforms, which in turn will protect and integrate cultural diversity, freedom of religion, and integration. The action plans adopted for the implementation of all commitments are based on strategic goals and recommendations. A country that tries to integrate into European structures has to ensure the respect for human rights and fundamental freedoms through the way of cooperation (including on religious grounds) and implement them following EU standards and recommendations.

At the same time, it is important to highlight that the inter-religious relations in the modern world put on the agenda the need to establish new global ethics, which pursues to create the conditions for the development of cultural and religious tourism since its content is based on the respect for the differences, cooperation, kindness, rejection of controversy, and the search for ways to cooperate and create a peaceful reality. For Georgia, as a country with multiethnic and religious diversity, an important component for the development of religious tourism is the compliance of state legislation concerning religious freedom.

The modern world is at its new and very important stage of development, when security, peace, solidarity, cooperation, mutual respect, and common well-being have been brought by the already existing extremes and following the state's interests. The twentieth century seeks to coexist peacefully by expanding and consolidating the community of states introducing democratic, legal principles.

According to the preamble to the UNESCO Charter, which stresses that "Peace must be established based on humanity's moral and intellectual solidarity" (UNESCO Constitution, 1945). Georgia was well aware of the statement and many measures were taken in the country to maintain a healthy situation. For instance, the Center for Tolerance has been operating and actively working in the county's Public Defender's Office since 2005, trying to promote the culture of tolerance and create an equal environment. Besides that The Center for Tolerance operates with active international support, it coordinates religious and ethnic minority councils that operate with the Public Defender's Office as well. It mostly deals with the protection of religious freedom, the promotion of multilateral dialogue between religious groups, the establishment of a tolerant and equal, peaceful environment, the social responsibility, and the involvement of religious minorities in the process of civic integration. At the same time, it follows the legal norms of religious freedom and equality in the frame of the Georgian legislation and international human rights treaties, guided by the principle of separation of state and religion, which is guaranteed by the Constitution of Georgia. Here we have to say that The Council of Religions guided by the Public Defender Office. It should be noted that at the initiative of this council, several important issues have been resolved in the frame of its commitments, ranging from promoting legislative changes to social cooperation.

Another important step can be considered the establishment of a special institution - the State Agency for Religious Issues - a specialized structure that implements religious policy based on national experience and modernity, caring for the country's image and aiming to deserve a certain place in the world arena.

Here we have to highlight that the institute was established in 2014 by the initiative of the government and the participation of qualified experts and scientists and is operating under the guidance Georgian the Prime Minister. Based on historical experience and universal values, professional research, and scientific analysis, the institute will develop recommendations in the field of religion and its challenges in the modern world.

Contributing Factors of the International Projects in the Development of Religious Tourism in Georgia

Generally, international tourism in Georgia has contributed significantly to the recovery and development of global policy cooperation. The increased number of international travelers is a source of promoting the process of common internationalization, which has a particular impact on socio-economic relations.

The sphere of tourism in large, successful countries is much more developed and efficient than in small countries. Being a small country, Georgia needs the support of the national policy for the development of international cooperation. Thus the Georgian government is trying to formulate a national strategy following international legal acts regulating the tourism industry. To gain an image of a prestigious developed touristic country, and to contribute to interstate religious and cultural tourism elaboration, Georgia is actively involved in the policy of sustainable development of countries and regions.

Georgia, being one of the most diverse religious and cultural countries pays special attention to the Universal Declaration on Cultural Diversity Development adopted by the UN General Assembly in 2001, declared May 21 to be the World Day for Cultural Diversity for Dialogue and Development, and held annually (UNESCO 2014).

Centuries long-established religious and cultural diversity in Georgia and the respect for different denominations, including communication, recognition of mutual knowledge are still important challenges for the government of Georgia, and consequently, are considered as one of its priorities.

Georgia, with its cultural diversity and distinctive tolerant skills, is a major challenge to international interest. It has been a member of the Council of Europe's Enlarged Partial Agreement since 2016 and has even greater potential for developing both local international cultural routes, integration into transnational routes, cooperating with the Council of Europe, and introducing European values.

It is important that Mariam Jashi, the chairman of the Parliamentary Committee on Education, Science and Culture while presenting cultural routes program to the Council of Europe spoke about the importance of the project and highlighted its role in the promoting of the Georgian heritage, knowledge and attitude and the readiness to bring them in full conformity with the EU countries, to implement the taken obligations in practice and gain the world recognition.

She reaffirmed full readiness of the Georgian government to support the program, and addressed the organizers of the event, saying, "Georgia has a huge historical treasury of the 17th century Christianity. The sacred relics of St. Maximus, the Confessor have been found in Tsageri, which has become a center for tourists' attraction. We have the greatest potential for cooperation and the discovery of new routes in this regard" (St. Maximus had immigrated to Georgia died on August 13, 662. Buried in St. In Tsageri).

Proof of the development of cultural diversity in Georgia is the fact that Georgia is among the certified cultural routes countries - the "European Route of Jewish heritage", also there are other cultural routes in Georgia: "the European Route of Thermal Cities", the "Prehistoric Rock Art Route", and the "Wine Route", which is operating in the House of Israel in Georgia. The Jewish Cultural Heritage Route tells the story of the Jews living in Georgia, particularly, the route passes through 21 cities and 43 points, where the Jewish cultural heritage is revealed. Throughout the country, Jewish cultural heritage includes synagogues, Jewish heritage sites, museums, tombs, and archeological sites.

It should be noted that this program is coordinated at the national level in Georgia by the Ministry of Education, Science, Culture, and Sports of Georgia, which is actively campaigning for its promotion and is a powerful stimulus for the development of tourism in the country. Georgia's history, culture, and cultural heritage are the main treasures of the country, which emphasizes that being a part of Europe, Georgia has made a great contribution to the creation of European culture. Besides, the program will play a major role in the development of local governments and businesses (Ministry of Education, Science, Culture, and Sports, 2020).

When discussing religious diversity and tolerance, it is worthy to mention several projects and events implemented in cooperation with the government of various organizations under the guidance of "Cultural Diversity and Friendly Bridge", which draws the interest in the field of its cooperation in the educational sphere.

There is a very interesting Department of Religious (Christian) Tourism at the Georgian Patriarchate, which plans and conducts both international and domestic tours. It is widely known that in the current year (2020) in Berlin, the Georgian capital won the LCD Berlin Awards at the joint ceremony Global Cultural Travel Network and VisitBerlin. The winners were selected by an international jury that included the editor of The European Culture Department of The New York Times and well-known representatives in the field cultures. At the event, Tbilisi was named the "Best Cultural City", even though the competitors were various world-famous touristic cities.

There were presented 50 cultural events from 28 countries at the award ceremony. 18 tourist destinations received various awards. We have to highlight this special award was given to the Embassy of Georgia in Germany. It is very vital for the development of Georgian tourism because the purpose of the event was to provide tourists with information about the latest travel destinations, which include cultural (religious) and adventure destinations.

It should be noted that important measures are implemented for the future development of tourism in Georgia, as Georgia is one of the fastest-growing and most attractive countries for the tourist market and indicates its right tourism policy; the country hosted more than 9,300,000 visitors in 2019.

As stated above Georgia won a new victory on the world tourist map, a very difficult competition held in Berlin, Germany (the duration was for the last 7-8 months, 09.05.2020), we have to highlight that 30 very strong and popular in the touristic sphere countries were participating in the tournament. The victory has been achieved by the country's right policy towards the issue, a safe environment, investing in tourism infrastructure and proactive development of the tourism industry, rich culture and traditions, and a liberal and transparent attitude, which gives uniqueness to the country. It is worth to say that it has been achieved due to the country's right, security policy, large investments in tourism infrastructure and proactive development of the tourism industry, rich culture and traditions of the country, the country's liberal attitude toward the multicultural and multi-faith policy and unique transparency. In addition to all mentioned political attitudes, the country won the price for its correct marketing position of the National Tourism Administration. Being awarded the name of "Best Cultural City", means that Georgia has received official confirmation from ITB Berlin, Germany's national tourism exhibition, that in 2022 it will be the host country of ITB Berlin the largest exhibition market in the tourism industry worldwide.

Hosting the ITB Berlin exhibition is an unprecedented opportunity for Georgia, which will shift Georgia to a higher level and present it as a country with special tourism opportunities. Host status is a signal to the whole world that Georgia has always been and will continue to be one of the world's best tourist destinations by 2022 and after it.

The preparation for this event was done by Georgian National Tourism Administration, which is a Legal entity of Public law, a part of the system of the Ministry of Economy and Sustainable Development of Georgia, independently conducting activity by state control.

The representatives of the tourism sector say that Georgia has the potential to develop religious tourism, however, the issue requires to be on the country's agenda constantly. In 2017, the National Tourism Agency published a catalog (in Georgian, Russian, English, Spanish, Portuguese, German) for Orthodox pilgrims. The leaflets include an overview of Georgia's cultural sights, ancient history, religion, temples and castles, rock-cut cities, traditional Georgian dances, and music. As well as it presents the basic information required for travelers in Georgia. We have to say that a catalog of Jewish heritage was published to facilitate Jewish tourists. However, this is not enough, since a very small percentage of tourists arriving in Georgia name religion as the aim of their visit.

Presidents of Georgia about the Religious Diversity in the Country

As soon as Georgia gain its independence it started following the Western course with the main goal to become a part of Euro Atlantic space and worked hard to fulfill the obligations taken in the frame of international agreements (UNESCO, the European Union's Eastern Partnership program and other) and ensured the promotion of tolerance, religious freedom, interreligious and intercultural dialogue (which is an important factor for the development of religious tourism). It is very actual and vital to observe the dynamic changes of Georgian government and presidents, to pay attention to the presidential speeches, issued various documents, the existing empirical materials, governmental interest and attitude toward religious tolerance, the country's commitments, and initiatives taken in that field.

Religious diversity in Georgia has always contributed to respect, cultural development, dialogue, and freedom of expression. As soon as Georgia sign the declaration of state independence, the Presidents of Georgia repeatedly stated their approach in their speeches and practices, We will only express the views of the first and current presidents, and I will add that the views of other presidents were similar, tolerant.

The first president of independent Georgia, Zviad Gamskhurdia, highlighted that the totalitarian regime and ideological dictatorship endangered the country's population, not only in the social, political and economic spheres but also in the spiritual and cultural spheres. He pointed out that the destruction of religion, the decline of morality, the forgetting of spiritual ideals, and the devaluation were the real culprits. Impairment was the real starting point for the crime. Consequently, the radical reform of the existing structures for the protection of cultural heritage was considered necessary. The State the restoration of independence and its welfare would not be possible without religious and moral revival, as it is evidenced by Georgia's past and present. He noted that the spiritual nature of the apostolic activity of the Church in itself precluded its direct participation in the political system of the country, but the separation of the Church from the political system is not considered to be mutually exclusive between the State and the Church. Zviad Gamsakhurdia was well aware of the role of the Orthodox Church in the history of Georgian state policy. The president in his speech said It his speech he said: "Freedom of conscience and religious liberty will be guaranteed by the state and to each person and the citizens' rights will not be violated on religious grounds. The Georgian state will remain committed to traditional religious tolerance" (Matsaberidze, 2007).

The current president of Georgia, Salome Zurabishvili, in her pre-election speech said that Georgia has always been a country of tolerance and respected all religions, and added "I will be a tolerant president of Georgia (Qeshelashvili, 2018).

She repeatedly highlighted the country's respect toward the faith, since "Religion is deeply rooted in the soul and culture of Georgian people. Georgia's religion and culture are intertwined".

"Salome Zurabishvili, in an interview with the Jerusalem bureau of TBC, said that Jews came to Georgia before Christianity and were welcomed with kindness. Georgia became their second home town it is a positive example of tolerance for the world (Potskhveria, 2020).

After examining several materials, we are convinced that all the presidents of Georgia tried to maintain the country's centuries-long religious tolerance. Every president repeatedly mentioned their attitude toward religious minorities and acts according to the taken international obligations. This is evidenced by a small example when Eduard Shevardnadze made a strict statement during one of the provocative incidents. He said: "I am against the escalation of religious extremism. Thus to fulfill the obliged commitments in the framework of the Council of Europe, the United Nations, and other international organizations to protect human rights, the persecution of religious minorities should be ended immediately". (Shevardnadze, 2003).

In Georgia, when we emphasize the coexistence of religious tolerance and peaceful life it is necessary to note that recently there have been cases when, unfortunately, the issue of Muslims became a matter of provocations (especially during the pre-election periods). Many scientists do not exclude the possibilities of extremism in that case. "If we observe the recent developments, we will see that there is a group of like-minded people who are engaged in organizing provocations, in this case, specifically can be considered the events of Minaret and Chela. The problems, which could have been solved without any noise or agitation, almost led to bloodshed" (Ivelashvili, 2018). Naturally, in any case, the Georgian government is trying its best to maintain a friendly relationship.

The Factor of Interreligious Dialogue in the Development of Religious Tourism

When discussing the issue of religious tourism, it is necessary to mention the special interest towards the religion in recent years, and its positive role in the peace-building process. Religious norms, common values, as a central aspect of cultural identity can lead people to cooperate, religious rituals are considered an important tool for promoting cooperation. Therefore, it is necessary not only to study religious traditions but also to make them visible to the public and society. This is a necessary condition for building trust between people of different religions.

Promoting interreligious dialogue in the modern era of globalization is considered as one of the most important preconditions for the harmonious coexistence of a peaceful world.

In our era, the main paradigm represents not the intense interactions between the groups with similar views and interests but in the cooperation of different groups or individuals with different religions and cultures. It is now almost impossible to see a homogeneous society whose members are followers of the same religion. Migration and tourism have shifted from the individual sphere to the mass movement. Consequently, the importance of inter-religious dialogue is growing steadily, and therefore, inter-religious dialogue is seen as an important new paradigm that plays a leading role in the peace-building process.

The Georgian Orthodox Church supports the principle of mutual respect and cooperation, and dialogue with representatives of different religions, - the statement issued by the Georgian Patriarchate (Statement of the Patriarchate, 2013). That is why the Georgian Patriarchate took the initiative in 2010 to create an inter-religious working group to discuss these issues. According to the statement, the Patriarchate welcomed the fact that the initiative was shared by religious associations and that the group, which is coordinated by the non-governmental organization Union "21st Century", has been functioning for years and unites the main religious spectrum operating in Georgia. Representatives of the country's legislative and executive branches are also taking part in the sessions. This is the first precedent of a full-fledged inter-religious dialogue in Georgia and we believe that the continuation of this activity and further deepening will greatly contribute to the further development of Georgia.

It is acknowledged that the best opportunity for interfaith dialogue and the promotion of a peaceful society lies in religious traditions. The study of these traditions is a necessary condition for overcoming negative stereotypes and building trust between different religions. Thus the State Agency for Religious Issues has been holding an annual conference in Georgia since 2016 entitled "Interreligious Dialogue for Peace", which is attended by representatives of all religions operating within the country. Realizing the centuries-old experience of their cohabitation and speeches are always presented with the idea of peaceful cooperation, (the conferences about Interreligious Dialogue for Peace were also held in 2017-2018-2019). A peaceful environment for its turn is an important factor in promoting religious tourism. For the Georgian reality, being a small country, even very small details are important in shaping its future agenda. To celebrate the International Convention on the protection of intangible cultural heritage, on 17 October 2003 in Paris, which is very important for the world's globalization processes and social transformation, creating the conditions for renewed dialogue among communities, and at the same time is a serious source for degradation, disappearance, and destruction. Therefore, the intangible cultural heritage is under threat, especially due to a lack of resources to protect it.

Since the beginning of 2020, the World Tourism Organization (UNWTO) has claimed that a new coronavirus pandemic has destroyed the international tourism and there emerges a new reality - a social distance. Tourist countries around the world plan to take measures to restore tourism. Georgia, as a member of the top five fast-growing countries, is slowly trying to open its tourism industry since it is a great source for the economy, emphasizing the safe travel corridors. On May 8, 2020, a video conference was held between the Secretary-General of the United Nations World Tourism Organization, Zurab Pololikashvili and Georgian Prime minister, Giorgi Gakharia to discuss the establishment of international standards and guidelines. It is noteworthy to say that in September of the current year, the ministers of more twenty countries will gather in Tbilisi, Georgia to discuss the prospects of the world tourism industry. The ministerial will be held under the guidance of the World Tourism Organization. In this regard Z. Pololikashvili noted: "I pledged our full support to the Prime Minister of Georgia Giorgi Gakharia in coordinating a strong and effective response to the neutralization of the influence of the new coronavirus in the region's tourism industry" (Pololikashvili, 2020).

CONCLUSION / RECOMMENDATIONS

Based on the empirical material studied and discussed in the presented chapter, the state's readiness for the elaboration of tourism in the country is obvious, however, the development initiatives specifically for religious tourism, despite the available wide range of opportunities, are less obvious.

Based on the study of several documents, it becomes clear that the democratic development of a society can only be achieved by focusing on the common welfare having equal rights and responsibilities.

It is noteworthy to say that the "Georgia Tourism Strategy 2025" is developed based on research, analysis, and consultation. To achieve its goals, in creating world-class tourism products, and offer the necessary high quality, diverse tourist products and the best impression for travelers' both international and domestic demands, the challenges and opportunities for tourism development are assessed. Indeed, the strategy and other state documents, a great deal with the preservation of cultural heritage monuments and other cultural assets, their accessibility, the improvement of existing services and interpretive infrastructure; both the international and domestic markets are focusing on increasing awareness of the country's tourism potential (Georgian National Tourism Administration, 2015), and even though religion is an integral part of Georgian culture, the emphasis on religious tourism in particular, despite its potential, is not stated in strategy or other documents. The high opportunities for the development of religious tourism in Georgia are based on ancient Christian foundations, preserved by religious relics, monuments of ancient culture along with the best geographical environment. Existing religions in the country since ancient times have never been violated and are always allowed to develop without any restrictions (except during the period of communist regime).

RECOMMENDATIONS

1. Under the auspices of the Georgian National Tourism Administration, it is necessary to consider the cooperation of professional tourism organizations in Georgia, both with each other and various organizations, including foreign organizations and as well as with educational institutions (which include the followings): preparation relevant literature, maps and illustrations in various major languages for tour guides; having high-level custom-preparation gives the possibility to create special vocational training centers - to train guides in the field of religious knowledge, nutritions for preparation of food according to religious demands, for instance for those who are fasted, etc.;
2. To cooperate with the construction companies for the development of religious tourism, to restore and arrange appropriate construction facilities, infrastructure in active places of pilgrimage.
3. Reconstruction of an insignificant building in the vicinity of any historical monument should not take place without the extensive discussion and permission of specialists.
4. Georgia needs to create a positive image of the country to establish itself in the world market of religious tourism, which means:
    - Proper presentation of religious freedom in the country.
    - Wide propaganda about the peaceful coexistence of different religions in the country.
    - The government has to ensure a good awareness of the citizens, about the requirements of religious tourism development.
    - To demonstrate the potential of religious tourism the government has to deal with several following factors: to protect the cultural heritage at the appropriate level; arrange/provide certain marketing steps, and spread the information about the field of religious tourism.

REFERENCES

Alasania, G., & Chkhartishvili, M. (2014). *The knowledge of Georgian Sources*. Publishing Hause Universal.

Bubulashvili, E. (2011). *The Life of St. Nino*. Retrieved 11 10, 2019, from Library Church: http://library.church.ge/index.php?option=com_content&view=article&id=424%3A2011-09-16-10-53-21&catid=47%3A2010-03-11-12-05-46&Itemid=67&lang=ka&fbclid=IwAR3YQqLiP9md-H6Djaz-59pRlcHXpA8F5N7NtLLL2wMISVR8f7lZpyuPba4k

Bulgarian Orthodox Church in the list of Georgia's praying places. (2019). Retrieved from The Official Web Page of Georgian Parliament.

Christian shrines preserved in Georgia. (2020). *Karibche*.

Decree of the Government of Georgia #1740. (2015). Retrieved from gov.ge: http://gov.ge/files/410_50997_650312_N1740.pdf

Devidze, E. (2012). *Cultural Heritage and Problems of Imereti Tourism Development* (Dissertation). Tbilisi State University.

Gambashidze, G. (2014). Proceedings Ministry of Culture and Monument Protections. Geprgian Jewish Historical Museum, 28.

Georgian National Tourism Administration. (2015). *The Georgian Tourism Strategy-2025*. Author.

Grdzelishvili, N. (2018). *Religious Tourism*. Publishing House Kentavri.

Ivelashvili, T. (2018). *Religious Minorities in Samtskhe-Javakheti.* Akhaltsikhe: Publishing House University Akhaltsikhe.

Kiknadze, Z. (2008). *Religions in Georgia*. Georgian Public Defender's Library.

Matsaberidze, M. (2007). *Inauguration Speeches of Georgian President*. Akhali Azri.

Meparidze, G., & Chachanidze, N. (2019). The Share of Religious Minorities in Georgian History. Tbilisi: Published by CIPDD and OSGF.

Natsvlishvili, N. (2019). *Catholic Churches in Georgia: History and Architecture*. Publishing House Ilia State University.

Ninikashvili, S. (2017). *A Great Synagogue of Tbilisi – Interesting Facts from History*. The Tbilisi Times.

Okrotsvaridze, A., Vadachkoria, M., & Okrotsvaridze, L. (2011). Tourism and Host Management. Academic Press.

Papashvili, M. (1995). *Georgian- Rome Relations in VI-XX Centuries*. Publishing House Aghmashenebeli.

Papashvili, M. (1998). *Georgian-Roman Relations*. Publishing House Sazani.

Pavlenishvili, E. (2019). *Catholicism in Georgia (XIX century)* (Dissertation). St. Andreas Georgian University.

Pololikashvili, Z. (2020). *The video conference held between him and the prime minister of Georgia, Giorgi Gakharia*. Retrieved from interpressnews.ge: https://www.interpressnews.ge/ka/article/598753-zurab-pololikashvili-giorgi-gaxarias-sruli-mxardachera-agvutkvi-turizmis-industriashi-koronavirusis-gavlenis-ganeitralebisken-mimartuli-epekturi-reagirebis-koordinaciashi

Potskhveria, G. (2020). *We need a positive example of Tolerance for the World, The coexistence of Georgian People, and Jewish is more than a Tolerance*. Retrieved from 1tv.ge: https://1tv.ge/news/salome-zurabishvili-chven-gvchirdeba-tolerantobis-pozitiuri-magalitebi-msofliostvis-qartvelebisa-da-ebraelebis-tanackhovreba-ki-tolerantobaze-metia/

Qeshelashvili, A. (2018). *Salome Zourabishvili: I am not a Catholic, I am an Orthodox*. Journal Liberal.

Saneblidze, N. (2016). *The State of Cultural Heritage of Religious Minorities in Georgia*. Published by Human Rights Center.

Shevardnadze, E. (2003). *Shevardnadzeadze demands the prevention of religious violence*. Retrieved from civil.ge: https://civil.ge/ka/archives/129134

Sikharulidze, A., Urushadze, M., Khmaladze, G., & Chichinadze, R. (2016). *Islam in Georgia: Politics and Integration*. Publishing House Caucasian House.

Songulashvili, M. (2017). Global Dynamics of Interreligious Dialogue: The Road from Peace Complaint to Ecumenical Charter. *Journal of Tolerance and Diversity Institute*.

The Constitution of Georgia. Article 9. (1995). Tbilisi.

UNESCO. (2014). *Fifth International Festival of Traditional Cultures and Crafts "ETHNORADUGA" devoted to the World Day of Cultural Diversity for Dialogue and Development*. Kaluga Region.

UNESCO Constitution, Constitution of the United Nations Educational, Scientific and Cultural Organization. (1945). Retrieved from http://portal.unesco.org/en/ev.php-URL_ID=15244&URL_DO=DO_TOPIC&URL_SECTION=201.html

Vardosanidze, S. (2017). *Georgian Orthodox Apostolic Church in the XIX-XX centuries"(Catholicos-Patriarch of All Georgia His Holiness and Beatitude Ilia II)*. Publishing House Sezani.

KEY TERMS AND DEFINITIONS

Collaboration: Working together to achieve a common, pre-determined goal.

Culture: Created by human hands, minds, and feelings. The division of culture takes place materially (the world of things) and spiritually (the world of knowledge, values, symbols).

Georgia: An independent, sovereign state in the South Caucasus.

Historical Monuments: A remarkable place, or architectural structure, erected by man or created by nature and related to a certain culture (architectural, ethnographic) facts or events.

Interreligious Dialogue: Sharing the views of people of different faiths for freedom and mutual respect.

Religion: Faith, belief, worship, and acknowledgment of God.

Religious Diversity: The human environment, with different religions, different views, values, traditions.

Religious Relics: Preserved religious objects, sacred parts and particles, and objects which belong to the saints, as a special memorial.

Religious Tourism: Travel for religious purposes, pilgrimage (pilgrimage) to holy places, and visit the religious buildings and shrines for cognitive purposes.

Tolerance: A fair approach to people of different faiths, respect their values and views.

Tourism: Travel for different purposes (time allocation, spending).

Chapter 13
Media and Religious Tourism:
A Thematic Review

Rizwan Shabbir
https://orcid.org/0000-0003-0859-7465
Government College University, Faisalabad, Pakistan

Aysha Batool
Government College University, Faisalabad, Pakistan

ABSTRACT

This chapter aims at presenting a thematic analysis of 64 research articles on religious tourism published from 2009 to 2020 by adopting a systematic literature review method. The results indicate that prominent topics discussed were amplification of concepts, tourist motivations, and experience. The evolution of religious tourism concept through nexus of diverse scholarly terms, related domains, and religious practices needs a comprehensive literary debate to refine the subject. The socio-economic impact of religious tourism also calls for the enrichment of topics on scholarly and practical grounds. Issues such as host and tourist behavior, social interaction, visitor management, marketing components, and the role of media need to be explored for the progression of religious tourism in line with sustainability. The chapter contributes to the literature by proposing a three-dimensional model focusing on the role of media and potential research domains for further exploration.

INTRODUCTION

Human interactions and exchanges are extremely important for satisfying social needs to belong and relate to others (Bernstein, Zawadzki, Juth, Benfield, & Smyth, 2018). Tourism is an important source of social interaction that enhances positive emotions and releases stress (Vittengle & Holt, 2000) through travel experiences as the tourist visits other places and meets people during his journey (H. Lin, Zhang, Gursoy, & Fu, 2019). The tourism and travel (T&T) business are considered one of the fastest-growing industries in the world with huge economic and developmental impacts on the host countries by providing employment, businesses, foreign exchange, and global recognition (Shahzad, Shahbaz, Ferrer, & Kumar,

DOI: 10.4018/978-1-7998-5792-1.ch013

2017). In 2019, the total contribution of the T&T industry to the world's GDP was 8.9 trillion US dollars, with the industrial growth rate of 3.5%, which was above the world economy of 2.5% (WTTC, 2020).

Tourism and religions have long been connected since the followers of different faiths used to travel to diverse geographical sites for the fulfillment of their religious obligations and peace-seeking, e.g., pilgrimage, etc. (El-Gohary, 2016; Fleischer, 2000; Hashim, Murphy, & Hashim, 2007). Researchers have investigated multiple interventions of tourism with religious creeds and activities such as spirituality and pilgrimage (J. Huang & Chu, 2019; Jafari & Scott, 2014; Nilsson & Tesfahuney, 2016). This religious travel to holy places is prevalent in the modern era too. The religious tourists and travelers are not only motivated to perform pilgrims, but their travel inclinations also include the quest for visiting sacred and historical sites (Buzinde, Kalavar, Kohli, & Manuel-Navarrete, 2014; Olya, Lee, Lee, & Reisinger, 2019; Wang, Chen, & Huang, 2016). This phenomenon has led to an increased scholarly and economic interest in religious tourism, and the industry is attracting enormous growth and positive developments globally (Hu & Huang, 2019; Ozcan, Bişkin, & Şimşek, 2018).

The development of religious tourism can also help to achieve sustainable development goals in countries with cultural and religious heritages. This industry creates a positive economic and social impact on societies, but many of the developing countries have not still utilized it to their gifted potentials such as Pakistan and Bangladesh (UNWTO, 2020). According to WTTC (2020), 1 out of 10 jobs are created by the tourism industry in 2019, and it is the third sector with the highest GDP growth after information & communication and financial services.

The evolution of new technologies and contemporary tourism services has transformed the tourism experience. With the advancement of the internet, smart devices, and media technologies, religious travel is also being influenced. Researchers are suggesting the marketers utilize social media as a tool of online marketing because the internet is a powerful and widely used means for providing quick and dynamic information about global customer experiences, including transport, residence, food, and other interrelated services. While making travel decisions, people use to search for information about the destination places, services, and other tourists' experiences (Qureshi, Humayun, & Hummayun, 2015). Hence, different media channels are impacting the travel decisions of tourists through visual and online content and electronic word of mouth (eWOM) from other tourists (Molinillo, Ximénez-de-Sandoval, Fernández-Morales, & Coca-Stefaniak, 2016). Customers extensively engage in social media for the pre and post-travel eWOM because it enhances their understanding of the desired tourism places. The massive use of smart media technologies globally has provoked the popularity of posting experiences on social media at all stages of the travel. An increasing number of people now involved in social media for pre-visit, during the visit and post-visit content sharing. Thus, social networking through modern internet technologies is modifying the social experience of tourism, including religious travels also (Muslim, Harun, Ismael, & Othman, 2020).

The technological influence is not limited to any specific religion. Novel research shows that followers of different religions either in the western or eastern part of the world are transforming their travel experiences through social media networks (Muslim et al., 2020; Qurashi & Sharpley, 2017). However, research about the use of social media in tourism is mostly concentrated to western countries (Miguéns, Baggio, & Costa, 2008) and social media role with respect to religious tourism and eastern part of the world is relatively less explored.

The potential of growth in the religious tourism industry has captured the attention of Asian countries as well as scholars, and major advancements are being done at governmental levels to promote this sector. Asia is the origin of major religions, civilizations, cultures, and hundreds of languages in the world and hub of their followers, including Hinduism, Islam, Buddhism, Sikhism, Christianity, and Judaism. The region not only has enormous holy places for spiritual travelers, including mosques, temples, cultural heritage, but it is also filled with immense natural beauty (UNWTO, 2011, 2020).

This chapter presents a review of 64 research papers that were published on religious tourism between 2009 and 2019, including early cites of 2020 in 12 leading journals of 5 renowned publishers. The literature review approach provides important insights into the thematic evolution in religion-oriented tourism and traveling research. The review also discusses the importance of technology, i.e., media, and its influence on religious tourism. The chapter identifies common research themes that have been explored in reviewed articles, i.e., motivation, experience, and investigates how, with the passage of time, integration of various research concepts in religious tourism has refined the subject. The conclusion part presents a definition of religious tourism and outlines future research directions for exploring the role of media under the attract-manage-delight model.

METHODOLOGY

A systematic literature review is a good approach for methodical assembly of research papers, and it provides collective and in-depth understandings about a specific research topic by reviewing the literature through following an explicit search criterion (Tranfield, Denyer, & Smart, 2003). The relevant papers for this study were searched via the Google Scholar search engine, which is a well-known database of academic research papers and journal articles. A large number of peer-reviewed articles on religious tourism are covered by Google Scholar.

Table 1. No. of articles by publishers

Publisher	No. of Articles
Elsevier	23
Emerald	3
SAGE	6
Taylor & Francis	17
Wiley	15
Total	**64**

Source: Author's calculation

The keywords used for identification of research papers were; 1) religious tourism, 2) religion and tourism, 3) pilgrimage, 4) religious tourism, and media or social media. Religion specific articles on tourism were also considered for review, such as Islamic, Buddhism, etc. The time period for review was set as from the year 2009 to 2019, and early cites of the year 2020 were also included. For inclusion and exclusion criteria, the titles and abstracts of the papers were manually analyzed by the authors to decide whether the research article was relevant to be included in the subject review or not. After analyzing, the authors shortlisted a total of 64 articles for the current work, which were published by five well-known publishing groups named Elsevier, Emerald, Wiley, Taylor & Francis and SAGE (summary presented in Table.1).

Among a number of tourism and hospitality domain journals owned by these publishers, only the top twelve were considered as cited by Kim, Kim, and King (2019) to restrict the review spectrum. These journals were also listed on the social sciences citation index (SSCI). The highest number of articles on religious tourism appeared in journals published by Elsevier, Taylor & Francis, and Wiley, i.e., 23, 17 & 15, respectively. The reviewed articles were also listed in chronological order according to citation year. This was done to avoid ambiguity in first-time publication date and citation year. The number of articles published each year in selected leading journals depicts 2016 as the most prominent with 14 papers (Fig 1).

Figure 1. Distribution of reviewed articles as per citation year

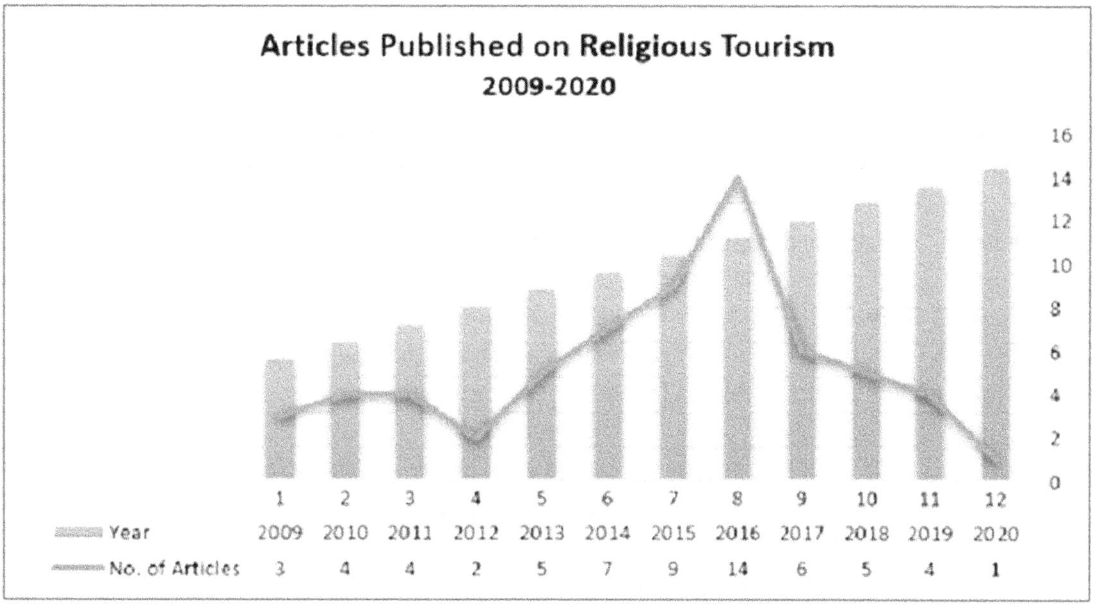

Asia embraces almost 60% of the world population and is considered the hub of hundreds of languages and religions. The research statistics about reviewed religious tourism articles also represent similar results (Table.2). Among 64 SSCI articles published in major tourism journals from 2009-2020, 30 were focusing on Asian countries. Only three studies exhibited a multi-regional perspective, whereas six articles discussed religious tourism as a scholarly debate without any regional focus, i.e., review, etc.

THEMATIC FINDINGS

Religious tourism refers to such kind of tourism, which has some religious associations (Rinschede, 1992). It has been elaborated with the combination of various religious activities and tourists' associated

Table 2. Distribution of Reviewed Articles as per Region

Region	No. of Studies
Africa	3
Asia	30
Australia	1
Europe	19
Europe, Africa, Asia	3
General	6
North America	2
Grand Total	**64**

Source: Author's calculation

motivations and benefits with such tours. For example, religious tourism, under Hinduism literature, is also termed as Ashram Tourism, Hindu pilgrimage, Yatra; under Islamic literature as Islamic tourism, a pilgrimage or pilgrim tourism, Halal tourism, Umrah tourism, Hajj tourism, Muslim tourism, Sufi tourism; under Buddhism research as Buddhist tourism, temples tourism; in Christianity focused articles as Christian tourism, Holy land tourism; and apart from any specific religion, in growing literature, religious tourism is also called by numerous interchanging names such as sacred tourism, sacred journeys, pilgrimage tourism, pilgrimage, spiritual tourism, spiritual journeys, secular tourism, heritage tourism, cultural tourism, secular pilgrimage tourism, etc.

The review of 64 articles publishes from 2009 to 2020 (listed in Table 3) reveals common themes that have been mostly discussed in the last decade. Among these, motivation and experience has remained topics of concentration by the scholars. The study focus statistics can be seen in figure 2.

For the purpose of identifying relatively unexplored topics, the articles were divided into 14 major groups according to the focus or purpose of the article. These themes were named as behavior, brand equity, concepts, economic effects, experience, loyalty, market characteristics, marketing components, motivation, satisfaction, service quality, socio-economic effects, value, and visitor management. Among 64 articles, 31 were investigating religious tourism under broad areas of tourist experience (11), subject concepts (10), and tourist motivation (10), whereas other aspects were relatively less explored. The detailed list is provided in table 3, while major themes are discussed further.

Motivations

Motivations are the reasons or factors that drive people to act or behave in a certain way. Tourists' motivations for this specific type of tourism, i.e., religious tourism, has been extensively studied in the literature. The reasons or motives for religious tourism widely differ among people and are greatly influenced by tourists' religious norms, socio-demographics, and experiences.

Religious and Non-Religious Motivations

Although the concept of religious tourism apparently demonstrate the notion that the tourists/pilgrims specifically visit sacred sites for the religious purpose only (Kruger & Saayman, 2016), the emerging

Figure 2. Distribution of reviewed articles according to study focus

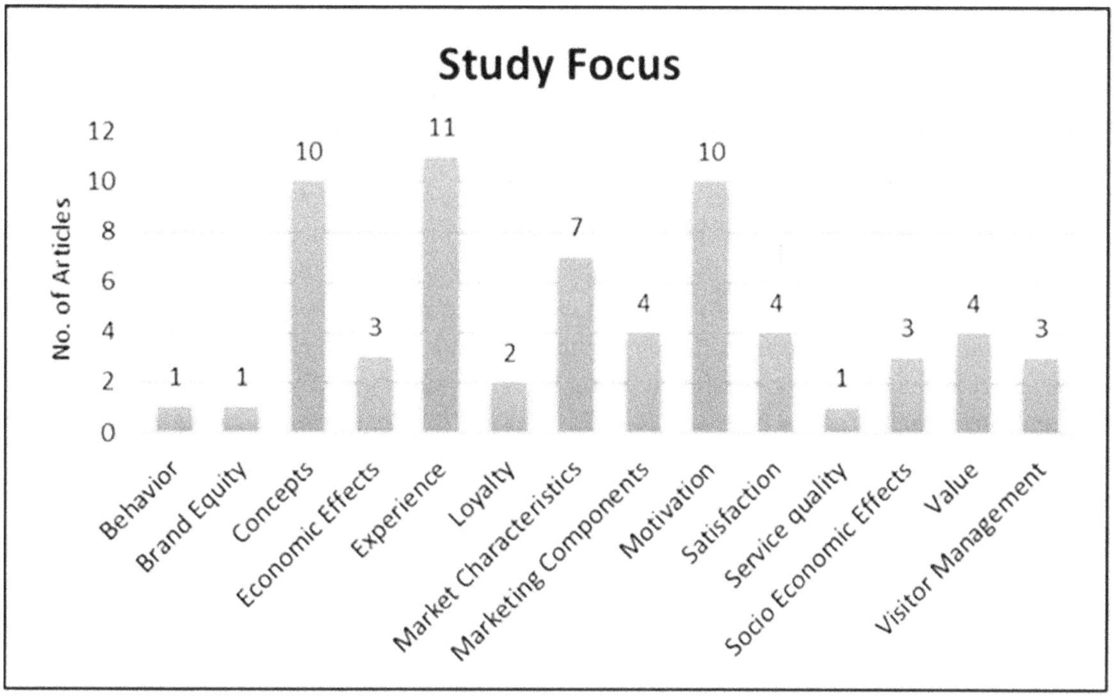

concepts in this field of study are somehow evolving this belief. In the present era of technology and dynamic changes in social events, religious activities, such as pilgrimage, are not being done under the sole influence of religious sentiments. Instead, the spiritual aspects of finding better change in life are impacting the progress of these practices (Amaro, Antunes, & Henriques, 2018). People tend to visit religious sites by inspiring religious and non-religious factors. Religious motivations include religious belief, norms, obligations, love for religion, quest to obey and find God, respect for religious activities, etc. Whereas, non-religious motivations may include but are not limited to reasons such as; to gain spiritual impetuses and novel experiences by meeting others, sports, visiting sacred places, cultural enjoyment, and mental relaxation, etc. Past experiences and degrees of religious belief also play an important role in enhancing the motivations to visit religious places (Wang et al., 2016).

The quest to learn about new things and people through social interactions is a long driven human instinct. During religious journeys, educational and learning desires can augment the underpinning travel motivations of gaining religious, environmental, spiritual, and cultural exposure. This novel dimension of education or learning pursuit can be utilized by triggering a positive emotional response of the tourists through educating them about the ancient heritage and destination attributes via conversation and interactions with site representatives, guides and local people (Robina Ramírez & Fernández Portillo, 2018).

Experience

Religious as well as non-religious antecedents are also important players in forming a religious tourist's experience. People who visit religious places also feel driven to materialism during their trips under the desire to escape from their daily routines. Other than being emotionally connected to religious activities, tourists also enjoy shopping, visiting religious and cultural sites, sightseeing natural sceneries, and historical heritage for joy, peace of mind, fun-seeking, and networking, etc. Hence, the maintenance and promotion of such places should also be focused on better targeting and marketing activities. The hotel managers, local public, and administration staff can be the means of facilitating social interactions with and among the tourists in order to give them better services, a sense of affiliation, and a feeling of community (Taheri, 2016). Researchers have found that most of the people are in search of an experience which fulfills their spiritual excursion in a way that is more reflective of life and linked to their own selves (Nilsson & Tesfahuney, 2016). While doing pilgrimage travel, pilgrims obtain spiritual exploration and the stimulus of social interaction with other tourists from diverse global backgrounds. Moreover, some of the values received from pilgrim experience also resemble those of holiday tourists. Religious tourists share a similar inspiration for fun-seeking tourists and find personal happiness while visiting and appreciating natural beauty (Kim, Kim, & King, 2016). Managers can enhance the on-site experience of religious tourists by recognizing and interpreting tourists' preferences. Faith, aesthetic and architectural features, and working community are recognized as important aspects for the marketers to grab the socio-economic benefits of the religious tourism industry (Ballantyne, Hughes, & Bond, 2016).

Emerging Concepts

With the passage of time, as theories, practices, and implications evolve, various concepts are emerging in religious tourism too. Researchers are exploring new dimensions and terminologies for this specific subject. Among these changing perspectives, literary tourism of visiting the graves of literary writers is also seen as a form of religious tourism. Brown (2016) describes the motivations of such tourists as the desire to feel related, paying homage, or to express their love and gratitude towards these writers, etc. Researchers are also trying to explore diverse aspects of secular and dark tourism concepts and their connection with religious tourism. The evolution of pilgrimage and religious tourism notions calls for a detailed examination and reconsideration of the concepts with the marketing perspective of the product life cycle. The PLC viewpoint can help in modifying the concepts by adapting novel ideas with changing customer requirements. Researchers are digging up the linkages between different expressions such as secular tourism, dark tourism, and pilgrimage or religious tourism to provoke the inclusion of novel terminologies in order to represent new generation phenomena (Collins-Kreiner, 2016b).

Association between religious and archaeology tourism is also appealing. Both disciplines focus on religious and archaeological heritage. The coalitions and conflicts in the nexus of two broad areas are being debated, and visitors' characteristics are also being explained. Such emerging concepts invite a scholarly and administrative focus on overlapping topics to help develop these areas in a productive way (Koren-Lawrence & Collins-Kreiner, 2019). As the branches of this field are expanding, new leaves for scholarly investigations are growing, e.g., motivations, experiences, satisfaction, behavior, values, benefits, etc., which will considerably differ as to the tourist perspective of the tour and destination changes.

It is also pertinent to discuss that most of the research on religious tourism is concentrated on developed nations. There is a need to focus on the growth of the religious tourism industry in developing countries as well. The immense cultural, religious, and historical heritage, as well as these country's potential for religious tourism, challenges, and opportunities, call for scholarly debate. The literary society has the responsibility for an in-depth investigation of the field from the perspective of developing nations. Implications for the policymakers and managers should be described to enhance the positive image of these countries among neighboring nations and at the international level (Heydari Chianeh, Del Chiappa, & Ghasemi, 2018).

Other Important Themes

With the purpose of highlighting other important themes for scholarly and practical attention, some of the topics have been discussed under this heading.

Tourist Identity, Behavior, and Social Interactions

The interplay of a pilgrim's role, identity, and environment evolves his behavior and experience at different levels (Damari & Mansfeld, 2016). The findings indicate that there is a dearth of research about religious tourist behavior and how it is influenced by tourist's cognitive attributes and social interactions. The study of C. F. Lin and Fu (2020) was the only detailed investigation found among 64 reviewed articles that explored this phenomenon through the lens of social network and means-end chain (MEC) theory. The experiences of self-expression with a sense of play while traveling and communicating with others during journeys and religious practices give tourists the freedom to express, and they also feel respect towards others' interests. This non-religious attribute of religious tourism fulfills their pursuit of escaping from daily materialistic life and engaging in memorable activities.

Thus, the importance of non-spiritual elements in religious tourism transforms tourists' experiences through providing pleasing moments, self-gratifying attributes, constructive emotional responses, group interactions, escape, and other social and hedonic benefits (Lochrie et al., 2019). Apart from performing religious activities and spiritual engagement, re-negotiation of religious practices is a particular act that involves interaction with other tourists, groups, and people. It brings feelings of joy and affiliation through social networking and helping others.

Among the values and attributes, religious tourists seek during their visit to sacred sites, the contentment of religious values is the most important. The personal and spiritual values of joy and self-satisfaction are seen as secondary critical reasons while the basic tourism values of fun-seeking and social empathy etc. are considered the least imperative. (Kim & Kim, 2019). Thus, motivations are largely constructed by tourists' personalities, needs, and their own previous experiences as well as those of others (Terzidou, Scarles, et al., 2018).

Economic and Socio-Economic Factors

Religious tourism is also seen through the lens of its social and economic impact on countries (Sánchez et al., 2017). In order to rationally utilize the socio-economic elements and benefits of religious tourism, a balanced approach between religious sanctity and commercialization is the key to safeguard religious heritage and cultural assets (Hung et al., 2017). In this process, the consensus between tourists and local residents is a crucial element that may demonstrate that religious tourism has positive economic and social impacts on their lives as it is a source of increased morality, sense of proud and socialization for them (Dafuleya et al., 2017). Literature is evident that people use to travel to such places where familiar faith and cultural atmosphere is prevalent. Religious affiliations impact the flow of tourism in the world. People tend to visit those places which have similar religions as their own country. The cultural, historical, and religious resemblance influences destination preferences and increases tourists' associa-

Table 3. List of Reviewed Articles (N=64)

Citation Year	Intext-Citation	Publisher	Region	Study Focus (Major Groups)
2009	(Belhassen & Ebel, 2009)	Taylor & Francis	Asia	Concepts
2009	(Andriotis, 2009)	Elsevier	Europe	Experience
2009	(Shuo, Ryan, & Liu, 2009)	Elsevier	Asia	Loyalty
2010	(Gutic, Caie, Clegg, & Wiley, 2010)	Wiley	Europe	Motivation
2010	(Shinde, 2010)	Wiley	Asia	Market Characteristics
2010	(Triantafillidou, Koritos, Chatzipanagiotou, & Vassilikopoulou, 2010)	Emerald	Asia	Marketing Components
2010	(Zamani-Farahani & Henderson, 2010)	Wiley	Asia	Socio-Economic Effects
2011	(Andriotis, 2011)	Elsevier	Europe	Market Characteristics
2011	(Battour, Ismail, & Battor, 2011)	Wiley	Asia	Market Characteristics
2011	(Henderson, 2011)	Wiley	Asia	Visitor Management
2011	(Hyde & Harman, 2011)	Elsevier	Asia	Motivation
2012	(Della Dora, 2012)	Elsevier	Europe	Concepts
2012	(Zamani-Farahani & Musa, 2012)	Elsevier	Asia	Market Characteristics
2013	(Krešić, Mikulić, & Miličević, 2013)	Wiley	Europe	Satisfaction
2013	(Linderman, 2013)	Emerald	Asia	Marketing Components
2013	(Scaffidi Abbate & Di Nuovob, 2013)	Taylor & Francis	Europe	Motivation
2013	(Hughes, Bond, & Ballantyne, 2013)	Elsevier	Europe	Marketing Components
2013	(Wong, McIntosh, & Ryan, 2013)	Elsevier	Asia	Market Characteristics
2014	(Barajas, Lago-Peñas, & Sanchez, 2014)	SAGE	Europe	Economic Effects
2014	(Battour, Battor, & Bhatti, 2014)	Wiley	Asia	Satisfaction
2014	(Buzinde et al., 2014)	Elsevier	Asia	Experience
2014	(Cheng & Chen, 2014)	Taylor & Francis	Asia	Market Characteristics
2014	(Jafari & Scott, 2014)	Elsevier	General	Concepts
2014	(Kirillova, Gilmetdinova, & Lehto, 2014)	Elsevier	North America	Concepts
2014	(Saayman, Saayman, & Gyekye, 2014)	Wiley	Africa	Economic Effects
2015	(Bond, Packer, & Ballantyne, 2015)	Emerald	Europe	Experience
2015	(Hung, 2015)	Taylor & Francis	Asia	Experience
2015	(Choe, Blazey, & Mitas, 2015)	Wiley	North America	Motivation
2015	(Eid, 2015)	SAGE	Europe, Africa, Asia	Loyalty
2015	(Eid & El-Gohary, 2015a)	Elsevier	Europe, Africa, Asia	Value
2015	(Eid & El-Gohary, 2015b)	Taylor & Francis	Europe, Africa, Asia	Satisfaction
2015	(Hung, Wang, & Tang, 2015)	Taylor & Francis	Asia	Service quality
2015	(Lee, Fu, & Chang, 2015)	Elsevier	Asia	Experience
2015	(Nyaupane, Timothy, & Poudel, 2015)	Taylor & Francis	Asia	Motivation
2016	(Damari & Mansfeld, 2016)	Elsevier	General	Experience

Source: Author's contribution (Continued)

Table 4. List of Reviewed Articles (N=64) Continued

Citation Year	Intext-Citation	Publisher	Region	Study Focus (Major Groups)
2016	(Ballantyne et al., 2016)	Wiley	Europe	Marketing Components
2016	(Brown, 2016)	Taylor & Francis	Europe	Motivation
2016	(Collins-Kreiner, 2016a)	Taylor & Francis	General	Concepts
2016	(Collins-Kreiner, 2016b)	SAGE	General	Concepts
2016	(Fourie, Rosselló, & Santana-Gallego, 2016)	Elsevier	General	Concepts
2016	(Kim et al., 2016)	Wiley	Europe	Value
2016	(Kruger & Saayman, 2016)	Elsevier	Africa	Visitor Management
2016	(Nilsson & Tesfahuney, 2016)	Elsevier	Europe	Concepts
2016	(Olsen & Wilkinson, 2016)	Taylor & Francis	Asia	Concepts
2016	(Suntikul & Dorji, 2016)	Taylor & Francis	Asia	Market Characteristics
2016	(Taheri, 2016)	Taylor & Francis	Asia	Experience
2016	(Wang et al., 2016)	SAGE	Asia	Motivation
2016	(Wong, McIntosh, & Ryan, 2016)	Taylor & Francis	Asia	Visitor Management
2017	(Battour, Ismail, Battor, & Awais, 2017)	SAGE	Asia	Satisfaction
2017	(Dafuleya, Gyekye, Oseifuah, Nethengwe, & Sumbana, 2017)	Wiley	Africa	Socio-Economic Effects
2017	(Hung et al., 2017)	SAGE	Asia	Socio-Economic Effects
2017	(Sánchez, Fernández, & Lara, 2017)	Taylor & Francis	Europe	Economic Effects
2017	(Shafaei, 2017)	Elsevier	Asia	Brand Equity
2017	(Terzidou, Scarles, & Saunders, 2017)	Taylor & Francis	Europe	Experience
2018	(Terzidou, Stylidis, & Terzidis, 2018)	Elsevier	Europe	Experience
2018	(Amaro et al., 2018)	Elsevier	Europe	Motivation
2018	(Terzidou, Scarles, & Saunders, 2018)	Elsevier	Europe	Motivation
2018	(Tkaczynski & Arli, 2018)	Elsevier	Australia	Motivation
2018	(Albayrak et al., 2018)	Taylor & Francis	Asia	Experience
2019	(Kim & Kim, 2019)	Wiley	Europe	Value
2019	(K. Huang, Pearce, Guo, & Shen, 2019)	Wiley	Asia	Experience
2020	(C. F. Lin & Fu, 2020)	Elsevier	Asia	Behavior
2019	(Shtudiner, Klein, Zwilling, & Kantor, 2019)	Taylor & Francis	Asia	Value
2020	(Luz, 2020)	Publisher	General	Concepts

Source: Author's contribution

tion with host countries. Moreover, cultural relations are also backed by religious norms (Fourie et al., 2016). Thus, developed as well as developing countries with globally widespread dominant religions and cultures can attract religious tourists by promoting their precious sacred and historical legacies.

The positive social and economic impacts of tourism and tourists on the host areas and residents are obvious, but an interesting phenomenon occurs when these effects mirror. For instance, the religious and cultural norms of local people in Bhutan are considered the major factor in forming positive attitudes towards tourists on sacred festivals. Due to religious moralities, people feel proud of the sense of cultural recognition while welcoming visitors. Thus, religious customs play an important role in the development of religious tourism. These values form people's attitudes in a positive way, which is beyond the materialistic intentions of economic gains from tourists. People do not perceive the visitors as a threat to their cultural heritage but see them as an opportunity for acknowledgment of their religious traits (Suntikul & Dorji, 2016).

CONCLUSION AND FUTURE RESEARCH DIRECTIONS

The chapter has presented a review of 64 research articles with the objective of exploring common research agendas in religious tourism, which have been explored during the last years from 2009 to 2020 in twelve leading journals of five publishers. On the basis of critical analysis of reviewed articles and research findings, the chapter proposes a definition of religious tourism and a three-dimensional framework with the role of media.

Religious Tourism Defined

Religion is a system of belief, faith, spirituality, and worship, so it should be the religious tourism. It is unfair to consider religious tourism a matter of '*visiting sacred sites*' only because the subject religious tourism does not represent any specific religion but such practices of all religions which involve travel and tourism. Followers of some religions also travel to various national and international places, which are not typically the 'sacred' or 'religious sites.' For instance, preaching groups in Islam travel among villages and cities to preach people about their religion. Similarly, people also use to travel and tour for motives which, as a whole or part, are not to perform religious worships directly but are indirectly backed by religious norms, obligations, and practices. Such as visiting graveyards to pay homage to loved ones by praying etc. Thus, it would not be inapt to say that "*religious tourism involves all such travels and tours which are directly or indirectly motivated by religious reasons of any faith.*"

Implications for Marketers and Researchers

Implications for marketers and researchers are discussed as under:

Religious Tourist Segmentation

Based on the religious and non-religious motivations, tourists can be segregated in different clusters in order to serve their needs better. The purpose of visit, behavioral, and socio-demographic characteristics of tourists can open new doors of opportunities for the industry and marketers. For example, these classes of visitors can be termed as devotees, explorers, and seeker. The presence of diverse clusters in religious tourists implies a customized focus on the segregation of this target industry by negating the idea of considering religious tourism as a homogenous market. These insights also create room for managerial and scholarly investigations as to whether the pilgrims should be considered as tourists or not?

Role of Media

Among various explored attributes to understand the complexity of tourist motivations and engagement, it is also exhibited that pilgrim motivations are a combination of identity, religiosity, experiences, and visual media. The transformation in human behavior and lifestyle brought by media plays an influential role in altering and fabricating the tourists' motivation to travel to sacred sites. Media is used to give awareness about and inspire people through fascinating tourism places. However, sometimes, it also serves as a communication channel to pose the myths and forging the realities about religious and historical sites. Tourists' involvement in religious sites tourism enhances brand awareness because they use to collect information about destinations and get inspired by others through various media channels. This is due to the fact that tourists love to talk and share information about religious places with friends, and in this process, media plays an important role in engaging tourists in positive word of mouth (Shafaei, 2017). Considering the importance of media channels, especially social media, the chapter proposes a conceptual three-dimensional model of Attract-Manage-Delight for sustainable development of the religious tourism industry. Practical implications and future research directions are discussed under each dimension.

Attract-Manage-Delight Model

Figure 3. Attract-Manage-Delight Model of religious tourism
Source: Authors' own source

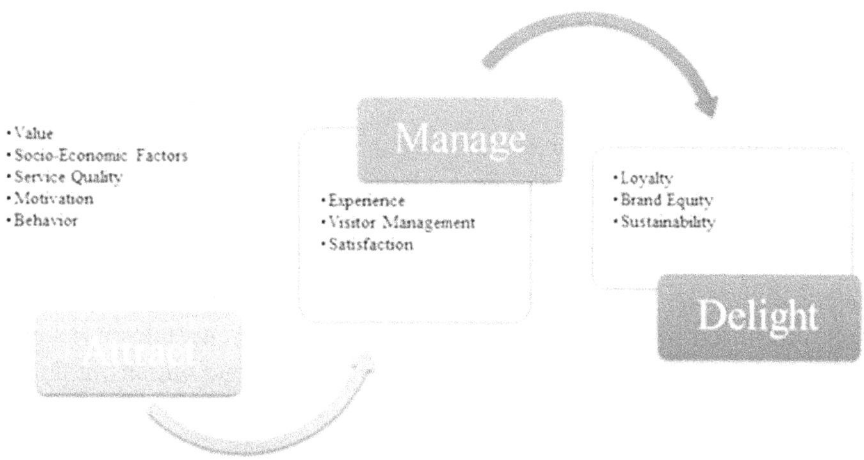

Attract Religious Tourist

Value

Novel research shows that tourists following different religions associate different values to the same things (Shtudiner et al., 2019). The religious, spiritual, destination, and tourism values that tourist desires can vary according to the religious and cultural norms, tourist needs, behavioral attributes, economic elements, and socio-demographics. Tourists' perceived values from a religious tour require in-depth analysis and literary debates with specific as well as generalized focus in order to find points of differences and similarities. Empirical studies need to be conducted to know the important sets of values that are crucial to be delivered to enhance religious tourism. A value analysis of religious tourists can also open the doors for new market segments to better serve the tourists according to their desired religious and non-religious values from the travel, destination, visit, and related activities. In this era of technological research transformation, market insights through big data and marketing analytics can also help enhance the understanding of religious tourists' perceived values.

Socio-Economic Factors

Tourism is a booming industry worldwide with a mega share in global GDP, and religious sites of a country can be a source of positive social and economic impact for it (Dafuleya et al., 2017; Saayman et al., 2014). The growing world population, expansion of urban areas, and busy work life make religious tours a way to inner peace and escape. These tours generate foreign exchange, employment opportunities, investments, cultural recognition, and social developments through interactions and values exchanges. The awareness and promotion of socio-economic benefits of religious tours for both hosts and guests can boost the sector by mitigating attitude barriers and providing assurance of innocuous tours. People feel an affiliation to such religious places where similar cultural and religious norms are present. So, developing countries, which are blessed with religious and cultural heritages, should promote such sites through media campaigns and give awareness to the related communities about their supporting culture and religion and should also educate local people about religious tourism benefits. Such as, Pakistan has promoted its asset of Sikhism origin, Kartarpura, by advertising and facilitating the Sikh community living across the globe. Researchers and marketers should explore the socio-economic benefits and the ways to avail these for attracting religious tourists.

Service Quality

Although service quality is a major topic of interest in the service industry, the usage of this tool in religious tourism is seen less evidently. This might be since religious sites and journeys are considered socially sensitive issues, and people feel it inappropriate to question the service quality at a temple, mosque or church, etc. However, as discussed, these sites and related tours are a source of huge socio-economic impact. Billions of people belonging to diverse faiths go for religious tours and eat food; live in hotels, hostels; hire travel services; give donations; shop vigorously; use telecom and other technology-oriented services and devices; devote days and months; hence, spend billions of dollars and precious time at religious sites and for religious tours. At every phase of their tour, service quality is being judged by them, and experiences are being shared through word of mouth. It is essential to identify desired service quality standards and elements that may attract people to hassle-free religious tourism. This will not only help in attracting more people to such tours but also support the sustainability of religious and historical heritages.

Motivation

There has been an extensive scholarly debate on tourists' motivations for religious tours since the beginning of this specific type of tourism research. But this area still needs further exploration with changing technological impacts. From numerous classifications of religious tourists' motivations, e.g., religious, spiritual, cultural, faith, relaxation, recreational, leisure, sightseeing, etc., it will not be inappropriate to divide religious tourists' motivation in two separate blocks, i.e., religious and non-religious motivations. Thus, the utilization of these blocks should also be focused on separate grounds as well. Scholars and marketers can attract religious tourists by targeting them upon these two types of motivations, motives or desires, etc. Media channels can present both the aspects of religious sites for inspiring tourists such as, the people who are more driven towards religious elements, can be attracted through advertising that side of the tour or destination. Similarly, people who are in search of non-religious benefits as well might be targeted through promoting other than pure religious attributes of the marketing components and market characteristics such a landscape, shopping, networking, local people, culture, etc. Keeping in mind the growth of the religious tourism industry, this topic demands more literary and managerial attention to further divide and utilize the religious and non-religious dimensions of motivation.

Behavior

Tourist's cognitive attributes, psychological states, personality traits, and social interactions are the most important and initial factors which affect a tourist's decision for religious tourism. In this aspect, opinion leaders and society members play an important role in forming people's behaviors (C. F. Lin & Fu, 2020). Similarly, during tours, tourists get inspired and motivated to see other pilgrims, respect others' interests, and create memories with strangers who, although they have diverse backgrounds yet come together for religious activities. Their interaction while traveling, eating and talking, etc., all become sources of happiness for them. Such group interactions motivate tourists to share each other's problems and create feelings of solidarity, which are unrestricted by any discrimination, ethnicity, and social status. (Terzidou et al., 2017). Tourist behavior with respect to personality and social interactions needs further exploration. Moreover, marketers should keep in mind the crucial impact of social interactions on decision making also. They should make marketing and promotional strategies for attracting people towards religious tourism in three aspects, i.e., tourist psychological attributes, tourist-to-society, and tourist-to-tourist interaction. Media channels, especially social media, should be utilized to develop effective marketing plans and generate positive eWOM.

Manage Religious Tourist

Experience

Religious tourist experience with respect to particular sites have been studied in the literature, and numerous attributes are identified which tourists seek during their tour or on a specific sacred place. The influence of media (television news and documentaries) on shaping a tourist's experience is vivid. Similarly, the visual expression of holy sites through enthralling tv news, presentation of political personalities, pictures, videos, etc., refine the destination image in people's minds and positively influence their expectations, willingness to visit and enhance the on-site experience of the tourists (Terzidou, Stylidis, et al., 2018). Awareness and familiarity about the travel process, activities, religious and non-religious destinations, cultural norms, and folks of the host places can enhance a tourist experience and confidence. Marketing campaigns and promotional tools can be used to reward religious tourists through positive spiritual, cognitive, and religious experiences.

Visitors Management:

Visitors' management at religious sites is relatively a less explored phenomena in literature as well as in practice. At some religious places such as a Buddhist sacred heritage in China, visitors are recognized under different categories by worshipers of these sites and local guardians. This is done to manage them according to their visit purposes while also safeguarding the religious activities and responsibility towards sacred sites. However, visitor management is a collective matter of respect and official administration (Wong, McIntosh, & Ryan, 2014). A well-organized pilgrimage, religious activity, and tour can not only positively enhance the associated values, benefits, and experience, but it can also support the sustainability of sacred sites and religious practices. By exploring and implementing visitor management strategies, the religious tourism industry can be flourished, and joyful experiences can be generated. Researchers can identify visitor management themes through analyzing religious site managers' interpretation of tourists and online content that visitors post about a particularly religious place.

Satisfaction

Research has shown that religion can moderate the relationship between religious tourists' perceived value and satisfaction. Similarly, religious norms can modify the tourist experience and level of satisfaction. Marketers and policymakers should manage to provide such products and services which are in line with the religious norms of that sacred site, e.g., in some religions, pork and wine are not allowed. Here again, media channels, especially social media, can be a convenient and less expensive source to understand the satisfaction and behavior influencing factors and to develop effective positioning strategies.

Delight Religious Tourist

Loyalty

Identification of desired values can help attract religious tourists by motivating them and influencing behaviors. Through acquiring a better understanding of religious tourists, tourism organizations, as well as policymakers, can build effective interactions, responses, and communication, which can, in turn, enhance satisfaction, loyalty, and visitor retention. By providing superior experiences, religious tourists can be delighted, and revisit intention can be enhanced. People use to post about their positive and negative experiences on social websites, which become a source of information for other tourists also. Through online content analysis, factors influencing religious tourists' word of mouth, decision making, revisit intention, and loyalty can be explored.

Brand Equity

By considering brand equity themes from media and religious tourism perspective, researchers and marketers can create a brand (sacred site) meaning through understanding the perception of religious tourists, e.g., pilgrims, about target religious destinations and tours. By promoting the spiritual imagery attributes and performance features, delightful experiences can be created. Moreover, pilgrims feeling associated with religious sites can be explored through their comments and narrative about that place. These sentiments can be written in the form of reviews and judgments etc. The marketers can implement such attributes for creating brand responses and unique proprietary assets, e.ge. The competitive advantage of religious tours.

Sustainability

Owing to the socio-economic impact of the industry, it is pertinent to develop such administrative and marketing strategies, which can increase the positive values while sustaining the religious heritages and attributes without harming the essence. Through media channels, visitors can be educated about the sensitive religious norms, worshiping activities and methods, the importance of the sites, host areas people and cultural values, etc. It can help in minimizing the negative impacts of commercializing religious tourism activities and can also create a positive image of the industry. The 2030 agenda of United Nations sustainable development goals (SDGs) also promote the benefits of the tourism industry and its impact on the economy, society, and environment. To achieve these SDGs, the religious tourism industry should be made sustainable by investigating the approaches and strategies which can make this industry sustainable by creating solidarity and a sense of responsibility among the beneficiaries.

REFERENCES

Albayrak, T., Herstein, R., Caber, M., Drori, N., Bideci, M., & Berger, R. (2018). Exploring religious tourist experiences in Jerusalem: The intersection of Abrahamic religions. *Tourism Management*, *69*(March), 285–296. doi:10.1016/j.tourman.2018.06.022

Amaro, S., Antunes, A., & Henriques, C. (2018). A closer look at Santiago de Compostela's pilgrims through the lens of motivations. *Tourism Management*, *64*, 271–280. doi:10.1016/j.tourman.2017.09.007

Andriotis, K. (2009). Sacred site experience. A Phenomenological Study. *Annals of Tourism Research*, *36*(1), 64–84. doi:10.1016/j.annals.2008.10.003

Andriotis, K. (2011). Genres of heritage authenticity. Denotations from a pilgrimage landscape. *Annals of Tourism Research*, *38*(4), 1613–1633. doi:10.1016/j.annals.2011.03.001

Ballantyne, R., Hughes, K., & Bond, N. (2016). Using a Delphi approach to identify managers' preferences for visitor interpretation at Canterbury Cathedral World Heritage Site. *Tourism Management*, *54*, 72–80. doi:10.1016/j.tourman.2015.10.014

Barajas, A., Lago-Peñas, S., & Sanchez, P. (2014). On the regional economic effects of the Pope's journeys: Evidence from the visit to Santiago de Compostela in 2010. *Tourism Economics*, *20*(2), 279–299. doi:10.5367/te.2013.0270

Battour, M., Battor, M., & Bhatti, M. A. (2014). Islamic attributes of destination: Construct development and measurement validation, and their impact on tourist satisfaction. *International Journal of Tourism Research*, *16*(6), 556–564. doi:10.1002/jtr.1947

Battour, M., Ismail, M. N., & Battor, M. (2011). The impact of destination attributes on a Muslim tourist's choice. *International Journal of Tourism Research*, *13*(6), 527–540. doi:10.1002/jtr.824

Battour, M., Ismail, M. N., Battor, M., & Awais, M. (2017). Islamic tourism: An empirical examination of travel motivation and satisfaction in Malaysia. *Current Issues in Tourism*, *20*(1), 50–67. doi:10.1080/13683500.2014.965665

Belhassen, Y., & Ebel, J. (2009). Tourism, faith, and politics in the Holy Land: An ideological analysis of evangelical pilgrimage. *Current Issues in Tourism, 12*(4), 359–378. doi:10.1080/13683500802469342

Bernstein, M. J., Zawadzki, M. J., Juth, V., Benfield, J. A., & Smyth, J. M. (2018). Social interactions in daily life: Within-person associations between momentary social experiences and psychological and physical health indicators. *Journal of Social and Personal Relationships, 35*(3), 372–394. doi:10.1177/0265407517691366

Bond, N., Packer, J., & Ballantyne, R. (2015). Exploring visitor experiences, activities, and benefits at three religious tourism sites. *International Journal of Tourism Research, 17*(5), 471–481. doi:10.1002/jtr.2014

Brown, L. (2016). Tourism and pilgrimage: Paying homage to literary heroes. *International Journal of Tourism Research, 18*(2), 167–175. doi:10.1002/jtr.2043

Buzinde, C. N., Kalavar, J. M., Kohli, N., & Manuel-Navarrete, D. (2014). Emic understandings of Kumbh Mela pilgrimage experiences. *Annals of Tourism Research, 49*, 1–18. doi:10.1016/j.annals.2014.08.001

Cheng, T. M., & Chen, M. T. (2014). Image Transformation for Mazu Pilgrimage and Festival Tourism. *Asia Pacific Journal of Tourism Research, 19*(5), 538–557. doi:10.1080/10941665.2013.764911

Choe, J., Blazey, M., & Mitas, O. (2015). Motivations of non-Buddhists visiting Buddhist temples. *Current Issues in Tourism, 18*(1), 70–82. doi:10.1080/13683500.2013.771627

Collins-Kreiner, N. (2016a). Dark tourism as/is a pilgrimage. *Current Issues in Tourism, 19*(12), 1185–1189. doi:10.1080/13683500.2015.1078299

Collins-Kreiner, N. (2016b). The lifecycle of concepts: The case of 'Pilgrimage Tourism.'. *Tourism Geographies, 18*(3), 322–334. doi:10.1080/14616688.2016.1155077

Dafuleya, G., Gyekye, A. B., Oseifuah, E. K., Nethengwe, T., & Sumbana, F. (2017). Religiously motivated travel and rural tourism in the Vhembe District of South Africa. *Tourism Economics, 23*(4), 911–918. doi:10.5367/te.2016.0564

Damari, C., & Mansfeld, Y. (2016). Reflections on pilgrims' identity, role and interplay with the pilgrimage environment. *Current Issues in Tourism, 19*(3), 199–222. doi:10.1080/13683500.2014.957660

Della Dora, V. (2012). Setting and Blurring Boundaries: Pilgrims, Tourists, and Landscape in Mount Athos and Meteora. *Annals of Tourism Research, 39*(2), 951–974. doi:10.1016/j.annals.2011.11.013

Eid, R. (2015). Integrating Muslim customer perceived value, satisfaction, loyalty and retention in the tourism industry: An empirical study. *International Journal of Tourism Research, 17*(3), 249–260. doi:10.1002/jtr.1982

Eid, R., & El-Gohary, H. (2015a). Muslim Tourist Perceived Value in the Hospitality and Tourism Industry. *Journal of Travel Research, 54*(6), 774–787. doi:10.1177/0047287514532367

Eid, R., & El-Gohary, H. (2015b). The role of Islamic religiosity on the relationship between perceived value and tourist satisfaction. *Tourism Management, 46*, 477–488. doi:10.1016/j.tourman.2014.08.003

El-Gohary, H. (2016). Halal tourism, is it really Halal? *Tourism Management Perspectives*, *19*, 124–130. doi:10.1016/j.tmp.2015.12.013

Fleischer, A. (2000). The tourist behind the pilgrim in the holy land. *International Journal of Hospitality Management*, *19*(3), 311–326. doi:10.1016/S0278-4319(00)00026-8

Fourie, J., Rosselló, J., & Santana-Gallego, M. (2016). Research note: Which God is good for tourism? *Tourism Economics*, *22*(1), 163–169. doi:10.5367/te.2014.0412

Gutic, J., Caie, E., Clegg, A., & Wiley, J. (2010). Visitors to an English Cathedral. *International Journal of Tourism Research*, *760*(June), 750–760. doi:10.1002/jtr.790

Hashim, N. H., Murphy, J., & Hashim, N. M. (2007). Islam and online imagery on malaysian tourist destination websites. *Journal of Computer-Mediated Communication*, *12*(3), 1082–1102. doi:10.1111/j.1083-6101.2007.00364.x

Henderson, J. C. (2011). Religious tourism and its management: The hajj in Saudi Arabia. *International Journal of Tourism Research*, *13*(6), 541–552. doi:10.1002/jtr.825

Heydari Chianeh, R., Del Chiappa, G., & Ghasemi, V. (2018). Cultural and religious tourism development in Iran: Prospects and challenges. *Anatolia*, *29*(2), 204–214. doi:10.1080/13032917.2017.1414439

Hu, X., & Huang, R. (2019). The impact of development of religious tourism on the host community - a case study of Mount E'mei in China. *Managing Religious Tourism*, 144–154. doi:10.1079/9781786393197.0144

Huang, J., & Chu, J. (2019). Tourist Experience at Religious Sites: A Case Study of the Chinese Visiting the Sheikh Zayed Grand Mosque. *Journal of China Tourism Research*, *00*(00), 1–25. doi:10.1080/19388160.2019.1658678

Huang, K., Pearce, P., Guo, Q., & Shen, S. (2019). Visitors' spiritual values and relevant influencing factors in religious tourism destinations. *International Journal of Tourism Research*, (October), 1–11. doi:10.1002/jtr.2337

Hughes, K., Bond, N., & Ballantyne, R. (2013). Designing and managing interpretive experiences at religious sites: Visitors' perceptions of Canterbury Cathedral. *Tourism Management*, *36*, 210–220. doi:10.1016/j.tourman.2012.11.022

Hung, K. (2015). Experiencing Buddhism in Chinese Hotels: Toward the Construction of a Religious Lodging Experience. *Journal of Travel & Tourism Marketing*, *32*(8), 1081–1098. doi:10.1080/10548408.2014.959632

Hung, K., Wang, S., & Tang, C. (2015). Understanding the normative expectations of customers toward Buddhism-themed hotels: A revisit of service quality. *International Journal of Contemporary Hospitality Management*, *27*(7), 1409–1441. doi:10.1108/IJCHM-12-2012-0264

Hung, K., Yang, X., Wassler, P., Wang, D., Lin, P., & Liu, Z. (2017). Contesting the commercialization and sanctity of religious tourism in the Shaolin Monastery, China. *International Journal of Tourism Research*, *19*(2), 145–159. doi:10.1002/jtr.2093

Hyde, K. F., & Harman, S. (2011). Motives for a secular pilgrimage to the Gallipoli battlefields. *Tourism Management*, *32*(6), 1343–1351. doi:10.1016/j.tourman.2011.01.008

Jafari, J., & Scott, N. (2014). Muslim world and its tourisms. *Annals of Tourism Research*, *44*(1), 1–19. doi:10.1016/j.annals.2013.08.011

Kim, B., & Kim, S. (2019). Hierarchical value map of religious tourists visiting the Vatican City/Rome. *Tourism Geographies*, *21*(3), 529–550. doi:10.1080/14616688.2018.1449237

Kim, B., Kim, S., & King, B. (2019). Religious tourism studies: Evolution, progress, and future prospects. *Tourism Recreation Research*, *0*(0), 1–19. doi:10.1080/02508281.2019.1664084

Kim, B., Kim, S. S., & King, B. (2016). The sacred and the profane: Identifying pilgrim traveler value orientations using means-end theory. *Tourism Management*, *56*, 142–155. doi:10.1016/j.tourman.2016.04.003

Kirillova, K., Gilmetdinova, A., & Lehto, X. (2014). Interpretation of hospitality across religions. *International Journal of Hospitality Management*, *43*, 23–34. doi:10.1016/j.ijhm.2014.07.008

Koren-Lawrence, N., & Collins-Kreiner, N. (2019). Visitors with their 'Backs to the archaeology': Religious tourism and archaeology. *Journal of Heritage Tourism*, *14*(2), 138–149. doi:10.1080/1743873X.2018.1503279

Krešić, D., Mikulić, J., & Miličević, K. (2013). The factor structure of tourist satisfaction at pilgrimage destinations: The case of Medjugorje. *International Journal of Tourism Research*, *15*(5), 484–494. doi:10.1002/jtr.1891

Kruger, M., & Saayman, M. (2016). Understanding the Zion Christian Church (ZCC) Pilgrims. *International Journal of Tourism Research*, *18*(1), 27–38. doi:10.1002/jtr.2030

Lee, T. H., Fu, C. J., & Chang, P. S. (2015). The support of attendees for tourism development: Evidence from religious festivals, Taiwan. *Tourism Geographies*, *17*(2), 223–243. doi:10.1080/14616688.2014.997280

Lin, C. F., & Fu, C. S. (2020). Cognitive implications of experiencing religious tourism: An integrated approach of means–end chain and social network theories. *International Journal of Tourism Research*, *22*(1), 71–80. doi:10.1002/jtr.2319

Lin, H., Zhang, M., Gursoy, D., & Fu, X. (2019). Impact of tourist-to-tourist interaction on tourism experience: The mediating role of cohesion and intimacy. *Annals of Tourism Research*, *76*(March), 153–167. doi:10.1016/j.annals.2019.03.009

Linderman, M. C. (2013). Royal accommodations: Pilgrim rest houses in early colonial South India. *International Journal of Contemporary Hospitality Management*, *25*(2), 187–203. doi:10.1108/09596111311301595

Lochrie, S., Baxter, I. W. F., Collinson, E., Curran, R., Gannon, M. J., Taheri, B., Thompson, J., & Yalinay, O. (2019). Self-expression and play: Can religious tourism be hedonistic? *Tourism Recreation Research*, *44*(1), 2–16. doi:10.1080/02508281.2018.1545825

Luz, N. (2020). Pilgrimage and religious tourism in Islam. *Annals of Tourism Research*, *82*(April), 102915. doi:10.1016/j.annals.2020.102915

Miguéns, J., Baggio, R., & Costa, C. (2008). Social media and Tourism Destinations: TripAdvisor Case Study. *Advances in Tourism Research*, *26*(28), 26–28.

Molinillo, S., Ximénez-de-Sandoval, J. L., Fernández-Morales, A., & Coca-Stefaniak, A. (2016). Hotel assessment through social media: The case of TripAdvisor. *Tourism & Management Studies*, *12*(1), 15–24. doi:10.18089/tms.2016.12102

Muslim, A., Harun, A., Ismael, D., & Othman, B. (2020). Social media experience, attitude and behavioral intention towards umrah package among generation X and Y. *Management Science Letters*, *10*(1), 1–12. doi:10.5267/j.msl.2019.8.020

Nilsson, M., & Tesfahuney, M. (2016). Performing the "post-secular" in Santiago de Compostela. *Annals of Tourism Research*, *57*, 18–30. doi:10.1016/j.annals.2015.11.001

Nyaupane, G. P., Timothy, D. J., & Poudel, S. (2015). Understanding tourists in religious destinations: A social distance perspective. *Tourism Management*, *48*(1), 343–353. doi:10.1016/j.tourman.2014.12.009

Olsen, D. H., & Wilkinson, G. (2016). Are fast pilgrims true pilgrims? The Shikoku pilgrimage. *Annals of Tourism Research*, *61*, 228–230. doi:10.1016/j.annals.2016.08.004

Olya, H. G. T., Lee, C. K., Lee, Y. K., & Reisinger, Y. (2019). What are the triggers of Asian visitor satisfaction and loyalty in the Korean heritage site? *Journal of Retailing and Consumer Services, 47*(November), 195–205. doi:10.1016/j.jretconser.2018.11.002

Ozcan, C. C., Bişkin, F., & Şimşek, Ç. (2018). *Regional Economic Effects and Marketing of Religious Tourism*, (August), 250–274. doi:10.4018/978-1-5225-5730-2.ch014

Qurashi, J., & Sharpley, R. A. (2017). *SMART Media Technologies impact on the Spiritual Experience of Hajj Pilgrims*. Academic Press.

Qureshi, F., Humayun, S., & Hummayun, S. F. (2015). Influence of Retail Atmospherics on Consumer Perception in Specialized Department Stores. *Management and Administrative Sciences Review*, *4*(5), 758–765.

Rinschede, G. (1992). Forms of religious tourism. *Annals of Tourism Research*, *19*(1), 51–67. doi:10.1016/0160-7383(92)90106-Y

Robina Ramírez, R., & Fernández Portillo, A. (2018). What role does touristś educational motivation play in promoting religious tourism among travellers? *Annals of Leisure Research*, *5398*. Advance online publication. doi:10.1080/11745398.2018.1561309

Saayman, A., Saayman, M., & Gyekye, A. (2014). Perspectives on the regional economic value of a pilgrimage. *International Journal of Tourism Research*, *16*(4), 407–414. doi:10.1002/jtr.1936

Sánchez, V. L., Fernández, M. D., & Lara, J. Á. S. (2017). Economic impact of a religious and tourist event: A Holy Week celebration. *Tourism Economics*, *23*(6), 1255–1274. doi:10.1177/1354816616675996

Scaffidi Abbate, C., & Di Nuovob, S. (2013). Motivation and personality traits for choosing religious tourism. A research on the case of Medjugorje. *Current Issues in Tourism*, *16*(5), 501–506. doi:10.1080/13683500.2012.749844

Shafaei, F. (2017). The relationship between involvement with travelling to Islamic destinations and Islamic brand equity: A case of Muslim tourists in Malaysia. *Asia Pacific Journal of Tourism Research*, *22*(3), 255–271. doi:10.1080/10941665.2016.1232741

Shahzad, S. J. H., Shahbaz, M., Ferrer, R., & Kumar, R. R. (2017). Tourism-led growth hypothesis in the top ten tourist destinations: New evidence using the quantile-on-quantile approach. *Tourism Management*, *60*, 223–232. doi:10.1016/j.tourman.2016.12.006

Shinde, K. A. (2010). Entrepreneurship and indigenous enterpreneurs in religious tourism in India. *International Journal of Tourism Research*, *12*(5), 523–535. doi:10.1002/jtr.771

Shtudiner, Z., Klein, G., Zwilling, M., & Kantor, J. (2019). The value of souvenirs: Endowment effect and religion. *Annals of Tourism Research*, *74*(October), 17–32. doi:10.1016/j.annals.2018.10.003

Shuo, Y., Ryan, C., & Liu, G. M. (2009). Taoism, temples and tourists: The case of Mazu pilgrimage tourism. *Tourism Management*, *30*(4), 581–588. doi:10.1016/j.tourman.2008.08.008

Suntikul, W., & Dorji, U. (2016). Local Perspectives on the Impact of Tourism on Religious Festivals in Bhutan. *Asia Pacific Journal of Tourism Research*, *21*(7), 741–762. doi:10.1080/10941665.2015.1069365

Taheri, B. (2016). Emotional Connection, Materialism, and Religiosity: An Islamic Tourism Experience. *Journal of Travel & Tourism Marketing*, *33*(7), 1011–1027. doi:10.1080/10548408.2015.1078761

Terzidou, M., Scarles, C., & Saunders, M. N. K. (2017). Religiousness as tourist performances: A case study of Greek Orthodox pilgrimage. *Annals of Tourism Research*, *66*, 116–129. doi:10.1016/j.annals.2017.06.009

Terzidou, M., Scarles, C., & Saunders, M. N. K. (2018). The complexities of religious tourism motivations: Sacred places, vows and visions. *Annals of Tourism Research*, *70*(February), 54–65. doi:10.1016/j.annals.2018.02.011

Terzidou, M., Stylidis, D., & Terzidis, K. (2018). The role of visual media in religious tourists' destination image, choice, and on-site experience: The case of Tinos, Greece. *Journal of Travel & Tourism Marketing*, *35*(3), 306–319. doi:10.1080/10548408.2017.1304316

Tkaczynski, A., & Arli, D. (2018). Religious tourism and spiritual leadership development: Christian leadership conferences. *Journal of Hospitality and Tourism Management*, *35*, 75–84. doi:10.1016/j.jhtm.2018.03.004

Tranfield, D., Denyer, D., & Smart, P. (2003). Towards a methodology for developing evidence-informed management knowledge by means of systematic review. *British Journal of Management*, *14*(3), 207–222. doi:10.1111/1467-8551.00375

Triantafillidou, A., Koritos, C., Chatzipanagiotou, K., & Vassilikopoulou, A. (2010). Pilgrimages: The "promised land" for travel agents? *International Journal of Contemporary Hospitality Management*, *22*(3), 382–398. doi:10.1108/09596111011035963

UNWTO. (2011). Religious Tourism in Asia and the Pacific. In Religious Tourism in Asia and the Pacific. doi:10.18111/9789284413805

UNWTO. (2020). Compendium of Tourism Statistics. *Data (København)*, 2014–2018.

Vittengle, J. R., & Holt, C. S. (2000). Getting Acquainted: The Relationship of Self-Disclosure and Social Attraction to Positive Affect. *Journal of Social and Personal Relationships*, *17*(1), 53–66. doi:10.1177/0265407500171003

Wang, W., Chen, J. S., & Huang, K. (2016). Religious Tourist Motivation in Buddhist Mountain: The Case from China. *Asia Pacific Journal of Tourism Research*, *21*(1), 57–72. doi:10.1080/10941665.2015.1016443

Wong, C. U. I., McIntosh, A., & Ryan, C. (2013). Buddhism and tourism. Perceptions of the monastic community at pu-tuo-shan, china. *Annals of Tourism Research*, *40*(1), 213–234. doi:10.1016/j.annals.2012.09.004

Wong, C. U. I., McIntosh, A., & Ryan, C. (2014). Visitor Management at a Buddhist Sacred Site. *Journal of Travel Research*, *55*(5), 675–687. doi:10.1177/0047287514563164

Wong, C. U. I., McIntosh, A., & Ryan, C. (2016). Visitor management at a Buddhist sacred site. *Journal of Travel Research*, *55*(5), 675–687. doi:10.1177/0047287514563164

WTTC. (2020). *Travel & tourism: Economic of all net new jobs created*. Retrieved from https://wttc.org/en-us/Research/Economic-Impact

Zamani-Farahani, H., & Henderson, J. C. (2010). Islamic tourism and managing tourism development in islamic societies: The cases of Iran and Saudi Arabia. *International Journal of Tourism Research*, *12*(1), 79–89. doi:10.1002/jtr.741

Zamani-Farahani, H., & Musa, G. (2012). The relationship between Islamic religiosity and residents' perceptions of socio-cultural impacts of tourism in Iran: Case studies of Sare'in and Masooleh. *Tourism Management*, *33*(4), 802–814. doi:10.1016/j.tourman.2011.09.003

KEY TERMS AND DEFINITIONS

Pilgrimage: A journey to a religious place that sometimes also involves circular movement around a sacred site.

Religious Tourism: Religious tourism involves all such travels and tours which are directly or indirectly motivated by religious reasons of any faith.

Chapter 14
Religious Tourism in Zimbabwe: A Stakeholders` Perspective

Zibanai Zhou
https://orcid.org/0000-0002-7918-2200
Midlands State University, Zimbabwe

ABSTRACT

The impact of religious tourism on destination economies is clearly an important question for public policymakers wishing to develop a robust tourism industry. This chapter aimed to offer insights into the contemporary state of religious tourism in Zimbabwe. Using interpretive qualitative data obtained through observation and in-depth interviews conducted with 30 key informants purposively selected from religious tourism stakeholders, the study established that although teeming with bright prospects, as manifested through the availing of employment opportunities to communities around religious tourism sites, religious tourism in Zimbabwe remains on the back foot. Findings revealed that religious tourism is hamstrung by the absence of a proper regulatory framework, spike in stampede cases jeopardising the safety and security of pilgrims, congestion and environmental pollution. Policy recommendations were proffered.

INTRODUCTION

Religious tourism has been well known for its huge economic gains and development of communities across the globe; however, there is dearth of literature on the impact of religious tourism activities on host communities in Zimbabwe. Nyikana (2017, p.1) noted that "... religious tourism has been in existence for centuries, there is still a dearth of academic focus on this tourism niche area especially in the developing world context".

Religious tourism is one of the least explored types of tourism in Zimbabwe despite its huge resources, opportunities and perceived attendant benefits on the socio-economic front. A perusal of available literature shows that religious tourism has not been adequately discussed in Zimbabwe`s academia; with a glaring paucity of information on its contemporary state, socio-economic significance, and as well as its underlying challenges.

DOI: 10.4018/978-1-7998-5792-1.ch014

There has been profound developments in Zimbabwe`s religious tourism landscape prompting authorities to have keen interest on religious tourism regarding its potential impact on the broader travel and tourism industry (Sunday Mail, 11 March, 2018). However, the extent of religious tourism development in contemporary times, its contribution to the social and economic spheres, and the attendant challenges are still not clear at policy making level in the context of a developing nation state like Zimbabwe.

This chapter addresses an infrequently researched aspect of religious tourism as a contemporary issue and seeks to establish how it intersects with the main stream tourism industry in Zimbabwe. The present and primary concerns of this chapter is to provide a rigorous empirical analysis of the niche tourism market, adding to the limited information around this subject matter within Zimbabwe`s tourism discourse. The study relied upon interpretive qualitative data obtained through in-depth interviews and observation, since there is hardly documentation readily available on religious tourism in Zimbabwe.

Religious tourism has enormous economic benefits for the tourism industry as it is a catalyst to local economic and social development (Kruger and Saayman, 2014; Yolal et al., 2016; Getz and Page, 2015). Religious tourism is significant in so many ways to the Zimbabwean economy which has not been performing well. Zimbabwe`s economy is currently characterised by high unemployment in the formal sector, shrinking job opportunities, and endemic poverty (Mkono, 2012, Nyahunzvi, 2015, Zhou, 2016). It is against a back drop of a significant slump of international tourism traffic into the country stemming from the bad publicity the country has been subjected to at the international scene, that religious tourism is rekindling hope for a quick turn- around of the tourism sector. Concerted efforts are underway to scale up tourism development through policy interventions like granting the sector an export status, designating certain areas special economic zones status, as well as characterising tourism as a low hanging fruit that can quickly revitalise the moribund economy (Ministry of Finance and Economic Development; Ministry of Environment, Climate, Tourism and Hospitality Industry, 2018).

Extant scholarship on tourism in Zimbabwe discusses various aspects of tourism from different perspectives (Mkono, 2012, Nyahunzvi, 2015, Zhou, 2018) but seldom explored religious tourism. A wave of religious tourism has engulfed the Zimbabwean tourism landscape lately, but the understanding of religious tourism by Zimbabwean policy makers and the academia in the broader sphere of tourism development trajectory is limited. This situation is further compounded by the scarcity of rigorous studies on religious tourism by the academia in Zimbabwe. Although Duran-Sanchez et al., (2018) acknowledged that publications on religious tourism are currently in an exponential growth stage, it is the United States that is at the forefront of religious tourism research, which is in sharp contrast on Zimbabwe.

Resultantly, the chapter seeks to contribute towards this. To this effect, the chapter seeks to: establish the current state of religious tourism development in Zimbabwe, amplifying on measures which the government of Zimbabwe (GOZ) and the Zimbabwe Tourism Authority (ZTA) have put in place to support further development of religious tourism; determine the socio-economic significance of religious tourism in Zimbabwe; and identify challenges associated with religious tourism in Zimbabwe. It is envisaged that a nuanced understanding of religious tourism in Zimbabwe can be gleaned from interrogating these objectives and the stance taken by the ZTA and Ministry of Environment, Climate, Tourism and Hospitality industry (MECOTHI) in supporting the development of religious tourism.

While it must be acknowledged that religious tourism has generated euphoria among Zimbabwean tourism operators, it still remains that very little is known about how it is regulated and managed; what are its socio-economic contribution, what are its attendant challenges, what is its contemporary state, are there any interventions to be made by Zimbabwean authorities if any, and at what point? Going forward these are critical questions that need to be unpacked. The chapter is aimed at addressing these issues hence contributing towards a better understanding of religious tourism governance and management issues.

The decline in agriculture and mining sectors respectively, can be filled up by tourism which has been on the rebound at global level, yielding benefits to local communities. At international level, tourism traffic was forecast to hit the 1.6 billion mark in 2020 and further increase to 1.8 billion in 2030 if the United Nations World Tourism Organisation (UNWTO) projections are anything to go by, barring the adverse impacts of the covid-19 pandemic (UNWTO, 2019; WTTC, 2018). Religious tourism consume a multiplicity of services at the destination area, and consistent with this thesis Akama (2002) posits that tourism`s economic justification is underpinned by its ability to generate foreign currency, contributing to the nation`s economic growth and replenish balance of payment (BOP) account. Ramukumba (2016) weighed in stating that tourism creates employment, has positive impact on upstream and downstream economic sectors, spur infrastructural development, and foster global peace.

To this end, religious tourism continue to grow steadily every year across the globe with nearly 10 million Muslims thronging Mecca in Saudi Arabia during the annual pilgrimage; 23 million visiting the Tirupati temple in India (Koldowski and Martin, 2008), and millions of pilgrims attend religious festival at mount Moria annually in South Africa (Kruger and Saayman, 2014). In this milieu, Zimbabwe`s aggregate tourist arrivals at national level stand around 2.5 million, with scope to increase tourist inflows if it harness religious tourism. Zimbabwe is drawing inspiration from countries such as Saudi Arabia, Nigeria, and India which have done extremely well on religious tourism development front.

The Concept of Religious Tourism

Wright (2009) explains religious tourism as travelling to a religious destination site, with a spiritual intent and leisure travel with fellowship intent. Religious tourism motivation comes from the desire for spiritually invigorating formative and cultural experience in which they grow and became enlightened by what they encounter at a destination. Rinsched (1992) and Shinde (2007b) illuminated further by defining religious tourism as visiting places of religious importance or pilgrimage sites where visitors aim to fulfil religious needs and recreational needs.

Turner and Turner (1979) framed religious tourism theoretically as comprising of three cardinal elements which are motivation, journey and destination. When a journey has an exclusively religious character with a motivation for spiritual fulfilment, a visitor is deemed to be a pilgrim. Motivations can be obligatory or non obligatory, the former being travel to performing life cycle ritual at the sites of family deities- such sites are sanctioned with the faith and religious system to which one belongs. George (2014) argues further that religious tourism is a type of tourism driven by a sense of obligation; it is undertaken largely for fellowship, pilgrimage or missionary. A similar narrative is proffered by Rinschede (1992) who argues that religious tourism is any trip motivated either exclusively or partly by religious reasons. The other one has to do with sentimental reasons, spiritual uplifting and finding solutions to mundane problems. However, other scholars like Kim and Kim (2018); Wang et al., (2016); Amarv et al., (2018); Abbate and Nuove (2013) contend that there is a battery of factors which motivate people to visit religious sites or events, ranging from cultural, spiritual, and traditional factors. In essence, tourist motivation is multi-dimensional.

It therefore suffices that religious tourism is mainly tourism engaged not only with the intention to visit places of worship, but also holy places and shrines (Adesina, 2003; Okpoko et al., 2008; Odum, n.d). Andriotis (2009) and Nicolaides (2016) put forth a compelling argument to say that people are motivated to participate in religious tourism because they want to be spiritually uplifted and closer or counted to the Almighty.

Literature Review

Overview of religious tourism

Mankind have been travelling for religious purposes since time immemorial (Chigozie et al., 2018). Literature documented that around 400 million people travel across the globe in pursuit of religion on the basis of the UNWTO`s estimate that 25 per cent of over a billion of contemporary international tourism traffic is for religious purposes with potential for further growth (UNWTO, 2017). If the UNWTO`s permutations are anything to go by, it implies that out of the forecast 1.6 billion travellers for 2020, it means not less than 400 million people will be travelling for religious purposes.

As a growing phenomenon, and given its impact on the immediate host communities, religious tourism has justifiably caught the attention of academics, business people and politicians. As for the business people, it represents future commercial value or opportunities for additional income and increased employment (Oslen, 2012; Simone-Charteris and Boyd, 2010).

Extant literature is replete with nation state, territories and jurisdictions which have morphed into renowned citadels of religious tourism destinations that entice multitudes of followers annually, for example, SCOAN in Nigeria, Enlightened Christian Group (ECG) of Bushiri in South Africa, United Family International Church (UFIC) in Zimbabwe, Saudi Arabia`s Mecca, St Peters in The Vatican City in Italy, Israel and Varamisa in India (Fourie et al., 2014). In particular, Mecca in tourism circles is the metaphorical word employed to refer to a destination that is visited by volumes of people.

It is against this background that understanding the interplay between the current status of religious tourism development in Zimbabwe, its socio-economic significance, and the attendant challenges is necessary especially in the context of policy makers and political leadership`s quest to develop effective strategies that not only tap into this niche market but also inform policy and plan appropriately so that the country benefit the most out of the religious phenomenon.

The Zimbabwean religious landscape is punctuated by pluralism in line with the country`s constitution. Zimbabwe is predominantly a Christian nation state with 80 per cent of its citizenry being ardent believers of Christianity (ZimStats, 2012). The constitution of Zimbabwe guarantees freedom of worship, co-existence and pluralism of religious faith (Zimbabwe Constitution, 2013). The recent proliferation of Pentecostal churches is testimony to a very liberal constitution of the republic of Zimbabwe. The existence of a triumvirate of mainline churches, traditional indigenous African churches and Pentecostal cum charismatic churches that attract worshippers from within and outside Zimbabwe has been a boon for religious tourism. To buttress this, there are basically three most dominant categories of religious organisations in Zimbabwe, which are the so- called traditional mainline churches, traditional indigenous African churches and Pentecostal cum charismatic churches.

The main line churches are essentially relics of colonisation, and therefore very conservative, and quite dominant in Zimbabwe`s religious landscape. They have a very wide geographically foot print spread across the country. Examples encompass Roman Catholic, Anglican, Lutheran, Salvation Army, Methodist, Reformed Church, Seventh Day Adventist (SDA), and Jehovah`s Witness to mention but a few.

The second category constitutes religious organisations firmly rooted in African indigenous traditional values and culture. These are apostolic churches commonly known as white garment apostolic sects. These in addition to being firmly rooted in African indigenous traditional values and culture, are conspicuous by their white garment trade mark. They seldom use the Bible and anchored in spiritism. They are dominant in both urban and rural areas. Examples encompass such as Zion Christian Church (ZCC), Johane Masowe and its variant break away sects, and The African Apostolic church led by Paul Mwazha, Zion Christian Church (ZCC) and other apostolic sects.

Then the third category is known as Pentecostal or charismatic religious organisations, which in most of the cases are off-shoots of mainline churches. These coalesce around a charismatic leader or personality, and are anchored in gospel of prosperity, miracles and healing. They are mainly concentrated in urban areas, and currently command an exceptionally huge following in Zimbabwe. Examples include Zimbabwe Assemblies of God in Africa (ZAOGA), Apostolic Faith Mission (AFM), United Family International Church (UFIC), Prophetic, Healing and Deliverance (PHD), Family of God (FOG), Spirit Embassy, Heartfelt International Ministries and so on.

Major religious events in Zimbabwe

Zimbabwe being a predominantly Christian society as highlighted earlier, observes Easter holiday. Ordinarily, it then follows that most major religious events are concentrated during the Easter holidays. The whole Easter week is packed with religious festivities across the country, with a hive of activity among the Christian community, ranging from mainline, indigenous and Pentecostal churches: judgement night, conventions, holy communions, mass celebrations, visiting sacred centres /places of worship like Mbungo, Defe Dopota, Bernard Mzeki, and camp meetings are some of the key events undertaken during this period.

Of late there have been so many religious activities lined up in Zimbabwe during Easter holidays. High profile religious events like the UFIC`s judgement night, PHD ministries night of turn around and cross over night and all night prayers, Jehovah`s Witness convention, SDA` camp meetings, Roman Catholic, Anglican, ZAOGA, ZCC, and Johane Masowe gatherings which keep on attracting an avalanche of pilgrims from as far afield as the UK, USA, China, South Africa, and Botswana. Prominent business people and sports personalities, politicians, musicians and celebrities have of late been attending these religious events or gatherings (Sunday Mail, 12 February, 2017). There is a narrative that political leaders and sitting presidents often attend such church gatherings as an endorsement to churches and also as a foothold for their future political gain when they seek political office. Most religious tourism attendees in Zimbabwe are from within, SADC region, Europe and America. Most are mainly Zimbabwe nationals working in the Diaspora.

Jackowski (2000); Olsem, (2006); and Rojo, (2007) concurred that religious tourism has grown significantly over the past few decades, and there is potential for religious tourism to play a far much more significant role on nation state`s economic arena if the niche form of tourism is given due attention and support by political leadership. On a comparative note, Oslen (2006) posit that unlike other niche tourism markets such as sport tourism, religious tourism is still very much in its infancy. This therefore raises concern on the quantum of the socio-economic spin-offs that nation states are losing in potential revenue, hence the need to come up with mechanisms which accurately captures this. Henderson (2011) observed that there is a conflation between tourism and religion, as the former motivates people to travel for the benefit of the latter. Religious tourism can generate significant income for the host cities and communities, in addition to a range of other positive impacts like sprucing of public utility spaces, cultural exchange and enhancing community cohesion (Tassiopoulos, 2011; Getz and Page, 2015; Yolal et al., 2016; Bama and Tichaawa, 2015).

The Zimbabwean tourism industry generated slightly above a US$1 billion in 2018, and efforts are currently underway to increase the contribution to 15 per cent of the country`s gross domestic product (GDP), by targeting 5 million tourists that would generate US$5 billion (Ministry of Tourism and Hospitality Industry, 2018).

Since the attainment of self rule in 1980, Zimbabwe`s economy has been anchored on mining and agriculture as the mainstay of the economy, with tourism coming on third position in importance. Tourism has been represented as the proverbial low hanging fruit, known for generating foreign currency, spurring infrastructural development, creating employment among a battery of other favourable impacts which transcend the socio-economic-ecological divide (Zhou, 2017). Zimbabwe` tourism sector has been predicated on Victoria Falls, medieval Great Zimbabwe monuments both of which are UNESCO recognised world heritage sites, lake Kariba, and The Eastern Highlands, inclusive of wildlife (the big five) (Mkono, 2012). On the religious tourism front, Zimbabwe trails behind South Africa but does well as compared to the rest of other southern Africa countries. South Africa boasts of 10 million visitors, and Zimbabwe boasts of 2.5 million (SATOUR, 2018; ZTA, 2018). There is heighted religious tourism in South Africa, courtesy of the ZCC church with its annual pilgrimage to Moria. South Africa could mutate into southern Africa region`s religious tourism power house. Kruger and Saayman (2014) and Muller (2015) documented that during Easter holiday millions of people troop to and gather in Moria for religious purposes, and some ZCC members end up sleeping in tents, buses and cars at Moria in Limpopo owing to unavailability of sufficient accommodation space at the site. Zimbabwe therefore aspires to be at the level of Saudi Arabia`s city of Mecca, in which 10 million pilgrims visit the city annually for religious purposes; and even at a better position like India` Kashmir in which 17 million visit for religious purposes.

Methods

The study utilised a combination of in-depth interviews with key informants and direct site observation. Fieldwork comprised of detailed interviews with representatives from key groups of social actors including government officials, religious functionaries, visitors, priests, organisers, and hotels. Interviewees were chosen on the basis of their being knowledgeable about religious tourism. Interview questions revolved around social and economic importance of religious tourism, challenges of religious tourism, and the state of religious tourism in Zimbabwe.

A qualitative approach was employed in this study as it is strongly recommended when meanings and individual interpretations are sought (Bloomberg and Volpe, 2008). A qualitative approach was also a clear choice as this study was exploratory (Denzin and Lincoln, 2003); given that there has not been an examination of Zimbabwe`s religious tourism. An exploratory study is recommended when there is limited prior research (Denzin and Lincoln, 2003). The present study was based on predominantly a series of open-ended, semi-structured interviews conducted with 18 female, 12 male tourism stakeholders that were found at the site during the research.

The religious tourism attendees were based at multiple sites, that is, national sports stadium, Waterfalls` PHD headquarters, Mbungo in Masvingo, Celebration Centre in Harare, Bernard Mzeki in Mutoko, Roman Catholic, Paul Mwazha, Marange, Bocha`s Johane Masowe, SDA, Jehovah`s Witness, ZAOGA, AFM`s Rufaro Conference centre. These religious tourism sites and venues have been operational quite for some time, and have became centres of attractions for parishioners in drawing huge crowds in Zimbabwe`s religious calendar, hence, it was deemed they were ideal sites that would yield richer data on religious tourism. These study sites were well established and represent the key religious tourism events in Zimbabwe, attended by multitudes of people coming from outside Zimbabwe.

Two preliminary visits were made in February 2019 facilitated by the ZTA to acquire the necessary permissions from religious organisation leadership of the sites/area before the interviews were conducted over a week in April 2019. The field visits facilitated the acquisition of contextual understanding and establishing rapport as well as gaining ethical approval. Furthermore, a pre-testing study was undertaken among another group of attendees that had a religious convention in Gweru (central Zimbabwe) led to the refinement of the data collection instrument and procedures. The pre-testing study was restricted to willing attendees.

Resultantly, 30 semi-structured interviews were conducted at the religious tourism events/sites. Semi-structured interviews were chosen because they allowed the researcher to probe for additional information and to seek clarification. The above dove tails with Stake and Torrance`s (2005, p. 33) remark that semi-structured interviews offer insights "into respondents" memories, and explanations of why things have come to be what they are, as well as descriptions. The interviews held were confined to people who gave their informed consent and each interview session lasted an average 30 minutes. The research participants were asked a broad range of questions in order to gain an understanding of the immediate impact and implication of religious tourism on Zimbabwe`s tourism development trajectory. The line of questioning explored, among other aspects, the perceived socio-economic benefits of religious tourism, its current state of development and growth prospects, strategies to develop it further, and challenges associated with it. In addressing these questions, the study builds upon and updates the existing literature, (for example, Rojo, 2007; Henderson, 2011; Nyikana, 2017) around religious tourism in developing countries.

The interpretation and analysis of the qualitative materials was an on-going process that proved inseparable from the fieldwork. In the field, after each interview, the data was analysed to inform subsequent interviews and to capture preliminary emergent themes as recommended by Mertens (2005). After the fieldwork, the interview material was later thematically analysed to reveal emergent themes following Veal`s (2006) guidelines. In line with the dictates of qualitative research, interviews were discontinued when theoretical saturation or data redundancy was reached (Jennings, 2005); henceforth 9 out of 30 interviews held contained no new themes. The interviews were complemented by site observation. The next section provides a detailed and concurrent presentation and discussion of the dominant themes that emerged during the interactions with research participants.

Findings and Discussion

Five themes were identified in line with objectives in the analysis of the qualitative materials that were obtained from the interview and personal observations. These were contemporary state of religious tourism, measures put in place to support the development of religious tourism, religious tourism`s socio-economic importance, and challenges of religious tourism. Each of these five themes is discussed below.

Current State of Religious Tourism Development in Zimbabwe

Results showed that religious tourism by the measure of tourist numbers and supportive infrastructural facilities, has not yet fully developed even though authorities have now recognised the potential of this type of tourism as a way to diversify the tourism product base, and are now beginning to take it seriously by embedding it in relevant tourism statutes and Acts, such as the national tourism sector development strategy (NTSDS). The appointment of a religious tourism advisor coupled with a battery of incentives supportive of the development and growth of religious tourism like land donations to religious tourism organisations, attending religious gatherings and events, organising national day of prayer where religious leaders are given a prominent role to play interfacing with state institutions underlines the government`s commitment in developing religious tourism. According to a government tourism development official, Informant B, (personal interview, 2 February, 2019), responsible for tourism policy formulation, a lot of consultations were being made with relevant stakeholders, and a lot of groundwork had already been covered behind the scenes, and they were finalising a detailed all encompassing policy framework specifically for religious tourism. All stakeholders were in unison. It was stated that there was now a religious tourism advisor at ministry level, which showed the government`s commitment in taping into the promising religious tourism niche market. It was also revealed that plans were under way to partner with institutions of higher learning to commissioning research on religious tourism so that there is a better understanding of this phenomenon supported by empirical research for the collective benefit of the country.

Measures Put in Place to Support Religious Tourism Development

The informant highlighted that as a way of supporting religious tourism the GOZ had provided state land in cities for religious organisation`s related activities, waived operating permits on commuter omnibus during religious tourism events so that people are easily ferried to religious events, offered land to build lodges, conference facilities, sanctioned the use of open spaces adjacent to urban centres as places of mass gathering, provision of security services during events, provision of bigger state owned venues, for example, the national sports stadium, to accommodate large church gatherings, for example, UFI, Jehovah`s Witness convention held at the NSS, representation of government at ministerial level where not only the minister of tourism attending churches` religious events but also other political elites, high ranking government officials and prominent business people. Government has also conferred religious tourism site status to some church building like the Celebration centre in Harare as a way of promoting religious tourism.

Socio-Economic Significance of Religious Tourism in Zimbabwe

Narratives from across the spectrum of religious tourism stakeholders gave the impression that religious tourism was very much significant on the socio-economic front. Respondents outlined the importance of religious tourism towards infrastructural development, its linkages with other sectors of the economy, projecting a positive image about Zimbabwe, enhancing of social cohesion and peace; and provision of free medical information through wellness expos.

i) *Infrastructural developments and amenities*

An official representing the association of religious organisations in Zimbabwe, Informant A (personal interview, 2 February 2020), responsible for public relations, stated that churches were making huge investment in the country`s tourism infrastructure through religious tourism. Religious organisations have built the Celebration Centre in the elite Borrowdale suburb which is a very big convention centre. Other accommodation facilities built by churches cited were PHD`s Yadah lodge, UFIC`s mount Hampden Family Centre, Chitungwiza`s UFI centre, ZCC`s Mbungo Estate. Such state of the art accommodation space and convention centres add to the nation`s total accommodation stock, and provide the much needed investment in the accommodation and conferencing sub-sector. These infrastructural facilities are a seed bed for further development of tourism. These findings conform to what was established by Getz and Page (2015); and Yolal et al., (2016) that religious tourism help in infrastructural developments. This is very significant in that at national level, Zimbabwe has got a serious deficit of accommodation and not only that but also the brutal reality that the accommodation sub-sector has suffered from under investment for a very long period of time. The informant impressed upon the fact that these utility spaces set up by churches for religious purposes can be hired out by any other stakeholder when there are no religious activities taking place, which is quite commendable.

ii) *Event tourism*

Results showed that religious tourism stimulate tourism business through religious activities lined up on the religious calendar. There are many religious events which have become permanent features in Zimbabwe. These were identified as: ZCC Defe Dopota in Gokwe and Mbungo annual gatherings/ festivals in Masvingo, Roman Catholics at Mutemwa in Mutoko, Anglicans at Bernard Mzeki shrine in Marondera, Paul Mwazha, Johane Masowe`s Gandanzara shrine, ZAOGA`s prayer mountain in Bindura, AFM`s Rufaro conference centre, PHD`s cross over might, UFIC`s Judgement, Spirit Embassy etc. These religious events are held at a time when the tourism sector is on low season. Furthermore, religious tourism through hosting of mega events provides tourism operators with hard facts or reality on the ground in relation to the exact facilities they have, in terms of deficits and excesses. This would then direct operators to introspect and work towards increasing investment in conventions, accommodation facilities, inspire product development, tailor make a tourism product that best meet the needs of the religious tourism market. Religious events create demand from outside the host communities for tourism products resulting in religious tourism events attendees boosting local economies. For example, when Jehovah`s Witness held their almost week long convention at the national sports stadium, a sport facility owned by Harare city council generating revenue for the local authority in particular and government in general through taxes. In addition, residents in suburbs like Warren Park, Westlea, Belvedere, Kambuzuma, and Dzivarasekwa literally camped at the site selling airtime and other necessities to people. To this end, the local communities benefited through entrepreneurial opportunities presented by the convention in the national sports stadium. A similar pattern or trend is noticed when UFIC hold its religious event at the same venue. Equally the same, residents of Highfields, Waterfalls, Glen Norah, Southerton, Glen View and Mbare throng PHD ministries headquarters when there is an event to cash-in through entrepreneurial activities.

iii) *Upstream and down- stream linkages during event tourism*

Respondents were unanimous that religious tourism affect almost every sector of the economy positively especially when there is a religious tourism event thus generating employment. The interconnectedness with various economic sectors prior to and during the religious tourism event result in every key player in the tourism value chain benefiting in one way or another. Religious tourism therefore creates a multiplicity of employment despite their short term nature. In essence, religious tourism attendees would have used airline carriers, long distance buses/coaches, commuter omnibuses, taxis, which facilitate in the transportation of religious tourism attendees from point A to point B; and the majority of them staying in formal accommodation facilities, whilst a few others would stay with friends and relatives, they will still use local catering organisations for food and beverages provisions especially fast food outlets and other outside catering companies, for example, Chicken inn, Chicken Slice, TN Grill; and tent making companies like Rooneys benefit through supplying food, tents and chairs respectively to the event attendees. Local taxis ferry people from the city centre to the venue where the event takes place, vendors benefit through selling airtime, bottled water, soft beverages, bracelets, bangles, handkerchiefs, stickers, posters and other religious related paraphernalia.

In addition, security services companies are contracted to control the crowds, whilst bottling water companies, soft drinks, bread and confectionary sector tend to benefit as well. All these commercial business entities are given a new lease of life during the period of the event. Whilst a few pilgrims would use tents which they would have pitched, they would contribute in a small way towards the local economy by injecting some income. Through interactions with respondents, most couples had intentions to visit the Victoria Falls, Gonarezhou and The Eastern Highlands after the religious event, whilst a fair number of pilgrims indicated that had already visited such prime tourist centres prior to the commencement of the religious event taking advantage of them being in Zimbabwe courtesy of the religious event. Such a chain of tourism activities are quite impactful on the broad travel and tourism sector at macro level. The above shows that a multiplicity of actors benefit in one way or the other courtesy of religious tourism events. These findings are in synch with Ramukumba (2016); Oslen (2012) and Akama (2002) who praised the multiplier effect of tourism in creating employment in various sectors of the economy.

To this end, religious tourism is very significant to Zimbabwe since it has the potential to create employment opportunities and generates revenue in a way as no other kind of tourism does. It is critically important to also note that religious tourism`s sphere of influence transcends social, cultural, political, economic and ecological divide. In addition, as Akama (2002); Ramukumba (2016) argued that the economic argument for tourism is predicated on its ability to generate foreign currency, BOP, employment, favourable upstream and downstream impacts on other economic sectors, incentivise infrastructural developments among a host of other positive ramifications at micro and macro level. This is equally applicable to Zimbabwe, and this is something that is so significant to Zimbabwe given that the country is currently in the throes of scarcity of foreign currency, high unemployment, in need of investment in critical sectors of the economy, characterised by a polarised society and a host of other socio-economic challenges.

iv) *Image propping- marketing destination Zimbabwe`s tourist attractions*

Another equally significant contribution of religious tourism to Zimbabwe is in the area of image building. Results indicated that religious tourism can be a cost effective vehicle to spruce up and up market destination Zimbabwe. According to the national regulatory authority official, Informant C, and a journalist from the print media, Informant A (personal interviews, 2 February 2020), responsible for MICE tourism and covering events respectively, there are pre and post religious event tours, which religious leaders engage in accompanied by journalists and political leadership. These are excursions to prime tourist resort areas, for example, when TB Joshua visited Zimbabwe, he proceeded to the Victoria Falls with his advance team and entourage, and that is huge business for the local tourism operators. This enable religious leaders, who are opinion leaders to sample Zimbabwe`s tourism product and have the correct perception. During such tours filming and video shooting of tourist attractions take place which would be later on used in the production of documentaries for marketing Zimbabwe, showcasing the tourism product, and thus reaching a wider audience.

Respondents also stated that during the religious tourism event, media houses would be present and such events are given wider coverage in newspapers, radio, and TVs, and at times may be broadcast live reaching millions of people across the globe. The overall net effect is to correct the negative perception about Zimbabwe, amplifying on the peace and tranquillity prevailing in the country, and its hospitable and peace loving Zimbabweans contrary to misrepresentations and distortions made in some hostile media outlets.

The understanding is that when religious tourism event attendees finally return to their countries they will market Zimbabwe as a tourist destination through word of mouth (WOM), thus they become free ambassadors. It has been highlighted earlier on that the country has gone through a rough political and economic patch that generated negative publicity, which subsequently caused the issuance of negative travel advisories; this could be an opportunity to deconstruct the adverse image tag through religious tourism. In this vein, religious tourism provides a platform for the publicity and marketing of Zimbabwe`s tourism product, through viral marketing.

v) *Social cohesion and peace*

Findings confirmed that religious tourism events help to cement social cohesion, entrench good relations between countries, nationalities, communities, and people of different races. An official with the Zimbabwe Council of Churches, Informant B (personal interviews, 2 February, 2020) said:

One good about religious tourism is that it brings together people of different backgrounds, nationalities, social and cultural orientation, social and economic standing, political persuasions under one roof, so it breaks all artificial societal, geographical, economic and political barriers.

This has the potential to reshape and change certain pre-conceived stereo types, and racial prejudice. The finding sits well with what Bama and Tichaawa (2015) found that religious tourism enhances family and community cohesion. Tourism thrives when there is peace, harmony and tolerance among and between nations, races, and citizens. Given that Zimbabwe is a highly polarised society characterised by hatred stemming from political orientation especially between the urban areas and rural areas, religious tourism can help in this regard for people in reaching out to one another. It is therefore clear that in spite of the economic significant of religious tourism, the diverse nationalities of attendees coming from rural and urban areas, having communion under one roof, provide considerable diversity and pluralism and thereby creating fertile ground for promoting peace, tolerance, co-existence and cohesion.

vi) *Health/wellness expos*

This was an interesting and unique finding in which respondents revealed that religious tourism helped the socio-economically disadvantaged members of society through provision of some basic services. An official from one religious organisation, Informant C (personal interviews, 3 February 2020) responsible for organising church events, stated that during their organisation`s religious event, they have a deliberate wellness program session which normally run concurrently with the religious tourism event. According to the informant this is meant to assist the under privileged and the general public on matters regarding their health especially lifestyle diseases. The religious organisation has a dedicated team of medical doctors who would provide medical information free of charge. This is also part of the church`s benevolent gesture of ploughing back to the community, that is, social corporate responsibility. However, it was stressed that these services are only offered during the course of the religious event or festivities in which there is free consultation, screening, free medical information, for example, SDA. An attendee was livid that he had been screened of cancer during the course of the festival, Informant A (personal interview, 20 February 2020) had this to say:

I am very excited and appreciate the medical procedure that has been administered on me. The wellness sessions are very helpful for our health. It is my prayer that these sessions would be given more time in future, so that many people can benefit.

Also as if taking a cue from such a noble deed, of late government departments like the registry in charge of the issuance of birth certificate records have also been visiting religious gatherings to facilitate issuance of birth records especially for those disadvantaged groups or communities, for example, in Marange community, they target Johane Masowe gatherings, and ZCC taking advantage of such huge gathering for religious tourism. One pilgrim or attendee, Informant F (personal interview, 31 January 2020), who had come to attend a religious event was very happy and livid saying:

I would like to thank the government for setting up an office here; I now have a birth certificate for my child. I could not travel from Gokwe since I did not have bus fare, but I came for the church gathering using transport hired by the church and I now have the document.

This is quite commendable given that such communities have challenges in travelling to centres where such services can be accessed.

Challenges of Religious Tourism

Results revealed that despite the numerous positive impacts spawned by religious tourism, there are some grey areas of concern as highlighted by respondents, which were a disconnect between tourism and religion in the aspect of regulation, unaccountability of some of the revenue generated during religious events, environmental pollution, breach of community peace and tranquillity, laxity in safety and security systems resulting in injuries and even death caused by stampedes, traffic congestion, non use of formal accommodation facilities, as well as short term nature of employment opportunities generated by religious events.

i) *Disconnect between tourism regulation and religion.*

The main issue of concern which was even admitted by government and religious organisation representatives relates to the lack of a proper legal regulatory framework for religious tourism due to its embryonic state. This aspect poses a dilemma to stakeholders in case of disease outbreak, death and permanent injuries caused by stampedes. There has been ugly and unpalatable incidents involving death of pilgrims before, during or soon after religious tourism gatherings, for example, in West Africa in Nigeria a church guest lodge collapsed at SCOAN and there was loss of life. This was followed by litigation of the church by aggrieved parties citing poor workmanship as the main cause of the collapse of the building and subsequent loss of lives. In neighbouring South Africa, in October 2015, there was a stampede during an ECG religious event at a stadium and scores of people were injured. In the same vein, in Zimbabwe in the town of Kwekwe, there was a stampede during PHD ministries` night religious event or vigil, and there was needlessly loss of life. These are incidents frowned upon since no one is held accountable for such loss of life. To date no one has been held accountable on cases of stampedes due to laxity of the security and policing system. The above makes a compelling case for the need of a very clear legal framework to handling such incidences, so that there is closure on the part of the grieving family members who would have lost their beloved ones or who would have been injured whilst attending religious events. There are also reported cases where children just go missing during religious events and really this is a cause of concern. Churches` protocol departments act as information gate keepers when it comes to any news deemed adverse for the religious organisation or event. The media is also equally complicit in this as it under report such incidences. The findings support what Shinder (2018) framed as institutional vacuum between religious tourism and the civilian institutions with the mandate to regulate the affairs of tourism industry.

ii) *Partial accountability of the potential revenue generated through religious tourism*

Another challenge relates to some of the potential revenue generated through religious tourism which is not accountable, or rather escapes the formal channels particularly for those attendees who stay with families, relatives and friends. The same can be said with regards to donations to pastors, tithes, and related gifts all of which do not directly get into the formal financial system for accountability so that one can have a true picture relating to the real economic value and impact of religious tourism. It is a fact that some religious tourism participants do not use formal accommodation facilities, however, they tend to stay with their acquaintances like friends, relatives and family members. This finding corroborates what Kruger and Saayman (2014) also found at Moria that some ZCC members slept in tents, buses and cars, since there are no other tourism related infrastructures or amenities during the annual pilgrimage at the site in Limpopo. This category of participants does not make any contribution to the tourism sector financially. The challenge is on how to make sure that the tourism sector benefits the most out of religious tourism events and income generated is accounted for.

iii) *Breach of community peace and environmental pollution through littering*

This is a challenge that communities in the vicinity where a religious tourism event is taking place have to endure unfortunately without any recourse. The use of public address system generates noise that disturbs communities. Another area of concern relates to environmental pollution through littering and excess carbon foot print in and around the religious event venue, for example, plastics, cans etc if there is no provision of proper disposal facilities, some towns/cities may be overwhelmed during the religious tourism event and fail to cope in terms of service delivery. It is the immediate communities surrounding the religious tourism event venue that bears the brunt of such ecological malfasceance. Furthermore, while religious actors actively participate and promote religious tourism they hardly shoulder responsibilities of addressing the negative environmental impacts. Shinder (2018) intimated that there is an institutional vacuum in dealing with both direct and indirect impacts of religious tourism, which is a cause for concern. Religious tourism conveners should behave as good corporate citizens to protect the environment and abiding by the regulations imposed by the secular authorities.

iv) *Congestion*

To add on, religious tourism is synonymous with congestion especially so during the UFI, PHD festivities. Traffic congestion at the national sports stadium (NSS) and at Mbudzi traffic circle in Harare, become permanent features disrupting the normal flow of traffic and other people engaging in their day to day chores are a cause for concern.

CONCLUSION AND RECOMMENDATIONS

This chapter has contributed in the area of religious tourism where there is still limited rigorous academic scrutiny in the context of Zimbabwe. Furthermore, the chapter sought to deepen our understanding of religious tourism as a new form of tourism with huge potential to significantly redefine the contours of Zimbabwe`s tourism industry. An argument has been advanced that the repercussions of religious tourism transcend the ecological, political, socio-economic and cultural divide, hence it can be deemed an agent of peace in the context of Zimbabwe. Zimbabwe destination authorities are now better informed in relation to how to formulate appropriate policies aimed at accelerating further development and nursing the religious tourism aspect for the greater good of the Zimbabwean economy. Areas of concern that need policy intervention relates to the regulation of religious tourism activities, so that they are aligned to the tourism master plan, and minimise ecological foot print among others.

In exploring the interplay between current status of religious tourism development in Zimbabwe, its socio-economic significance and attendant challenges, this chapter has glossed over many issues which can be deemed as limitations but also opportunities for further research. The chapter has been primarily concerned with Christianity yet there is a fair share of the Muslim community in Zimbabwe, which has been excluded in this study. However, it is vital to note that despite the above limitation of the current study, it can be considered indicative for sketching the contours of religious tourism in Zimbabwe. It is prudent to explore the same aspect of religious tourism in the context of the Muslim community for a comprehensive understanding of religious tourism. In light of this, the following recommendations were proffered. Collaboration between ZTA and religious organisations when planning events, expedite the crafting of a legal regulatory framework to regulate religious tourism, and develop a mechanism that captures religious tourism potential income so that its true economic value or significance is ascertained. Religious tourism has bright prospects of significantly contributing to the economic fortunes of Zimbabwe as country embarks on an ambitious plan to attain an upper middle income status by 2030.

REFERENCES

Abbate, C., & Nuovo, S. (2013). Motivation and Personality Traits for Choosing Religious Tourism. A Research on the case of Medjugorje. *Current Issues in Tourism*, *16*(5), 501–506. doi:10.1080/13683500.2012.749844

Adesina, B. (2003). *Ife Museum, Enuwa, Ife-Ife*. National Commission for Museum and Monuments.

Akama, J. S. (2002). The Role of government in the development of tourism in Kenya. *International Journal of Tourism Research*, *4*(1), 1–13. doi:10.1002/jtr.318

Amaro, S., Angela, A., & Carla, H. (2018). A Closer Look at Santiago de Compostela's Pilgrims through the Lens of Motivations. *Tourism Management*, *64*, 271–280. doi:10.1016/j.tourman.2017.09.007

Andriotis, I. C. (2009). Sacred sites experience: A phenomenological study. *Annals of Tourism Research*, *36*(1), 64–84. doi:10.1016/j.annals.2008.10.003

Bama, H. K. N., & Tichaawa, T. M. (2015). Major sporting events as responsible tourism: Analysis of the 2012 Africa Cup of nations (AFCON) tournament in Port Elizabeth, South Africa. *African Journal for Physical Health Education. Recreation and Dance*, (December), 205–219.

Bloomberg, L. D., & Volpe, M. (2008). *Completing your questionnaire dissertation: A roadmap from beginning to end*. London: Sage.

Chigozie, O. J., Uchema, O., & Ngozi, E. (2018). Impact of Religious Tourism on Host Communities: The Case of Awhum Monastery. *American Journal of Social Sciences*, *6*(3), 39–47.

Constitution of the Republic of Zimbabwe. (2013). *Zimbabwe Constitution*. https://parlzim.gov.zw/component/k2/download/1290_da927a81557040d47c3a2c27012f6e1

Denzin, N. K., & Lincoln, Y. S. (2003). Introduction: The discipline and practice of qualitative research. In N. K. Denzin & Y. S. Lincoln (Eds.), *The landscape of qualitative research. Themes and issues* (pp. 1–45). Sage.

Duran-Sanchez, A., Alvarez-Garas, J., Rio-Rama, M. de la. C., & Oliveira, C. (2018). *Religious Tourism and Pilgrimage. Bibliometric Overview*. MDPI.

Fourie, J., Rosello, J., & Santana-Gallego, M. (2014). *Religion, Religious Diversity and Tourism*. Stellenbosch Economic Working paper, November 9/14.

George, R. (2014). *Marketing Tourism in South Africa. Cape Town*. Oxford University Press.

Getz, D., & Page, S. (2015). *Progress and prospects for event tourism research*. Tourism Management. doi:10.1016/.tourman.2015-.03.007

Henderson, J. C. (2011). Religious Tourism and its management; the Hajj in Saudi Arabia. *International Journal of Tourism Research*, *13*(6), 541–552. doi:10.1002/jtr.825

Jackowski, A. (2000). Religious Tourism: Problems with terminology. *Peregrimus Craaviensis*, *1*, 64–74.

Jennings, G. R. (2005). Interviewing: A focus on questionnaire techniques. In B. W. Ritchie, P. Burns, & C. Palmer (Eds.), *Tourism research methods* (pp. 99–118). Wallingford: CABI.

Karar, A. (n.d.). *Impact of Pilgrimage in Tourism at Haridwa. Anthropological survey of India.* www.krepublishers.coml

Mertens, D. M. (2005). *Research Methods in education and psychology: Interpreting diversity with quantitative and qualitative approaches* (2nd ed.). Sage.

Ministry of Environment, Climate Change, Tourism and Hospitality Industry. (2018). http://www.environment.gov.zw

Ministry of Finance and Economic Development. (2019). *Incentives to Boost Tourism.* https://www.zimtreasury.gov.zw

Mkono, M. (2012). Zimbabwe's Tourism Woes in the Last Decade: Hindsight Lessons for African Tourism Planners and Managers. *Tourism Planning & Development*, *9*(2), 205–210. doi:10.1080/21568316.2011.630749

Muller, R. (2015). The Zion Christian church and Global Christianity: Negotiating a tightrope between localisation and globalisation. *Religion*, *45*(2), 174–190. doi:10.1080/0048721X.2014.992111

Nicolaides, A. (2016). Moria Tourism: Eastern Orthodox and Roman Catholic Pilgrimage. *African Journal of Hospitality, Tourism and Leisure*, *5*(4).

Nyahunzvi, D. K. (2015). Negotiating livelihoods among Chivi curio traders in a depressed Zimbabwe trading environment. *Anatolia- An International Journal of Tourism and Hospitality Research, 26*(3), 397-407. 10. 1080/13032917.2014.974065

Nyikana, S. (2017). Religious Tourism in South Africa: Preliminary analysis of a major festival in Limpopo. *African Journal of Hospitality, Tourism and Leisure*, *6*(1), 1–9.

Odum, C.J. (n.d.). *Religious tourism. The Nigerian perspective: A survey of religious tourism in Nigeria.* Academic Press.

Okpoko, P. U., Okonkwo, E. C., & Dimiyi, C. (2008). *Understanding Tourism.* Nsukka University of Nigeria Press.

Olsen, D. (2012). Negotiating Identify at Religious Sites: A Management Perspective. *Journal of Heritage Tourism*, *7*(4), 359–366. doi:10.1080/1743873X.2012.722642

Oslen, D. H. (2006). Management issues for religious heritage attractions. In D. J. Timothy & D. H. Oslen (Eds.), *Tourism, Religion and Spiritual Journey* (pp. 104–118). Routledge.

Ramukumba, T. (2016). The role of government in support of tourism businesses: A perspective from guest houses in the Eastern district region. *African Journal of Hospitality, Tourism and Leisure*, *5*(1), 1–11.

Rinschede, G. (1992). Forms of religious tourism. *Annals of Tourism Research*, *19*(1), 51–67. doi:10.1016/0160-7383(92)90106-Y

Rojo, D. M. (2007). *Religious tourism: the way to Santiago. MA European Tourism Management*. Bournemouth University.

SATOUR. (2018). *South Africa Tourism*. https://www.statista.com

Shinde, K. (2018). Governance and Management of Religious Tourism in India. *International Journal of Religious Tourism and Pilgrimage*, *6*(1), 7. Doi.org/10.21427/D7K42m

Shinde, K. A. (2007b). Visiting sacred sites in India: religious tourism or pilgrimage. In R. Raj & N. D. Morpeth (Eds.), *Religious Tourism and Pilgrimage Festivals Management: An International Perspective* (pp. 186–199). CABI. doi:10.1079/9781845932251.0184

Simone-Charteris, M., & Stephen, B. (2010). The Development of Religious Heritage Tourism in Northern Ireland: Opportunities, Benefits and Obstacles. *Turizam: medunarodni znanstveno-strucnicasopis, 58*, 229-257.

Stake, S., & Torrance, H. (2005). Case study. In B. Somekh & C. Lewin (Eds.), *Research methods in the social sciences* (pp. 33–40). Sage.

Tassiopoulos, D. (2011). *New Tourism Ventures* (2nd ed.). Juta.

The Sunday Mail. (2017, Feb. 12). Religious tourism booms in Zimbabwe. *The Sunday Mail.*

The Sunday Mail. (2018, Mar. 11). Religious tourism racks in big bucks. *The Sunday Mail.*

Turner, V., & Turner, E. (1978). *Image and Pilgrimage in Christian Culture*. Columbia University Press.

UNWTO. (2011). *Religious Tourism in Asia and the Pacific: United Nations world Tourism Organisation*. UNWTO.

UNWTO. (2017). *Tourism Highlight 2017*. https://planeta.com/unwto-tourism-highlights-2017/

UNWTO. (2019). *International Tourism Highlights*. https://www.unwto.org/publications/international-tourism-highlights-2019-edition

Veal, A. J. (2006). *Research methods for leisure and tourism: A practical guide* (3rd ed.). Pearson Education.

Wang, W., Chen, J., & Huang, K. (2016). Religious Tourist Motivation in Buddhist Mountain: The Case from China. *Asia Pacific Journal of Tourism Research*, *21*(1), 57–72. doi:10.1080/10941665.2015.1016443

WTTC. (2018). *Travel and Tourism economic impact 2018 world*. https:www.wttc.org/-/media/files/reports/2018/power-and-performance-rankings-2018.pdf

Yolal, M., Gursoy, D., Uysal, M., Kim, H., & Karacaoglu, S. (2016). Impacts of festivals and events on residents' well-being. *Annals of Tourism Research*, *61*, 1–18. doi:10.1016/j.annals.2016.07.008

Zhou, Z. (2016). Post 2010 evaluation of Zimbabwe as a preferred tourist destination. *African Journal of Hospitality, Tourism and Leisure*, *5*(1), 1–15.

Zhou, Z. (2016). Implications of China' Approved Tourism Destination Status (ADS) on Zimbabwe's Tourism Environment: Insights from Local and Government Tourism Groups in Zimbabwe. *International Journal of Hospitality and Tourism Systems*, *9*(1), 44–52.

Zhou, Z. (2018). The Tourism Sector: A bright light on Zimbabwe's depressed economic environment. *African Journal of Hospitality, Tourism and Leisure*, *7*(1), 1–15.

Zimbabwe National Statistical Agency. (2012). *Census 2012 National Report*. www.zimstat.co.zw/sites/default/files/img/publications/Population/NationalReport.pdf

ZTA. (2018). *Zimbabwe Tourism Authority Annual Statistics Report*. ZTA.

KEY TERMS AND DEFINITIONS

Christianity: A religion based on the life and teachings of Jesus Christ.

Mainline Church: Mainstream church which was started by missionaries from another continent.

Mainstream Tourism Industry: An economic sector concerned with travelling of the public away from the places of origin utilising hotel accommodation facilities, food, and drink.

Muslims: People who follow or practice Islam.

Pentecostal or Charismatic Church: A church which share with Pentecostals and emphasise on the gift of the spirit who remain a part of a main line church.

Pilgrimage: A trip or journey to a religious event, site or shrine.

Religious Festival: Refers to a period of special importance marked by adherents to a particular religious sect.

Religious Shrine: Is a holy or sacred site dedicated to a specific deity, hero, martyr, saint or similar figure of respect, where they are venerated and worshipped.

Religious Tourism: A special type of tourism that is strongly motivated by religious reasons. Its variants are faith tourism or spiritual tourism.

Spiritual Pilgrimage: A physical journey toward a place of sacred or religious significance.

Tourism Stakeholder: Any group or individuals who can affect or is affected by the activities and operations of the tourism sector.

Traditional Indigenous African Church: An African initiated church, is a Christian church independently started in Africa by Africans rather than by missionaries from another continent.

Wellness Tourism: Travel associated with the pursuit of enhancing one`s health and well-being through physical, psychological or spiritual activities.

Chapter 15
Use of Social Media by Tourists at Religious Tourism Destinations in India

Lakhvinder Singh
Government College, Kaithal, India

Dinesh Dhankhar
Kurukshetra University, India

ABSTRACT

The chapter assesses the social media usage by tourists during their visit to religious tourism destinations and evaluates rationale and challenges faced by religious tourism establishments and other stakeholders in using social media as tools for marketing of religious tourism. The study used secondary data sources for getting a theoretical understanding of social media and religious tourism. Primary data conveniently collected through field survey from tourists and tourism establishment at religious tourism destination to know the extent of social media usage. The chapter indicating increased use of social media by tourists in searching, planning, booking, and sharing experiences about different products and services. The study identifies a crucial role of social media in the marketing and development of religious tourism along with certain challenges related to using as a mechanism to promote destination. Finally, implication and recommendations for different stakeholders were given to ensure adequate presence on social in more effective manner.

DOI: 10.4018/978-1-7998-5792-1.ch015

INTRODUCTION

Religious places are among the most ancient of travel destinations such as holy shrines, temples, rivers, and other sacred sites. This has become a new stem of the tourism industry, and become a preferred form of travel throughout the globe since time immemorial (Rodrige, 2012). The areas covered by the form of tourism not just limited to visits to the sacred places, but also encompass all kinds of travel that are motivated by traveling to perform pilgrimage, attending religious meetings or discourses, and traveling to participate in religious fairs and festivals (Chattopadhyay, 2006).

Taking into account of these discussions, more than 300 million people have been traveling to religious visits every year (UNWTO, 2011). Religious tourism has assumed an increasing significance in a country like India where cultural, religious, and spiritual diversity prevalent and people give utmost importance to visit sacred places (Jha, 1978). Religious tourism places in India are spread throughout the breadth and width of the country and major destination includes Badrinath in Uttarakhand, Dwarika in Gujarat, Puri in Odhisha and Rameshwaram in Tamilnadu, Kumbha fair destinations, Golden Temple in Amritsar, Bodh Gaya in Bihar, Mount Abu in Rajasthan, Ajmer Sharif (Rajasthan) and churches of Basilica of Bom Jesus and Se Cathedral in Goa (Poddar, Goswami & Shastri, 1951). In a survey more than 100 million tourists were noticed for religious and pilgrimage purposes with eight out of the top ten destinations of the country being religious tourism sites (Koldowski & Martin, 2008).

This growing phenomenon of religious tourism also gave birth to increasing demands and expectations among tourists regarding better products and services. This paradigm shift forced concerned stakeholders to come up with suitable strategies to successfully market the destination for increasing survival and revenue generation at the destination (Rinsched, 1992; Sharma, 2010). The infrastructural development such as information & communication technology and computerized reservation system (CRS) marched the growth of religious tourism in the country (Sharma, 2010). Moreover inventions of internet facilities have further added new dimensions to the marketing of religious tourist places (Eck, 2012). This paradigm shift has influenced the way destination lanners market their products and services and consumption of facilities and services by tourists. Therefore, the marketing of religious destinations through the application of digital platforms such as the internet, website, and electronic media is required for transforming the business of tourism (Jauhari & Sanjeev, 2010; Kumar & Singh, 2015).

The emerging technology-based marketing especially 'social media' that appeared after the developments in Internet technology are thought to be useful tools in the field of tourism and hospitality. The use of social media is directly related to the purchase of decisions of present and potential tourism consumers. This relationship can be used as an efficient support in promoting and marketing of destinations in the religious tourism industry. The travelers during religious visits also used to share their past travel experiences with friends and relatives on social media. This tool can be increasingly used for travel plans of others, from choosing destinations to hotels and restaurants (Yesil, 2013). Hence, it is important to examine the impact of social media on religious tourism destinations from the perspective of its different stakeholders including tourists and service providers and destination planners.

This brings to notice that, despite the increasing role of social media in religious tourism, the research area remains unexplored. Moreover, there is no particular attention has been given by researchers for getting insights on these aspects about a specific type of topic related to its effect on religious tourism. The present chapter considers this research gap and critically assesses the relationship between religious tourism and social media. To begin with, the chapter evaluates the way tourists are using social media platforms for making their travel arrangements. Similarly, whereas tourism businesses adopt social media to enhance their marketing activities, from this point of view, this study evaluates the factors that contribute to reliance on social media when engaged in travel research and planning.

THEORETICAL BACKGROUND ON SOCIAL MEDIA AND RELIGIOUS TOURISM NEXUS

The present section of the chapter described the interrelationship between social medial and religious tourism. The section first elaborates on the role of social media in the tourism industry to provide a deep insight into social media concepts and the need to apply this technology in the tourism sector. The later part of the section defines the interrelationship between social media and religious tourism as a sub-sector of the tourism industry. The analysis of the section helps to provide a strong theoretical background for the study.

Tourism is considered as the fastest growing industry throughout the world with its consequential benefits to destinations including economic and other benefits (Pforr & Hosie, 2009). In this regard, it is critical to observe that apart from its tremendous economic growth, the tourism industry also experiences a high rate of competition among firms. As a result, businesses that operate in the industry, they need to rely on innovation to not only reduce the prices of their products and services but also offer new, high-quality products that can attract the attention of the target customers (Alsos, Eide & Madsen, 2014). Hence, there is a need for transformations in the process of serving to present and potential customers by tourism & hospitality firms during the stay of tourists at the destination. Kotler and Gertner (2002) observed, in this case, by marketing itself as a brand, a country can focus on promoting products, services, and experiences that better attract tourists rather than using a general approach to tourism marketing. With this in mind, it is essential to have an insight of different factors of technology in marketing itself as a brand. This chapter will therefore, examine the role of social media in the religious tourism industry in India.

Unlike traditional media, there are a variety of definitions of social media, the content of which prepared by the users. This is described as a virtual environment where that provide follower and user with opportunity of content marketing and share it with other people (Kaplan & Haenlein, 2010;_Carneiro, Eusébio, and Caldeira, 2018). After this initiative, a large number media outlets, that can be called social media, began operating in a virtual environment, until the foundation of all-time record-breaking usage of social media such as Facebook, Twitter, LinkedIn, Reddit, MySpace, YouTube, Newsvine, StumbleUpon, Digg ve Xing are considered to be the top players of all social media networks.

The digital platform of social media influences the various businesses including tourism in different spheres such as promotion, marketing, management, and research (Zeng & Gerritsen, 2014; Liana, 2017). Travelers through this technology can search, organize, share, and annotate their travel stories and experiences (Leung, Law, Van, & Buhalis, 2013; Shaista & Khan,2019) through a large number of different types of social media sites, which a formidable number of people make use of during their trip planning process.

Tourism practitioners therefore, use social media to create interest in their properties, introduce promotions, drawing the media attention, and creating awareness, recognition, and promotion of their brands (Murphy, Chen & Cossutta,2016). This emerging technology extends several benefits to tourists also such as searching, booking, and sharing about tourism products, hence, it is imperative to consider the social media behavior of people in terms of a tourist destination (Green, 2009: Tafveez,2017).

Religious tourism is growing as an important segment of the tourism industry throughout the world. Thus, about other touristic products it was determined that the religious phenomenon remains scarcely explored. The specialists consider that both the religious tourism market and marketing through using digital technology such as social media in the field are less studied. It is just recently that researchers, travel agencies, and governments noticed the great number of travelers starting, from religious reasons, towards sacred locations. Thus, as a direct result of the promotion of religious sites, the number of those that start has increased (Cristea, Apostol & Dosescu, 2015).

The emerging religious tourism market had to cooperate with the technology and media constantly, to the survival of the destination. Information and comments in the media about destinations can be effective on the decisions of people who are planning to travel to sacred places (Yesil, 2013). Growing numbers of social media use by travelers for search and consume travel information created by other travelers for their travel planning and then share their experiences when they return from their trips. Given the experiential nature of tourism, the information created by other travelers is even more important and influential in the search and decision-making process than when considering other types of purchases (Gretzel, 2011). Thus, religious tourism planners also should make use of social media networks which may be considered the most developed and sophisticated promotional media tool of the digital age as it allows service providers to form one-on-one relationships with their clientele in a competitive environment.

Fotis, Buhalis, and Rossides (2012) have also reported social media as effective, the possibility, for people, to make changes in travel plans increases. That's why, it can be said that religious tourism which addresses a growing market, should cooperate closely with social media networks to promote her travel programs to people who make travel plans according to the comments published in social media networks.

The purpose of this study is to find out the importance of the role that social media can play in the promotion of religious tourism and possible use by tourists that may arise from this interaction.

Since the concepts of religious tourism and social media are new compared with the traditional ones, there are not many studies in which we can examine the interaction process. Some studies examine the relationships between the tourism sector and social media (Fotis et al., 2012; Ranjha, 2010). There are also a lot of studies on the tourism sector and traditional media relations in general (Scott, 2007; Gonzales, 2008; Harrett & Benjamin, 2009; Bingqui, Bing, Taolong & Ruofei, 2011; Popescu & Corboş, 2012). As the chapter in the above discussion the existence of a research gap on social media use in religious tourism and the probable application and hurdles that may arise from this interaction.

METHODOLOGY

To arrive at research findings that truly answer the research questions and as a result, meet the objectives of the study, there is a need to utilize a specific methodology. Speaking from this point of view, this research will first analyze the past research and study findings on how social media has been used in the past in tourism marketing. For this the study accessed and reviewed books, articles, dissertations, government reports, and online data links from websites of different religious tourism shrines, destination websites, websites of Ministry of Tourism, Government of India, and many other sources. Similarly, the research used primary data responses collected from tourist and destination stakeholders including religious tourism shrine board, religious tourism service providers. The primary data for the study collected by using field surveys and through online tools such as Google docs, email, Facebook. For enhancing the validity and as such the reliability of the data collected, the study strongly adhered to the standards of creating, testing, and sending out questionnaires to respondents. Regarding the collection of data responses, the study identified small sample groups of tourists and other respondents as a true representation of the larger population. To achieve this, the study opted to use convenient sampling technique, since targeted a samples who have social media account(s) and using this platform before to plan their travel in case of tourists and using social media for management, planning, marketing, dissemination of information, delivering services in case of restaurants, hotels, resorts and shrine boards, religious destination establishments. The collected raw data was organized to find out the completed survey forms to use for the analysis through Microsoft excel sheets and data entry sheets of statistical package for social science(SPSS). The organized data was analyzed by using different statistical tools to determine whether it supports or opposes the secondary research findings. The collected data were tabulated and classified in a spreadsheet and data entry sheet of statistical package for social science (SPSS) by giving separate coding to every question of the survey form. The data were used to get the empirical findings through the application of descriptive statistical tools. The author specifically carried out percentage analysis, frequency distribution, and average score analysis to comply with the objectives and aims of the chapter. The values of empirical data analysis were presented through tables to make it easy to understand for its readers.

MAIN FOCUS OF THE CHAPTER

The present section of the chapter describes the interrelationship of social media and religious tourism followed by the use of social media tools by tourists during their visit to religious tourism destinations from the demand-side perspective of tourism business. From a supply-side perspective, the section also describes the potential factors that force religious tourism suppliers to adopt social media in religious tourism operations and challenges faced by them in using social media.

Use of Social Media by Tourists in Religious Tourism

The development of social media provides an opportunity for its users to produce their content and send it without geographical and time restrictions. Millions of users and businesses have become active members of social media networks to make use of this unique opportunity. Besides its use by customers for searching and purchasing, social media also helps companies to promote products and services. Kaplan and Haenlein (2010) apprehended the issue and described that the use of social media is not an easy task and may require new ways of thinking, but the potential gains are far from being negligible.

Social media such as Facebook and Myspace, provide instant ways to reach thousands or millions of people, Twitter is a fast way to get a few words out, and social media, such as LinkedIn connect people interested in business relationships (Tourism & More, 2011; Mariani, Felice, & Mura, 2016). Taking advantage of the opportunities offered by social media, a lot of people who have become users of this media share their experiences with friends and colleagues around the world. Some studies on social media users report that sharing experiences with others plays an important role in the decision of people, who live in different regions or countries, to purchase any products and services or to give up buying.

Groups of friends sharing their work and travel experiences in social media are becoming effective on consumers decisions. This influence is also true in the field of travel and tourism (Gretzel, 2011; Tourism & More, 2011; Gonzalo, 2012). With similar findings to the studies mentioned above Fotis et al. (2012) further elaborated that social media are used during all stages of the holiday planning process (before, during, and after holidays). At this point, a question may arise that influence of reviews and comments on the decision to purchase and plan travel holiday. The study of Gretzel (2011) reached findings that confirm the above-discussed supposition: trust levels are very high and that a considerable number of people even prefer the opinions of unknown others over opinions of friends and relatives. This also noticed that individuals are pretty good at detecting false reviews, and this ability increases with the length of the review.

The chapter findings noticed that out of total tourists surveyed, 49.47 percent were female, and remaining were male tourists. The percentage of male and female respondents was almost equal (approximately 50%). Respondents in the age group of less than 20 and above 60 were almost equal to 11 percent meaning they were less represented. All age groups make quite a good representation in each age group. The groups being so much represented indicates that they are the ones familiar with the internet in their daily routine life, as the majority of them started to use it. As regards the occupations, most of the respondents were employed, but students were also included in the survey along with 25 percent retired persons as part of the sample. Data signifies that a lot many travelers these days, whether, young or old, student or self-employed or retired, resorted to using social media in their choice of destination selection, and selection of products or services at the destination. The country of the respondents was also checked-in and found most of them were of Indian origin covering more than 90 percent and the rest of the respondents were from foreign countries.

Social media habits of consumers for religious tourism were also investigated in the chapter. The religious tourism destination planners need to know how potential tourists get information when they want to decide to be on a religious tourism trip. By doing so, companies can reorient their social media strategies for the potential customers.

The chapter also highlights the percentage analysis of consumer respondents to use social media to search for the destination. The tabular presentation below shows that before finalizing the trip 41% of the respondents consulted their friends/family members, 72% checked-in at internet accommodation websites, 70% checked-in at internet hotel booking websites, 33% respondents pondered over the website of the destination, 7% checked-in at general social media, 22% respondents reviewed the customer

Table 1. Sources used for search about religious destination

Source of Information	Yes (%)	No (%)
Family and Friends	41	59
Shrine Board/Temple Trust	72	28
Online Booking Portal	70	30
Destination Website	33	67
General Social Media	7	93
Review Social Media	22	78
Books and Magazines on Destination	21	79
Travel Fair and Exhibition	3	97
Travel Intermediaries	2	98
Other	1	99

Source: Authors' Creation

comments on social media, 21% referred to books/guides about the destination. Very little percentage of respondents visited the travel fairs or consulted their travel agents.

According to Hudson (2008), reference groups such as family, friends, colleagues and other travelers have a direct or indirect influence on a person's behavior. Thus e-presence is extremely important for the destination planners. For 84% of the respondents, tourism products and services must be available for booking through social media websites of the religious destinations.

Table 2. Which Social Media Do You Use

Uses of social media	Yes (%)	No (%)
Facebook	81	19
Twitter	27	77
Pinterest	10	90
Google+	33	67
Trip Advisor	31	69
Instagram	1	99
Blogger	2	98
LinkedIn	1	99
Trave Buddy	1	99
I don't use any social media	10	90

Source: Author's Creation

As far as the actual usage of social media is concerned, the below table presents the percentage usage of social media by the respondents. Facebook was highly relied upon. The respondents shared information and/or opinion about their trip with friends, and even shared pictures from their trip with friends on social media.

Consumers were also asked as to for what purpose they use social media. Many consumers were using social media for their reservation of attraction and other services at religious destinations and solution of queries too. Questionnaire focused on finding as to which social media were used to book religious destinations attractions and services and 93% believed that it is important for a religious tourist attractions site to have a booking page on social media. 43 percent of the respondents used social media to write reviews about the quality of different services and facilities after the trip, 60 percent of the respondents shared pictures with friends about the trip, and 24 percent of them prefer to share information and/or outlook with friends about their religious tourism trip. All these outcomes display that social media were really popular at the time of planning and, during and after the trips. Moreover, social media is used for feedback while writing the reviews at sites like Google+, Trip Advisor, etc.

Therefore, it seems that there is still a long way to critically evaluate the use of social media in marketing and promotion of different products and services.

Table 3. Use of social media during the travel process

Uses of social media	Yes (%)	No (%)
Before a trip to get the opinion of other peoples regarding religious tourism places	44	56
Before a trip to get information about different services and facilities	46	54
Reservation of attraction and another service through website, Expedia	93	7
Before a trip to access reviews and comments of people of those already visited religious destinations	52	48
Used during trip to share picture of religious tourism places	49	51
Use during a trip to share information and opinion about religious tourism destination	36	64
To share information and opinion to friends about religious tourism trip	43	57
To share pictures of religious tourism trips to friends	60	40
To write a review about different services consumed on religious tourism trip	24	76

Source: Authors' Creation

Social Media for Marketing and Promotion of Religious Tourism Destination

The marketing communication became an intrinsic necessity in the superior valorization of the religious touristic product. The success of this pilgrimage is owed to the concerted effort of competent authorities (religious, and local authority), directly interested in developing the touristic flows and local economy. In the modern world religious tourism reached the highest peak, when the chronics of those times record over millions of religious tourism visits. An entire industry was developed as a result of the great number of believers coming in pilgrimage and different infrastructural facilities were built to ensure quality experience to tourists (Timothy & Olsen, 2006). Social media is useful for every company but for tourism companies it is essential. Social media use made it possible to promote travel destinations and the programs, such as travel packages and tours with very low prices. Some studies on social media are encouraging tourism companies to use social media in their businesses social media has made a huge impact on essentially every major industry across the world, and the business of travel and hospitality has reaped the rewards perhaps as well as any other commercial venture. Travel and tourism depends heavily on the use of word-of-mouth to spread opinions and recommendations, and social platforms such as Twitter and Facebook allow customers to easily share tips and suggestions, which can be enormously valuable when positive (Bennett, 2012).

The research is based upon the religious tourism destinations of India, and it tracks their visibility in the social media. Due to technological development several initiatives were taken to provide elaborative information about attractions and other facilities on websites of destination such as online booking portals, websites, and social media, and so on. The social media page of different religious tourism sites or shrine boards contains complete information about the destination. The information concerns the access, historical data, and the accommodation possibilities for each destination.

Rationale Behind Using Social Media

The description of data relating to the area of operation of selected religious tourism destination shrine boards and organizations including shrine board and service providers, the location of the firm, and segment type of organizations is presented first. This indicates that 41 out of one hundred and twenty-four were providing services in the global tourism market, followed by 30 covering regional tourism market, 22 were domestic tourism organizations and 21 organizations were providing services only in the local tourism market. As far as the location of religious tourism destinations planners and organizations are concerned the most of the organizations which comprise 39 were located in north India, 37 were located in west India followed by 23 in south India, and the remaining 15 were located in east India. It is also observed that every segment of the tourism & hospitality sector covered under selected one hundred and twenty-four organizations including 43 travel agency, 35 tour operators, 26 accommodation units, and 20 were transport and ground service providers included in the sample. The demographic data of the employee respondents such as the age of the respondents, employment status of the respondents, and the total number of years for which the respondents have been working in the tourism industry also presented in the chapter. Further, out of total respondents, 52 were marketing personnel, the majority, 41 were sales personnel, 18 were accounting personnel and 13 were either owners or general managers. Selected tourism firms and destination organizations in the field of religious tourism were d asked to specify the total number of years for which they have been working in the tourism industry. The question was necessitated as it was assumed that the number of years one is employed in the tourism industry, it increases their assessment capability to express whether social media marketing usage impacted the technical and financial profitability in their organizations. 48 respondents indicated that they were employed for between one and three years, 32 respondents had and experience of four to six years, whilst 34 were employed for seven to ten years and 10 respondents were in employment for more than ten years. The results indicate that selected destination shrine boards and organizations respondents were capable enough to visualize and provide information about the benefits of using social media tools in their respective organizations.

Table 4. Effects of Social Media in Religious Tourism Marketing

Social Media Effect on Religious Tourism Marketing	Mean Score
Social media helps a religious tourism establishment in disseminating information faster	3.51
Social media is a popular tool among consumers in sharing personal experiences, comments and opinions	3.47
Social Media has been proven to be an excellent strategy rather than any promotional campaign	3.04
Hospitality management can also use social media as a means to analyze performance and consumer satisfaction	3.21
Social media can be an advantage to reach a global audience even in remote corners of the world	3.3
Social media can also be the ultimate tool for word of mouth	3.18
Social Media gives opportunities for hospitality management to improve service recovery	3.22
Social Media provides opportunities to manage and present business contents	3.22
Social Media can easily ruin a business reputation	3.16
Social Media helps to overcome the distance barriers and other communication obstacles	3.16
Social media can disclose confidential information	2.91
Social media can easily expose confidential information and could go viral resulting to loss of privacy	3.06
Social media can spread bad comments easily and go viral which can harm a business	3.18

Source: Authors' Creation

Social media is characterized by accessibility, low costs allowing us to concentrate the information that the issuer wishes to convey. Such advantages render this means of communication to be preferred by both issuer and receiver. The online media is a dynamic and interactive environment, allowing the receiver to send feedbacks employing comments, numbers of views and likes, thus contributing to the improvement of marketing communication. In this regard, social media pages were created to encourage the users in sharing their experiences connected to the consumption of the touristic product. The social media promotion of the touristic offer is differently performed by organizations, according to the objectives in mind. Thus, the religious tourism shrine boards through their websites or social media pages providing detailed information about the entire internal touristic offer, including a detailed presentation of each program. The religious tourism service providers have included in their offer touristic packages with having the presence of social media platforms. The visibility on such pages created by local and county administrations contains general data and they have links containing cultural-religious information, possibilities of recreation and they promote the representative touristic receiving structures in the area.

It was found out that the overall assessment was agreed with a composite mean of 3.20. Social media helps a religious tourism establishment in disseminating information faster ranked first with a weighted mean score of 3.51 and rated strongly agree. It was followed by Social media is a popular tool among consumers in sharing personal experiences, comments, and opinions with a weighted mean of 3.47. From the perspective of the researcher, Social Media tools are changing the way people communicate, so the result did not come as a surprise. The researchers, themselves as tourism students are aware of those benefits that social media can contribute to the tourism and hospitality industry. Social media is not just for social interaction, but most probably becoming the best form of a promotional campaign for all the hospitality establishments wherein they can review all the comments and opinions that will help in improving their services to satisfy customers. One good thing that social media can provide is that customers no longer need to go to a certain place or destination before deciding whether to stay or not.

On the other hand, social media has been proven to be an excellent strategy rather than any promotional campaign (3.04) and can disclose confidential information (2.91) that can goes viral very quickly got the lowest mean value. Some religious tourism boards in India such as shrine board of Goddess Vaishno in Jammu & Kashmir, Tirumala Tirupati Shrine Board in Andhra Pradesh, Goddess Naina temple trust in Himachal Pradesh and many more are already starting using social media in promoting the destination, due to ease of attracting tourists through social media. Such temple trusts and shrine boards need to further utilize the continuous usage of social media. That is the reason why social media has been proven to be an excellent strategy rather than a promotional campaign in terms of religious tourism in India.

Challenges Ahead Using Social Media

The introduction of social media, which provides an easy and cost-effective promotion opportunity, was met with great interest in the tourism industry, too. Hurdles integral in the execution of various social media practices in the religious tourism industry were look over through the literature and were put before the respondents for their view on a five-point scale starting from extreme barrier to not a barrier. Dhankhar (2016) conducted a study and identified that some challenges also associated with the use of social media in the tourism and hospitality sector. These challenges can be noticed as the lack of physical contact with the customers, poor technical skills among service providers, perceived security threats related to payment, the unwillingness of management, reliability of information, and uncertain

Table 5. Barrier encountered during the use of social media

Barriers	Mean Score
Social media can be possibly abused by the tourists and employees in exposing an employer's trade secrets or business proprietary information	3.01
Social media can also lead to unfair negativity, inaccurate information and undue criticism	2.9
Social media can spread false information about destination	2.88
In Social media, consumers are not always presented a good representation of the opinions, thoughts, and expressions of consumers because only the popular social media sites are represented well in the internet	3.25
Companies find ways to abuse social media without anyone truly knowing and without any real consequences	2.86
Companies can have bad publicity of using social media	2.74
Social media holds your business as you go online	3.04
The trend toward social networking offers an extraordinary opportunity for hospitality firms to garner a competitive edge by analyzing submitted feedback, contemplating constructive advice, and determining how best it is	3.21
Web sites catering to industry employees can be harmful to the success of a single firm or industry segment	2.73
Spammers can make the social network a problem for new users and people new to the social media scene	3.08

Source: Authors' Creation

returns (Dhankhar, 2016). Thus overall social media was considered too expensive to implement, and it is required to give more attention to consider social media marketing from some empirical evidence.

The responses collected from different stakeholders of religious tourism in India revealed the major challenges faced that need to consider for the efficient execution of social media marketing. The above table shows the problem encountered in using social media in religious tourism in India. It was found out that the over-all assessment was agreed with a composite mean of 2.97. Among the problems mentioned, Social media consumers are not always presented a good representation of the opinions, thoughts, and expressions of consumers because only the popular social media sites are represented well on the internet, with a weighted mean score of 3.25 and rated agree.

From the researchers' point of view, who always use different social networking sites such as Facebook, Twitter, Instagram, and Blogs, it is true that sometimes using social media, users experience or received some bad representation of information, ideas, comments, that in result can harm one individual's credibility and reputation. Social media is open to the public and no one can control what people say. One reason why customers and guests experience those negative effects of social media, is because they only depend on those popular websites and networks like Facebook and Twitter. On the other hand, respondents believed that companies can have bad publicity of using social media with a weighted mean of 2.74, being second to the lowest and that websites catering to industry employees may not be too harmful to the success of a single firm or industry segment, with a weighted mean of 2.73 got the lowest mean rank. Therefore, the study shows that everyone cannot expect that the outcome of social media is always positive, sometimes it can lead to false assumptions, critiques, and opinions, so there's still a need to consider different sorts of aspects.

According to these findings, social media play an important role in marketing, and it is a useful tool in reaching the customer and making face to face talk but it is a harmful tool at the same time if it is not used properly. Preventive measures are not taken due to improper use of social media companies, and hazards such as loss of reputation. Senior executives and directors do not make use of and do not trust the information coming from social media to use for their companies. Companies do not have any infrastructure which to process the information they gather from social media and do not know how to use this information.

SOLUTIONS AND RECOMMENDATIONS

The concept of religious tourism is becoming the most important segment of tourism in a country like India. Hence, it needs publicity and marketing of religious destination for survival. Henderson (1998) corroborates the issue of needs and marketing those religious places to reduce the mismatch between expectations of visitors and service providers' strategies. Kotler (2002) also suggested that to satisfy the target market's needs, wants, and demands, marketers must consider that their products and services will be successful only if they deliver value to customers (Kotler, 2002).

Hence, it is appropriate that marketers must pay close attention to the trends and developments in these environments and make timely adjustments to their marketing strategies by applying different initiatives and technologies assuming that consumers want widely available, affordable products; the product concept assumes that consumers want products with the most quality, performance, or innovative features. Keeping this concept in mind, smart companies started adding combination of technology, globalization, and deregulation is influencing customers, service providers and other stakeholders in a variety of ways.

Henderson (1998) in case of religious products and services recommended this is essential to market through latest innovation and efforts. Hence, the organization (Religious tourism shrine board) will do their best when they choose to attain their goals of reaching out more and winning to their sides their target markets. The 'value' is the Organization's offering (message, product, and service) that is of benefit to the chosen target markets', that is, to the public – internal to her and external. By application, the religious Organization(s) that seek to reach her chosen target markets must know the needs of customers and seek to satisfy the same while using integrated marketing to reach them at a minimal cost.

To eliminate a bad representation of opinions, thoughts, and expressions caused by the social media users, the findings of the chapter suggested that there are certain procedures that omit unnecessary comments, photos, and videos that can affect your reputation. For ensuring the smooth usage of social media in marketing of religious tourism, the prevention of spammers from being a cause of delayed transaction and operations of the tourism business activities is recommended. There are certain ways to avoid spammers such as the use of filter folder wherein spam messages will be directed to it. To be aware and cautious in using social media for their business activities, the study suggested for Implementation of policy among the employee on how to use internet or social media. To evade from abusing the use of social media caused by exposing employers' trade secrets and business proprietary information, the authors focus religious tourism establishments to provide a seminar and orientation to all the staff of an organization regarding social media usage. To prepare for the negative comments or information and undue criticism that may arise while using social media. All employers should implement a strict policy on internet/ social media use and individual and establishment should be advised not to easily accept a friend request particularly from people you do not know.

FUTURE RESEARCH DIRECTIONS

The present chapter investigated the use of social media by tourists and its influence in promoting religious tourism sector in India. The findings of the study indicated that different social media platforms are not equally applicable in promoting tourism. Further research could investigate the influence of specific social media platforms in promoting tourism business. Despite of the high adoption of social media by tourism consumers and suppliers in terms of tourism, the successful practice of manipulating and managing social media in promoting religious tourism still remains largely unknown to both practitioners and scholars. Further research could investigate how to successfully manage and manipulate the business opportunity presented by various social media platforms such as Twitter, Facebook, and Pinterest etc. Current chapter also provided a suitable window for future research targeting to mobile social media and tourism relationships may be the future research subject. This research can work as seed for future researches in the social media marketing and its relation with profitability of firm is still debatable in tourism sector. It is suggested on the basis of findings, that future research could observe the impact of different social media on business profitability in religious and other tourism segment as comparative study is the need of the hour.

CONCLUSION

The chapter intends to find out the uses social media by tourists and by religious tourism industry from the viewpoint of role that social media played in the development and promotion of religious tourism followed by challenges associated with use of social media. From the chapter it can be ascertained that customers have considered social media very important in terms of tourism industry. Also in order for better business outcomes religious tourism destination organization and service providers need to continually use social media for monitoring, analyzing and evaluating customer reviews. The findings of the study indicated increasing use of social media by tourists in searching, planning, booking and sharing experiences about different products and services consumed during their visit to religious tourism sites in India. In order to avoid inaccurate information and negative publicity, social media should be embraced with proper planning and rigorous guidelines for maximum success. Outcomes stated that in addition to searching and sharing, many consumers were using social media for reservation and solution of queries too. Presence of destination on social media pages found tremendously famous for this purpose. As regards the actual usage of social media, Facebook was highly relied upon. The respondents shared information and/or opinion about their trip with friends, and even shared pictures from their trip with friends on social media. Nearly half of the respondents surveyed indicated use of social media to write reviews about quality of their trip to religious destination after the trip, shared pictures with friends about the trip, and prefer to share information and/or outlook with friends about their visit. All these outcomes displays that Social Media were really popular at the time of planning and, during and after the trips. Moreover, social media is used for feedback while writing the reviews at sites like Google+, Trip Advisor etc. Furthermore, the chapter produced enough evidence to explain that social media plays an effective role in the development of religious tourism sector as in the case of other sub-sectors of tourism industry. Social media applications served as an avenue to disseminate the information faster especially for the religious tourism establishments with lesser cost. The study also notices careful handling of customer queries and complaints over social media as potential buyers' sturdily consider reviews and comments or post submitted on social media for destination and other services. Further in this line, the use of social media by tourism industry, it came to experience some challenges also in marketing and promotions of religious destinations. Improper use of social media can cause bad image and reputation of destination among potential tourists. The main problem encountered in using social media is that, customers' opinions, thoughts and expressions are not well presented that leads to bad impressions and unfair criticism. The destination organizations and service providers must be familiar with different social media applications to protect the image of company among customers. Therefore, the findings of the chapter confirm that improper use of social media brings more benefits than costs. Finally, the chapter indicates there is a long way to go for most of the companies to learn effective use social media.

REFERENCES

Alsos, G. A., Eide, D., & Madsen, E. L. (2014). *Handbook of research on innovation in tourism industries*. Edward Elgar Publishing. doi:10.4337/9781782548416

Bennett, S. (2012). *The impact of social media on travel and tourism*. Retrieved from https://www.mediabistro.com/alltwitter/social-media-travelhospitality_b24679

Bingqui, F., Bing, D., Taolong, X., & Ruofei, F. (2011). Place media events and place image communication: The role of festivals and special events from the perspective of media relations. *Catalan Journal Of Communication & Cultural Studies*, *3*(2), 159–173. doi:10.1386/cjcs.3.2.159_1

Carneiro, M. J., Eusébio, C., & Caldeira, A. (2018). The influence of social contact in residents' perceptions of the tourism impact on their quality of life: A structural equation model. *Journal of Quality Assurance in Hospitality & Tourism*, *19*(1), 1–30. doi:10.1080/1528008X.2017.1314798

Chattopadhyay, M. (2006). Influence of Religion on Tourism: Implications for India's Tourism Policy. *The ICFAI Journal of Consumer Behavior*, *1*(3), 59–67.

Cristea, A. A., Apostol, M. S., & Dosescu, T. (2015). The Role of media in Promoting Religious Tourism in Romania. *Procedia: Social and Behavioral Sciences*, *188*, 302–306. doi:10.1016/j.sbspro.2015.03.398

Dhankhar, D. (2016). *E Marketing in Hospitality Industry: Practice, Performances and Service Quality* (Unpublished Thesis). Maharishi Dayanand University, Rohtak, India.

Eck, D. L. (2012). India: a Sacred Geography. New York: Harmony (Random House Group).

Fotis, J., Buhalis, D., & Rossides, N. (2012). *Social media use and impact during the holiday travel planning process*. Springer-Verlag. doi:10.1007/978-3-7091-1142-0_2

Gonzales, H. (2008). Tourism, mass media, and the making of visual culture in the greater Yucatan Peninsula. *Journal of Film and Video*, *60*(2), 50–58. doi:10.1353/jfv.0.0003

Gonzalo, F. (2012). *How Travelers Use Online Sources for Travel Decision-Making*. Retrieved from https://fredericgonzalo.com/2012/09/11/how-travelers-use-online-sources-for-travel-decision-making/

Green, C. E. (2009). *Sensible marketing in tough times*. Hospitality Upgrade Magazine.

Gretzel, U. (2011). *Do social media influence our travel decisions? The Institute for Innovation in Business and Social Research (IIBSoR) Faculty of Commerce*. Retrieved from https://uowresearch.wordpress.com/2011/03/27/do-social-media-influence-our-travel-decisions/

Harrett, J., & Benjamin, T. (2009). Travel with a time Lord: Using media to enhance literacy. *Literacy*, *43*(3), 134–142. doi:10.1111/j.1741-4369.2009.00510.x

Henderson, D. W. (1998). *Culture Shift: Communicating God's Truth to our changing World*. Baker Books. http://www.tourismandmore.com/tidbits/using-the-social-media-for-tourism

Hudson, S. (2008). *Tourism and hospitality marketing: a global perspective*. Sage. doi:10.4135/9781446280140

Jauhari, V., & Sanjeev, G. M. (2010). Managing Customer Experience for Spiritual and Cultural Tourism: An Overveiw. *Worldwide Hospitality and Tourism Themes*, *2*(5), 467–476. doi:10.1108/17554211011090094

Jha, M. (1978). *Aspects of Great Traditional City in Nepal*. Kishore Vidya Niketan.

Kaplan, A. M. V., & Haenlein, M. (2010). Users of the world, unite! The challenges and opportunities of Social Media. *Business Horizons*, *53*(1), 59–68. doi:10.1016/j.bushor.2009.09.003

Koldowski, J. M., & Martin, O. (2008). Emerging Market Segments: Religious and Medical Tourism in India. In C. Roland & B. Martin (Eds.), *Trends and Issues in Global Tourism*. Springer.

Kotler, P. (2002). *Marketing Management*. Pearson Custom Publishing.

Kotler, P., & Gertner, D. (2002). Country as brand, product, and beyond: A place marketing and brand management perspective. *Journal of Brand Management*, *9*(4), 249–261. doi:10.1057/palgrave.bm.2540076

Kumar, R. B., & Singh, L. (2015). Expectation & Satisfaction of Hindu Pilgrims in North-West India-A Case Study of Naina Devi Shrine. *African Journal of Hospitality, Tourism and Leisure*, *4*(2), 1–18.

Leung, D., Law, R., Van, H. H., & Buhalis, D. (2013). Social *Media in Tourism and Hospitality: A Literature Review. Journal of Travel & Tourism Marketing*, *30*(1-2), 3–22. doi:10.1080/10548408.2013.750919

Liana, S. (2017). *The role of social media in the tourism industry in Austria, How does Austria Implemented the Social Media to Attract Tourists?* (Bachelor Thesis). Module University Vienna.

Mariani, M. M., Felice, M., & Mura, M. (2016). Facebook as a destination marketing tool: Evidence from Italian regional Destination Management Organizations. *Tourism Management*, *54*(1), 321–343. doi:10.1016/j.tourman.2015.12.008

Murphy, H. C., Chen, M. M., & Cossutta, M. (2016). An investigation of multiple devices and information sources used in the hotel booking process. *Tourism Management*, *52*, 44–51. doi:10.1016/j.tourman.2015.06.004

Pforr, C., & Hosie, P. (2009). *Crisis management in the tourism industry: Beating the odds?* Ashgate Publishing, Ltd.

Poddar, H., Goswami, C., & Shastri, M. A. (1951). *Kalyan: Tirthank*. Gita Press.

Ppoescu, R.I., & Corboş, R. A. (2012). The role of Festivals and Cultural Events in the Strategic Development of Cities. Recommendations for Urban Areas in Romania. *Informatica Economică*, *16*(4).

Ranjha, A. (2010). *Promoting tourism in Abu Dhabi using social media* (Unpublished Master Thesis). Royal Institute of Technology of Stockholm.

Rinschede, G. (1992). Forms of Religious Tourism. *Annals of Tourism Research*, *19*(1), 51–67. doi:10.1016/0160-7383(92)90106-Y

Rodrigo, N. N. (2012). *Influence of religious factors on tourism in Maramures*. Retrieved from http://writepass.co.uk/journal/2012/12/influence-of-religious factors-on-tourism-in-maramures-northern-part-of-romania/

Scptt, D. W. (2007). Constructing sacred history: Multi-media narratives and the discourse of "museumness" at Mormon Temple Square. *Journal of Media and Religion*, *6*(3), 201–218. doi:10.1080/15348420701530098

Shaista, A., & Khan, M. M. (2019). Role of Social Media in Promoting Tourism in Pakistan. *Journal of Social Sciences and Humanities*, *1*(1), 101–113.

Sharma, V. (2010). *Pilgrim Satisfaction - A Study of Selected Shrines of Northern India* (Unpublished Ph.D. Thesis). Shri Mata Vaishno Devi University, Katra.

Tafveez, M. (2017). Role of social media in tourism: A literature review. *International Journal for Research in Applied Science and Engineering Technology*, *5*(11), 633–635. doi:10.22214/ijraset.2017.11099

Thomas, D. R. (2006). A general inductive approach for analyzing qualitative evaluation data. *The American Journal of Evaluation*, *27*(2), 237–246. doi:10.1177/1098214005283748

Thomas, J. (2013). *Social media effects on tourism sector*. Retrieved from https://socialnetworking.lovetoknow.com/Social_Media_Effects_on_Tourism

Timothy, D. J., & Olsen, D. H. (2006). *Tourism, religion and spiritual journeys*. Routledge. doi:10.4324/9780203001073

Tourism & More. (2011). *Using the social media for tourism*. Academic Press.

United National World Tourism Organization. (2011). *Religious Tourism in Asia and the Pacific*. UNWTO. Retrieved from http://pub.unwto.org/epages/Store.sf/?ObjectPath=/Shops/Infoshop/Products/1509/SubProd ucts/1509-1

Yesil, M. M. (2013). The Social Media Factor in the Development and Promotion of Tourism Destination. *International Periodical for the Languages. Literature and History of Turkish or Turkic*, *8*(7), 733–744.

Zeng, B., & Gerritsen, B. (2014). What do we know about social media in tourism? A review. *Tourism Management Perspectives*, *10*, 27–36. doi:10.1016/j.tmp.2014.01.001

ADDITIONAL READING

Ell, A., & Bokhari, R. (2013). Key quality factors affecting users' perception of social networking websites. *Journal of Retailing and Consumer Services*, *20*(1), 120–129. doi:10.1016/j.jretconser.2012.10.013

Kong, L. (2001). Religion and technology: Refiguring place, space, identity and Community. *Area*, *33*(4), 404–413. doi:10.1111/1475-4762.00046

MacWilliams, M. W. (2002). Virtual pilgrimages on the Internet. *Religion*, *32*(4), 315–335. doi:10.1006/reli.2002.0408

Minazzi, R. (2015). *Social media marketing in tourism and hospitality*. Springer. doi:10.1007/978-3-319-05182-6

Molz, J. G. (2012). *Travel connections: Tourism, technology, and togetherness in a mobile world*. Routledge. doi:10.4324/9780203123096

Pike, S. (2016). *Destination marketing: Essentials*. Routledge.

Raj, R., & Morpeth, N. D. (2007). *Religious Tourism and Pilgrimage management. An International Perspective*. CABI. doi:10.1079/9781845932251.0000

Rodriguez, I. L. (2009). *Social Media in Tourism Behavior*. Bournemouth University.Université de Savoie.

Sparkler, 2013. *Facebook Travel: Near and Now*. London: Sparkler.

Yoo, K. H., & Gretzel, U. (2012). Use and Creation of Social Media by Travellers. In M. Sigala, E. Christou, & U. Gretzel (Eds.), *Social Media in Travel, Tourism and Hospitality: Theory, Practice and Cases* (pp. 189–206). Ashgate Publishing Limited.

KEY TERMS AND DEFINITIONS

Blog: Blogs are usually maintained by an individual or a business with regular entries of content on a specific topic, descriptions of events, or other resources such as graphics or video.

Chat: A chat is an online conversation with one or more people.

Comment: A comment is a response that is often provided as an answer or reaction to a blog post or message on a social network.

Computer Reservation Systems (CRS): Computerized reservation systems used for inventory management by airlines, hotels and other facilities and allow direct access to check availability, make reservations and print tickets.

Content Marketing: Content marketing is the practice of attracting and retaining customers through the creation and distribution of original, valuable content such as videos, whitepapers, guides, and infographics.

Follower: In a social media setting, a follower refers to a person who subscribes to your account in order to receive your updates.

Platform: The term platform is often used to mean a set of software tools that help SMMs organize their social media accounts.

Post: A post refers to any social media status update, photo, or video, or an item shared on a blog or forum.

Share: This is what your goal is when you're using social media - to spread your message, products, brand awareness, thoughts, and company voice with users.

Viral: To go viral on social media is to have a particular post bring in an unusually large number of engagements.

Chapter 16
Religious Tourism:
Sanctuary Cities as the Primary Destinations

Luciana Aparecida Barbieri da Rosa
https://orcid.org/0000-0001-9240-0236
Universidade Federal de Santa Mari, Brazil

Maria Carolina Martins Rodrigues
https://orcid.org/0000-0003-2575-8611
CinTurs, Portugal

Waleska Yone Yamakawa Zavatti Campos
https://orcid.org/0000-0001-5050-1557
IAG Business School, Pontifical Catholic University of Rio de Janeiro, Brazil

Maria José Sousa
https://orcid.org/0000-0001-8633-4199
ISCTE, Instituto Universitário de Lisboa, Portugal

Martius Vicente Rodriguez y Rodriguez
https://orcid.org/0000-0001-8270-7488
Universidade Federal Fluminense, Brazil

Larissa Cristina Barbieri
Faculdade de Educação São Luis, Brazil

ABSTRACT

The present study has as its general objective to analyze the differences in behavior and culture among religious tourists who visit sacred sites. The study is descriptive through research. The quantitative instrument was adapted of Irimiás and Michalkó and Irimiás, Mitev, and Michalko. The results show a heterogeneous sample that includes different age groups. Among the data it appears that the majority of

DOI: 10.4018/978-1-7998-5792-1.ch016

respondents are female and have higher education. A large percentage were born in a religious cradle and thus are considered religious. In addition, a high number of religious tourism participants are part of a religious community. Thus, it is suggested for future research to expand the analyzed countries in order to compare results, with the aim of analyzing other contents different from those analyzed. Among other limitations of this study we have: the information presented is limited to the articles selected only with Brazilian and Portuguese tourists; the research was carried out with the snowball technique, being limited only to the sample for convenience.

INTRODUCTION

Religious tourism, seeking to live moments of faith and devotion, has grown worldwide. However, this demonstration of faith and veneration of sacred places has existed since antiquity. It is known that this practice emerges from different aspects, either in the fulfillment of a promise or in the desire to visit places of a historical-religious character. From this perspective, different studies on the subject have been published (Savola, 2003; Nyaupane, Timothy, & Poudel, 2015; Abumanssur, 2017; among others).

Some studies indicate that religious architectural aspects attract tourists from different perspectives, reflecting on different behaviors (Nolan & Nolan, 1992; Prazeres & Carvalho, 2015). Others highlight the cultural desire combined with historical political and educational reasons (Digance, 2003; Baggio, Parise, and Sanchez, 2016.

In this scenario, the United Nations World Tourism Organization (2014) highlights that about 330 million tourists visit sacred places around the world. Based on this aspect, religious tourism has some specificities such as pilgrimage, pilgrimage, penance, religious events, and visits to sacred places (Macedo, 2011). Collins-Kreiner, (2010) reinforces in his study that, religious places of different spheres have become tourist destinations for people who wished to know the culture of the place or to know the religious history.

According to the authors Andrade (2000) and Oliveira (2000), studies on religious tourism have motivated the growing attention of researchers from different disciplinary areas. It is an emerging theme that, according to Silveira (2007) and Vilas Boas (2012), has evolved with the authors' lack of unanimity on some theoretical issues such as the concept of religious tourism, with a plurality of theoretical perspectives and definitions.

In this perspective, tourism, especially religious tourism, characterized as a "symbolic exchange economy" connection, is "linked to the religious calendar of the locality receiving the tourist flow, that is, religious destinations are linked to sacred places, where there are reports of the manifestation of a deity, or of the lives of people who have been beatified or sanctified" (Tendeiro, 2010, p. 5).

According to Prazeres & Carvalho (2015), sanctuary cities are epicenters of religious tourism, defined as a complex and dynamic phenomenon. The relevant issues related to religious tourism are essential to frame and understand the role of sanctuary cities in the action of welcoming their visitors (p.1147).

The population is looking for pilgrimages to sanctuaries, as they are places that allow more magnificent "meditation" with God. These trips to the Shrines gave rise to religious tourism, which in addition to the religious aspect, can also have the entertainment aspect, as is the case of parties, processions, pilgrimages.

According to Oliveira (2004), pilgrimages, parties, processions, pilgrimages, provide a break in the daily life of the population, that is, a renewal of strength and faith for a specified period. Religious tourism emerges as a new option for segmenting the tourist market, where tourists seek spiritual peace, pay promises, obtain graces, and others.

In this perspective, several places have stood out for the practice of religious tourism, such as the Basilica of Senhor do Bonfim, in Bahia, the city of Aparecida in São Paulo, both in Brazil, Sanctuary of Nossa Senhora do Rosario de Fátima, in Fátima, and Sanctuary of Nossa Senhora do Sameiro, in Braga, both in Portugal.

Based on such inferences and knowing the emerging importance of the theme, the research problem was elaborated: "What are the differences in behavior and culture that exist between religious tourists who visit sacred places?". Thus, the present study has the general objective: To analyze the differences in perception of behavior and cultures among religious tourists who visit sacred sites in both countries.

In the analysis of the articles, we sought to incorporate the research of authors with works with scientific recognition, as well as recent articles that have not yet had the time necessary to be cited by researchers in scientific works.

The present study is a survey, characterized as the approach as a quantitative descriptive. The questionnaire applied contains 10 questions and was developed to obtain global radiography of the perception of the behavior of religious tourism, differences in behavior, and culture among religious and cultural tourists who visit sacred sites.

Some dichotomous questions, multiple options, multiple answers, as well as a 5-point Likert scale. For data processing, the spreadsheet Excel 2016 for Windows and SPSS version 22 for Windows were used.

It is in the last chapter of this work that the conclusions of the study will be presented, taking into account the theoretical review and the empirical results obtained, in order to answer the questions raised in the work methodology and taking into account the question defined at the outset. The study's limitations and difficulties are also presented, as well as suggestions for further investigations.

The work will be completed in two phases. The first phase focuses on the theoretical objectives of the investigation, defined in the methodology chapter, and analyzed in the theoretical chapters of the literature review. The second phase presents the conclusions of the empirical part of the study, which were obtained by analyzing the results of the questionnaires.

LITERATURE REVIEW

Religious Tourism on Different Perspectives

Studies on religious tourism, according to Lochrie, et al. (2019), have focused on the relevance of involvement, diversity, socialization and experimentation of local culture for pilgrim tourists or visitors.

In this sense, religious tourism has faith as its fundamental motivation. This type of tourism is more comprehensive because it can translate cultural motives into knowing other religious manifestations or others such as visiting religious spaces, parties, shows and activities, involving the relationship and the exploration of a destination (Kim, Kim, & King, 2019).

Thus, two issues must be considered: the first one prevents religious destinations from being visited by a large number of tourists, whether for religious or cultural reasons, which raises concerns about the possible negative impacts of mass tourism (Wong et al., 2013); the second evidence is that the identity of religious places has been transformed, so that it becomes common that the reason for the trip is not just religious tourism, with a consequent increase in cultural reasons although there is a spiritual connection between the tourist and the destination (Kim, Yilmaz, & Ahn, 2019; Wong et al., 2013).

From an economic point of view, religious commerce, as well as secularization, are challenges imposed on religious tourism destinations due to commercial and secular activities, including parties, shows, purchases of items based on brands, restaurants and hotels. luxury hotels close to religious sites can affect local identity (Huang & Pearce, 2019; Singh, 2004).

Secular travelers and pilgrims differ because for the former, religious tourism appears as a consequence of religious motivation, and for the latter, it aims to base the faith, and have common needs as a tourist, in terms of accommodation, visits, meals, purchase of souvenirs, etc. (Ebron, 1999; Ambrósio, 2006; Henderson, 2011).

Martins-Rodrigues (2018) defines sustainability as an opportunity that companies have to add value and differentiation to their processes, products and services, becoming more competitive in the market.

In relation to sustainability, mass tourism to religious places poses challenges in the social, economic and environmental dimensions, such as: high number of visitors, using hotel establishments, restaurants, transport. Relationships are also created with the local populations that receive them, creating impacts of a socio-cultural, economic and environmental nature.

At an economic level, tourism companies and services are the essential part of the system, with a basic part of tourism production: provision of accommodation, restaurants, distribution, shows and other services. According to Cunha (2007), these private and public organizations have to guarantee the entire functioning of the system.

Thus, it is necessary to develop tourism with the objective of preservation and development, based on sustainable actions on cultural, historical and environmental heritage, transforming the place into a suitable and attractive environment for all (Bulatović & Stranjančević, 2019; Huang & Pearce, 2019).

Another important aspect, regarding sustainability, is related to the sustainability of the local culture. Therefore, actions are needed to protect the religious identity of sacred sites, by valuing the accumulated knowledge, minimizing purely commercial promotion that works as an attraction to visitors instead of loyal ones and minimizing environmental impacts such as, for example, those measured by footprints. carbon (El Hanandeh, 2013; Hung et al., 2017).

Regarding the socio-cultural potential of religious tourism, this type of tourism acts as a social practice capable of positioning man in the world, influencing motivation and spiritual awareness (Kurmanaliyeva et al., 2014).

From the point of view of visitors, the perceived motivation, safety and quality of services are able to positively impact the likelihood that they will be satisfied with the trip, which, in turn, favors the recommendation and revisit (Henderson, 2011; Verma & Sarangi, 2019).

This is of fundamental importance for public policy managers, who must observe local needs, aligning them with economic, social and environmental development. Visitors' motivation emerges from affective commitments to the sacred, and thus, the justification for visits is nourished by religious texts, verbal reports, practices and experiences (Scaffidi & Nuovob, 2013; Terzidou, Scarles, & Saunders, 2018).

The memories of religious groups are caused by three possible types: events, people and geographic locations, the latter being, due to greater materiality, which allow the more appropriate promotion of religious tourism (Follis, 2018).

Considering the spatial context, urban planning is a crucial tool for the better organization of the economic, social and environmental objectives of cities that host religious tourism. They must also take into account mass tourism and the possible local and regional development brought by pilgrims (Chadha & Onkar, 2016; Shinde, 2017). In addition, issues of rapid demographic growth and construction challenge municipal master plans in an attempt to organize commercial establishments, hotels, infrastructure and pollution (Shinde, 2017).

In relation to the promotion of religious tourism, the media have a preponderant role in preserving the image of cities related to religious tourism, disseminating through the internet, adequately disseminating the benefits of each destination (Cristea, Apostol, & Dosescu, 2015; Huang & Pearce, 2019).

There are other aspects such as the quality of information, means of transport, accessibility, parking facilities, cleaning and organization, which encourage religious tourism and the intention to return, (Prazeres & Carvalho, 2015).

Main Sanctuary Cities in the World

When we travel, we are not just going to places to observe exotic landscapes, beaches, gastronomy or different cultures, but we will also see unique and fascinating buildings, markets, museums, churches and Cathedrals. Therefore, we will describe some of the largest churches and cathedrals in the world, places that we must admire, and to enjoy these architectural wonders it is not essential to be religious. Churches and cathedrals, in addition to points of spirituality and peace, are tourist places of religious, architectural and historical interest.

It is evident that there are countless mystical places of religious tourism in the world, such as the Cathedral of San Sava, in Belgrade, capital of Serbia, the largest orthodox church in the world, located on the plateau of Vracar; Poland, with the Basilica of Our Lady of Lichen, one of the most modern in Europe, built between 1994 and 2004, has an impressive architecture. It was the old chapel of the Convent of the Sisters of the Mother of God of Mercy, where Saint Faustina Kowalska, the seer of Divine Mercy, lived. We can observe the relics of this Saint, to whom the Lord of Divine Mercy appeared and transmitted the well-known prayer of the Rosary of Mercy. We also observed in this Basilica the spiritual treasures the "cornerstone", which was taken from Golgotha (the Calvary Mount) and placed at the entrance.

In the European context, the Fátima sanctuary, located in Portugal, attracts approximately seven million of Portugal's ten million religious tourists per year, and its population has more than doubled in 45 years (Prazeres & Carvalho, 2015; Ambrósio et al., 2019). The Basilica of Nossa Senhora do Rosário, one of the most important marian shrines for Catholics (Orth, 2015), where are the tombs of the three shepherds (Lúcia, Francisco and Jacinta) and the Basilica of the Holy Trinity with a modern architecture, with decorations inspired by Byzantine art. The sanctuary of Fátima is an international reference in terms of religious offerings, information, transport and mobility, parking, organization and cleaning mechanisms (Prazeres & Carvalho, 2015).

In Spain, the Cathedral of Santiago de Compostela was built in 1075 on a small mausoleum where the remains of the apostle São Tiago are located, with various architectural styles. It is one of the most important religious tourism sites in Europe, together with Rome, configuring a center of Christian pilgrimage since the Middle Ages (Ambrósio 2006; De La Torre & Pérez, 2017).

In the United Kingdom, the second most visited sanctuary at the national level is Liverpool Cathedral, a monument to the city of the Beatles, a red sandstone building with a neo-Gothic style that stands out for its austerity and the impressive dimensions of the 31 ton bells, known as Bartlett Bells; the Basilica of the Sacred Heart in Belgium, known as Koekelberg by the inhabitants of the Belgian capital, built throughout the 20th century, is one of the most beautiful buildings in Brussels. With Art-Deco architecture, its construction was interrupted by the two World Wars (Ambrósio, 2006).

According to Urien-Lefranc (2020), in the union of archaeological and religious characteristics, one of the main destinations in the world is Jerusalem, the so-called Holy Land, in Israel, which since the 19th century has been preferentially visited at Easter. In view of this, there is concern with the cultural care of the so-called Samaritans, populations seen as "relic peoples" from biblical times, and should be preserved. In addition to the religious heritage, Israel carries archaeological treasures, which fosters religious tourism in that region, but also imposes the need to care for the archaeological heritage.

In Asia, there are a multitude of religions, such as Buddhism, Confucianism, Christianity, Hinduism, Islam, Jainism, Sikhism, Shinto and Taoism (Lin & Fu, 2020; Shinde, 2015). Therefore, without the intention of exhausting the places and their beliefs, it is important to emphasize that religious tourism is of great importance for pilgrims and the community. The Basilica of Nossa Senhora da Paz, on the Ivory Coast, the main church in Africa, erected between 1986 and 1989, has beautiful architecture and the largest extension of stained glass in the world. In Kyoto in Japan, the Kiyomizu-dera Temple or "Temple of Pure Waters", listed by UNESCO, is next to Otowa Mountain, where pilgrims seek the sacred waters. On the other side of the city, gold leaves cover the Zen Buddhist temple Kinkaku-ji, originally built in the 15th century.

In India, according to Singh (2004), we can observe historical archaeological sites, whose trip aims at the Hindu practice of mass bathing in rivers or reservoirs, or to pray in sacred places. The impacts of mass travel for religious tourism in sacred places must be analyzed from an anthropological (Singh, 2004), sociological (Kim & Kim, 2019) perspective, with a focus on sustainability through the triple bottom line perspective of social, economic and environmental development (Buckley, 2012; Paiano, Crovella, & Lagioia, 2020), and with interdisciplinary orientation of marketing studies (Hosany, Ekinci, & Uysal, 2006; Muriuki, Bururia, & Mutegi, 2018).

In the American continent the Basilica of Nossa Senhora Aparecida, in Brazil also known as the National Shrine of Nossa Senhora Aparecida, is a catholic religious temple located in Aparecida, in the interior of the state of São Paulo, it is the largest sanctuary in Brazil and worldwide second largest (Oliveira 2018). The sanctuary is visited annually by approximately 12 million pilgrims from all parts of Brazil. In November 2016, by decree of Pope Francis, the basilica was elevated to the dignity of a cathedral church of the Archdiocese of Aparecida, title transferred from the church of Santo Antônio, in Guaratinguetá.

Another sanctuary widely visited in North America, more precisely in Mexico, the Basilica of the Virgin of Guadalupe that was inaugurated in 1976 and the Tilma of San Juan Diego is preserved there, where the image of Our Lady of Guadalupe is preserved (Orth, 2015). In the USA, the Cathedral of Saint John the Divine, New York, the largest Anglican cathedral in the world, was built in less than 50 years, between 1892 and 1941, originally in Byzantine-Romanesque style in the late 19th century, but the its architecture was changed to gothic style in the 70s. it has a capacity for 20 thousand people. Located in the heart of Manhattan.

METHODOLOGY

The present study is quantitative and exploratory based on a survey, in order to obtain specific information for the research that is not possible to obtain from public databases. The questionnaire was adapted from the study by Irimiás, Mitev & Michalko (2016), composed of ten questions that sought to identify the perception of the behavior of individuals in relation to religious tourism, as can be seen in table 1.

In the process of applying the questionnaire, a pre-test was carried out with a group of twenty people, with no difficulty in understanding the questions. For grading the answers, a five-point Likert scale was used, based on the agreement or disagreement with the question (1 = strongly disagree to 5 = strongly agree).

The ten questions seek to identify information such as: the link between religious tourism and national identity, the purpose of visiting tourist cities with a bias towards religion, health issues, the magnitude of donations to sacred visits, souvenir purchases, and cultural attractions.

Table 1. Research Model

Demographic variables related to religion	References
1. Visiting one or more churches is usually part of the journey.	Irimias, Mitev and Michalko (2016)
2. I consider churches primarily for their architectural and historical value.	
3. I usually buy some souvenirs when I attend religious events or when visiting a sacred place.	
4. When traveling, I am also interested in visiting non-religious places.	
5. I think it is important that my family and friends know the significant historical places of Christianity.	
6. I usually travel to destinations where I can visit religious attractions.	
7. Through a donation, I support the church I visited as a tourist.	
8. I prefer to visit churches when I travel abroad.	
9. Visiting sacred places reinforces my national identity.	
10. I visited a sacred place to improve my health conditions or ask my family.	

Source: Adapted Irimias, Mitev and Michalko (2016)

In order to obtain a greater number of responses, snowball sampling was used. The sampling was initiated by researchers from public and private universities so that it was possible to send it to their contact networks, as we seek to analyze the behavior of tourists. The parameters for them to participate in the study was at least to have traveled to religious events or religious tourism in the last 3 years. Thus, the study was composed of 396 responses from Brazilians and Portuguese. The data were analyzed using SPSS Software, version 22.

The results of the study will be explained and discussed below.

DISCUSSION OF RESULTS

Sample Characterization

The study was applied to a sample of 396 individuals, of Portuguese and Brazilian nationality.

The questionnaire consisting of the respondents' sex, nationality, age, marital status, professional status and religious roots are shown in Table 1. It is important to note that individuals answered the questionnaire based on their knowledge of the topic and experience. Thus, the characterization of respondents will be presented below.

It is noteworthy that there was a heterogeneity regarding the respondents' age range of the total sample, it appears that 66.2% are female and 33.8% male and with a level of Literacy, 40.2% have higher education (undergraduate, postgraduate, masters and doctorates), they have a maturity in the answers and experiences about the object of the study. In addition, it is evident that the predominance regarding the age group is between 31 to 65 years.

Still in accordance with the data, the respondents' predominant marital status is married, followed by single. In this perspective, the professional situation is expressed as employees and approximately 86% have Catholic religious roots, contributing to the objective of this study. These data are very important for the purpose of the research that studies religious tourism that emphasizes Nolan and Nolan (1992) of the relevance of studying the religious bias aimed at individuals who are religious, as well as tourists who have other religions or even who visit places sacred due to cultural history. This survey highlighted the respondents' significant experiences.

Following it is presented the characterization of the sample:

Based on the data, it is possible to infer that the respondents of this study have the profile to refute the established and compatible questions with the objective of this research. The results show that individuals, whether Catholics or from different religions, when visiting religious places, contemplate the importance of the place.

The results of the research on religious tourism are presented below.

Religious Tourism

This section presents the results regarding the religious tourism behavior of the respondents in the study sample. It is worth mentioning that the respondents were asked to agree on the ten variables highlighted below in table 2, being based on the agreement or disagreement with the question (1 = strongly disagree to 5 = strongly agree).

As can be seen, some variables have higher agreement averages such as "I consider churches mainly for their architectural and historical value", "When traveling, I take the opportunity to visit non-religious places" and "Visiting one or more churches is usually part of the journey" as well as with the smallest averages "Through a donation, I support the church I visited as a tourist", "I prefer to visit churches when I travel abroad" and "Visiting sacred places reinforces my national identity".

As can be seen in the table above, the variable "I consider churches mainly for their architectural and historical value" obtained the highest average among respondents in the study. This indication is due to the fact that the majority have a high level of knowledge, are retired or employed and are Catholics. In this context, the aspiration to opulent mastery over the cultural and historical contextualization of sacred places is of fundamental importance. In this sense, Prazeres & Carvalho (2015) indicate that religious architectural aspects attract tourists from different perspectives, reflecting on different behaviors.

However, another variable that deserves to be highlighted "When traveling, I take the opportunity to visit non-religious places", leading to the reflection that in addition to the sacred places, visitors seek to learn about the culture of the country visited, be it local cuisine, other environments that do not belong to religion individual, as well as visiting local businesses to purchase souvenirs (Clarke & Raffay, 2015).

Table 2. Distribution by Demographic Data

Gender	Frequency	%
Feminine	264	66,2
Male	134	33,8
Total	**396**	**100**
Nationality	**Frequency**	%
Portugal	202	51,0
Brazil	193	49,0
Total	**396**	**100**
Age	**Frequency**	%
until 18	08	2,0
18 a 24	11	2,8
25 a 30	30	7,6
31 a 40	110	27,8
41 a 50	85	21,5
51 a 65	93	23,7
more 65	58	14,6
Total	**396**	**100**
Education	**Frequency**	%
Elementary School	30	7,5
High School	101	25,5
University Graduate	106	26,8
Postgraduate	159	40,2
Total	**396**	**100**
Marital status	**Frequency**	%
In Union	38	9,6
Single	90	22,5
Widow	30	7,6
Divorced	38	9,8
Married	200	50,5
Total	**396**	**100**
Professional situation	**Frequency**	%
Employee	276	69,7
Unemployed	54	13,6
Retired	66	16,7
Total	**396**	**100**
Religious roots	**Frequency**	%
Catholic	344	86,9
Evangelical	13	3,3
Catholic and Spiritist	2	0,5
Kardecist	1	0,3
Spiritist	3	0,8
Protestant	3	0,8
Muslim	1	0,3
Ubandista	1	0,3
Orthodox	2	0,5
No religion	25	6,3
Total	**396**	**100**

Source: Processed by author´s

In this context, according to table 3, the third variable with the highest average was "Visiting one or more churches is usually part of the journey", it is shown that religious trips provide visits to religious places in that particular tour package. In this way, visitation to sacred environments influences the perspicacity of tourists with regard to the objective of this study, the behavior of religious tourists.

Table 3. Data from the survey on religious tourism

	N	Minium	Maxium	Mean	STDEV	Var
4.2.1 Visiting one or more churches is usually part of the journey.	396	1	5	**3,70**	1,244	1,546
4.2.2 I consider churches mainly for their architectural and historical value.	396	1	5	**3,91**	1,157	1,339
4.2.3 I usually buy souvenirs when I participate in religious events or when visiting a sacred place.	396	1	5	2,78	1,346	1,811
4.2.4 When traveling, I take the opportunity to visit non-religious places.	396	1	5	**3,90**	1,197	1,434
4.2.5 I think it is important that family and friends know the significant historical places of Christianity.	396	1	5	3,60	1,221	1,491
4.2.6 I usually travel to destinations where I can visit religious attractions.	396	1	5	2,60	1,177	1,385
4.2.7 Through a donation, I support the church I visited as a tourist.	396	1	5	**2,19**	1,293	1,671
4.2. 8 I prefer to visit churches when I travel abroad.	396	1	5	**2,25**	1,091	1,189
4.2.9 Visiting sacred places reinforces my national identity.	396	1	5	**2,40**	1,305	1,704
4.2.10 I like to watch movies with religious themes.	396	1	5	3,10	1,250	1,562
Total	**396 responses**					

Source: Processed by author´s

However, it can be detailed that three variables obtained low averages among respondents "Through a donation, I support the church I visited as a tourist", "I prefer to visit churches when I travel abroad" and "Visiting sacred places reinforces my national identity ".

With regard to the variable "Through a donation, I support the church I visited as a tourist" (2,19), it is noteworthy that this attitude is not essentially carried out by respondents when participating in tourist visits.

Still in this perspective, the second variable "I prefer to visit churches when I travel abroad" (2.25), emphasizes the importance of religious tourist places so that they have an excellent structure in receiving these tourists. It appears that they do not visit churches only when they travel and it can be concluded that they visit religious places assiduously.

And finally, "Visiting sacred places reinforces my national identity" (2.40) pointing out that the identity, for the respondents, is not related to religiosity, but to the environment in which it is inserted. In this perspective, religious tourism goes beyond pilgrimage or resolving individual and religious issues, involving the relationship and the exploration of a place (Kim, Kim, & King, 2019).

However, religious places focused on tourism are positively related to the sagacity of the behavior of individuals in religious tourism. Thus, the particularities found in this study assist Tourist Managers in the way of better guaranteeing quality in the provision of services aimed at religious tourism.

FINAL CONSIDERATIONS

The popularity of religious tourism, the importance of Sanctuaries as tourist destinations and the motivation of tourism in people's lives are the inspiration to carry out this study of the motivations and behaviors for visiting contemporary Sanctuaries in the World. For the sustainable development of tourist spots, religious and pilgrimage tourism, priority must be given to the diversity of attributes of religious tourists.

New challenges will continue to appear in the management of religious tourism destinations, including sacred sites, as the global travel market continues to evolve. For researchers, this requires an understanding of current changes in pilgrimages and religious tourism, as well as trends and ways of traveling. For tour operators, it means finding appropriate resources and offers to meet the multifaceted needs of visitors with motivations such as: religious, cultural, leisure or tourism. The managers of the sanctuaries, must have an adequate planning to the diverse expectations of the visitors, guaranteeing their own standard for the comfort of the tourist. The purpose of the activities is to have visitors satisfied, both pilgrims and non-pilgrims, so that everyone wishes to return and maintain a sustainable level of use of the Shrines.

The study aimed to analyze the behavior and cultural differences between religious tourists who visit sacred sites. The results showed that we are facing a heterogeneous sample, which included different age groups, the majority of respondents being female and having higher education as academic qualifications. A large percentage was born in a religious cradle and is therefore considered religious. In addition, a high number of religious tourism participants are part of a religious community. Regarding to the findings of the research the most important activities of the respondents regarding religious tourism are: visiting one or more churches is usually part of the journey; consider churches mainly for their architectural and historical value; and they visit non-religious places, when traveling for sanctuary cities.

The main value of this study is to confirm and reinforce the idea that sanctuary cities are one of the most important destinations for tourists, and that churches are the main interests, nevertheless other type of monuments are also of interest and compose the visit to the cities. This is important information for tourist guides, as they could conceive mixed tourist travel packages, not only focused on religious artifacts and monuments.

Among other limitations of this study, we highlight: the information presented is limited only to Portuguese and Brazilian tourists; the research was carried out with the snowball technique, limited to the sample for convenience.

The contribution of this article to the studies in Management is due to the resulting indicators of the motivation of the secular travelers and pilgrims to visit or pilgrimage to Sanctuaries.

For future research, it is suggested that the analyzed others countries be expanded in order to compare results and to analyze other content different from those analyzed, for example with a range of different religious beliefs or non-religious beliefs and convictions.

REFERENCES

Abumanssur, E. S. (2017). A força do mito: tensões e acomodações narrativas em torno das imagens de devoção. In Academia Marial de Aparecida. Aparecida, 300 anos de fé e devoção. Aparecida: Santuário/ Academia Marial de Aparecida.

Ambrósio, V. (2006). *Turismo Religioso: Desenvolvimento das Cidades – Santuário. (Tese de Doutoramento).* Universidade Nova de Lisboa.

Ambrósio, V., Fernandes, C., Krogmann, A., Leitão, I., Oremusová, D., & Šolcová, L. (2019). From church perseverance in authoritarian states to today's religious inspired tourism. PASOS. *Revista de Turismo y Patrimonio Cultural*, *17*(5), 1065–1078. doi:10.25145/j.pasos.2019.17.074

Andrade, J.V. (2000). *Turismo: Fundamentos e dimensões*. São Paulo: Ática.

Baggio, F., Parise, P., & Sanchez, W. L. (2016). *Mobilidade humana e identidades religiosas*. Paulus.

Bourdieu, P. (2005). *A Economia das trocas simbólicas* (6th ed.). Perspectiva.

Bulatović, I., & Stranjančević, A. (2019). Integrated model of religious tourism in underdeveloped destinations. *Journal of the Geographical Institute Jovan Cvijic SASA*, *69*(1), 53–66. doi:10.2298/IJGI1901053B

Cabrera, H. (2011). Cartografía regional de las fiestas tradicionales. In Regionalización cultural del Uruguay. Manosanta Desarrollo Editorial.

Chadha, H., & Onkar, P. (2016). Changing Cities in the Perspective of Religious Tourism – A Case of Allahabad. *Procedia Technology*, *24*, 1706–1713. doi:10.1016/j.protcy.2016.05.200

Clarke, A., & Raffay, Á. (2015). Religion, Local Produce and Sustainability at Religious Sites in Hungary. *International Journal of Religious Tourism and Pilgrimage*, *3*(2), 33–47.

Collins-Kreiner, N., & Kliot, N. (2000). Pilgrimage tourism in the holy land: The behavioral characteristics of Christian pilgrims. *GeoJournal*, *501*(1), 55–67. doi:10.1023/A:1007154929681

Cristea, A. A., Apostol, M. S., & Dosescu, T. (2015). The Role of media in Promoting Religious Tourism in Romania. *Procedia: Social and Behavioral Sciences*, *188*, 302–306. doi:10.1016/j.sbspro.2015.03.398

Cunha, L. (2007). *Introdução ao Turismo* (3rd ed.). Verbo.

De La Torre, G. M. V., & Pérez, L. M. (2017). Religious tourism in different geographical areas in Spain: The tourist profile. *Boletín de la Asociación de Geógrafos Españoles*, (75), 29–54.

Digance, J. (2003). Pilgrimage at contested sites. *Annals of Tourism Research*, *30*(1), 143–159. doi:10.1016/S0160-7383(02)00028-2

Ebron, P. A. (1999). Tourists as Pilgrims: Commercial Fashioning of Transatlantic Politics. *American Ethnologist*, *26*(4), 910–932. doi:10.1525/ae.1999.26.4.910

El Hanandeh, A. (2013). Quantifying the carbon footprint of religious tourism: The case of Hajj. *Journal of Cleaner Production*, *52*, 53–60. doi:10.1016/j.jclepro.2013.03.009

Follis, R. (2018). Turismo religioso, adventismo e lugares de memória. *HORIZONTE - Revista de Estudos de Teologia e Ciências Da Religião, 16*(49), 38–65.

Henderson, J. C. (2011). Religious tourism and its management: The hajj in Saudi Arabia. *International Journal of Tourism Research, 13*(6), 541–552. doi:10.1002/jtr.825

Hosany, S., Ekinci, Y., & Uysal, M. (2006). Destination image and destination personality: An application of branding theories to tourism places. *Journal of Business Research, 59*(5), 638–642. doi:10.1016/j.jbusres.2006.01.001

Huang, K., & Pearce, P. (2019). Visitors' perceptions of religious tourism destinations. *Journal of Destination Marketing & Management, 14*(June), 100371. doi:10.1016/j.jdmm.2019.100371

Hung, K., Yang, X., Wassler, P., Wang, D., Lin, P., & Liu, Z. (2017). Contesting the Commercialization and Sanctity of Religious Tourism in the Shaolin Monastery, China. *International Journal of Tourism Research, 19*(2), 145–159. doi:10.1002/jtr.2093

Irimias, A., & Mitev, A., & Michalko, G. (2016). Demographic Characteristics Influencing Religious Tourism Behaviour: Evidence form a Central-Eastern-European country. *International Journal of Religious Tourism and Pilgrimage, 4*, 19–32.

Kim, B., & Kim, S. (2019). Hierarchical value map of religious tourists visiting the Vatican City/Rome. *Tourism Geographies, 21*(3), 529–550. doi:10.1080/14616688.2018.1449237

Kim, B., Kim, S., & King, B. (2019). Religious tourism studies: Evolution, progress, and future prospects. *Tourism Recreation Research, 0*(0), 1–19.

Kim, H., Yilmaz, S., & Ahn, S. (2019). Motivational landscape and evolving identity of a route-based religious tourism space: A case of Camino de Santiago. Sustainability (Switzerland), 11(13).

Kurmanaliyeva, R. S., Duissenbayeva, A., & Izmailov, I. (2014). Religious Tourism as a Sociocultural Phenomenon of the Present "The Unique Sense Today is a Universal Value Tomorrow. This is the Way Religions are Created and Values are Made.". *Procedia: Social and Behavioral Sciences, 143*, 958–963. doi:10.1016/j.sbspro.2014.07.534

Lin, C. F., & Fu, C. S. (2020). Cognitive implications of experiencing religious tourism: An integrated approach of means–end chain and social network theories. *International Journal of Tourism Research, 22*(1), 71–80. doi:10.1002/jtr.2319

Lochrie, S., Baxter, I. W. F., Collinson, E., Curran, R., Gannon, M. J., Taheri, B., Thompson, J., & Yalinay, O. (2019). Self-expression and play: Can religious tourism be hedonistic? *Tourism Recreation Research, 44*(1), 2–16. doi:10.1080/02508281.2018.1545825

Macedo, L. (2011). La dimensión territorial del turismo religioso. In Turismo espiritual: Una visión iberoamericana. Centro Universitario de los Altos de la Universidad de Guadalajara.

Martins-Rodrigues, M. C. (2018). *Empresas incubadoras em Portugal e o seu efeito na sustentabilidade das incubadas*. Chiado Books.

Muriuki, L. M., Bururia, D., & Mutegi, J. (2018). Determinants of religious tourism branding tactics in Kenya: A cross sectional study. *Tourism and Hospitality Research*, *18*(4), 456–469. doi:10.1177/1467358416670937

Nolan, M., & Nolan, S. (1992). Religious sites as tourism attractions in Europe. *Annals of Tourism Research*, *19*(1), 1–17. doi:10.1016/0160-7383(92)90107-Z

Oliveira, C. D. M. (2000). *Viagens a santuários: uma modalidade de turismo religioso ou de religiosidade turística?* Boletim Turístico de Administração Hoteleira, Out.

Oliveira, C. D. M. (2004). *Turismo Religioso*. Aleph.

Oliveira, C. D. M. (2018). Matergrafia e patrimônio: Santuários Marianos como espaço simbólico e vetorial da Latinidade. *Ateliê Geográfico*, *12*(3), 170–194. doi:10.5216/ag.v12i3.47188

Orth, M. (2015). The world's most powerful woman. *National Geographic*, (December), 32–59.

Paiano, A., Crovella, T., & Lagioia, G. (2020). Managing sustainable practices in cruise tourism: the assessment of carbon footprint and waste of water and beverage packaging. *Tourism Management, 77*(August), 104016.

Pinto, S. O. A. (2011). Turismo Religioso – potencial de desenvolvimento da Vila de Arcozelo, Vila Nova de Gaia (Dissertação de Mestrado). Universidade Lusófona de Humanidades e Tecnologias, Lisboa, Portugal.

Prazeres, J., & Carvalho, A. (2015). Turismo Religioso: Fátima no Contexto dos Santuários Marianos Europeus. *PASOS Revista de Turismo y Patrimonio Cultural*, *13*(5), 1145–1170. doi:10.25145/j.pasos.2015.13.078

Prazeres, J., & Carvalho, A. (2015). Turismo Religioso: Fátima no Contexto dos Santuários Marianos Europeus. *PASOS Revista de Turismo y Patrimonio Cultural*, *13*(5), 1145–1170. doi:10.25145/j.pasos.2015.13.078

Savola, C. (2003). EI turismo religioso ya mueve millones por el país y eI mundo. Clarín. *Cuadernos Sociedad, Buenos Aires*, *22*(sep), 3.

Scaffidi, A. C., & Nuovob, S. (2013). Motivation and personality traits for choosing religious tourism. A research on the case of Medjugorje. *Current Issues in Tourism*, *16*(5), 501–506. doi:10.1080/13683500.2012.749844

Shinde, K. A. (2015). Religious tourism and religious tolerance: Insights from pilgrimage sites in India. *Tourism Review*, *70*(3), 179–196. doi:10.1108/TR-10-2013-0056

Shinde, K. A. (2017). Planning for urbanization in religious tourism destinations: Insights from Shirdi, India. *Planning Practice and Research*, *32*(2), 132–151. doi:10.1080/02697459.2016.1198197

Silveira, E. J. S. (2004). Turismo religioso popular? Entre a ambiguidade conceitual e as oportunidades de mercado. *Revista de Antropologia Experimental*, *4*, 1–16.

Singh, S. (2004). Religion, heritage and travel: Case references from the Indian Himalayas. *Current Issues in Tourism*, *7*(1), 44–65. doi:10.1080/13683500408667972

Tendeiro, I. L. G. S. (2010). *A igreja de Santo António de Lisboa e o turismo religioso italiano*. Dissertação de Mestrado da Escola Superior de Hotelaria e Turismo de Estoril.

Terzidou, M., Scarles, C., & Saunders, M. N. K. (2018). The complexities of religious tourism motivations: Sacred places, vows and visions. *Annals of Tourism Research*, *70*(February), 54–65. doi:10.1016/j.annals.2018.02.011

Urien-Lefranc, F. (2020). From religious to cultural and back again: Tourism development, heritage revitalization, and religious transnationalizations among the samaritans. *Religions*, *11*(2), 86. doi:10.3390/rel11020086

Verma, M., & Sarangi, P. (2019). Modeling attributes of religious tourism: A study of Kumbh Mela, India. *Journal of Convention & Event Tourism*, *20*(4), 296–324. doi:10.1080/15470148.2019.1652124

Vilas Boas, N. F. S. (2012). *A Pastoral do turismo: da peregrinação ao Santuário. Dissertação de mestrado*. Universidade Católica Portuguesa.

Wong, C. U. I., Ryan, C., & McIntosh, A. (2013). The Monasteries of Putuoshan, China: Sites of Secular or Religious Tourism? *Journal of Travel & Tourism Marketing*, *30*(6), 577–594. doi:10.1080/10548408.2013.810996

Chapter 17
Religious Tourism in Africa's Global South:
Indigenous African Traditional Spirituality

Unathi Sonwabile Henama
https://orcid.org/0000-0002-1111-0729
Department of Tourism Management, Tshwane University of Technology, South Africa

Lwazi Apleni
https://orcid.org/0000-0001-8521-6174
Department of Recreation and Tourism, University of Zululand, South Africa

ABSTRACT

International tourist arrivals are projected to surpass 1.8 billion by 2030 on the back of rapid growth in emerging tourism economies. Tourism has emerged as an economic messiah for a plethora of countries in Sub-Saharan Africa. It has emerged as a cost-effective means by which countries can diversify their economies, especially countries with low economies that depend on agricultural products to diversify their economies. Religious tourism can contribute to deeper economic benefit for a destination. The synthesis of literature adds to the paucity of academic gaze on religious tourism in Southern Africa. The synthesis takes the reader on a religious tourism journey that includes African spirituality, Pentecostal Christianity, and the interface between Africans spirituality and Christianity. These areas are neglected in the academic gaze and are outside the tourism beaten track, and these forms of religious tourism bring in much needed economic activities for areas on the tourism fringe.

INTRODUCTION

Tourism is the world's largest industry, therefore making big business, located within the tertiary sector of the economy. It is for this reason, where almost all countries have jumped on the tourism bandwagon, seeking to benefit from the positive economic impacts associated with tourism. "Tourism destinations attract tourism because of the positive economic impacts such as labour-intensive jobs, tourism acting

DOI: 10.4018/978-1-7998-5792-1.ch017

as a catalyst for other industries, the attraction of foreign exchange, and foreign direct investment that comes with tourism" Henama (2016: 1). The tourism industry, which is a service, depends on human production, which is an experience, which is inseparable. This leads to tourism being labour-intensive and the majority of the value-adding happens to occur at the destination area, which is the site of the tourism consumption. "Tourism is labour-intensive, therefore any growth in tourism is associated with growth in employment" Henama (2014: 1). An increase in the number of tourists arrivals, will lead to an increased demand for more labour, therefore of the labour intensive nature of tourism consumption. In the case of South Africa, it is accepted that one permanent job is created with the arrival of eight international tourists.

This makes it an attractive means of creating job opportunities for a developing country such as South Africa, which has a high rate of unemployment according to Henama (2017). Tourism more than any other industry (Tauoatsoala, Henama & Monyane, 2015) has the potential to create jobs and the sharing of economic benefits. The vast majority of businesses within the tourism value chain are small and medium enterprises (SMMEs), which are known and recognised engines of job creation and economic growth. The continued and sustained growth of tourism, has meant that it had been integrated into the economic growth policies of almost all countries, jumping on the tourism bandwagon. The arrival of tourists at a destination has the impact that government may speed up the development of infrastructure and super-structure. The tourism industry can motivate a destination country to create infrastructure that are shared by the locals and the tourists. "The tourism industry is beneficial to support infrastructure for tourism, such as transport, financial, telecommunications, energy and distribution services" Henama (2013: 230).

As a result, tourism has a positive impact on the Quality-of-Life of citizens. The tourism industry that is owned and managed by locals, is more sustainable as it would retain the majority of the tourism expenditure within the local economy. Spreading tourism's developmental benefits are imperative to ensure that urban and rural areas benefit from tourism expenditure. Tourism must be developed in such a way that it is locally owned, and that it is not concentrated in urban areas, but tourism must experience geographic spread if it is to become a tool for development. "These locally owned enterprises would not only increase the retention of the tourism multiplier effect, this would lead to lower economic leakages and increase local employment and asset base of the locals. Local ownership of tourism product offerings would assist in reducing the levels of poverty, and inequality that are a key feature of South African society" Henama (2012: 78). African countries have adopted tourism as an economic panacea, to diversify the economies of many African countries, which are still dominated by agriculture and poor exports. African countries still attracts less than 10% of international tourist arrivals, whereas tourism arrivals are disproportionally important for African countries. Tourism remains an attractor of foreign exchange par excellent, and in South Africa the growth of tourism as a sector has led it to be called the 'new gold' as it has surpassed gold exports in attracting foreign exchange.

The growth trajectory of tourism is sustained during times of economic decline. This proves that the tourism industry is very resilient, bouncing pack to its perennial growth rates after a slight dip, caused by either economic decline or global health emergencies. The emergence of fast growing economies, and better connectivity through aviation and the wide spread availability of transportation have promoted tourism's growth. This is because the tourism industry is transport-intensive, linking tourists with places they call home, and the tourism destination area. "The demand for tourism is almost unending, as more and more customers are getting introduced to tourism consumption" Henama, Janse van Rensburg & Nicolaides (2016: 2). "Another advantage is that tourism is a product that has potential to be consumed over the lifetime of an individual as new destinations are visited and old ones are revisited. The institutionalisation of vacations as a normal part of society as people move up the social strata is imperative as it has increased the number of tourists as many countries experienced rapid economic growth, wealth and associated normal middle class consumption that includes tourism" Henama (2014a: 2). "Tourism consumption must be made to be fashionable for society in a language and images that will appeal within the prevalent consumer culture. Once it is assimilated as part of a status symbol, this may assist in the institutionalisation of holiday taking especially for the growing middle class in South Africa" Henama & Sifolo (2017: 50).

Theoretical Framework: Special Interest Tourism (SIT)

The tourism interests and motivations are very much diverse and this has led to the development of Special Interest Tourism (SIT), tourists that share similar consumer behaviour patterns and are grouped into SIT markets. Special interest tourism has been closely associated with the growth of a form of "new tourism" Henama (2014: 3). These are forms of tourism that in contrast to mass tourism, and show alternative consumption patterns when compared to mass tourism. For example, business tourism is a form of SIT, which can be consumed at a destination area during the valley season at a destination. According to Ingle (2013), the advent of "niche tourism" has seen a move away from traditional "mass tourism" towards a predilection for remote, "authentic" rural regions and their small towns. Forms of SIT include business tourism, cultural tourism, medical tourism, justice tourism, educational tourism, rural tourism, adventure tourism and etc. SIT is markedly different and divergent from the mass tourism market, associated with sand, sea and sun tourists that had, at times represented a plantation economy in the tourism industry.

"Alternative tourism generally consists of small-scale dispersal and low-density developments often located in and organised by villages or communities, where it is hoped than more meaningful interactions will be fostered between tourists and local residents, resulting in less social and cultural disruption than in the case of tourism enclave development. Secondly, ownership patterns in alternative tourism is weighed towards local, often family-owned, relatively small-scale business, rather than foreign-owned tourism capital. Thirdly, alternative tourism encourages community participation in planning and seeks to strengthen institutions designed to enhance local participation and give priority to local needs and local enterprises" Visser & Baker (2004: 229). Van Der Merwe & Saayman (2013) noted that two factors that motivates individuals to travel, first, tourists travel as a response to what is lacking yet desired (that is, a need) and secondly, tourists travel as a response to individual needs. These are several markets segments that are set to grow at an exponential growth in the years to come in Africa.

Literature Review: Religious Tourism

"Religious tourism, chiefly a form of tourism whereby people travel to pilgrimage, missionary, fellowship purposes is growing worldwide" Mushava (2015: 1). "Religious travel is probably the oldest type of tourism in the world. Today, faith tourism is a $18 billion global industry with 300 million travellers" Seyer & Muller (2011: 45-46). "Religious tourism is that type of tourism whose participants are motivated either in part or exclusively for religious reasons" Rinschede (1992: 52). ''Religious tourism is a form of travel with the main interest of experiencing religious forms or the products they induce like art, traditions and architecture. Two aspects of this form of tourism can also be listed as: (a) the faith of a tourist, who has a spiritual attachment to the destination in line with his religious belief, and (b) the new spiritual experience of a tourist, belonging to another belief or a different religion and for whom the destination and the religious practices have the dimensions of newness'' Jude, Uchenna & Ngozi (2018:40). ''Religious tourism is not only a visitation to a particular holy destinatuion, but may also be travel for a humanitarian cause, for reasons of friendship or even as a form of leisure. Religious tourism can be the primary reason for a trip but it can also be part of a trip and provide a destination with additional attractions'' Henama & Sifolo (2018: 105).

Due to the large faith-based populations, and the definitive trend towards people purchasing products and services related to their beliefs, religious tourism has experienced a remarkable increase in recent years, and especially during the past decade. The majority of academic gaze on religious tourism and pilgrimage has been dominated by a focus on such destinations located in the Global North. These include the Vatican in Vatican City, which is the smallest country in the world. Countries such as Saudi Arabia, Israel and Nigeria draw significant revenue from religious tourism according to Mushava (2015). Saayman (2013) noted that St Paul's in London, St Peter's in Rome, the Blue Mosque in Istanbul and the Taj Mahal in India are religious tourism attracts that attract visitors from other religions. Destinations in the Global South, defined as developing countries are sites of perennial religious tourism events that have experienced as paucity of academic gaze. According to Kastarlak & Barber (2012), the activities associated with religious pilgrimage, include:

- Praying (e.g., in church, temples, mosques, shrines, or other spiritual places)
- Meditating (e.g., in sacred places)
- Participating (e.g., pilgrimage, processing, rituals)
- Other (e.g., buying religious artefacts, bringing offerings to shrines or holy places).

"Religious tourism is also considered by the government as a strategic activity for employment and generation of value to host communities" Gyekye et al. (2014: 2). Religious tourism gaze in South Africa is dominated by the Zion Christian Church (ZCC), which has headquarters in Zion Moria City, outside Polokwane in the Limpopo Province of South Africa. The peak season for religious tourism in South Africa is the Easter season, when millions of South Africans travel to various places of pilgrimage and religious festivities. ''In South Africa, the highlight on the religious tourism calendar is the Zion Christian Church's ZCC annual Easter religious gathering at Moria in the Limpopo province. Millions of people gather together at Moria, and the worshipping lasts for several days. Some worshippers come from the neighbouring countries such as Lesotho, Botswana, Namibia, Zimbabwe and Mozambique'' Apleni & Henama (2020:3). Religious tourism is not only located in urban areas, it has a presence in rural and peri-rural areas. Furthermore, Gyekye et al. (2014) noted that there is a paucity of religious tourism studies in South Africa. This would be a welcomed addition to the limited body of knowledge on religious tourism and African traditional spirituality, within a South African context. The religious tourism areas gazed as part of this body of knowledge are located outside the main tourism regions of South Africa, in areas that need tourism as a means of creating employment and availing various small scale business opportunities.

African Spirituality and Pilgrimage

According to Chiorazzi (2015) indigenous African religions refer to the indigenous or native religious beliefs of the African people before the Christian and Islamic colonization of Africa are by nature plural, varied, and usually informed by one's ethnic identity, where one's family came from in Africa. African spirituality is very much holistic. African spirituality believes that those cease to exist on earth, enter the spirit world as ancestors. '' The role of ancestors in the African cosmology has always been significant. Ancestors can offer advice and bestow good fortune and honor to their living dependents, but they can also make demands, such as insisting that their shrines be properly maintained and propitiated. And if these shrines are not properly cared for by the designated descendant, then misfortune in the form of illness might befall the caretaker. A belief in ancestors also testifies to the inclusive nature of traditional African spirituality by positing that deceased progenitors still play a role in the lives of their living descendants'' Chiorazzi (2015:2). African spirituality is steeped in spiritual ethos that is sacred, and that that leave the world are able to interface with a higher power in the eternal world.

Mantsopa (Modderpoort), Free State Province, South Africa

Sacred sites, particularly caves have been used as sites for religious tourism, as they attract pilgrims, which are a form of special interest tourism. Ngobese (2018) noted that the sacred cave could be regarded as African university for ancestral insight, strength and power to reveal divince healing for various pandemic diseases. The cave is steeped in African traditional religion and embodied to the African belief system which identifies the existence of ancestors or *badimo* who are regarded as people, who have passed on, and exist in the afterlife. African traditional beliefs systems has a set of beliefs, traditions, customs and values that must be performed. For example, in African traditional belief systems, the death of an African is not complete, as the one who dies, is regarded as still living in the afterlife, and therefore still part of the family. ''Cave dwellers hope that cave visits on the sacred space will reconnect with one's religious beliefs and reinforce cultural identities'' Ngobese (2018: 481). The cave is a pilgrimage site for worshipers from the Zion Christian Church and Anglican Church. ''The Zion Christian Church (ZCC) has in recent years, held services in the cave and erected a pulpit at the far end'' Coplan (2003: 989). The ZCC is an African Initiated Church (AIC), which is the largest church in Southern Africa with more than 16 million members.

According to Apleni & Henama (2020), the ZCC has headquarters at Zion City Moria, in Boyne outside Polokwane, which is the provincial capital city of the Limpopo Province. The Zion City Moria has the largest Christian gathering in the country during Easters, as ZCC pilgrims come from all over South Africa and neighbouring countries such as Zimbabwe, Zambia, Mozambique, Botswana, Malawi, eSwatini, Lesotho, Namibia and Malawi. ''Mantsopa's 'cave church', located in the old Anglican priory and mission of St. Augustine's at Modderpoort near Ladybrand, is surely the most famous of the cave shrines in the eastern Free State and possibly in South Africa'' Coplan (2003:988). Prophetess Makhetha Mantsopa used a cave in rural Ladybrand, in present day Free State Province as a place for refuge after she fled the Kingdom of Lesotho, which was ruled by her brother King Moshoeshoe. King Moshoeshoe formed the Basotho Nation after the Lifaqane. Hodgson (1995) noted that the refugees dislodged by the Zulu kingdom formed the Sotho kingdom, which had Thaba Bosiu as its capital. The cave has evidence of life since the Stone Ages and has been ancient San paintings, which were the first residents of present day South Africa. Makhetha Mantsopa had powers associated with her ancestral beliefs. The cave as an African traditional religion site, would be frequented by initiates who attend to the university to learn, and upon graduation, they can leave to go practice African religion and spirituality. The initiates can stay from several days to several months, and they decide to go to the university after receiving ancestral calling through dreams.

As noted by Ngobese (2018) the *ithwasa* (a trainee) had to consult an experienced diviner in order to establish the cause of the dreams. The ancestral calling can cause a decline in the health of a person who refuses to comply, which includes accepting the calling to practice as an African traditional doctor, using African spirituality. Some had to leave official employment to undertake the ancestor's call to become a *sangoma*, an African traditional doctor. This happens by accepting the ancestral calling or *ukuvuma idlozi*, a step towards training and receiving knowledge in divination and healing in African traditional beliefs. Prophetess Makhetha Mantsopa was a renowned healer, rainmaker and diviner who was born around 1759 in a place called Likotso or Ramakhetheng near present day Lesotho according to Scripps (2019). ''In 1859 Mantsopa prophesied that the Basotho would win a battle against the British troops led by Major Warden. She become renowned as a prophetess, living legend and advisor to Moshoeshoe I when her prophesies about successive wars were fulfilled…her role as an advisor to Moshoeshoe ended in 1868 when the British annexed Basutoland (Lesotho)'' Goitsemodimo (2019:3). In addition, Mantsopa was a famed rain maker. Hodgson (1995) noted that after Mantsopa settled in Modderpoort she began foretelling future events. ''Mantsopa's prophetic powers acquired mythic proportions extending over space and time. She is credited with foretelling the riderpest epidemic after Union, and World War I, warning Africans involvement. The sinking of the SS Mendi with the huge loss of African life is attributed to their disregard of this warning'' Hodgson (1995: 220).

She was baptised into Christianity by French missionary Theophile Jousse, and she then moved to Modderpoort outside Ladybrand. She moved to Modderpoort to join the new church after being exiled by Moshoeshoe. '' By the 1860s, Moshoeshoe had come to fear the power of this national prophetess. When she correctly, but unpatriotically, predicted the defeat of the Basotho by the Free State in the wars of 1865-1868, he 'exiled' her to her original home in Modderpoort'' Coplan (2003: 988). She continued to practice African traditional spirituality even through she was baptised into Christianity. Coplan (2003) noted that Makhetha Mantsopa was converted on 13 March 1870, the death after the death of her cousin Moshoeshoe. The cave provided refuge and shelter for the first Anglican missionaries in the late 1800s, and she converted to Christianity. According to Scripps (2019) the Priory of St Augustine's monastery is located next to the cave, established in 1871. The St Augustine monastery is built in Gothic style. The Anglicans acquired the track of land for the church and it is still farmed today still. Burgess (2007) noted that Anglican mission was established between 1868 and 1869, the buildings were rebuilt by Canon Frederick Beckett, James Douglas, Arthur Byrd, Bernard Garret, Henry Sanderson and James Carmichael. They made use of the cave as their temporary home, as they began the construction of the chapel, which is today the St Augustine's monastery. The Cave Chapel is the most important focal point of the St Augustine's complex.

During the Anglo-Boer War as noted by Hodgson (1995), the Modderpoort was occupied by the British and the cave was used to stable horses. Hodgson (1995) noted that cave was the site of the first 'native' school and church services were also held in the cave. There is a spring at the site, where people come to drink waters they believe to be sacred, holy and provide cleansing and can lead to the attainment of good fortune. The cave pilgrims use candles to light and illuminate the cave, and pray and seek healing from the African traditional spirituality. According to Coplan (2003), the 'Well of Mantsopa'' is a water well that was used by Mantsopa for devotion and prognostication. Goitsemodimo (2019) noted that she passed away aged 111 and buried at the Modderpoort cemetery alongside the Brothers of the Society of St Augustine and white residents of Modderpoort. She died at the year of 1906. ''Her grave, the cave where she prayed and the freshwater spring (reputedly pounted out to Mantsopa as a source of 'sacred' water by her ancestors) are major tourist attractions visited mostly by pilgrims who go for various reasons'' Goitsemodimo (2019:4). The grave of Makhetha Mantsopa is located in a gravesite on site, and pilgrims visit the grave, plant candles and pray towards the grave of Manstopa. ''Mantsopa's cave, grave and a spring at the base of Viervoet, where she spent most of her time, are still inundated with offerings left by pilgrims. These include slaughtering stock, crockery, food, money, lotto, cards, and written appeals left in the hope of blessing. Clay is often collected, believing it will bring good fortune'' Burgess (2007:3). The Free State Provincial Heritage Resource Authority declared the grave and the cave as provincial heritage sites. Hodgson (1995) noted that on Boxing Day, the altar is pilled high with cakes and sweets as gifts from children.

This site experiences the arrivals of pilgrims that want to practice African traditional spirituality and Christian pilgrims. The cave presents visitors with the spiritual site where both African spirituality and Christianity coexist. Burgess (2007) noted that Anglican church has transformed the old institution into a lodge, conference and wedding venue to generate an alternative income to farming. ''The Mantsopa movement has had no formal organisation or leadership and pilgrims from different language groups, especially Sotho and Tswana, come to Modderpoort throughout the year. Whilst Cave Sunday remains the highlight, Christmas, New Year and Easters are also special occasions which see busloads of pilgrims camping out to a week'' Hodgson (1995:228). ''The cave, home to the yearly St Augustine celebrations, has since been a symbol to Anglicans of the adversity these men had to endure'' Burgess (2007:3). This has made Modderpoort a site of mass pilgrimage. Hodgson (1995) noted that the majority of pilgrims comes from the resettlement camps of Onverwacht and Botshabelo.

Hodgson (1995) noted that the St Augustine's feast day (28 August) is annually attracts people from far and near for this thanksgiving ceremony, where as the thousands of African pilgrims are there to honour Mantsopa. Coplan (2003) noted that to make the sites tourist-friendly, invites forms of permanent settlement, casual sightseeing and commercial exploitation that are not in keeping with the character of the site. The heritage protection that has been given on this religious site is important to protect it. This site is a contested site between Christianity, African traditional spirituality and the African Initiated Churches, namely the ZCC. In the past, there may have been clashes, which have resulted in today's co-existence which allows for the different beliefs space in enjoying the religious and spiritual connection at the site. Ngobese (2018) noted cave dwellers do not have a problem with the influx of tourists provided they do honour the sacredness of the site and the ethical considerations attached to the site. The cave must be declared as a national heritage site, so that it can get heritage protection, as a form of cultural property. Scripps (2019) noted that 10-15 visitors per week visit the cave. The religious pilgrims that visit Modderpoort have diversified the agricultural nature and character of the Eastern Free State, and this niche tourism market attracts tourists that benefit the economy of the Eastern Free State.

Thathe Vondo and Lake Fundudzi, Limpopo Province, South Africa

Thathe Vondo is located in the Venda, region located in the Limpopo Province of South Africa.

The Limpopo Province is the far northern part of South Africa and is surrounded by the neigbouring countries of Botswana, Zimbabwe and Mozambique, and is therefore the gateway to other African countries. Beitbridge Border post is located within the Limpopo Province, which is the most busy land based border post within the southern hemisphere. The Limpopo Province shares borders with three South African provinces, namely the North West Province, Gauteng Province and Mpumalanga Province. Venda is rural in nature, is at the centre and heart of the Soutpansberg Mountains, and the people continue to live traditionally, in this remote and subtropical part of South Africa. The Thathe Vondo

Forest and Lake Fundudzi are located around the mountainous area of Tshiavha village. Thathe Vondo is a state forest, and it regarded by the local VhaVenda population as a holy and sacred forest. The sacred forest is the burial site of the VhaVenda royal family. The Thate Vondo Forest surrounds Lake Fundudzi, and the area is full of spirits, that the local VhaVenda people respect the forest. According to the State of the Rivers Report (2001), Lake Fundudzi has high cultural importance for the people of the area. South African Tourism (2016) noted that the lake's water is regarded as possessing healing properties. VanHunks (2016) noted that Fundudzi is described as zombie lake, where buried ancestors are said to come alive at nights and play drums from beneath the water. This is attributed to the village that was swallowed by the landslide.

''The lake and its origin form the focus of mythical and cultural interest to the VhaVenda people which goes back in their history to the period when their ancestors moved south from Zimbabwe. This lake is considered sacred and is alleged to be inhabited by the ancestral spirits of the Vhatavhatsindi people'' Stayt (1968:1). "Lake Fundudzi is defended by a Venda python god who lives in the mountain on a rock. The ancestral spirits who inhabit the lake are said to be guarded by a white crocodile" South Africa Tourism (2016: 1). MediaClubSouthAfrica (2016) noted that in Venda culture, the python is the god of fertility and maidens perform the famous *Domba*-python dance in the sacred lake to honour the god. ''The young women form a line in which they hold on to the elbows of maiden in front of them, and using their joint arms, make snake-like movements next to the lake to honour the god'' MediaClubSouthAfrica (2016: 3). Stayt (1968) identified the myths and superstitions around Lake Fundudzi include the following:

- The lake is respected by everyone approaching the lake for the first time by approaching the lake by bowing in a special way, facing away from the lake and viewing it by looking upside down through stretched legs.
- The *Thevula,* an annual thanksgiving ritual is performed by *Gota* Netshiavha, where traditional beer in a full calabash is poured into the lake.
- When the Motale River comes down in flood, its dark, clear water form a contrast with the silt-laden red murky waters of lake, and its believed that the waters never mix.

The myth around Thathe Vondo, are that there are two mythical creatures that keep guard over the holy forest, include the white lion (the spirit of Nethathe) and the Ndadzi bird which is regarded as thunder and lightning bird. VanHunks (2016) noted that the one eyed *zwidudwane (ditutwane)* protect the sacred area, and are half human, and are invisible to humans, residing within the spirit world. According to VanHunks (2016), on flat rock above the Phiphidi Falls, offerings are left by *zwidudwane,* and the gifts would be taken by the *zwidudwane* to the foot of the pool at the foot of the waterfall, on the Mutshindudi River. Mystical tales are that a white crocodile likes in the lake. Legend according to Sunday World (2014) has it that when a Venda kind dies, the crocodile coughs up a stone which the new king has to swallow it. Anyumba & Nkuna (2017) noted that it has a reasonable population of 'normal' crocodiles, and there has no record in living memory of any attacks on human beings. ''It is said that the sacred water of the lake will not mix with normal water. Because there is no obvious outlet, the lake was believed to be bottomless as far as the human world is concerned'' VanHunks (2016:2). ''The lake is often associated with myth and legends of the VhaVenda people and is believed to be protected by a python god'' MediaClubSouthAfrica. (2016:2).

Anyumba & Nkuna (2017) noted the VhaVenda people are referred to as the 'Vhatatsindi' or People of the Pool. This is in reference to Lake Fundudzi. ''The People of the Pool have been part of Lake Fundudzi's conservation since their ancestors migrated here centuries ago. Chief Ntsandeni Netshiava, his father and grandfather before him, were the only people who could give permission to strangers to approach'' South Africa Tourism (2016:2). ''Even today, the foreign public is not allowed access to the lake. Only *Gota* (headman) Netshiavha can give permission to visit the late itself'' Van Der Waal (1997: 51). According to VanHunks (2016) permission to visit the lake shore is rarely granted to tourists. Anyumba & Nkuna (2017). MediaClubSouthAfrica (2016) noted that outsiders are not permitted to visit the lake, only those granted special permission by the Netshiava Royal Family, can visit the lake. According to Show Me (2017), the holy forest is protected by Nethane and the lighting bird, Ndadzi, and the area also contains the Lake Fundudzi, which is home of the Python God and the half-spirit, half-human predatory Ditutwane. The Thathe Vondo Forest surrounds Lake Fundudzi, and Lake Fundudzi is regarded as the most sacred place in the area, and the custom is that the first view of the lake must be between your legs. MediaClubSouthAfrica (2016) noted that this ritual is known as *ukudola*. ''Sacred Lake Fundudzi has a wealth of folklore and mythology. Upon a fuller awareness of the diversity of the spiritual meaning of the lake and the sacred Holy Forest of Thathe Vondo, one is surprised by the minimal tourism development some distance from the lake and the forest. In other countries, lodges and local craftmen would be competing to sell their wares to domestic and international visitors.

But this is not the case in the vicinity of Lake Fundudzi, a sacred area that is not really open to tourism development'' Tourism development is therefore impossible as the area is regarded as being sacred and used by the VhaVenda. According to Show Me (2017) when you pass graves by car, you may not come out of your car and to come out of your car. The dense forest has giant trees, a wide variety of ferns, creeps and giant mushrooms. According to Stayt (1968) Lake Fundudzi is a unique inland lake. It was formed by a landslide within the Soutpansberg Mountains, blocking part of the Mutale River. Anyumba & Nkuna (2017) noted that Fundudzi Lake is 5km in length and 3km in width, when full from seasonal rains. The lake is estimated to be atleast 10 000 years old, and may have drowned a village when it was formed by the landslide. Lake Fundudzi has been declared a national heritage site, and it is the only inland natural lake in South Africa. The Mutale River leads to the Phiphidi Falls, which have been developed by the government of South Africa to have accommodation and ablution blocks, creating them as a site for tourism consumption.The VhaVenda exhume the bodies of the VhaVhenda kings after 10 to 15 years, burn the bones and then releasing the ash into Lake Fundudzi.

According to Sunday World (2014), local people perform rituals on the shores of Lake Fundudzi to celebrate the new harvest or to ask for rain, making offerings to the river. This respected custom is threatened by the invasion of houses towards the lake and the tourists that do not respect the sacred lake and the holy forest, much to the disgust of the local population. Visitors to the holy forest encounter VhaVenda spirituality, and must observe traditional protocols, to ensure that they don't offend the host community. ''Visitors travelling to Limpopo, are able to get glimpse of the lake, from above, when travelling along surrounding mountains'' MediaClubSouthAfrica (2016: 3). Tshipala & Chauke (2020) noted that Thathe Vondo has adventure tourism potential whilst it remains an undeveloped destination. Thathe Vondo and Lake Fundudzi are sacred sites of the VhaVenda people and can therefore not be commercialised for tourism consumption. This does not however, stop the packaging of tours to gaze from an acceptable distance the mystery around these scared places. The translation of much of the oral history can richly benefit the preservation of the history around Thathe Vondo and Lake Fundudzi, not just for heritage, but also for history of the VhaVenda people. The definitng feature of these sacred sites is that local control and ownership has prevented tourism development, and this presents an opportunity to gaze a religious tourism destination that is anti-developmental.

CONCLUSION

The development of tourism at a destination is supposed to improve the standard of living and the Quality-of-Life of residents. Special interest tourism is associated with alternative tourism which has a less negative impact on a destination when compared to mass tourism, and more importantly, it allows for local economic participation by the host community. In examples of In the case of religious tourism, the existence of an important religious building such as the headquarters of a church has the unintended impact of attracting religious tourists to a destination. Religious tourists would leave their places of residents, undertake a journey and arrive at a tourism destination, the majority of time and financial expenditure would be undertaken at the destination. Religious tourism more than any form of tourism, is able to have a direct economic impact on a destination, with deep forward and backward linkages in local ownership. Religious pilgrimages have the opportunity to be attracted to "new" destinations that seek to catalyse their economies using tourism-led economic development. Many countries that have jumped on the tourism bandwagon have used mega-events as a means of entering the tourism economy.

There is a plethora of research that indicates that mega-events may not have a positive impact on a destination, especially developing countries. Religious pilgrimages are a safe and sustainable means of promoting high impact, tourism that has benefits citizens and locally owned businesses. Destinations must be creative in seeking to stimulate this market for drive their economies. The ECG weekly church services in the economically depressed region of Pretoria, attracts local and international religious tourists, and the highlight of the year, is the crossover service which is broadcast live and live streamed. Pretoria's tourism economy is closely linked to government business, as it's the seat of government. During weekends, the hospitality capacity within the CBD is satisfied by demand from ECG congregants. Religious tourism demand has kept the tourism industry robust in the CBD. The town of Ladybrand is an agricultural outpost in rural Eastern Cape attracts religious pilgrims which is one of the most important pilgrimage caves in South Africa, the Cave Church. In the absence of this market, Ladybrand would be exclusively dependent on agriculture to sustain its economy. Religious tourism is one of the oldest and largest forms of tourism and in the case of South Africa is one of the most geographically dispersed.

REFERENCES

Adogame, A., & Jafta, L. (1980). *African Initiated Churches*. Source Unknown.

Anyumba, G., & Nkuna, M. (2017). Lake Funfudzi: A Sacret Lake in South Africa that is not open for Tourism Development. *African Journal of Hospitality, Tourism and Leisure*, *6*(4), 1–20.

Apleni, L., & Henama, U. S. (2020). Visitor's perceptions of religious tourism in South Africa: The case of Moria in the Limpopo Province. *African Journal of Hospitality, Tourism and Leisure*, *9*(2), 1–19.

Burgess, M. (2007). Spiritual clash over land resolved. *Farmers Weekly*. Retrieved from: https://www.farmersweekly.co.za/rural-insight/

Chiorazzi, A. (2015). The spirituality of Africa. *The Harvard Gazette*. Retrieved from: https://news.harvard.edu/gazette/story/2015/10/the-spirituality-of-africa/

Chitando, E., & Biri, K. (2016). Walter Magaya's Phophetic Healing and Deliverance Ministries and Pentecostalism in Zimbabwe: A preliminary study with particular reference to Eumenism. *Studia Historiae Ecclesiasticae*, *42*(2), 72–85.

Coplan, D. B. (2003). Land from the Ancestors: Popular Religious Pilgrimage along the South Africa-Lesotho Border. *Journal of Southern African Studies*, *29*(4), 976–993. doi:10.1080/0305707032000135923

Goitsemodimo, G. A. (2019). *Mantsopa Makhetha*. National Museum Publications. Retrieved from: https://nationalibrarypublications.co.za/mantsopa-makhetha/

Gyekye, A., Oseifuah, E., Nethengwe, T., Sumbana, F., & Dafuleya, G. (2014). Religiously motivated travel and rural tourism in Vhembe District of South Africa. *SHS Web of Conference, 12*, 1-9. 10.1051hsconf/20141201057

Henama, U. S. (2012). Tourism accommodation in Mangaung: Current dynamics, immediate challenges and future prospects. *International Journal of Business. Economics and Law*, *1*, 76–84.

Henama, U. S. (2013). Attracting Indian Outbound Tourist to South Africa: A BRICS Perspective. *India Quarterly. Journal of International Affairs*, *69*(3), 229–247.

Henama, U. S. (2014a). Health tourism in South Africa: Opportunities and possibilities. *African Journal of Hospitality, Tourism and Leisure*, *3*(1), 1–16.

Henama, U. S. (2014b). The demise of 1Time airline and the reaction of various interest groups. *African Journal of Hospitality, Tourism and Leisure*, *6*(4), 1–16.

Henama, U. S. (2016). The low cost carrier bandwagon: Lessons for Skywise Airline. *African Journal of Hospitality, Tourism and Leisure*, *5*(2), 1–11.

Henama, U. S. (2017b). Marikana: Opportunities for Heritage Tourism. *African Journal of Hospitality, Tourism and Leisure*, *6*(4), 1–11.

Henama, U. S., Janse van Rensburg, M., & Nicolaides, A. (2016). Nkandla: The unexplored frontier of heritage tourism to Zululand, South Africa. *African Journal of Hospitality, Tourism and Leisure*, *5*(2), 1–16.

Henama, U. S., & Sifolo, P. P. S. (2017). Tourism migration in South Africa: Current dynamics, immediate challenges and future prospects. *International Journal of Innovation in the Digital Economy*, *8*(1), 47–60. doi:10.4018/IJIDE.2017010103

Henama, U. S., & Sifolo, P. P. S. (2018). Religious Tourism in South Africa: Challenges, Prospects and Opportunities. In Global Perspectives on Religious Tourism and Pilgrimage. IGI.

Herald. (2015). *Night of Turn Around experience*. Retrieved from: http://www.herald.co.zw/night-of-turn-around/

Hodgson, J. (1995). *Mantsopa: Popular religions and the Anglican Church in South Africa*. Source Unknown.

Ingle, M. (2013). *Counter urbanisation and the emergence of a post-productivist economy in South Africa's arid Karoo region, 1994-2010*. Source Unknown.

Jude, O. C., Uchenna, O., & Ngozi, E. (2018). Impact of Religious Tourism in Host Communities:The Case of Awhum Monastery. *American Journal of Social Sciences*, *6*(3), 39–47.

Jugmohan, S., & Giampiccoli, A. (2015). Cruise tourism and community-based tourism: Towards a possible relationship. *African Journal of Hospitality, Tourism and Leisure*, *4*(2), 1–10.

Kastarlak, B. I., & Barber, B. (2012). *Fundamentals of Planning and Developing Tourism*. Pearson.

Kings, S. (2017a). *Frogs vs angels: Prophet Bushiri's angel landing pad plan declared illegal*. Retrieved from: https://mg.co.za/article/

MediaClubSouthAfrica. (2016). *Sacred lake to heritage site*. Retrieved from: https://www.mediaclub-southafrica.com/culture/

Mill, R. C., & Morrison, A. M. (1985). *The Tourism System: An Introduction Text*. Prentice-Hall.

Mushava, S. (2015). *Zimbabwe: Mega-churches a boon for faith tourism*. Retrieved from: https://allafrica.com/stories/

Nengovhela, M., Tshipala, N., & Nyikana, S. (2017). Domestic tourism in practice: An investigation of Vhembe District Municipality residents' travel motivations. *African Journal of Hospitality, Tourism and Leisure*, *6*(4), 1–15.

News, P. M. (2015). *Nigeria's loss is South Africa's gain-TB Joshua birthday fallout*. Retrieved from: http://www.pmnews.co.ng/

Ngobese, D. H. D. (2018). A cultural appraisal of sacred sites of the Eastern Free State: Adjunct to African Traditional Religious and Belief Systems. *China-USA Business Review*, *17*(9), 474–486.

NTV. (2017). *Fury in Kigezi as politician takes over Punishment Island aka Akampene*. Retrieved from: http://www.ntv.co.ug/news/local/18/jan/2018/fury-kigezi-politician-takes-over-punishment-island-aka-akampene-22199#sthash.Y0Tuqo4D.dpbs

Nyikana, S. (2017). Religious tourism in South Africa: Preliminary analysis of a major festival in Limpopo. *African Journal of Hospitality, Tourism and Leisure*, *6*(1), 1–8.

Papathanassis, A. (2011). The Long Tail of Tourism: Holiday Niches, and their Impact in Mainstream Tourism. Gabler Verlang.

Poplak, R. (2016). *Planes, blames and autocracies: How TB Joshua will get away with murder*. Retrieved from: http:www.dailymaverick.co.za/article/

Premium Times. (2015). *South Africa threatens to jail Pastor Chris Oyakhilome*. Retrieved from: https://www.premiumtimesng.com/news/

Ramogale, M., & Galane, S. (1997). *Articles from the 1997 festival of American Folklore Programme Book*. Retrieved from: http://www.folklore.si.edu/resources/festival1977/faithin.htm/

ReinsteinD. (2014). Retrieved from: https://www.travelweekly.com/Middle-East-Africa-Travel/Insights/South-Africa-becoming-a-more-popular-stop-for-cruise-ships/

Rinschede, G. (1992). Forms of religious tourism. *Annals of Tourism Research*, *19*(1), 51–67. doi:10.1016/0160-7383(92)90106-Y

Rogerson, C.M., Simango, S. & Rogerson, J.M. (2013). From recreation to niche tourism: The transformation of birdwatching in South Africa. *African Journal of Physical, Health Education, Recreation and Dance*, 122-137.

Saayman, M. (2013). *En-route with tourism: An introductory text* (4th ed.). Juta.

Scripps, T. (2019). *Visiting An African Prophetess' Cave*. Retrieved from: https://thestarlitpath.com//

Seyer, F. & Muller, D. (2011). Religious tourism: Niche or mainstream? In *The Long Tail of Tourism: Holiday niches and their impact on mainstream tourism*. Gabler Verlag.

Show, Me. (2017). *Thathe Vondo-The Holy Forest, Soutpansberg, Limpopo*. Retrieved from: https://showme.co.za/tourism/thathe-vondo/

South Africa Tourism. (2016). *Limpopo's Lake Fundudzi*. Retrieved from: https://www.southafrica.net/za/en/aticles/

State of the Rivers Report. (2001). *Letaba and Luvuvhu Rivers Systems*. Pretoria: Water Research Council.

Stayt, H. A. (1968). *The Bavenda*. Frank Cass.

Sunday World. (2014). *Tourism and housing encroaches on sacred lake*. Retrieved from: https://www.sundayworld.co.za/lifestyle/

Tauoatsoala, P., Henama, U. S., & Monyane, M. (2015). Service quality as a fundamental concernstone for visitor satisfaction: Lessons learned from Pretoria, South Africa. *African Journal of Hospitality, Tourism and Leisure*, *4*(1), 1–15.

The New Age. (2016). *South African still visit SCOAN despite 2014 incident*. Retrieved from: http://www.thenewage.co.za/south-africans-still-visit-scoan-despite-2014-incident/

Thornycroft, P. (2017). *TB Joshua visit a ''blessing'' for Zim's religious tourism*. Retrieved from: https://www.iol.co.za/news/special-features/zimbabwe/

Tribe, J. (2016). *The economics of recreation, leisure and tourism* (5th ed.). Routledge.

Tshipala, N., & Chauke, J. (2020). Thathe Vondo: Is this a missed opportunity for Adventure Tourism development? *African Journal of Hospitality, Tourism and Leisure*, *9*(1), 1–15.

Turner, M. (2015). *Trip Tips: Serenity meets dark history on Uganda's Lake Bunyonyi*. Retrieved from: https://uk.reuters.com/article/travel-uganda/trip-tips-serenity-meets-dark-history-on-ugandas-lake-bunyonyi-idUKL6N0UR2YG20150227/

Van Der Merwe, P., & Saayman, M. (2013). Who are the South African hunters and why do they hunt? *Journal of Hospitality Management and Tourism*, *4*(1), 9–18. doi:10.5897/JHMT2013.0085

Van Der Waal, B. C. W. (1997). Fundudzi, a unique, sacred and unknown South African lake. *Southern African Journal of Aquatic Sciences*, *23*(1), 42–55. doi:10.1080/10183469.1997.9631387

VanHunks. (2016). *Bewitched Forests and Waters of the VhaVhenda*. Retrieved from: http://www.van-hanks.com/lwveld1/venda1.html./

Visser, G., & Baker, C. (2004). A Geography of British Geographic Tourists in South Africa. *Geography (Sheffield, England)*, *89*(3), 226–239. doi:10.1080/00167487.2004.12094100

Waldo, M. (1976). *Travel guide to South Africa: Revised and Updated Edition*. MacMillan Publishing.

Zambia Eye. (2018). *Botswana govt shuts down Bushiri's church*. Retrieved from: https://zambianeye.com/botswana-govt-shuts-down-bushiris-church/

Zambia Observer. (2017). Retrieved from: http://www.zambianobserver.com/prophet-uebert-angel-refused-entry-to-zambia/

KEY TERMS AND DEFINITIONS

African Spirituality: The traditional African religions (or traditional beliefs and practices of African people) are a set of highly diverse beliefs that includes various ethnic religions. Generally, these traditions are oral rather than scriptural, include belief in an amount of higher and lower gods, sometimes including a supreme creator, belief in spirits, veneration of the dead, use of magic and traditional African medicine.

Free State Province: Is a centrally located province of South Africa. Its capital is Bloemfontein, which is also South Africa's judicial capital. Its historical origins lie in the Boer republic called the Orange Free State and later Orange Free State Province.

International Tourists: This refers to tourists that cross national borders.

Limpopo Province: Limpopo is the northernmost province of South Africa. It is named after the Limpopo River, which forms the province's western and northern borders. The capital and largest city in the province is Polokwane (formerly Pietersburg).

Quality of Life: An individual's perception of their position in life in the context of the culture and value systems in which they live and in relation to their goals, expectations, standards, and concerns. It is a broad ranging concept affected in a complex way by the person's physical health, psychological state, personal beliefs, social relationships, and their relationship to salient features of their environment.

Religious Tourism: Religious tourism is a type of tourism exclusively or strongly motivated by religious reasons. This is one of the oldest forms of tourism. It can also be referred to as faith tourism or spiritual tourism.

Special Interest Tourism (SIT): Are the provision of customised tourism activities that caters to the specific interests of groups and individuals. In this case, tourism is undertaken to satisfy a particular interest or need.

Tourism: Are the activities of people traveling to and staying in places outside their usual environment for leisure, business, or other purposes for not more than one consecutive year.

Tourists: People traveling to and staying in places outside their usual environment for not more than one consecutive year for leisure, business, and other purposes.

Chapter 18
Types of Religious Tourism

Aniqa Ijaz
University of Okara, Pakistan

ABSTRACT

Religious tourism is considered as one of the interesting and oldest areas of research in the field of tourism. The process of observing and practicing the religious activity during the visit of any sacred place combined with the zeal of faith in the heart of practitioner is categorized as religious tourism. Many of the researchers have explained its different types in their own perspectives. Within the boundaries of modern scientific society, modern religious tourism can be classified into two major types. One is pilgrimage and the other one is intellectual learning excursions. The second main category has two branches. These intellectual learning excursions are based on two mottos of the tourists. One is regarding the scientific know-how, and the second one is regarding attending the religious-cultist events.

Objectives of the Study

Religious tourism is gaining attention in today's globalized world rapidly. Though it is practiced for a long time as the Pilgrimage. But over time, it has been merged into various types, which are still to be explored. So, this study will investigate new types of religious tourism. The objectives of the study are as follow:

1. To put attention towards the significance and recognition of religious tourism in the field of tourism.
2. To define the types of religious tourism within the modern scientific society.
3. To explore how these other types of religious tourism were realized in the modern world.

INTRODUCTION

Religion and tourism are linked with each other closely. Many of the travelers execute religious voyages, including domestic and international tours. According to UNWTO (2011), almost six hundred religious tours take place every year, comprising domestic and international religious tours. Religious tourism

DOI: 10.4018/978-1-7998-5792-1.ch018

has been characterized in the following words: "It is a special tourist activity oriented by religious culture, with the help of specific eco-cultural environment, and it refers to such special tourist activities as worship, research, sightseeing and culture carried out by religious followers and lay tourists"(Raj et al., 2013). Religious voyage is a unique type of the tourism industry, which is totally or seriously roused for religious reasons.

One of the pioneer types of religious tourism is Pilgrimage. Religious tourism often referred the spiritual tourism (Tala & Padurean (2019); Smith et.al. (2010)), is the process of observing and practicing religious activity during the visit of any sacred place combined with the zeal of faith in the heart of the practitioner. Religious tourism is one of the ancient types of tourism, which is practiced by the faith keepers since the rise of religion. According to CBI[1] Ministry of Foreign Affairs, "Religious tourism is a type of tourism exclusively or strongly motivated by religious reasons. This is one of the oldest forms of tourism. It can also be referred to as faith tourism or spiritual tourism".

With time, the increase in globalization has given a force to push forward the phenomena of religious tourism (Mustafa 2010). Many tourists go out for tourism just for the sake of love for their God, or in search of knowledge of reality, or attend religious events as an obligatory part of their religion or just for innovation and knowledge gaining activity. Reasons behind the religious tourism can be many, as, to pray for their physical or spiritual motive, to pray for their selves or relatives, to gain peace and spiritual benefits, to visit the places just for the sake of curiosity, to visit the shrines or temples or sacred places to increase their knowledge.

By traveling towards the centers, mainly considered as prestigious, flawless, and pure, many travelers in modern societies expect to go through a pure and religious reality in which their shattered individuals gathered again (Allcock 1988; Cohen 1986, 1988 & 2006). Their attachment and scattered bounds may be explained as a diaspora of responsiveness – a dynamic or religious, despite physical, scattered in which individuals don't entirely present in the here and now; they are broken into groups, in the past and future (Cohen 1986; Cohen 2006). Travelers usually make these journeys in search of peace, serenity, and change their selves and "personal authenticity". (Desforges 1998; Noy 2004; Cohen 2006). Like the ancient way of pilgrims, the traveler wished to be altered in a more useful and better form. In the words of MacCannell (1976, 1992), tourism has been considered as "cognitive activity", but it affects the individual's behaviors and sentiments (Cohen 2006).

If we look some years back, religious tourism has become a widely used research area. Mukherjee et.al. (2020), investigated the importance of religious tourism and also posed a need to do more work in this field. Many of the researchers have explained the importance of religious tourism and its relation and impacts on other variables but its types are still needed to be understood, as no such research work has been done in this regard. By living in a more dynamic world and more scientific society, it is important to also throw some light on the types of religious tourism.

Religious Tourism can be categorized into two major types, one is Pilgrimage, and the other one is intellectual learning excursions. Pilgrimage is one of the oldest types and some researchers have also given the name of pilgrimage as religious tourism (Pusacasu 2015), it is one of the oldest types practiced by people of many ages. It is offered by faith keepers of many religions, to derive the sense of peace, serenity, forgiveness and to fulfill obligatory duties. The later one, intellectual learning excursions have emerged as a modern and more important type of religious tourism. With a more globalized world, this modern type has gained much value. These intellectual learning excursions may be further categorized into two more types. These excursions may be performed to seek scientific knowledge. The scientific knowledge, which tourists are looking for, may be to look for reality. Intellectual Learning Excursions are related to the desire to attend religious culturists' events like urs celebration at shrines by Muslims etc.

Literature Review

Globally, the tourism sector contributes positively and has direct and indirect relations with the economy (Rasul et.al., 2016). Religious tourism has been gaining more attention in the fields of tourism, which gained more attention over the years. (Annuwichannant et.al., 2019; Collins-Kreiner, N. (2020); Mukherjee et.al., 2020; Smith et.al., 2010; Seyer & Muller 2011).

Tala & Padurean (2019) tried to investigate the dimensions of religious tourism. By referring to religious tourism as spiritual tourism, they threw light on the significance, determinants, and facets of religious tourism. They explained, pilgrimage, a type of religious tourism, has been extensively adopted by the pilgrims (one who performs pilgrimage) for very long ages. They concluded that, at a worldwide level, the social receptiveness from the previous years is likewise favorable for religious tourism. The world is an intriguing religious mosaic: it could be a fascinating mosaic for religious tourism also.

Annuwichannant et.al., (2019), examined tourist's attitude towards religious tourism in Thailand. They investigated the elements impacting the fulfillment and future conduct patterns of religious tourists and Dhamma experts. Furthermore, they analyzed the contrasts among desires and discernments towards the travel practices. It was noticed that the perspectives towards merit-making inventions and Dhamma learning and rehearsing inventions were seen as the most remarkable determinants of overall fulfillment and future conduct trends. Lastly, desire and observation of religious tours were fundamentally unique.

Mukherjee et al. (2020), underlined the importance of the travel industry as there is a shortage of major scholarly examination and it is imperative to research in this new field of tourism (religious tourism). It has additionally explored the potential growth and proposed special procedures for the improvement of the tourism industry in terms of spiritual tourism in India.

Terzidou et. al. (2018), explained the complexity of religious tourism motivations, vows, and visions of sacred places. They found that the motivations behind religious tourism are related to believes, faith, or may be an attachment with any personality or ritual performances. This study was made only for the organized tours in Greece.

Luz (2020), attempted to investigate the account for the pilgrimage vs. tourism problem. He explained that under the heading of devout tourism, an integrator category was characterized among the thoughts that rose. In this paper, he contended a simply devout point to view journey in Islam, which is specified within the Quran; termed as the Hajj. He concluded that remaining devout Muslims persuaded ventures to fall in another type which may be named; religious tourism. He re-framed and re-conceptualized the Hajj as the devout journey. Further, a lot of religious tours and a variety of religious voyages have been explained in the paper.

Chianeh et al. (2018), explained the different anticipations of cultural and religious tourism progress and growth in Iran. He pointed out that, Iran has experienced a larger degree from the downward trending advertisement internationally through the media and tourism mediators when compared to Turkey, Egypt, and Malaysia. Further, he explained that these countries have many similarities so they should be equally promoted among international and western travelers. Many policy implications were also suggested like to change the entire way of tourism in a dynamic world, to maintain a good relationship with other foreign countries, improvement of the administrative bodies of embassies to issue visas earlier, to promote ICHTO (Iran Craft, Heritage and Tourism Organization, to provide improved tourism facilities, to invest on conservation of cultural heritage.

Rinschede (1992), investigated different forms of religious tourism. It was explained that religious tourism is not only confined to its one of the elderly known type but cultural and social factors also work as a motivational force behind religious tourism. He explained different forms of religious tourism as short-term, long term, and institutional tourism. He explored that due to increased mobility and resources people are now more motivated towards religious tourism.

Nolan & Nolan (1992), threw light on the tourism system of Europe by focusing on the attainment of the wants of the travelers varying from religious travelers to secular travelers at three centers, which are shrines for pilgrimage, devotional centers, places of religious festivals. Further, some management strategies were also presented to serve the interests of the pilgrims.

Blackwell (2007), investigated the desires behind religious tourism, religious festivals, Pilgrimage, and religious events. He explained these concepts with the help of two theories, i-e, content theories, and process theories. By analyzing these theories and by relating them with the concepts of devout travels, motivations and their strength were measured.

Cohen (2006), explained the want to flock together of the Diaspora in the wake of religion, caused tourists to attain a specific form of the tour, affecting the form of tourism to the religious centers. It was concluded that the educational implications of religious tourism area are linked to variations in the ways of imparting knowledge about religion, pre-assumptions for tours, and preferences in education.

Review of literature depicts the research work which has been conducted on religious tourism and only one type of religious tourism has been explained, i-e, Pilgrimage (the pioneering type of religious tourism). Further, the light was thrown on the more availability and improved mobility of tourists by Rinschede (1992). He has explained some of the forms of religious tourism. And Mukherjee et.al (2020) stressed the need to do more inventions in the field of religious tourism. By keeping n view the previous literature and the need and importance of religious tourism it is essential to carry out some useful analysis in it. Hence, other new emerging types of religious tourism still exist to be investigated, in this more scientific society.

Types of Religious Tourism

Religious tourism is classified into two main types, which are as follow

1. Pilgrimage
2. Intellectual learning excursions

Pilgrimage has been in practice by many religions. Islam and Christianity are the main religions of the world and after these Hinduism comes on the third number. The believers of the religion, Islam, label it as 'Hajj". It is the oldest type of religious journey. According to the teachings of Islam, Hajj is obligatory for its physically and financially capable believers at least one time in their life. Thus, a large number of Muslims set out to pay their obligation and gather in their center, i-e, Saudi Arabia. Like-wise, believers of other religions gather on their holy centers to perform their religious duties. Whereas, the second type of religious tourism is intellectual learning excursions. These excursions could be made for two main reasons which are, to gain scientific knowledge and, to attend religious-cultural rich events. Some of the tourists visit the sacred and holy places to achieve their scientific goals while others went

Figure 1.
Source: Author's analysis

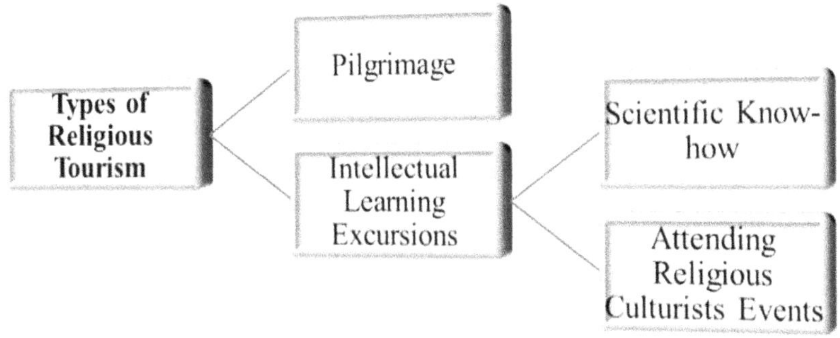

out to perform some rituals on their holy centers or to do or attend culture-related programs. The flow chart is given below clearly depicts the types of religious tourism explained above.

Pilgrimage

MacCannell (1973, 1976, 1992) was the first to gave the notion of tourism and explained one of its type pilgrimages. Pilgrimage is known to be the oldest type of religious tourism (Timothy & Olsen 2006) (Fang 2020). It is practiced by many people belonging to different religions. One who performs this sacred activity is often referred to as "Pilgrim", which is a Latin word derived from 'Peregrinus', which is also meant as a traveler or any unknown person. The word, peregrines, is also referred to as, "the idea of wandering over a distance" (Yeoman 2009) (Nieminen 2012) (Wisniewski 2019). After looking on the previous categorization of tourism and pilgrimage, (e.g. Apostolopoulos et al. 1996; Smith 1992; Cohen 1979; Cohen 2003; Cohen 2006; Collins & Kliot 2000; Jafari 1987; MacCannell 1976;), religious visits and pilgrimage are neither opposite forces nor edges of one rod, they can be explained as two axes of a pilgrimage and tourism model. Coleman (2002) has defined pilgrimage as religious phenomena which gives some useful insights and sacred religious places to visitors and is a source of providing a base to make theories by anthropologists. Pilgrimage is performed by many devotees, approximately 240 million people, belonging to their religions (Jackowski 2000). Every year many religious tourists go their holy Centers to perform their sacred duties, mainly Christians, Muslims, and Hindus. Christian pilgrimages may be a number of the following: having the urge to know the depth matters of Christian faith, urge to know about the importance of Christianity to civilization, to achieve inspiration from the well-known people or religious personalities of Christianity, getting informed about Christians within the previous times, to look for the spiritual or emotional peace or yo went out on holidays for a pilgrimage, for instance, Santiago de Compostela stream. Europe has provided approximately 20 centuries of heritage of Christians, along with the basics of many Christian churches. The attractive or sacred holy places to Europe for Christians are The Vatican, theologian, and the house of C.S. Lewis. (Wright 2008) Thus, in step with professionals, pilgrimages also keenly contribute to visit Santiago de Compostela or Mecca. Voyages of sacred centers, for instance, are said to be a pilgrimage. Such journeys are in line with the practices of St. Paul or theologian, at this place the travelers get to know about the places.

Muslims perform Pilgrimages according to Islamic shariah, according to which, pilgrims have to visit the Mecca and Madina (situated in the Kingdom of Saudi Arabia) and to perform it according to the divine orders of their God. According to the religion of Muslims, Islam, pilgrims are those who have the financial ability to perform the pilgrimage.

Within the boundaries of religion Islam, there exist many sectarian groups. Among these, the main two groups are Sunni and Shi'ia. These groups have formed different types and destinations of religious pilgrimage. There exist 87 to 90% Sunni Muslims, 10 to 13% Shi'ia and other groups have a small number of individuals in it. Most of the Muslims have resided in the 57 countries of OIC (Organization of Islamic Cooperation) (Zamani-Farahani & Eid, 2016).

According to (Esposito, 2004), the contemporary traditional setting is getting progressively convoluted against the increasing number of faith keepers of Islam living in European areas and Northern and Southern parts of America. In ongoing decades, Muslim immigrants have migrated in large number with an increasing trend. These immigrants become increasingly more significant on occasion as travelers to their nations of origin and to traditional Muslim locales just as rising holy Centers in their new nations (Xavier, 2018). This critical and exceptionally nuanced geological distribution is likewise answerable for the huge development of religious visitors to various holy centers that far outperform the number of Hajjis to Mecca. All religions have different ways to perform the pilgrimage. Individuals belonging to Hinduism, Jainism, Sikhism, and Buddhism also participate in performing the pilgrimages at the specified holy places according to their religions. Without a doubt, there exist a great many other religious excursions which are occurring mainly due to the faith-keepers of Islam but these excursions are not Hajj (Arjana 2017; Luz 2020). So, this also depicts the existence of another type of religious excursion, which we will discuss in the next section.

Now we will discuss the pilgrimage of one of the largest religious groups, Buddhism. The holy centers associated with this faith can be found in Bihar and the northern Indian states of Uttar Pradesh. Many decades back, this place was called Puratthima, nowadays it is known as Buddhist Circuit. In this religion, any religious activity, including Pilgrimage, gains its significance from the mental peace it provides. That's why the centers visited by individuals and Buddha (in his lifetime) itself has very important. (Singh 2003). The importance of the quote of Buddha for the tourists by the noble does not have equal centrality in the religion of Buddhism, like it contains, for instance, for Muslims to do Hajj at least one time in life. However, it is important here to explain that all Buddhists visit is not Pilgrimage, some are secular ones.

Intellectual Learning Excursions

Religious tours are not always belonging to spirituality. Yes, they are related to faith and religion but this will not be the same always. There are also some other dimensions and may be the aim to face any spiritual truth. Spiritual tours are linked with culture, scientific education, and nature. Thus it can be easily said that now there is not only spirituality i-e, seeking God but other preferences or other dimensions have also been added in the religious tours (Wisniewski 2018). The fame of religion, education related to religion, a trend towards scientific knowledge and religion, and the realization of informal education to develop the identity are all increasing the trend towards intellectual learning journeys, for religious purposes and educational basis (Cohen 2006).

Intellectual learning excursions may be practiced to gain any scientific knowledge and to attend any religious culture related event. In case of gaining scientific knowledge, any person enriched with the thirst of visiting the holy places to gain the scientific knowledge related to the phenomena of their desire may be categorized into one of the types of intellectual learning excursion. This type of excursion is fully based on the process of gaining scientific know-how. Mostly the people looking for the answers of scientific phenomena related to the holy places. They are mostly looking for the answers to questions that arise in their mind when they hear about the miracles of great people like saints etc, belonging to any holy place like shrines, darbars or other saintly places. Sometimes, these people are looking for the power hidden behind the phenomena. The experience cycle illustrated in Kelly's (1963) & Cohen (2006) hypothesis of "man as a scientist" might be utilized to comprehend and assess scientific knowledge excursions. The cycle comprises of expectation, interest in the occasion, experience, affirmation/ disconfirmation, and helpful correction. The various Mazars in Xinjiang is continually visited because they are accepted to have the ability to fix fertility issues (Dawut, 2009). Likewise, many other shrines and Mazars are associated with such types of cures. So, this situation creates idiosyncrasy, to unveil the reality and any power hidden or any scientific phenomena hidden behind it and generates the urge to attempt the intellectual learning excursions for scientific purposes.

Another component of the intellectual learning excursion is to attend the religious culturalist events. Every religion has some past experiences and memories of some of their great personalities. These experiences and memories of great personalities form some events. These events are celebrated on some fixed periods. Most of these events are celebrated yearly. Every devotee has a strong desire to attend these events to show their great affection to their religion or may be due to the great affection with sacred personalities related to their religion. But some religious tourists attend the culture related events to visit such places. Tourists who are interested in exploring the religious-cultural rituals and heritage are also part and parcel of this type of religious excursion. One who went out to visit the historical places, old architectural buildings, religious museums, and galleries, attend cultural plays can be referred to as the religious-cultural tourist. The tours coordinated at nature and physical action, saintly music, innovativeness, or religious activities are also accessible in shrines. As indicated by the Directory on Popular Piety and the Liturgy, a shrine can emerge due to many causes: as a celebration of an uncommon occasion and being an observer to devotion and appreciation of the unwavering. Similarly, there are many sacred places enriched with the mercy of God, because many people have experienced many marvels and cures. Shrines observe and commemorate God's works by assisting to remember that everything in the universe including earth, heaven, and humans are created by God. Besides, each place of worship has its individuality and magnetism, a blessing from God for all travelers who went on the holy journey, where they can get unique healings. The traveler's awareness and the beginning of a shrine rely upon the uncommon mystique of it. Most shrines follow their beginnings to healings or other significant occasions concerning faith (Wisniewski 2018). Such characteristics of shrines form a special power to travel to these places for religious and cultural excursion.

Religious vacations create mental harmony and also provide a chance to visit religious places. Along with the performance of religious obligations like Hajj, as mentioned in the Holy Quran there is an assortment of religious excursions. Surinder (1998) discussed that voluntary voyages are additionally offering us a typology of these religious journeys in various regions and bridle the regular term which is adequate among most researchers in the field, ziyarat. This has been practiced for quite some time been noticeable on the worldwide Islamic scene (Werbner, 2003). This kind of religious excursion is of great importance, not just because it is the thing that most Muslims do, yet additionally it allows knowing about religion in its full intricacy. Further, Wali-Sango is the nine blessed walis of Indonesia representing most of the Muslims, could never go on a journey to Mecca but instead would get the job done themselves with visits to nearby Islamic destinations (Lucking, 2014).

Many travelers have keen to visit the Islamic areas, Islamic events, and Islamic festivals by considering them as religious excursions (Haq & Yin Wong, 2010). The first person who went out for travel and tourism to visit most of the Islamic world was Ibn Battutah. It took about over thirty years. So, the concept of a religious excursion was notably rooted in history (Nursanty, 1945). Internationally, there are many famous and attractive religious sites for religious tourism, which are Mecca, Jerusalem, and Rome. Likewise, for Muslims, these attractive Holy Centers are the sacred places of Iran, Iraq, and Saudi Arabia. Other than performing pilgrimage to Saudi Arabia, called Hajj in Islamic shariah, travelers visit these sacred places as part of the religious rituals and in the commemoration of the sacred personalities of their religion. Many Muslims are attracted to such religious sites and do excursions in these cultural religious sites (Iravani & Mozaffari, 2013).

Hence, intellectual learning excursion to gain scientific know-how are made to any sacred or holy place to which the tourist want to visit and, at their desired holy centers, to which they feel attraction or to which they are attracted or want to attend some rituals or festivals related to culture and religion. These centers could be any place related to culture and religion. These tourists firstly perform their religious obligations or activities and after this, they went to visit the culture, heritage, and architecture of the holy place and are enriched with religious feelings. (Timothy & Olsen 2006; Jackson & Hudman 1995)

CONCLUSION

Within this scientific society, tourism has been widely practiced by many tourists. It tends to be easily said since the religious travel industry has achieved huge consideration recently. Religious and spiritual tourists have gotten across the board and advocated in late decades, possessing a significant fragment of worldwide tourism, has developed considerably in recent decades both in an absolute and proportional way. A proceeded with increment in this market portion is by all accounts a predictable pattern later on too (Holmberg 1993; Bywater 1994; Post et al. 1998; Singh 1998; Olsen and Timothy 1999; Russell 1999; San Filippo 2001; Timoothy & Olsen 2006). Increments in religious travel have harmonized with the development of the travel industry in the cutting edge time (Lloyd 1998), and even though the business and its "associated practices interact with religious life and the institutions of religion in virtually every corner of the world" (Bremer 2005: 9260).

Religious tourism is the oldest form of tourism and can be classified into two major types, which are pilgrimage and intellectual learning excursions. Pilgrimage, an ancient type of religious tourism, practiced by faith keepers attached to any religion, is been practiced from past to present and will be practiced in the future. These faith keepers move to the centers to offer their religious duties to show devotion to their religion. Mainly it is practiced by the three main religions of the world which are Islam, Christianity, and Hinduism. Most of the pilgrims are Christians because their large number exists in the world, one the second number is Muslim's pilgrimage i-e, Hajj. Muslim pilgrims visit their Holy Centers, Mecca, and Madinah, situated in Saudi Arabia to perform this obligation for financially able Muslims only within the limits of Islamic shariah. Christians are also associated with the strong desire of pilgrimages. Their Holy Centers include Jerusalem, Bethlehem, Sephoria, etc. Hindus also went towards their Holy Centres to gain purity and to feel near to God. The pilgrims visit Holy places and events to perform the pilgrimage at least one time in their life, their Holy Centres include Kumbh Mela at Ganges River, and many other rivers, mountains, and temples associated with their Gods and Goddesses. Likewise, other religious groups also perform pilgrims but they are not in very large numbers. Intellectual learning excursions may be for scientific knowledge purposes and, may be to attend religious culturists' events. A great number of tourists went out to such types of excursions. This type of excursions are usually short term and includes ziyarats, urs, to attend fairs on saint's tombs, shrines, voyages of religious and cultural places, to participate and attend ritual ceremonies and festivals, to visit and gain the scientific significance of such places and rituals, and the miracles attached with such places, and other sacred and religious culturally rich events and places. The sacred places which attract the tourists for the religious culture excursion are known for many miracles and rituals attached to these holy places. Intellectual learning excursions are made both at the national and international levels. Mostly such excursions at the international level are made in Iraq, Iran, and Saudi Arabia. While the religious-cultural excursion at the national level is experienced at many sacred places within the countries. So, it can be now easily concluded from afore stated research that religious tourism is emerging rapidly and is experienced by almost all individuals of the faith keepers. The oldest type of religious tourism is a pilgrimage, while the other one is Intellectual Learning excursions, which are made to achieve scientific knowledge and to attend the events related to culture and religion.

REFERENCES

Allcock, J. B. (1988). Tourism as a sacred journey. *Loisir et Société*, *11*(1), 33–48. doi:10.1080/07053436.1988.10715288

Anuwichanont, J., Serirat, S., Mechinda, P., & Archarungroj, P. (2020). Examining Tourists' Attitude Towards the Religious Tourism in Thailand. *Review of Integrative Business and Economics Research*, *9*(3), 55–66.

Apostolopoulos, Y., Leivadi, S., & Yiannakis, A. (Eds.). (1996). *The Sociology of Tourism*. Routledge.

Arjana, S. R. (2017). *Pilgrimage in Islam: traditional and modern practices*. Simon and Schuster.

Bhardwaj, S. (1998). Non-hajj pilgrimage in Islam: A neglected dimension of religious circulation. *Journal of Cultural Geography*, *17*(2), 69–87. doi:10.1080/08873639809478321

Blackwell, R. (2007). Motivations for religious tourism, pilgrimage, festivals and events. *Religious tourism and pilgrimage festivals management: An international perspective*, 35-47.

Bremer, T. S. (2005). Tourism and religion. In L. Jones (Ed.), *Encyclopedia of Religion*. Macmillan.

Bywater, M. (1994). Religious travel in Europe. *Travel & Tourism Analyst*, *2*, 39–52.

Canon Law Society of America, Catholic Church, Pope (1978-2005: John Paul II), & Libreria Editrice Vaticana. (1999). Code of Canon Law, Latin-English Edition: New English Translation. Canon Law Society of Amer.

Cohen, E. (1979). A phenomenology of tourist experiences. *Sociology*, *13*(2), 179–201. doi:10.1177/003803857901300203

Cohen, E. (1988). Authenticity and commoditization in tourism. *Annals of Tourism Research*, *15*(3), 371–398. doi:10.1016/0160-7383(88)90028-X

Cohen, E. H. (1986). Tourisme et identité. *Pardès*, *4*, 84–97.

Cohen, E. H. (2003). Tourism and religion, a case study: Visiting students in Israel. *Journal of Travel Research*, *42*(1), 36–47. doi:10.1177/0047287503254247

Cohen, E. H. (2006). Religious tourism as an educational experience. In *Tourism, religion and spiritual journeys* (pp. 94–109). Routledge.

Coleman, S., & Elsner, J. (1995). *Pilgrimage: Past and present in the world religions*. Harvard University Press.

Collins-Kreiner, N. (2020). Religion and tourism: A diverse and fragmented field in need of a holistic agenda. *Annals of Tourism Research*, *82*, 102892. doi:10.1016/j.annals.2020.102892

Collins-Kreiner, N., & Kliot, N. (2000). Pilgrimage tourism in the Holy Land: The behavioural characteristics of Christian pilgrims. *GeoJournal*, *50*(1), 55–67. doi:10.1023/A:1007154929681

Dawut, R. (2009). Shrine pilgrimage among the Uighurs. *Silk Road*, *6*(2), 56–67.

Desforges, L. (1998). Checking out the planet: global representations/local identities and youth travel. In T. Skelton & G. Valentine (Eds.), *Cool Places: Geographies of Youth Cultures*. Routledge.

Esposito, J. L. (2004). *Islam: The Straight Path*. Oxford University Press.

Fang, W. T. (2020). Religious Tourism. In *Tourism in Emerging Economies* (pp. 15–43). Springer. doi:10.1007/978-981-15-2463-9_2

Gilsenan, M. (2008). *Recognizing Islam: religion and society in the modern Middle East*. Tauris.

HALL. C. M. (2006). Buddhism, tourism and the Middle Way. In Tourism, Religion and Spiritual Journeys (pp. 188-201). Routledge.

Haq, F., & Wong, H. Y. (2010). Is spiritual tourism a new strategy for marketing Islam? *Journal of Islamic Marketing*, *1*(2), 136–148. doi:10.1108/17590831011055879

Heydari Chianeh, R., Del Chiappa, G., & Ghasemi, V. (2018). Cultural and religious tourism development in Iran: Prospects and challenges. *Anatolia*, *29*(2), 204–214. doi:10.1080/13032917.2017.1414439

Holmberg, C. B. (1993). Spiritual pilgrimages: Traditional and hyperreal motivations for travel and tourism. *Visions in Leisure and Business*, *12*(2), 4.

Iravani, M. R., & Mozaffari, A. M. (2013). Review The Status of Tourism from the Perspective of The Qur'an. *Int. J. Basic Appl. Sci*, *1*, 628–633.

Jackowski, A. (2000). Religious tourism: problems with terminology. *Peregrinus Cracoviensis*, (10).

Jackson, R. H., & Hudman, L. (1995). Pilgrimage tourism and English cathedrals: The role of religion in travel. *Tourism Review*.

Jafari, J. (1987). Tourism models: The sociocultural aspects. *Tourism Management, 8*(2), 151–159. doi:10.1016/0261-5177(87)90023-9

Kelly, G., Kelly, G. A., & Kelly, G. A. (1963). A theory of personality: The psychology of personal constructs (No. 152). WW Norton & Company.

Laviniaţală, M., & Pădurean, A. M. (2008). *Dimensions of religious tourism.* Academic Press.

Lloyd, D. W., & Tourism, B. (1998). *Pilgrimage and the Commemoration of the Great War in Britain.*

Lücking, M. (2014, April). Making'Arab'One's Own: Muslim Pilgrimage Experiences in Central Java, Indonesia. In *Internationales Asien Forum. International Quarterly for Asian Studies* (Vol. 45, No. 1/2, p. 129). Arnold Bergsträsser Institut.

Luz, N. (2020). Pilgrimage and religious tourism in Islam. *Annals of Tourism Research, 82*, 102915. doi:10.1016/j.annals.2020.102915

MacCannell, D. (1973). Staged authenticity: Arrangements of social space in tourist settings. *American Journal of Sociology, 79*(3), 589–603. doi:10.1086/225585

MacCannell, D. (1976). *The Tourist: A New Theory of the Leisure Class.* Schocken Books.

MacCannell, D. (1992). *Empty Meeting Grounds.* Routledge.

Mukherjee, S., Bhattacharjee, S., & Singha, S. (2020). *Religious to Spiritual Tourism-An Era of Paradigm Shift in India.* Available at SSRN 3546903

Mustafa, M. H. (2010). Tourism and globalization in the Arab world. *International Journal of Business and Social Science, 1*(1).

Nieminen, K. (2012). *Religious tourism-a Finnish perspective.* Academic Press.

Nolan, M. L., & Nolan, S. (1992). Religious sites as tourism attractions in Europe. *Annals of Tourism Research, 19*(1), 68–78. doi:10.1016/0160-7383(92)90107-Z

Noy, C. (2004). This trip has really changed me: Backpackers' narratives of self-change. *Annals of Tourism Research, 31*, 78–102. doi:10.1016/j.annals.2003.08.004

Nursanty, E. (2012). *Halal Tourism The New Product In Islamic Leisure Tourism And Architecture.* Department of Architecture.

Olsen, D. H., & Timothy, D. J. (1999). Tourism 2000: Selling the millennium. *Tourism Management, 20*(4), 389–392.

Post, P., Pieper, J., & Van Uden, M. (1998). *The modern pilgrim: Multidisciplinary explorations of Christian pilgrimage* (Vol. 8). Peeters Publishers.

Puşcaşu, V. (2015). Religious tourism or pilgrimage. *European Journal of Science and Theology, 11*(3), 131–142.

Raj, R., Griffin, K., & Morpeth, N. D. (Eds.). (2013). *Cultural tourism.* CAB International. doi:10.1079/9781845939236.0000

Rasul, F., Fatima, U., & Sohail, S. (2016). Religion Tourism and Entrepreneurial Development (A Case Study Hazrat Data Ganj Bakhsh Shrine). *South Asian Studies, 31*(1).

Rinschede, G. (1992). Forms of religious tourism. *Annals of Tourism Research, 19*(1), 51–67. doi:10.1016/0160-7383(92)90106-Y

Russell, P. (1999). Religious travel in the new millennium. *Travel & Tourism Analyst*, (5), 39–68.

San Filippo, M. (2001). The religious niche. *Travel Weekly*, *60*(18), 12.

Seyer, F., & Müller, D. (2011). Religious tourism. In *The Long Tail of Tourism* (pp. 45–56). Gabler. doi:10.1007/978-3-8349-6231-7_6

Singh, R. P. (2003). *Where the Buddha walked: A companion to the Buddhist places of India*. Indica Books.

Singh, S. (1998). Probing the product life cycle further. *Tourism Recreation Research*, *23*(2), 61–63. doi:10.1080/02508281.1998.11014839

Smith, M., Macleod, N., & Robertson, M. H. (2010). Religious and spiritual tourism. In *SAGE key Concepts: Key concepts in tourist studies* (pp. 140–145). SAGE Publications Ltd., doi:10.4135/9781446251027

Smith, V. L. (1992). Introduction: The quest in guest. *Annals of Tourism Research, 19*(1), 1–17. doi:10.1016/0160-7383(92)90103-V

Terzidou, M., Scarles, C., & Saunders, M. N. (2018). The complexities of religious tourism motivations: Sacred places, vows and visions. *Annals of Tourism Research*, *70*, 54–65. doi:10.1016/j.annals.2018.02.011

Timothy, D., & Olsen, D. (Eds.). (2006). *Tourism, religion and spiritual journeys*. Routledge. doi:10.4324/9780203001073

Werbner, P. (2003). *Pilgrims of love: The anthropology of a global Sufi cult*. Indiana University Press.

Wiśniewski, Ł. (2018). Religious tourism in Christian sanctuaries: The implications of mixed interests for the communication of the faith. *Church. Communication and Culture*, *3*(3), 199–220. doi:10.1080/23753234.2018.1537674

World Tourism Organization. (2011). *Religious Tourism in Asia and the Pacific*. UNWTO., doi:10.18111/9789284413805

Wright, K. (2008). *The Christian travel planner*. Thomas Nelson.

Xavier, M. S. (2018). *Sacred spaces and transnational networks in American Sufism: Bawa Muhaiyaddeen and contemporary shrine cultures*. Bloomsbury Publishing.

Yeoman, I. (2009). *Tomorrow's tourist: scenarios & trends*. Routledge. doi:10.4324/9780080878904

Zamani-Farahani, H., & Eid, R. (2016). Muslim world: A study of tourism & pilgrimage among OIC Member States. *Tourism Management Perspectives*, *19*, 144–149. doi:10.1016/j.tmp.2015.12.009

KEY TERMS AND DEFINITIONS

Faith: Strong reliance on religion.

Hajj: An obligation for financially stable people to perform it on the specified Holy Center by the faith-keepers of Religion, Islam.

Intellectual Learning Excursions: A short religious journey enriched with the sense to attain scientific knowledge or to attend some religious cultural events.

Pilgrimage: A religious journey according to some specified rules designed by the religion.

Religious Cultusrist Events: The events celebrated for individuals for the sake of culture and religion.

Religious Events: The events specifically associated with the religion performed by their faith-keepers.

Religious Journeys: A short or long traveling towards some religious places.

Tourism: A journey to a place for some specific purpose.

ENDNOTE

1 It is an agency of the Netherlands Ministry of Foreign Affairs, established in 1971, to promote imports from developing countries.

Compilation of References

"Title 36 – Patriotic and National Observances, Ceremonies, and Organizations". U.S. Government Publishing Office. Archived from the original on 12 May 2019. Retrieved 12 March, 2020. §302. National motto "In God we trust" is the national motto.

Abbasi, A. S. (2018). 15 Nankana Sahib as 'A Symbol of Religious Coexistence'. *Religious Tourism in Asia: Tradition and Change Through Case Studies and Narratives*, 138.

Abo-Sak, M. (1999). *US Involvement in Israeli-Palestinian Conflict:Solution or Complication.* Al-Hewar Center.

Abumanssur, E. S. (2017). A força do mito: tensões e acomodações narrativas em torno das imagens de devoção. In Academia Marial de Aparecida. Aparecida, 300 anos de fé e devoção. Aparecida: Santuário/Academia Marial de Aparecida.

Acar, A. (2012). Imam Ghazali and the Seljuks. In *Imam Ghazali at the 900th Year of Death, International Controversial Scientific Meeting 07 - 09 October 2011*. Istanbul: Marmara University Faculty of Theology Foundation Publications Nu: 271. Accessed from http://www.isav.org.tr/img/20150908__3924219012.pdf

Adesina, B. (2003). *Ife Museum, Enuwa, Ife-Ife*. National Commission for Museum and Monuments.

Adogame, A., & Jafta, L. (1980). *African Initiated Churches*. Source Unknown.

Agrawal, M., Choudhary, H., & Tripathi, G. (2010). Enhancing Buddhist tourism in India: An exploratory study. *Worldwide Hospitality and Tourism Themes*, *2*(5), 477–493. Advance online publication. doi:10.1108/17554211011090102

Ahmad, S. (2016). Tourism in Pakistan: Challenges, Prospects and Potential of Gilgit-Baltistan. *Pakistan Perspectives, 21*(2).

Ahmad, A. (2016). Gwadar: A Historical Kaleidoscope. *Policy Perspectives: The Journal of the Institute of Policy Studies*, *13*(2), 149–166. doi:10.13169/polipers.13.2.0149

Aitchison, C., MacLeod, N. E., Macleod, N. E., & Shaw, S. J. (2002). *Leisure and Tourism Landscapes: Social and Cultural Geographies*. Psychology Press.

Akaev, V. (1999). *Sufism and Wahhabism in the North Caucasus. Confrontation or compromise?* The Institute of History, Archeology of Ethnography.

Akaev, V. (2014). Religious and Political Elites in the Northern Caucasus: Formation, Ideological, Ideological Contradictions, and Practical Opposition. *Religion in Society, Central Asia and the Caucasus, 15*(1).

Akama, J. S. (2002). The Role of government in the development of tourism in Kenya. *International Journal of Tourism Research*, *4*(1), 1–13. doi:10.1002/jtr.318

Akram, M. (2015). *The Temples of Katasraj in Pakistan between Legend and History*. Academic Press.

Al-Amin, H. (2002). Religious Tourism in Islamic Heritage. *Islamic Tourism, 2*.

Alasania, G., & Chkhartishvili, M. (2014). *The knowledge of Georgian Sources*. Publishing Hause Universal.

Al-Aseli, K. C. (2002). Jerusalem: The Ottoman Period and After. In *Religious Foundation of Turkey [TDV] Encyclopedia of Islam*. Accessed from https://islamansiklopedisi.org.tr/kudusat 19.03.2020

Albayrak, T., Herstein, R., Caber, M., Drori, N., Bideci, M., & Berger, R. (2018). Exploring religious tourist experiences in Jerusalem: The intersection of Abrahamic religions. *Tourism Management*, *69*, 285–296. doi:10.1016/j.tourman.2018.06.022

Alecu, I. C. (2010). Epistemological aspects of religious tourism in rural areas. International Journal of Business. *Management and Social Sciences*, *2*(3), 59–65.

Alegre, J., & Garau, J. (2010). Tourist satisfaction and dissatisfaction. *Annals of Tourism Research*, *37*(1), 52–73. doi:10.1016/j.annals.2009.07.001

Ali, S., & Maharani, L. (2019). Development of Religious Tourism in Bandar Lampung. Academic Press.

Alikberov, A., Bobrovnikov, V., & Bustanov, A. (2019). *Russian Islam, Features of History and Cultures*. The Institute of Oriental Studies, Russian Academy of Science.

Alipour, H., Olya, H. G., & Forouzan, I. (2017). Environmental Impacts of Mass Religious Tourism: From Residents' Perspectives. *Tourism Analysis*, *22*(2), 167–183. doi:10.3727/108354217X14888192562285

Ali, Y., Awan, M. A., Bilal, M., Khan, J., Petrillo, A., & Khan, A. A. (2019). Risk assessment of China-Pakistan fiber optic project (CPFOP) in the light of multi-criteria decision making (MCDM). *Advanced Engineering Informatics*, *40*, 36–45. doi:10.1016/j.aei.2019.03.005

Allcock, J. B. (1988). Tourism as a sacred journey. *Loisir et Société*, *11*(1), 33–48. doi:10.1080/07053436.1988.10715288

Almaimoni, H., Altuwaijri, N., Asiry, F., Aldossary, S., Alsmadi, M., Al-Marashdeh, I., ... Alrajhi, D. (2018). Developing and Implementing WEB-based Online Destination Information Management System for Tourism. *International Journal of Applied Engineering Research*, *13*(10), 7541–7550.

Almuhrzi, H. M., & Alsawafi, A. M. (2017). Muslim perspectives on spiritual and religious travel beyond Hajj: Toward understanding motivations for Umrah travel in Oman. *Tourism Management Perspectives*, *24*, 235–242. doi:10.1016/j.tmp.2017.07.016

Alshattnawi, S. (2012). Effective use of QR codes in religious tourism. In *Proceedings - 2012 International Conference on Advanced Computer Science Applications and Technologies, ACSAT 2012*. 10.1109/ACSAT.2012.43

Alsos, G. A., Eide, D., & Madsen, E. L. (2014). *Handbook of research on innovation in tourism industries*. Edward Elgar Publishing. doi:10.4337/9781782548416

Alves, G. M., Sousa, B. M., & Machado, A. (2020). The Role of Digital Marketing and Online Relationship Quality in Social Tourism: A Tourism for All Case Study. In J. Santos & Ó. Silva (Eds.), *Digital Marketing Strategies for Tourism, Hospitality, and Airline Industries* (pp. 49–70). IGI Global. doi:10.4018/978-1-5225-9783-4.ch003

Amaro, S., Antunes, A., & Henriques, C. (2018). A closer look at Santiago de Compostela's pilgrims through the lens of motivations. *Tourism Management*, *64*, 271–280. doi:10.1016/j.tourman.2017.09.007

Ambrósio, V. (2006). *Turismo Religioso: Desenvolvimento das Cidades – Santuário. (Tese de Doutoramento)*. Universidade Nova de Lisboa.

Ambrósio, V., Fernandes, C., Krogmann, A., Leitão, I., Oremusová, D., & Šolcová, L. (2019). From church perseverance in authoritarian states to today's religious inspired tourism. PASOS. *Revista de Turismo y Patrimonio Cultural, 17*(5), 1065–1078. doi:10.25145/j.pasos.2019.17.074

Amy, S. (1997). *The Soul of Development: Biblical Christianity and Economic Transformation in Guatemala*. Oxford University Press.

Andrade, J.V. (2000). *Turismo: Fundamentos e dimensões*. São Paulo: Ática.

Andriotis, K. (2009). Sacred site experience. A Phenomenological Study. *Annals of Tourism Research, 36*(1), 64–84. doi:10.1016/j.annals.2008.10.003

Andriotis, K. (2011). Genres of heritage authenticity. Denotations from a pilgrimage landscape. *Annals of Tourism Research, 38*(4), 1613–1633. doi:10.1016/j.annals.2011.03.001

Anrew, P. (2014). *8 Reasons people embrace religion*. Retrieved from https://listverse.com/2013/06/13/8-reasons-people-embrace-religion/

Antunes Angela, A. S. (2016). Intentions to use a pilgrimage app: Which features really matter? A Pathway for the New Generation of Tourism Research. In F. Dias (Ed.), *Proceedings of the EATSA Conference* (pp. 525–535). Lisbon: APTUR.

Antunes, J., Barroco Novais, C., & Dias, H. (2016). A importância do turismo religioso no desenvolvimento das regiões: O caso do santuário da nossa senhora da lapa. *International Journal of Scientific Management and Tourism, 2*(1), 273–285.

Anuwichanont, J., Serirat, S., Mechinda, P., & Archarungroj, P. (2020). Examining Tourists' Attitude Towards the Religious Tourism in Thailand. *Review of Integrative Business and Economics Research, 9*(3), 55–66.

Anyumba, G., & Nkuna, M. (2017). Lake Funfudzi: A Sacret Lake in South Africa that is not open for Tourism Development. *African Journal of Hospitality, Tourism and Leisure, 6*(4), 1–20.

Apleni, L., & Henama, U. S. (2020). Visitor's perceptions of religious tourism in South Africa: The case of Moria in the Limpopo Province. *African Journal of Hospitality, Tourism and Leisure, 9*(2), 1–19.

Apostolopoulos, Y., Leivadi, S., & Yiannakis, A. (Eds.). (1996). *The Sociology of Tourism*. Routledge.

Arjana, S. R. (2017). *Pilgrimage in Islam: traditional and modern practices*. Simon and Schuster.

Arshad, M. I., Iqbal, M. A., & Shahbaz, M. (2018). Pakistan tourism industry and challenges: A review. *Asia Pacific Journal of Tourism Research, 23*(2), 121–132. doi:10.1080/10941665.2017.1410192

Asadi, M. M., Heidari, M., & Asadian, F. (2016). Forecasting religious tourism demand using artificial intelligence case study: Mashhad city. *International Journal of Humanities and Cultural Studies*, 1597-1617.

Ashraf, M. M. (2018). *Opening the Kartarpur Corridor*. Academic Press.

Aslam, N., & Bukhari, M. F. (2018). Dharmarajika Stupa: A Great Buddhist Heritage of Taxila (Pakistan). Journal of the Centre for Heritage Studies, *35*.

Asmelash, A. G., & Kumar, S. (2019). Assessing progress of tourism sustainability: Developing and validating sustainability indicators. *Tourism Management, 71*, 67–83. doi:10.1016/j.tourman.2018.09.020

Atlas of Islamic Thought (AIT). (2020). *Ghazzali*. Accessed from https://www.islamdusunceatlasi.org/timemaps/1

Avci, C. (2002). Jerusalem. In *Religious Foundation of Turkey [TDV] Encyclopedia of Islam*. Accessed from https://islamansiklopedisi.org.tr/kudus

Avci, İ. Y. (2018). *It is very easy to go to Jerusalem.* Accessed from https://www.gzt.com/mecra/kuduse-gitmek-cok-kolay-3390623

Babary, A., & Zeeshan, M. (n.d.). Reminiscent of Hinduism: An insight of Katas Raj Mandir. Explorer, 1(4), 121–124.

Backman, S., & Crompton, J. (1991). The usefulness of selected variables for predicting activity loyalty. *Leisure Sciences*, *13*(3), 205–220. doi:10.1080/01490409109513138

Badar, R., & Chakradeo, S. (2016). Sufism and tourism interface. *Space, 20*(1-2).

Badone, E., & Roseman, S. R. (2004). *Approaches to the Anthropology of Pilgrimage and Tourism.* University of Illinois Press.

Baggio, F., Parise, P., & Sanchez, W. L. (2016). *Mobilidade humana e identidades religiosas.* Paulus.

Baghdad Travel Guide. (2020). *Places to Visit in Baghdad.* Accessed from https://www.flypgs.com/sehir-rehberi/bagdat-gezi-rehberi

Bahadori, S. R., Farsani, N. T., & Shafiei, Z. (2019). Tourists' views on Islamic spiritual tours (case study: Yazd, Iran). *Journal of Islamic Marketing*, *10*(1), 323–335. doi:10.1108/JIMA-05-2017-0055

Baker, D., & Crompton, J. (2000). Quality, Satisfaction and Behavioral Intentions. *Annals of Tourism Research*, *27*(3), 785–804. doi:10.1016/S0160-7383(99)00108-5

Bal-El, Y., Durst, R., Katz, G., Zislin, J., Strauss, Z., & Knobler, H. (2000). Jerusalem Syndrome. *The British Journal of Psychiatry*.

Balfour Declaration. (1917). Retrieved from Israel Ministry of Foreign Affairs: https://mfa.gov.il/mfa/foreignpolicy/peace/guide/pages/the%20balfour%20declaration.aspx

Ballantyne, R., Hughes, K., & Bond, N. (2016). Using a Delphi approach to identify managers' preferences for visitor interpretation at Canterbury Cathedral World Heritage Site. *Tourism Management*, *54*, 72–80. doi:10.1016/j.tourman.2015.10.014

Baloch, Q. B., & Rehman, A. (2015). Regional integration of Pakistan tourism: Exploring prospects. *Abasyn University Journal of Social Sciences*, *8*(2), 405–415.

Bama, H. K. N., & Tichaawa, T. M. (2015). Major sporting events as responsible tourism: Analysis of the 2012 Africa Cup of nations (AFCON) tournament in Port Elizabeth, South Africa. *African Journal for Physical Health Education. Recreation and Dance*, (December), 205–219.

Barajas, A., Lago-Peñas, S., & Sanchez, P. (2014). On the regional economic effects of the Pope's journeys: Evidence from the visit to Santiago de Compostela in 2010. *Tourism Economics*, *20*(2), 279–299. doi:10.5367/te.2013.0270

Barber, T. (2012). Europe must confront crisis of legitimacy. *Financial Times.* https://www.ft.com/content/4e2c793c-8d50-11e1-8b49-00144feab49a

Barretto, M. (1995). *Manual de Iniciação ao Estudo do Turismo*. Papirus Editora.

Battour, M., Battor, M., & Bhatti, M. A. (2014). Islamic attributes of destination: Construct development and measurement validation, and their impact on tourist satisfaction. *International Journal of Tourism Research*, *16*(6), 556–564. doi:10.1002/jtr.1947

Battour, M., Ismail, M. N., & Battor, M. (2011). The impact of destination attributes on a Muslim tourist's choice. *International Journal of Tourism Research*, *13*(6), 527–540. doi:10.1002/jtr.824

Battour, M., Ismail, M. N., Battor, M., & Awais, M. (2017). Islamic tourism: An empirical examination of travel motivation and satisfaction in Malaysia. *Current Issues in Tourism, 20*(1), 50–67. doi:10.1080/13683500.2014.965665

Bayly, S. (2001). *Caste, Society and Politics in India from the Eighteenth Century to the Modern Age*. Cambridge University Press.

BBC. (n.d.). https://www.bbc.com/news/world-asia-india-50342319

Becker, J. (2010). U.S. Approved Business with Blacklisted Nations. *The New York Times*. https://www.nytimes.com/2010/12/24/world/24sanctions.html

Belhassen, Y., Caton, K., & Stewart, W. P. (2008). The search for authenticity in the pilgrim experience. *Annals of Tourism Research, 35*(3), 668–689. Advance online publication. doi:10.1016/j.annals.2008.03.007

Belhassen, Y., & Ebel, J. (2009). Tourism, faith, and politics in the Holy Land: An ideological analysis of evangelical pilgrimage. *Current Issues in Tourism, 12*(4), 359–378. doi:10.1080/13683500802469342

Benjamin, W. (2005). Capitalism as Religion [Fragment 74]. In Religion as Critique: The Frankfurt School's Critique of Religion. New York: Routledge.

Bennett, S. (2012). *The impact of social media on travel and tourism*. Retrieved from https://www.mediabistro.com/alltwitter/social-media-travelhospitality_b24679

Berger, P. (2004). *Max Weber is alive and well and living in Guatemala*. Paper prepared for the conference 'The Norms, Beliefs and Institutions of the Twenty-First Century' at Cornell University, Ithaca, NY.

Bernstein, A., & Rule, S. (2011). Flying under South Africa's Radar: The Growth and Impact of Pentecostals in a Developing Country. In P. Berger & G. Redding (Eds.), *The Hidden Form of Capital: Spiritual Influences in Societal Progress* (pp. 91–131). Anthem.

Bernstein, M. J., Zawadzki, M. J., Juth, V., Benfield, J. A., & Smyth, J. M. (2018). Social interactions in daily life: Within-person associations between momentary social experiences and psychological and physical health indicators. *Journal of Social and Personal Relationships, 35*(3), 372–394. doi:10.1177/0265407517691366

Bhardwaj, S. (1998). Non-hajj pilgrimage in Islam: A neglected dimension of religious circulation. *Journal of Cultural Geography, 17*(2), 69–87. doi:10.1080/08873639809478321

Bhonsale, M. (2019). *Religious Tourism as Soft Power: Strengthening India's Outreach to Southeast Asia, Observer Research Foundation, India Special Report No.97, Sep.2019*. https://www.orfonline.org/research/religious-tourism-as-soft-power-strengthening-indias-outreach-to-southeast-asia-55674/

Bigné, J., Andreu, L., & Gnoth, J. (2005). The theme park experience: An analysis of pleasure, arousal and satisfaction. *Tourism Management, 26*(6), 833–844. doi:10.1016/j.tourman.2004.05.006

Bingqui, F., Bing, D., Taolong, X., & Ruofei, F. (2011). Place media events and place image communication: The role of festivals and special events from the perspective of media relations. *Catalan Journal Of Communication & Cultural Studies, 3*(2), 159–173. doi:10.1386/cjcs.3.2.159_1

Blackwell, R. (2007). Motivations for religious tourism, pilgrimage, festivals and events. *Religious tourism and pilgrimage festivals management: An international perspective*, 35-47.

Bloomberg, L. D., & Volpe, M. (2008). *Completing your questionnaire dissertation: A roadmap from beginning to end*. London: Sage.

Bogdan-Constantin, I., & Alina, M. (2018). Territorial and marketing strategies related to religious tourism in Northern Moldavia, Romania. *CES Working Papers, 10*(1), 129–148.

Bohg, J., Hausman, K., Sankaran, B., Brock, O., Kragic, D., Schaal, S., & Sukhatme, G. S. (2017). Interactive perception: Leveraging action in perception and perception in action. *IEEE Transactions on Robotics, 33*(6), 1273–1291. doi:10.1109/TRO.2017.2721939

Bokare, M. G. (2009). *Hindu Economics* (2nd ed.). Dr. M.G. Bokare Memorial Foundation.

Bolay, S. H. (2019). *Comparison of Aristotle Metaphysics and Ghazali Metaphysics*. MEB Publications.

Bond, N., & Falk, J. (2013). Tourism and identity-related motivations: Why am I here (and not there)? *International Journal of Tourism Research, 15*(5), 430–442. Advance online publication. doi:10.1002/jtr.1886

Bond, N., Packer, J., & Ballantyne, R. (2015). Exploring Visitor Experiences, Activities and Benefits at Three Religious Tourism Sites. *International Journal of Tourism Research, 17*(5), 471–481. Advance online publication. doi:10.1002/jtr.2014

Bornhorst, T., Brent Ritchie, J. R., & Sheehan, L. (2010). Determinants of tourism success for DMOs & destinations: An empirical examination of stakeholders' perspectives. *Tourism Management, 31*(5), 572–589. Advance online publication. doi:10.1016/j.tourman.2009.06.008

Bourdieu, P. (2005). *A Economia das trocas simbólicas* (6th ed.). Perspectiva.

Bowen, J. (n.d.). *1967 War: Six days that changed the Middle East*. Retrieved from BBC News: https://www.bbc.com/news/world-middle-east-39960461

Bowen, D. Z., Zubair, S., & Altinay, L. (2017). Politics and Tourism Destination Development: The Evolution of Power. *Journal of Travel Research, 56*(6), 725–743. doi:10.1177/0047287516666719

Bozkaya, Z. (2019). *Tus City*. Accessed from https://irangezi.com/Tûs-kenti/

Bozkurt, A. (2014). Branding Adana as a Touristic Destination in the Eastern. In International Interdisciplinary Business-Economics Advancement Conference. University of South Florida M3 Center Publishing.

Bozkurt, N., & Kucukasci, M. S. (2003a). Mekka. In *Religious Foundation of Turkey [TDV] Encyclopedia of Islam*. Accessed from https://islamansiklopedisi.org.tr/mekke at 20.03.2020

Bozkurt, N., & Kucukasci, M. S. (2003b). Medina. In *Religious Foundation of Turkey [TDV] Encyclopedia of Islam*. Accessed from https://islamansiklopedisi.org.tr/medine

Bremer, T. S. (2005). Tourism and religion. In L. Jones (Ed.), *Encyclopedia of Religion*. Macmillan.

Brewer, P. S. (1947, Dec. 30). Irgun Bomb Kills 11 Arabs, 2 Britons. *The New York Times*.

Brown, D. (2001). Suez and the end of empire. *The Guardian*. https://www.theguardian.com/politics/2001/mar/14/past.education1

Browne, C. (2017). *Buddhist Economics: An Enlightened Approach to the Dismal Science*. Bloomsbury Press.

Brown, L. (2016). Tourism and pilgrimage: Paying homage to literary heroes. *International Journal of Tourism Research, 18*(2), 167–175. doi:10.1002/jtr.2043

Bubulashvili, E. (2011). *The Life of St. Nino*. Retrieved 11 10, 2019, from Library Church: http://library.church.ge/index.php?option=com_content&view=article&id=424%3A2011-09-16-10-53-21&catid=47%3A2010-03-11-12-05-46&Itemid=67&lang=ka&fbclid=IwAR3YQqLiP9md-H6Djaz59pRlcHXpA8F5N7NtLLL2wMISVR8f7lZpyuPba4k

Buhalis, D., & Amaranggana, A. (2015). Smart Tourism Destinations Enhancing Tourism Experience Through Personalisation of Services. In Information and Communication Technologies in Tourism 2015. doi:10.1007/978-3-319-14343-9_28

Buhalis, D., & Law, R. (2008). Progress in information technology and tourism management: 20 years on and 10 years after the Internet-The state of eTourism research. *Tourism Management*, *29*(4), 609–623. Advance online publication. doi:10.1016/j.tourman.2008.01.005

Bulatović, I., & Stranjančević, A. (2019). Integrated model of religious tourism in underdeveloped destinations. *Journal of the Geographical Institute Jovan Cvijic SASA*, *69*(1), 53–66. doi:10.2298/IJGI1901053B

Bulgarian Orthodox Church in the list of Georgia's praying places. (2019). Retrieved from The Official Web Page of Georgian Parliament.

Bulti, A. G., Ray, A., & Bhuyan, P. (2019). Smart tourism system architecture design using the internet of everything (IOE) over cloud platform. *International Journal of Innovative Technology and Exploring Engineering*.

Burgess, M. (2007). Spiritual clash over land resolved. *Farmers Weekly*. Retrieved from: https://www.farmersweekly.co.za/rural-insight/

Burke, J. (2012). UK government ends boycott of Narendra Modi. *The Guardian*. https://www.theguardian.com/world/2012/oct/22/uk-ends-boycott-narendra-modi

Bussard, S. (2018). En Italie, les élections de la peu. *Le Temps*. https://www.letemps.ch/opinions/italie-elections-peur

Butler, R., & Suntikul, W. (2013). *Tourism and War*. Routledge. doi:10.4324/9780203107706

Buttaeva, A. (2009). Islamic Revival in Russia. *Bulletin of the Dagestan Scientific Center of the Russian Academy of Sciences*.

Buzinde, C. N., Kalavar, J. M., Kohli, N., & Manuel-Navarrete, D. (2014). Emic understandings of Kumbh Mela pilgrimage experiences. *Annals of Tourism Research*, *49*, 1–18. doi:10.1016/j.annals.2014.08.001

Buzpinar, S. T. (2010). Damascus. In *Religious Foundation of Turkey [TDV] Encyclopedia of Islam*. Accessed from https://islamansiklopedisi.org.tr/sam--suriye#2

Bywater, M. (1994). Religious travel in Europe. *Travel & Tourism Analyst*, *2*, 39–52.

Cabrera, E. J., dela Vega, J. M., dela Vega, J. M., Dioquino Jr, A., Favor, C., & Navarro, L. (2019). Religious Tourism Assessment and Carrying Capacity: The Case of National Shrine of Saint Padre Pio in Batangas, Philippines. *Asia Pacific Journal of Multidisciplinary Research*, *7*(3).

Cabrera, H. (2011). Cartografía regional de las fiestas tradicionales. In Regionalización cultural del Uruguay. Manosanta Desarrollo Editorial.

Can, A., & Gulpınar, S. (2010). Baghdad, 1831 - 1869. *Seljuk University Journal of Faculty of Letters*, *23*, 209–217. Accessed from https://dergipark.org.tr/en/download/article-file/151651at

Canon Law Society of America, Catholic Church, Pope (1978-2005: John Paul II), & Libreria Editrice Vaticana. (1999). Code of Canon Law, Latin-English Edition: New English Translation. Canon Law Society of Amer.

Carneiro, M. J., Eusébio, C., & Caldeira, A. (2018). The influence of social contact in residents' perceptions of the tourism impact on their quality of life: A structural equation model. *Journal of Quality Assurance in Hospitality & Tourism*, *19*(1), 1–30. doi:10.1080/1528008X.2017.1314798

Carrillo-Hidalgo, I., & Pulido-Fernández, J. I. (2020). The Role of the World Bank in the Inclusive Financing of Tourism as an Instrument of Sustainable Development. *Sustainability*, *12*(1), 285. doi:10.3390u12010285

Cassidy, R., & Hawkins, B. (2017). *Interviewing key informants from the corporate sector*. Academic Press.

Causevic, S., & Lynch, P. (2013). Political (In)stability and its Influence on Tourism Development. *Tourism Management*, *34*, 145–157. doi:10.1016/j.tourman.2012.04.006

CBI Ministry of Foreign Affairs. (2018). *What are the opportunities for religious tourism from Europe?* Academic Press.

Census. (2019). https://www.census.gov/newsroom/facts-for-features/2019/holiday-season.html

Cesari, J. (2008). *Muslims in Europe and the Risk of Radicalism. Jihadi Terrorism and the Radicalisation Challenge in Europe*. Asgate.

Chadha, H., & Onkar, P. (2016). Changing Cities in the Perspective of Religious Tourism – A Case of Allahabad. *Procedia Technology*, *24*, 1706–1713. doi:10.1016/j.protcy.2016.05.200

Chaline, E. (2002). *Zen and the Art of Travel*. MQ Publications.

Chatri, S. (2015). Islam and socio-economic, political and cultural development of women in Iran. *Historical and Social Review*, 7-10.

Chattopadhyay, M. (2006). Influence of Religion on Tourism: Implications for India's Tourism Policy. *The ICFAI Journal of Consumer Behavior*, *1*(3), 59–67.

Chen, G.-h., & Huang, Y.-s. (2008). A study on the degree of international impact of China's tourism research: Comparative analysis and upgrading approach. *Luyou Xuekan*, *23*(5), 91–96.

Cheng, T. M., & Chen, M. T. (2014). Image Transformation for Mazu Pilgrimage and Festival Tourism. *Asia Pacific Journal of Tourism Research*, *19*(5), 538–557. doi:10.1080/10941665.2013.764911

Chigozie, O. J., Uchema, O., & Ngozi, E. (2018). Impact of Religious Tourism on Host Communities: The Case of Awhum Monastery. *American Journal of Social Sciences*, *6*(3), 39–47.

Chiorazzi, A. (2015). The spirituality of Africa. *The Harvard Gazette*. Retrieved from: https://news.harvard.edu/gazette/story/2015/10/the-spirituality-of-africa/

Chitando, E., & Biri, K. (2016). Walter Magaya's Phophetic Healing and Deliverance Ministries and Pentecostalism in Zimbabwe: A preliminary study with particular reference to Eumenism. *Studia Historiae Ecclesiasticae*, *42*(2), 72–85.

Choe, J., Blazey, M., & Mitas, O. (2015). Motivations of non-Buddhists visiting Buddhist temples. *Current Issues in Tourism*, *18*(1), 70–82. doi:10.1080/13683500.2013.771627

Christian shrines preserved in Georgia. (2020). *Karibche*.

Chufrin, G. (2001). *The Security of the Caspian Sea Region*. Oxford University Press, Inc.

CihinaS. C. (2019). Retrieved from http://eng.mod.gov.cn/Database/WhitePapers/: http://eng.mod.gov.cn/

Clarke, A., & Raffay, Á. (2015). Religion, Local Produce and Sustainability at Religious Sites in Hungary. *International Journal of Religious Tourism and Pilgrimage*, *3*(2), 33–47.

Cohen, P. (2014, October). *The Root Cause of the Never-Ending Conflict in Palestine; and How to Fix It*. Retrieved from Huffington Post: https://www.huffpost.com/entry/root-causes-in-palestine_b_5499127?guccounter=1&guce_referrer=aHR0cHM6Ly93d3cuZ29vZ2xlLmNvbS8&guce_referrer_sig=AQAAAGGfgtKWMMbxGDoS8EW3B75gVCCguodTDQ-0af0TjW0qqLOwqJYw0wNhDwwuxZBwjMxqnTQ5M6uK-tIzbmDsLA18P9beID-YSfo7ELWcQKhqD

Cohen, E. (1979). A phenomenology of tourist experiences. *Sociology*, *13*(2), 179–201. doi:10.1177/003803857901300203

Cohen, E. (1988). Authenticity and commoditization in tourism. *Annals of Tourism Research*, *15*(3), 371–398. doi:10.1016/0160-7383(88)90028-X

Cohen, E. H. (1986). Tourisme et identité. *Pardès*, *4*, 84–97.

Cohen, E. H. (2003). Tourism and religion, a case study: Visiting students in Israel. *Journal of Travel Research*, *42*(1), 36–47. doi:10.1177/0047287503254247

Cohen, E. H. (2006). Religious tourism as an educational experience. In *Tourism, religion and spiritual journeys* (pp. 94–109). Routledge.

Coleman, S., & Elsner, J. (1995). *Pilgrimage: Past and present in the world religions*. Harvard University Press.

Cole, S. (2005). Cultural tourism, community participation and empowerment. In M. Robinson & S. E. Phipps (Eds.), *Cultural Tourism in a Changing World* (Vol. 2). Channel View Publications.

Collection of decrees by the President of the Chechen Republic from 1 November 1991 to 30 June 1992. (n.d.). 4.

Collier, P. (2007). *The Bottom Billion: Why the Poorest Countries are Failing and What Can be Done About It*. Oxford University Press.

Collins Kreiner, N. (2009). Researching Pilgrimage: Continuity and Transformations. *Annals of Tourism Research*, *37*(2), 440–456. doi:10.1016/j.annals.2009.10.016

Collins-Kreiner, N. (2016a). Dark tourism as/is a pilgrimage. *Current Issues in Tourism*, *19*(12), 1185–1189. doi:10.1080/13683500.2015.1078299

Collins-Kreiner, N. (2016b). The lifecycle of concepts: The case of 'Pilgrimage Tourism.'. *Tourism Geographies*, *18*(3), 322–334. doi:10.1080/14616688.2016.1155077

Collins-Kreiner, N. (2020). Religion and tourism: A diverse and fragmented field in need of a holistic agenda. *Annals of Tourism Research*, *82*, 102892. doi:10.1016/j.annals.2020.102892

Collins-Kreiner, N., & Kliot, N. (2000). Pilgrimage tourism in the holy land: The behavioral characteristics of Christian pilgrims. *GeoJournal*, *501*(1), 55–67. doi:10.1023/A:1007154929681

Colvin, R. (2009). *Obama to reach out to Muslims in Egypt speech*. https://www.reuters.com/article/us-obama-muslims/obama-to-reach-out-to-muslims-in-egypt-speech-idUSTRE54754920090508?feedType=RSS&feedName=ObamaEconomy&virtualBrandChannel=10441

Constitution of the Republic of Zimbabwe. (2013). *Zimbabwe Constitution*. https://parlzim.gov.zw/component/k2/download/1290_da927a81557040d47c3a2c27012f6e1

Coplan, D. B. (2003). Land from the Ancestors: Popular Religious Pilgrimage along the South Africa-Lesotho Border. *Journal of Southern African Studies*, *29*(4), 976–993. doi:10.1080/0305707032000135923

Crabtree, S. (2010). *Religiosity Highest in World's Poorest Nations, United States is among the rich countries that buck the trend*. https://news.gallup.com/poll/142727/religiosity-highest-world-poorest-nations.aspx

Creutz, K. I.-M. (2019). The changing global order and its implications for the EU. Finnish Institute of International Affairs, 128.

Cristea, A. A., Apostol, M. S., & Dosescu, T. (2015). The Role of media in Promoting Religious Tourism in Romania. *Procedia: Social and Behavioral Sciences*, *188*, 302–306. doi:10.1016/j.sbspro.2015.03.398

Criteo. (2019). https://www.criteo.com/insights/ramadan-2019-advertising-trends

Crotti, R., & Misrahi, T. (2015). The travel & tourism competitiveness index 2015: T&T as a resilient contribution to national development. *The Travel & Tourism Competitiveness Report, 13.*

Cui, Q., Xu, H., & Yang, Y. (2014). Secular Pilgrimage; Tourist Experience in Tibet. *Luyou Xuekan*, *29*(2), 110–117.

Cunha, L. (2013). *Economia e Política do Turismo*. Lidel: edições técnicas Lda.

Cunha, L. (2007). *Introdução ao Turismo* (3rd ed.). Verbo.

Dafuleya, G., Gyekye, A. B., Oseifuah, E. K., Nethengwe, T., & Sumbana, F. (2017). Religiously motivated travel and rural tourism in the Vhembe District of South Africa. *Tourism Economics*, *23*(4), 911–918. doi:10.5367/te.2016.0564

Damari, C., & Mansfeld, Y. (2016). Reflections on pilgrims' identity, role and interplay with the pilgrimage environment. *Current Issues in Tourism*, *19*(3), 199–222. doi:10.1080/13683500.2014.957660

Dann, G. (1977). NoAnomie, ego enhancement and tourism. *Annals of Tourism Research*, 184–194.

Dann, G. (1981). Tourist motivation. An appraisal. *Annals of Tourism Research*, *8*(2), 187–219. doi:10.1016/0160-7383(81)90082-7

Dann, G., Nash, D., & Pearce, P. (1988). Methodology in Tourism Research. *Annals of Tourism Research*, *15*(1), 1–28. doi:10.1016/0160-7383(88)90068-0

Dario, C., Elahi, M., David, M., Stefano, M., Elena, N., & Francesco, R. V. A. (2018). Tangible Tourism with the Internet of Things. Information and Communication Technologies in Tourism. In Information and Communication Technologies in Tourism. Jönköping: Springer.

Dawut, R. (2009). Shrine pilgrimage among the Uighurs. *Silk Road*, *6*(2), 56–67.

De Ascaniis, S., Mutangala, M. M., & Cantoni, L. (2018). *ICTs in the tourism experience at religious heritage sites: a review of the literature and an investigation of pilgrims' experiences at the sanctuary of Loreto*. Church, Communication and Culture. doi:10.1080/23753234.2018.1544835

De La Torre, G. M. V., & Pérez, L. M. (2017). Religious tourism in different geographical areas in Spain: The tourist profile. *Boletín de la Asociación de Geógrafos Españoles*, (75), 29–54.

DeAscaniis, S., & Cantoni, L. (2016). Pilgrims in the Digital Age: a research manifesto. *International Journal of Religious Tourism and Pilgrimage*. doi:10.21427/D7Z120

Decree of the Government of Georgia #1740. (2015). Retrieved from gov.ge: http://gov.ge/files/410_50997_650312_N1740.pdf

Delcker, J. (2015). German anti-refugee movement sparks fear Officials warn a resurgent far-right has seized on the crisis to incite hatred and violence. https://www.politico.eu/article/asylum-pegida-refugees-far-right-protest-seekers-as-criminals-politicians-as-traitors/

Della Dora, V. (2012). Setting and Blurring Boundaries: Pilgrims, Tourists, and Landscape in Mount Athos and Meteora. *Annals of Tourism Research*, *39*(2), 951–974. doi:10.1016/j.annals.2011.11.013

Deloitte. (n.d.). https://www2.deloitte.com/content/dam/Deloitte/de/Documents/consumer-business/Deloitte%20Christmas%20Survey%202019%20-%20European%20results.pdf

Demircioglu, A. (2018). *Criticism and Doubt in Ghazali* (2nd ed.). Gece Publishing.

Denzin, N. K., & Lincoln, Y. S. (2003). Introduction: The discipline and practice of qualitative research. In N. K. Denzin & Y. S. Lincoln (Eds.), *The landscape of qualitative research. Themes and issues* (pp. 1–45). Sage.

Descartes. (2018). *Principles of Philosophy* (16th Edition). Istanbul: Say Publishing

Desforges, L. (1998). Checking out the planet: global representations/local identities and youth travel. In T. Skelton & G. Valentine (Eds.), *Cool Places: Geographies of Youth Cultures*. Routledge.

Devidze, E. (2012). *Cultural Heritage and Problems of Imereti Tourism Development* (Dissertation). Tbilisi State University.

Dhankhar, D. (2016). *E Marketing in Hospitality Industry: Practice, Performances and Service Quality* (Unpublished Thesis). Maharishi Dayanand University, Rohtak, India.

Difference on Destination Image and Travel Options: An Exploratory Text-Mining Study. (2018). In *2018 15th International Conference on Service Systems and Service Management (ICSSSM)* (pp. 1-5). IEEE.

Digance, J. (2003). Pilgrimage at contested sites. *Annals of Tourism Research, 30*(1), 143–159. doi:10.1016/S0160-7383(02)00028-2

Donnelly, M. (2009). *Building Customer Loyalty: A customer experience based approach in a tourism context.* Retrieved from http://repository.wit.ie/id/eprint/1396

Donnet, P. (1994). *Antoine*. Survival in Question. Oxford University Press.

Dumbaugh, K. (2008). *China-U.S. Relations: Current Issues and. Foreign Affairs, Defense, and Trade Division, Asian Affairs*. Congressional Research Service.

Duran-Sanchez, A., Alvarez-Garas, J., Rio-Rama, M. de la. C., & Oliveira, C. (2018). *Religious Tourism and Pilgrimage. Bibliometric Overview*. MDPI.

Durán-Sánchez, A., Álvarez-García, J., Río-Rama, D., De la Cruz, M., & Oliveira, C. (2018). Religious tourism and pilgrimage: Bibliometric overview. *Religions, 9*(9), 249. doi:10.3390/rel9090249

Durkheim, É., & Mauss, M. (2016). *Hobbes à lagrégation: Un cours dÉmile Durkheim suivi par Marcel Mauss*. Éditions de l'EHESS.

Easwaran, E. (2016). *Meditação*. Centro Lusitano de Unificação Cultural.

Ebron, P. A. (1999). Tourists as Pilgrims: Commercial Fashioning of Transatlantic Politics. *American Ethnologist, 26*(4), 910–932. doi:10.1525/ae.1999.26.4.910

Echtner, C., & Jamal, T. (1997). The disciplinary dilemma of tourism studies. *Annals of Tourism Research, 24*(4), 868–883. doi:10.1016/S0160-7383(97)00060-1

Eck, D. L. (2012). India: a Sacred Geography. New York: Harmony (Random House Group).

Ecommerce Guide. (n.d.). https://ecommerceguide.com/news/effect-ramadan-ecommerce-revealed/

Economic Times. (n.d.). https://economictimes.indiatimes.com/markets/stocks/news/investors-see-ingredients-for-stocks-revival-after-diwali/articleshow/71756398.cms

Eda, A., Gizem, K., & Cihan, Ç., K. E. (2019). The Potential Usage of Augmented Reality Applications in Tourist Guidance Education. In A. L. Cobanoglu Cihan (Ed.), *Global Congress On Smart Tourism* (pp. 79–81). Canakkale: Çanakkale Onsekiz Mart University.

Ed-Duri, A. (1991). Baghdad. In *Religious Foundation of Turkey [TDV] Encyclopedia of Islam.* Accessed from https://islamansiklopedisi.org.tr/bagdat#1-genel-bakisat

Eid, R. (2015). Integrating Muslim Customer Perceived Value, Satisfaction, Loyalty and Retention in the Tourism Industry: An empirical study. *International Journal of Tourism Research*, *17*(3), 249–260. Advance online publication. doi:10.1002/jtr.1982

Eid, R., & El-Gohary, H. (2015a). Muslim Tourist Perceived Value in the Hospitality and Tourism Industry. *Journal of Travel Research*, *54*(6), 774–787. doi:10.1177/0047287514532367

Eid, R., & El-Gohary, H. (2015b). The role of Islamic religiosity on the relationship between perceived value and tourist satisfaction. *Tourism Management*, *46*, 477–488. doi:10.1016/j.tourman.2014.08.003

El Hanandeh, A. (2013). Quantifying the carbon footprint of religious tourism: The case of Hajj. *Journal of Cleaner Production*, *52*, 53–60. doi:10.1016/j.jclepro.2013.03.009

El-Gohary, H. (2016). Halal tourism, is it really Halal? *Tourism Management Perspectives*, *19*, 124–130. doi:10.1016/j.tmp.2015.12.013

Embacher, J., & Buttle, F. (1989). A repertory grid analysis of Austria's image as a summer vacation destination. *Journal of Travel Research*, *27*(3), 3–7. doi:10.1177/004728758902700302

Enongene, V., & Griffin, K. (2019). Tackling the problems of deficient data when planning for religious tourism management. *Managing Religious Tourism*, 79.

Eric, D. (2015). *Myth, Magic, and Mysticism in the Age of Information.* Retrieved from https://lareviewofbooks.org/article/myth-magic-mysticism-age-information/

Esposito, J. L. (2004). *Islam: The Straight Path.* Oxford University Press.

Europe's Growing Muslim Population. (2017). *Pew Research Center.* https://www.pewforum.org/2017/11/29/europes-growing-muslim-population/

European Citizenship. (2017, November). *Standard Eurobarometer*, *88*, 117.

Fang, W. T. (2020). Religious Tourism. In *Tourism in Emerging Economies* (pp. 15–43). Springer. doi:10.1007/978-981-15-2463-9_2

Femenia-Serra, F., Neuhofer, B., & Ivars-Baidal, J. A. (2019). Towards a conceptualisation of smart tourists and their role within the smart destination scenario. *Service Industries Journal*, *39*(2), 109–133. Advance online publication. doi:10.1080/02642069.2018.1508458

Ferreira, J., & Sousa, B. (2020) Experiential Marketing as Leverage for Growth of Creative Tourism: A Co-creative Process. In Advances in Tourism, Technology and Smart Systems. Smart Innovation, Systems and Technologies, (vol 171. pp 567-577). Springer. doi:10.1007/978-981-15-2024-2_49

Fleischer, A., & Felsenstein, D. (2000). *Support for rural tourism: Does it make a difference?* Academic Press.

Fleischer, A. (2000). The tourist behind the pilgrim in the holy land. *International Journal of Hospitality Management*, *19*(3), 311–326. doi:10.1016/S0278-4319(00)00026-8

Follis, R. (2018). Turismo religioso, adventismo e lugares de memória. *HORIZONTE - Revista de Estudos de Teologia e Ciências Da Religião, 16*(49), 38–65.

Ford, P. (2005). Europe's rising class of believers: Muslims. *Christian Science Monitor.* https://www.csmonitor.com/2005/0224/p10s01-woeu.html

Forster, D. (2018). *Hydrodynamic fluctuations, broken symmetry, and correlation functions.* CRC Press. doi:10.1201/9780429493683

Fotis, J., Buhalis, D., & Rossides, N. (2012). *Social media use and impact during the holiday travel planning process.* Springer-Verlag. doi:10.1007/978-3-7091-1142-0_2

Fourie, J., Rosello, J., & Santana-Gallego, M. (2014). *Religion, Religious Diversity and Tourism.* Stellenbosch Economic Working paper, November 9/14.

Fourie, J., Rosselló, J., & Santana-Gallego, M. (2016). Research note: Which God is good for tourism? *Tourism Economics, 22*(1), 163–169. doi:10.5367/te.2014.0412

Freeman, D. (2012). The Pentecostal Ethic and the Spirit of Development. In D. Freeman (Ed.), *Pentecostalism and Development: Churches, NGOs and Social Change in Africa* (pp. 1–38). Palgrave Macmillan. doi:10.1057/9781137017253_1

Freston, P. (2013). Pentecostals and Politics in Latin America. In D. E. Miller, K. H. Sargeant, & R. Flory (Eds.), *Spirit and Power: The Growth and Global Impact of Pentecostalism* (pp. 3–36). Oxford University Press. doi:10.1093/acprof:oso/9780199920570.003.0005

Friedman, M., & Friedman, R. (1980). *Free to Choose.* University of Chicago Press.

Fujita, H., Gaeta, A., Loia, V., & Orciuoli, F. (2018). Resilience analysis of critical infrastructures: A cognitive approach based on granular computing. *IEEE Transactions on Cybernetics, 49*(5), 1835–1848. doi:10.1109/TCYB.2018.2815178 PMID:29994107

Future of Europe. (2018). *Special Eurobarometer, 479,* 118.

Gambashidze, G. (2014). Proceedings Ministry of Culture and Monument Protections. Geprgian Jewish Historical Museum, 28.

Garayeva, N., & Naumov, N. (2016). Religious sites as tourism attractions: The case of Azerbaijan. *International Hospitality and Tourism Student Journal, 8*(1), 8–14.

Geary, D. (2018). India's Buddhist circuit(s): A growing investment market for a "rising" Asia. *International Journal of Religious Tourism and Pilgrimage, 6*(1), 47–57. doi:10.21427/D7PT46

Genner, S., & Süss, D. (2017). *Socialization as Media Effect. The International Encyclopedia of Media Effects.* Zurich University of Applied Sciences.

George, R. (2014). *Marketing Tourism in South Africa. Cape Town.* Oxford University Press.

Georgian National Tourism Administration. (2015). *The Georgian Tourism Strategy-2025.* Author.

Getz, D. (1986). Models in Tourism Planning: Towards Integration of Theory and Practice. *Tourism Management, 7*(1), 21–32. doi:10.1016/0261-5177(86)90054-3

Getz, D., & Page, S. (2015). *Progress and prospects for event tourism research.* Tourism Management. doi:10.1016/.tourman.2015-.03.007

Gewali, B. R. (2013). Isha-Upanishad and Economics Consideration: An Elucidation. *Economic Literature, 11*(35-39).

Gezender. (2018). *Kingdom of Heaven Jerusalem: Travel Notes*. Accessed from https://www.gezender.com/kudus-gezi-rehberi/at

Gezimanya. (2020a). *Things to Know About Baghdad*. Accessed from https://gezimanya.com/bagdatat

Gezimanya. (2020b). *Things to Know About Nishabur*. Accessed from https://gezimanya.com/iran/nisabur-hakkinda-bilinmesi-gerekenler

Gezimanya. (n.d.a). *Damascus: General Information / Guide*. Accessed from https://gezimanya.com/samat

Gezimanya. (n.d.b). *Hebron General Information / Guide*. Accessed from https://gezimanya.com/filistin/hebron

Gezimanya. (n.d.c). *Jerusalem - Israel*. Accessed from https://gezimanya.com/kudus-israil

Gezimanya. (n.d.d). *Things to Know About Medina*. Accessed from https://gezimanya.com/medineat

Ghani, G. M. (2019). Tourist Arrivals to Muslim Countries: Is Religion Important? *Tourism, Culture & Communication*, *19*(3), 209–214. doi:10.3727/194341419X15554157596128

Ghazali. (1994). Mişkatü'l Envar (Lighthouse of Lights). Bedir Publishing.

Ghazali. (2019). *Ihya'u Ulumi'd Din* (3rd Edition). Istanbul: Bedir Publishing.

Ghazali. (2020). *El Munkız Mine'd-Dalal (From Dalalet to Hidaday)*. Istanbul: Cagdas Book.

Gillberg, C. (2006). *Warriors of Buddhism: Buddhism and violence as seen from a Vajrayana Tibetan Buddhist perspective*. Retrieved from https://www.researchgate.net/: https://www.researchgate.net/publication/326888242_Warriors_of_Buddhism_Buddhism_and_violence_as_seen_from_a_Vajrayana_Tibetan_Buddhist_perspective

Gilsenan, M. (2008). *Recognizing Islam: religion and society in the modern Middle East*. Tauris.

GizemK. (2019). *PinctadaAR*. Retrieved from https://play.google.com/store/apps/details?id=com.SMK.PinctadaAR&hl=tr

Goitsemodimo, G. A. (2019). *Mantsopa Makhetha*. National Museum Publications. Retrieved from: https://nationalibrarypublications.co.za/mantsopa-makhetha/

Gonzales, H. (2008). Tourism, mass media, and the making of visual culture in the greater Yucatan Peninsula. *Journal of Film and Video*, *60*(2), 50–58. doi:10.1353/jfv.0.0003

Gonzalo, F. (2012). *How Travelers Use Online Sources for Travel Decision-Making*. Retrieved from https://fredericgonzalo.com/2012/09/11/how-travelers-use-online-sources-for-travel-decision-making/

Government of Chechen Republic. (2012). *Social and Economic Development Strategy Chechen Republic until 2025*. Ministry of Economic Development and Trade.

Gozzoli, R. B., Chen, K. T. N., & Talawanich, S. (2019). Religious Tourism and Heritage Knowledge at Phra Pathom Chedi, Thailand. *Tourism, Culture & Communication*, *19*(2), 135–152. doi:10.3727/109830419X15536971539416

Grdzelishvili, N. (2018). *Religious Tourism*. Publishing House Kentavri.

Greece elections: Centre-right regains power under Kyriakos Mitsotakis. (2019). *DDC news*. https://www.bbc.com/news/world-48902766

Green, C. E. (2009). *Sensible marketing in tough times*. Hospitality Upgrade Magazine.

Gretzel, U. (2011). *Do social media influence our travel decisions? The Institute for Innovation in Business and Social Research (IIBSoR) Faculty of Commerce*. Retrieved from https://uowresearch.wordpress.com/2011/03/27/do-social-media-influence-our-travel-decisions/

Gretzel, U., Sigala, M., Xiang, Z., & Koo, C. (2015). Smart tourism: Foundations and developments. *Electronic Markets, 25*(3), 179–188. Advance online publication. doi:10.100712525-015-0196-8

Gretzel, U., Werthner, H., Koo, C., & Lamsfus, C. (2015). Conceptual foundations for understanding smart tourism ecosystems. *Computers in Human Behavior, 50*, 558–563. Advance online publication. doi:10.1016/j.chb.2015.03.043

Griffin, K., Raj, R., & Yasuda, S. (2018). Religious tourism and sacred sites in Asia. *Religious Tourism in Asia: Tradition and Change Through Case Studies and Narratives*, 1.

Griffin, K. (2017). The importance of religious tourism and pilgrimage: Reflecting on definitions, motives and data. *International Journal of Religious Tourism and Pilgrimage, 5*(3), 2.

Grygiel, J. (2016). The Return of Europe's Nation-States. The Upside to the EU's Crisis. *Foreign Affairs*, 94–101.

Gurung, R. (2016). *Marketing religious tourism destination. Case study: the birthplace of Gautama Buddha*. Centria University.

Gutic, J., Caie, E., Clegg, A., & Wiley, J. (2010). Visitors to an English Cathedral. *International Journal of Tourism Research, 760*(June), 750–760. doi:10.1002/jtr.790

Guzel, A. (1988). *An Essay on the Bibliography of Ghazali, Abu Hâmid Muhammed el-Gazzali (1058-1111)*. Kayseri: Erciyes University Gevher Nesibe Medical History Institute Publications. Accessed from http://isamveri.org/pdfdrg/D017726/1988/1988_GUZELA2.pdf

Gyekye, A., Oseifuah, E., Nethengwe, T., Sumbana, F., & Dafuleya, G. (2014). Religiously motivated travel and rural tourism in Vhembe District of South Africa. *SHS Web of Conference, 12*, 1-9. 10.1051hsconf/20141201057

Haber 7. (2020). *Four mausoleums belonging to famous scientists and scientists in the city of Nishabur in Khorasan*. Accessed from https://www.haber7.com/dunyayi-geziyorum/haber/2907764-horasanin-nisabur-sehrinde-unlu-ilim-ve-bilim-insanlarina-ait-4-turbe

Hadden, J. K. (1995). Religion and the quest for meaning and order: Old paradigms, new realities. *Sociological Focus, 28*(1), 83–100. Advance online publication. doi:10.1080/00380237.1995.10571040

Hafner, R. W. (2006). Islamic Economics and Global Capitalism, at the Symposium: Visible Hands, Religion and the Market. *Transaction Social Sciences and Modern Society, 44*(1).

Haj. (n.d.). https://www.haj.gov.sa/en/News/Details/12324

Halacoglu, Y. (1991). Baghdad. In *Religious Foundation of Turkey [TDV] Encyclopedia of Islam*. Accessed from https://islamansiklopedisi.org.tr/bagdat#1-genel-bakis

Halbach, U. (n.d.). *Islam in the North Caucasus*. Retrieved from http:// journals.openedition.org/assr/18403

HALL. C. M. (2006). Buddhism, tourism and the Middle Way. In Tourism, Religion and Spiritual Journeys (pp. 188-201). Routledge.

Hall, J. A. (1985). Religion and the Rise of Capitalism. *European Societies, 26*(2), 193–223. doi:10.1017/S0003975600004422

Halpenny, E. (2010). Pro-environmental behaviors and park visitors: The effect of place attachment. *Journal of Environmental Psychology, 30*(4), 409–421. doi:10.1016/j.jenvp.2010.04.006

Hammond, J. (2013, August). Washington Report on Middle East Affairs. In *The Role of the U.N in Creating the Israeli-Palestinian Conflict.* American Educational Trust. Retrieved from https://www.wrmea.org/013-august/the-role-of-the-u.n.-in-creating-the-israeli-palestinian-conflict.html

Hanahu, R. A. (1997). *Traditional culture of the North Caucasus: the challenges of time: (Socio-philosophical analysis).* Maikop.

Handriana, T., Yulianti, P., & Kurniawati, M. (2019). Exploration of pilgrimage tourism in Indonesia. *Journal of Islamic Marketing, 11*(3), 783–795. doi:10.1108/JIMA-10-2018-0188

Haq, F., & Wong, H. Y. (2010). Is spiritual tourism a new strategy for marketing Islam? *Journal of Islamic Marketing, 1*(2), 136–148. doi:10.1108/17590831011055879

Hardy, J. (2018). *Se eu não cuidar de mim, quem cuidará?* (1ª Edição). Pergaminho.

Harman, O. F. (2002). Jerusalem. In *Religious Foundation of Turkey [TDV] Encyclopedia of Islam.* https://islamansiklopedisi.org.tr/kudus

Harmsen, R. (2005). *Euroscepticism: Party Politics, National Identity and European Integration.* Rodopi.

Harrett, J., & Benjamin, T. (2009). Travel with a time Lord: Using media to enhance literacy. *Literacy, 43*(3), 134–142. doi:10.1111/j.1741-4369.2009.00510.x

Hashim, N. H., Murphy, J., & Hashim, N. M. (2007). Islam and online imagery on malaysian tourist destination websites. *Journal of Computer-Mediated Communication, 12*(3), 1082–1102. doi:10.1111/j.1083-6101.2007.00364.x

Heath, R. (2018). Europeans love the EU (and populists too). *Politico.* https://www.politico.eu/article/europeans-love-the-eu-and-populists-too/

Heelas, P., & Woodhead, L. (2005). *The spiritual revolution: Why religion is giving way to spirituality.* Blackwell Publishing.

Heidari, A., Yazdani, H. R., Saghafi, F., & Jalilvand, M. R. (2017). Developing strategic relationships for religious tourism businesses: A systematic literature review. *EuroMed Journal of Management, 2*(1), 77–98. doi:10.1504/EMJM.2017.084273

Heidari, A., Yazdani, H. R., Saghafi, F., & Jalilvand, M. R. (2018). The perspective of religious and spiritual tourism research: A systematic mapping study. *Journal of Islamic Marketing, 9*(4), 747–798. Advance online publication. doi:10.1108/JIMA-02-2017-0015

He, M., Li, J., Li, J., & Chen, H. (2019). A comparative study on the effect of soundscape and landscape on tourism experience. *International Journal of Tourism Research, 21*(1), 11–22. doi:10.1002/jtr.2237

Henama, U. S., & Sifolo, P. P. S. (2018). Religious Tourism in South Africa: Challenges, Prospects and Opportunities. In Global Perspectives on Religious Tourism and Pilgrimage. IGI.

Henama, U. S. (2012). Tourism accommodation in Mangaung: Current dynamics, immediate challenges and future prospects. *International Journal of Business. Economics and Law, 1,* 76–84.

Henama, U. S. (2013). Attracting Indian Outbound Tourist to South Africa: A BRICS Perspective. *India Quarterly. Journal of International Affairs, 69*(3), 229–247.

Henama, U. S. (2014a). Health tourism in South Africa: Opportunities and possibilities. *African Journal of Hospitality, Tourism and Leisure, 3*(1), 1–16.

Henama, U. S. (2014b). The demise of 1Time airline and the reaction of various interest groups. *African Journal of Hospitality, Tourism and Leisure, 6*(4), 1–16.

Henama, U. S. (2016). The low cost carrier bandwagon: Lessons for Skywise Airline. *African Journal of Hospitality, Tourism and Leisure*, *5*(2), 1–11.

Henama, U. S. (2017b). Marikana: Opportunities for Heritage Tourism. *African Journal of Hospitality, Tourism and Leisure*, *6*(4), 1–11.

Henama, U. S., Janse van Rensburg, M., & Nicolaides, A. (2016). Nkandla: The unexplored frontier of heritage tourism to Zululand, South Africa. *African Journal of Hospitality, Tourism and Leisure*, *5*(2), 1–16.

Henama, U. S., & Sifolo, P. P. S. (2017). Tourism migration in South Africa: Current dynamics, immediate challenges and future prospects. *International Journal of Innovation in the Digital Economy*, *8*(1), 47–60. doi:10.4018/IJIDE.2017010103

Henderson, D. W. (1998). *Culture Shift: Communicating God's Truth to our changing World*. Baker Books. http://www.tourismandmore.com/tidbits/using-the-social-media-for-tourism

Henderson, J. C. (2011). Religious tourism and its management: The hajj in Saudi Arabia. *International Journal of Tourism Research*, *13*(6), 541–552. doi:10.1002/jtr.825

Henley, J. (2019). Denmark's centre-left set to win election with anti-immigration shift. *The Guardian*. https://www.theguardian.com/world/2019/jun/04/denmark-centre-left-predicted-win-election-social-democrats-anti-immigration-policies

Henrion-Dourcy, I. (2016, January). Tibet. In *The Wiley-Blackwell Encyclopedia of Race, Ethnicity and Nationalism*. Retrieved from https://www.researchgate.net

Herald. (2015). *Night of Turn Around experience*. Retrieved from: http://www.herald.co.zw/night-of-turn-around/

Hernández-Martín, R., Rodríguez-Rodríguez, Y., & Gahr, D. (2017). Functional zoning for smart destination management. *European Journal of Tourism Research*.

Herz, E. (2007). *Mandate of Palestine: The Legal Aspects of Jewish Rights*. Myths and Facts.

Heslam, P. S. (2015). in his article: The Rise of Religion and the Future of Capitalism. *De Ethica. A Journal of Philosophical. Theological and Applied Ethics*, *2*, 3.

Heydari Chianeh, R., Del Chiappa, G., & Ghasemi, V. (2018). Cultural and religious tourism development in Iran: Prospects and challenges. *Anatolia*, *29*(2), 204–214. doi:10.1080/13032917.2017.1414439

Hidalgo, M., & Hernandez, B. (2001). Place attachment: Conceptual and empirical questions. *Journal of Environmental Psychology*, *21*(3), 273–281. doi:10.1006/jevp.2001.0221

Hisar Tourism. (2019a). *Visit Places and Mubarak Places in Makkah Al Mukarramah*. Accessed from https://hisarturizm.com.tr/mekke-i-mukerremedeki-ziyaret-yerleri-ve-mubarek-mekanlar/

Hisar Tourism. (2019b). *Visit Places and Mubarak Places in Medina-i Münevvere*. Accessed https://hisarturizm.com.tr/medine-i-munevveredeki-ziyaret-yerleri-ve-mubarek-mekanlar/

Hodgson, J. (1995). *Mantsopa: Popular religions and the Anglican Church in South Africa*. Source Unknown.

Hoffmann, H. (2018). *Religions of Tibet* (Vol. 4). Routledge. doi:10.4324/9780429441486

Holmberg, C. B. (1993). Spiritual pilgrimages: Traditional and hyperreal motivations for travel and tourism. *Visions in Leisure and Business*, *12*(2), 4.

Hop. (2017). *5 things you need to know about Smart Destinations*. Retrieved from https://hopu.eu/5-things-you-need-to-know-about-smart-destinations/

Hopkin, J. P., & Paolucci, C. (1999). The business firm model of party organisation: Cases from Spain and Italy. *European Journal of Political Research*, *35*(3), 307–339. doi:10.1111/1475-6765.00451

Hosany, S., Ekinci, Y., & Uysal, M. (2006). Destination image and destination personality: An application of branding theories to tourism places. *Journal of Business Research*, *59*(5), 638–642. doi:10.1016/j.jbusres.2006.01.001

Hoseinipor, S. E., Dolatabadi, H. R., & Kazemi, A. (2019). Effect of decision-making style on travel intention of religious tourism (Qom as religious tourism destination in Iran). *International Journal of Applied Management Science*, *11*(2), 113–123. doi:10.1504/IJAMS.2019.098825

Hou, J., Lin, C., & Morais, D. (2005). Antecedents of attachment to a cultural tourism destination: The case of Hakka and non-Hakka Taiwanese visitors to Pei-Pu, Taiwan. *Journal of Travel Research*, *44*(2), 221–233. doi:10.1177/0047287505278988

Hu, X., & Huang, R. (2019). The impact of development of religious tourism on the host community - a case study of Mount E'mei in China. *Managing Religious Tourism*, 144–154. doi:10.1079/9781786393197.0144

Huang, C. D., Goo, J., Nam, K., & Yoo, C. W. (2017). Smart tourism technologies in travel planning: The role of exploration and exploitation. *Information & Management*, *54*(6), 757–770. Advance online publication. doi:10.1016/j.im.2016.11.010

Huang, J., & Chu, J. (2019). Tourist Experience at Religious Sites: A Case Study of the Chinese Visiting the Sheikh Zayed Grand Mosque. *Journal of China Tourism Research*, *00*(00), 1–25. doi:10.1080/19388160.2019.1658678

Huang, K., & Pearce, P. (2019). Visitors' perceptions of religious tourism destinations. *Journal of Destination Marketing & Management*, *14*, 100371. doi:10.1016/j.jdmm.2019.100371

Huang, K., Pearce, P., Guo, Q., & Shen, S. (2019). Visitors' spiritual values and relevant influencing factors in religious tourism destinations. *International Journal of Tourism Research*, (October), 1–11. doi:10.1002/jtr.2337

Hudson, S. (2008). *Tourism and hospitality marketing: a global perspective*. Sage. doi:10.4135/9781446280140

Hughes, K., Bond, N., & Ballantyne, R. (2013). Designing and managing interpretive experiences at religious sites: Visitors' perceptions of Canterbury Cathedral. *Tourism Management*, *36*, 210–220. Advance online publication. doi:10.1016/j.tourman.2012.11.022

Hung, K. (2015). Experiencing Buddhism in Chinese Hotels: Toward the Construction of a Religious Lodging Experience. *Journal of Travel & Tourism Marketing*, *32*(8), 1081–1098. doi:10.1080/10548408.2014.959632

Hung, K., Wang, S., & Tang, C. (2015). Understanding the normative expectations of customers toward Buddhism-themed hotels: A revisit of service quality. *International Journal of Contemporary Hospitality Management*, *27*(7), 1409–1441. doi:10.1108/IJCHM-12-2012-0264

Hung, K., Yang, X., Wassler, P., Wang, D., Lin, P., & Liu, Z. (2017). Contesting the commercialization and sanctity of religious tourism in the Shaolin Monastery, China. *International Journal of Tourism Research*, *19*(2), 145–159. doi:10.1002/jtr.2093

Huntington, S. P. (1993). The Clash of Civilizations? *Foreign Affairs Magazine*. https://www.foreignaffairs.com/articles/united-states/1993-06-01/clash-civilizations

Huntington, S. P. (1996). *The Clash of Civilizations and the Remaking of World Order*. New York, NY: Simon and Schuster.

Hwang, S., Lee, C., & Chen, H. (2005). The relationship among tourists' involvement, place attachment and interpretation satisfaction in Taiwan's national parks. *Tourism Management*, *26*(2), 143–156. doi:10.1016/j.tourman.2003.11.006

Hyde, K. F., & Harman, S. (2011). Motives for a secular pilgrimage to the Gallipoli battlefields. *Tourism Management*, *32*(6), 1343–1351. doi:10.1016/j.tourman.2011.01.008

Idrees, I. A., Vasconcelos, A. C., & Ellis, D. (2018). Clique and elite: Inter-organizational knowledge sharing across five-star hotels in the Saudi Arabian religious tourism and hospitality industry. *Journal of Knowledge Management*, *22*(6), 1358–1378. doi:10.1108/JKM-07-2017-0276

Idris, J. (2019). Spiritual motivation for religious tourism destinations. *Spiritual and Religious Tourism: Motivations and Management*, 59.

Ingle, M. (2013). *Counter urbanisation and the emergence of a post-productivist economy in South Africa's arid Karoo region, 1994-2010*. Source Unknown.

Inside Arabia. (n.d.). https://insidearabia.com/muslim-hajj-religious-ritual-lucrative-business/

International Crisis Group. (2012). *Europe Report: N°220. The North Caucasus: The Challenges of Integration (I), Ethnicity and Conflict.* Author.

International Tourism Highlights 2019 Edition. (n.d.). https://www.e-unwto.org/doi/pdf/10.18111/9789284419470

Investor Guide to the Chechen Republic. (2016). PricewaterhouseCoopers.

Iqbal, J., Khan, Y., Haq, Z., & Hesseln, H. (2017). Estimation of Economic Value of an Archaeological Site: A Case Study of Takht-i Bahi. *Ancient Pakistan, 28*.

iQvizyon. (2020). *Internet of Things*. Retrieved from http://iqvizyon.com/?page_id=223

Iravani, M. R., & Mozaffari, A. M. (2013). Review The Status of Tourism from the Perspective of The Qur'an. *Int. J. Basic Appl. Sci*, *1*, 628–633.

Irimias, A., & Mitev, A., & Michalko, G. (2016). Demographic Characteristics Influencing Religious Tourism Behaviour: Evidence form a Central-Eastern-European country. *International Journal of Religious Tourism and Pilgrimage*, *4*, 19–32.

Isim Bulutu. (2020). *"Sapur" item*. Accessed from https://www.isimbulutu.com/%C5%9Fapur-isminin-anlami-nedir at 14/03/2020

Israeli Letter of July 10, 1. t.-G. (1967, July). Brief History of Israel and the Jewish People. *Israel Science and Technology Directory, 6*, 846-850.

Ivelashvili, T. (2018). *Religious Minorities in Samtskhe-Javakheti*. Akhaltsikhe: Publishing House University Akhaltsikhe.

Jackowski, A. (2000). Religious tourism: problems with terminology. *Peregrinus Cracoviensis*, (10).

Jackowski, A. (2000). Religious Tourism: Problems with terminology. *Peregrimus Craaviensis*, *1*, 64–74.

Jackowski, A., & Smith, V. A. (1992). Polish pilgrim-tourists. *Annals of Tourism Research*, *19*(1), 92–106. doi:10.1016/0160-7383(92)90109-3

Jackson, R. H., & Hudman, L. (1995). Pilgrimage tourism and English cathedrals: The role of religion in travel. *Tourism Review*.

Jafari, J. (1987). Tourism models: The sociocultural aspects. *Tourism Management*, *8*(2), 151–159. doi:10.1016/0261-5177(87)90023-9

Jafari, J., & Aaser, D. (1988). Tourism as the subject of doctoral dissertations. *Annals of Tourism Research*, *15*(3), 407–429. doi:10.1016/0160-7383(88)90030-8

Jafari, J., & Scott, N. (2014). Muslim world and its tourisms. *Annals of Tourism Research, 44*, 1–19. Advance online publication. doi:10.1016/j.annals.2013.08.011

Jafar, S., & Badaruddin, M. (2011, July). Factors Impact on Religious Tourism Market: The Case of Palestinian Territories. *International Journal of Business and Management, 6*, 254–260.

Jamal, T. (2006). Collaborative Networks and Partnerships for Integrated Destination Management. Tourism Management Dynamics. doi:10.1016/b978-0-7506-6378-6.50028-7

Jauhari, V., & Sanjeev, G. M. (2010). Managing Customer Experience for Spiritual and Cultural Tourism: An Overveiw. *Worldwide Hospitality and Tourism Themes, 2*(5), 467–476. doi:10.1108/17554211011090094

Jawabreh, O. A. (2017). Distinction of Jordan as a destination for religious tourism. *Journal of Environmental Management and Tourism, 8*(22), 1171–1182.

Jelacic, I. (2017). *Knowledge gap identification as a driver of learning and knowledge reuse in product development.* Academic Press.

Jennings, G. R. (2005). Interviewing: A focus on questionnaire techniques. In B. W. Ritchie, P. Burns, & C. Palmer (Eds.), *Tourism research methods* (pp. 99–118). Wallingford: CABI.

Jeong, M., & Shin, H. H. (2019). Tourists' Experiences with Smart Tourism Technology at Smart Destinations and Their Behavior Intentions. *Journal of Travel Research.* Advance online publication. doi:10.1177/0047287519883034

Jha, M. (1978). *Aspects of Great Traditional City in Nepal.* Kishore Vidya Niketan.

Johnston, R. (2018). Summary of Terrorist Attacks in Israel. *Johnston's Archive.* Retrieved from http://www.johnstonsarchive.net/terrorism/terrisraelsum.html

Jones, S. (1968). The Status of Jerusalem: Some National and International Aspects. *Law and Contemporary Problems, 33*(1), 177. doi:10.2307/1190849

JPost. (n.d.). https://www.jpost.com/Israel-News/Christmas-in-Israel-16500-tourists-expected-10-percent-increase-from-2018-611955

Jude, O. C., Uchenna, O., & Ngozi, E. (2018). Impact of Religious Tourism in Host Communities:The Case of Awhum Monastery. *American Journal of Social Sciences, 6*(3), 39–47.

Jugmohan, S., & Giampiccoli, A. (2015). Cruise tourism and community-based tourism: Towards a possible relationship. *African Journal of Hospitality, Tourism and Leisure, 4*(2), 1–10.

Kamadan, Y. S. (2018). *Gazali's Tomb Awaits Interest.* Accessed from https://www.gzt.com/mecra/gazzlnin-mezari-ilgi-beklenen-3424638 at 16/03/2020

Kamboh, A. Z., Kamal, Y., Nisar, Q. A., & Kamboh, M. A. (2018). Kartarpur Corridor: A New Role for Political Marketing. *Journal of Indian Studies, 4*(2), 233–238.

Kaplan, A. M. V., & Haenlein, M. (2010). Users of the world, unite! The challenges and opportunities of Social Media. *Business Horizons, 53*(1), 59–68. doi:10.1016/j.bushor.2009.09.003

Karar, A. (n.d.). *Impact of Pilgrimage in Tourism at Haridwa. Anthropological survey of India.* www.krepublishers.coml

Karlan, D., & Appel, J. (2011). *More Than Good Intentions: Improving the Ways the World's Poor Borrow, Save, Farm, Learn, and Stay Healthy.* Penguin.

Karliga, B. (1997). *Gazali (Works). Religious Encyclopaedia of Islam.* Turkey Religious Foundation Publications.

Kartal, B., Tepeci, M., & Atli, H. (2015). Examining the religious tourism potential of Manisa, Turkey with marketing perspective. *Tourism Review, 70*(3), 214–231. doi:10.1108/TR-09-2013-0048

Kastarlak, B. I., & Barber, B. (2012). *Fundamentals of Planning and Developing Tourism*. Pearson.

Katz, R. (2009). The Cartel Party Thesis: A Restatement. *Perspectives on Politics*, 753-766.

Katz, R. M., & Mair, P. (1995). Changing models of party organization and party democracy. *Party Politics, 1*(1), 5–28. doi:10.1177/1354068895001001001

Kawsar, M. D., & Raj, R. (2017). *Customer Relationship Marketing Tool for Growth of Religious Tourism Business: An Islamic perspective*. Academic Press.

Kayar Gizem, Ç. C. (2019). How to attract young generation to museums using the idea of gamification and an augmented reality based mobile application. In *Global Conference of Business and Economics* (pp. 133–135). Istanbul: Academic Press.

Kelly, G., Kelly, G. A., & Kelly, G. A. (1963). A theory of personality: The psychology of personal constructs (No. 152). WW Norton & Company.

Kessler, K., & Raj, R. (2018). 2 The Rise of Heritage and Religious Tourism to Sacred Sites in Oman. *Religious Tourism in Asia: Tradition and Change Through Case Studies and Narratives*, 10.

Keytours. (n.d.). https://blog.keytours.com/top-religious-pilgrimages-and-christian-tours-in-europe-and-mediterranean

Khan, A. U., Zaman, G., & Baloch, Q. B. (2019). Promoting Religious Tourism in Pakistan: Empirical Evidence from Sikh Pilgrims Visiting Pakistan. *City University Research Journal, 9*(4).

Khanbabaev, K. (2007). *Islamic Radicalism in the North Caucasus. Ideology, Goals, Ways of Financing*. Journal Free Thought.

Khan, M. A., Arif, S., Butt, A., Pervaiz, A., & Arif, M. (2014). Excavation at Badalpur Monastery, District Haripur (Khyber Pakhtunkhwa), Pakistan: A Preliminary Report of Season 2014. *Journal of Asian Civilizations, 37*(1), 85.

Kiknadze, Z. (2008). *Religions in Georgia*. Georgian Public Defender's Library.

Kim, H., Yilmaz, S., & Ahn, S. (2019). Motivational landscape and evolving identity of a route-based religious tourism space: A case of Camino de Santiago. Sustainability (Switzerland), 11(13).

Kim, B., & Kim, S. (2019). Hierarchical value map of religious tourists visiting the Vatican City/Rome. *Tourism Geographies, 21*(3), 529–550. doi:10.1080/14616688.2018.1449237

Kim, B., Kim, S. S., & King, B. (2016). The sacred and the profane: Identifying pilgrim traveler value orientations using means-end theory. *Tourism Management, 56*, 142–155. doi:10.1016/j.tourman.2016.04.003

Kim, B., Kim, S., & King, B. (2019). Religious tourism studies: Evolution, progress, and future prospects. *Tourism Recreation Research, 0*(0), 1–19. doi:10.1080/02508281.2019.1664084

Kim, B., Kim, S., & King, B. (2019). Religious tourism studies: Evolution, progress, and prospects. *Tourism Recreation Research*, 1–19.

King James version of Bible Online. (n.d.). https://www.kingjamesbibleonline.org/

Kings, S. (2017a). *Frogs vs angels: Prophet Bushiri's angel landing pad plan declared illegal*. Retrieved from: https://mg.co.za/article/

Kirillova, K., Gilmetdinova, A., & Lehto, X. (2014). Interpretation of hospitality across religions. *International Journal of Hospitality Management*, *43*, 23–34. doi:10.1016/j.ijhm.2014.07.008

Knausgaard, K. O. (2015). The Inexplicable, Inside the mind of a mass killer. *The New Yorker*. https://www.newyorker.com/magazine/2015/05/25/the-inexplicable

Kolbi, O. (2012). *Developing a high value added tourism in. University of Lausanne, Switzerland*. University of Lausanne.

Koldowski, J. M., & Martin, O. (2008). Emerging Market Segments: Religious and Medical Tourism in India. In C. Roland & B. Martin (Eds.), *Trends and Issues in Global Tourism*. Springer.

Koren-Lawrence, N., & Collins-Kreiner, N. (2019). Visitors with their 'Backs to the archaeology': Religious tourism and archaeology. *Journal of Heritage Tourism*, *14*(2), 138–149. doi:10.1080/1743873X.2018.1503279

Korstanje, M. E., & George, B. P. (2018). 7 Religious Tourism in the Ideological Framework of Chinese Tourism Education. *Religious Tourism in Asia: Tradition and Change Through Case Studies and Narratives*, 58.

Korstanje, M., Raj, R., & Griffin, K. (2018). *Risk and Safety Challenges for Religious Tourism and Events*. CAB International. doi:10.1079/9781786392282.0000

Kotler, P. (2002). *Marketing Management*. Pearson Custom Publishing.

Kotler, P., & Gertner, D. (2002). Country as brand, product, and beyond: A place marketing and brand management perspective. *Journal of Brand Management*, *9*(4), 249–261. doi:10.1057/palgrave.bm.2540076

Kozak, M. (2001). Repeaters' behavior at two distinct destinations. *Annals of Tourism Research*, *28*(3), 784–807. doi:10.1016/S0160-7383(00)00078-5

Kozak, M. (2003). Measuring Tourist Satisfaction with Multiple Destination Attributes. *Tourism Analysis*, *7*(3/4), 229–240. doi:10.3727/108354203108750076

Kozak, M., & Rimmington, M. (2000). Tourist satisfaction with Mallorca, Spain, as an off-season holiday destination. *Journal of Travel Research*, *38*(3), 260–269. doi:10.1177/004728750003800308

Krešić, D., Mikulić, J., & Miličević, K. (2013). The factor structure of tourist satisfaction at pilgrimage destinations: The case of Medjugorje. *International Journal of Tourism Research*, *15*(5), 484–494. doi:10.1002/jtr.1891

Kroupenev, A. (2007). *Application of Poliheuristic Methodology to a Sequence of Boris Yeltsin's Decisions During the 1995 Hostage Crisis in Budyonnovsk*. Retrieved 04 02, 2020, from https://papers.ssrn.com/sol3/papers.cfm?abstract_id=1027054

Kruger, M., & Saayman, M. (2016). Understanding the Zion Christian Church (ZCC) Pilgrims. *International Journal of Tourism Research*, *18*(1), 27–38. doi:10.1002/jtr.2030

Kucukasci, M. S. (2003). Mecca: The Ottoman Period. In *Religious Foundation of Turkey [TDV] Encyclopedia of Islam*. Accessed from https://islamansiklopedisi.org.tr/mekke

Kumar, R. B., & Singh, L. (2015). Expectation & Satisfaction of Hindu Pilgrims in North-West India- A Case Study of Naina Devi Shrine. *African Journal of Hospitality, Tourism and Leisure*, *4*(2), 1–18.

Kurmanaliyeva, R. S., Duissenbayeva, A., & Izmailov, I. (2014). Religious Tourism as a Sociocultural Phenomenon of the Present "The Unique Sense Today is a Universal Value Tomorrow. This is the Way Religions are Created and Values are Made.". *Procedia: Social and Behavioral Sciences*, *143*, 958–963. doi:10.1016/j.sbspro.2014.07.534

Kurtulus, R. (1993). Curcan. In *Religious Foundation of Turkey [TDV] Encyclopedia of Islam*. Accessed from https://islamansiklopedisi.org.tr/curcan at 17/03/2020

Kurtulus, R. (2012). Tus. In *Religious Foundation of Turkey [TDV] Encyclopedia of Islam*. Accessed from https://islamansiklopedisi.org.tr/Tûs

Lamsfus, C., & Alzua-Sorzabal, A. (2013). Theoretical Framework for a Tourism Internet of Things: Smart Destinations. *TourGUNE Journal of Tourism and Human Mobility*.

Lane, J.-E. M. (1996). *Political Data Handbook: OECD Countries (Comparative European Politics)*. Oxford University Press. doi:10.1093/019828053X.001.0001

Lapidoth, R. (2011). *Some Legal Issues*. The Jerusalem Institute for Israel Studies.

Larsen, S. (2007). Aspects of a psychology of the tourist experience. *Journal of Hospitality and Tourism*, *7*(1), 7–18. doi:10.1080/15022250701226014

Laviniaţală, M., & Pădurean, A. M. (2008). *Dimensions of religious tourism*. Academic Press.

Law, R., Buhalis, D., & Cobanoglu, C. (2014). Progress on information and communication technologies in hospitality and tourism. *International Journal of Contemporary Hospitality Management*, *26*(5), 727–750. Advance online publication. doi:10.1108/IJCHM-08-2013-0367

Ledford, J. R., & Gast, D. L. (2018). *Single case research methodology: Applications in special education and behavioral sciences*. Routledge. doi:10.4324/9781315150666

Lee, T. H., Fu, C. J., & Chang, P. S. (2015). The support of attendees for tourism development: Evidence from religious festivals, Taiwan. *Tourism Geographies*, *17*(2), 223–243. doi:10.1080/14616688.2014.997280

Leggett, T. (2017). Survey development: Creating intended consequences. *Radiologic Technology*, *88*(5), 568–571. PMID:28500097

Lekesiz, O. (2016). *Al-Halil: The Story of the Captive City*. Accessed from http://www.gercekhayat.com.tr/yazarlar/el-halil-esir-sehrin-hikayesi/

Leppakari, M., & Griffin, K. A. (Eds.). (2016). *Pilgrimage and Tourism to Holy Cities: Ideological and Management Perspectives*. Cabi.

Leung, D., Law, R., Van, H. H., & Buhalis, D. (2013). Social *Media in Tourism and Hospitality: A Literature Review. Journal of Travel & Tourism Marketing*, *30*(1-2), 3–22. doi:10.1080/10548408.2013.750919

Lewis, B. (1990). The Roots of Muslim Rage. *Policy*, *17*(4). https://www.cis.org.au/app/uploads/2015/04/images/stories/policy-magazine/2001-summer/2001-17-4-bernard-lewis.pdf

Liana, S. (2017). *The role of social media in the tourism industry in Austria, How does Austria Implemented the Social Media to Attract Tourists?* (Bachelor Thesis). Module University Vienna.

Li, Chu, Zhao, & Wang. (2015). Tibetian tourism hotspots:co-word cluster analysis of English blogs. *Luyou Xuekan*, *30*, 35–43.

Lin, C. F., & Fu, C. S. (2020). Cognitive implications of experiencing religious tourism: An integrated approach of means–end chain and social network theories. *International Journal of Tourism Research*, *22*(1), 71–80. doi:10.1002/jtr.2319

Linderman, M. C. (2013). Royal accommodations: Pilgrim rest houses in early colonial South India. *International Journal of Contemporary Hospitality Management*, *25*(2), 187–203. doi:10.1108/09596111311301595

Lin, H., Zhang, M., Gursoy, D., & Fu, X. (2019). Impact of tourist-to-tourist interaction on tourism experience: The mediating role of cohesion and intimacy. *Annals of Tourism Research*, *76*(March), 153–167. doi:10.1016/j.annals.2019.03.009

Liszewski, S. (2000). Pilgrimages or religious tourism. *Peregrinus Cracoviensis, 10*, 47–51.

Liu, A., & Pratt, S. (2017). Tourism's vulnerability and resilience to terrorism. *Tourism Management, 60*, 404–417. doi:10.1016/j.tourman.2017.01.001

Li, Y., Hu, C., Huang, C., & Duan, L. (2017). The concept of smart tourism in the context of tourism information services. *Tourism Management, 58*, 293–300. Advance online publication. doi:10.1016/j.tourman.2016.03.014

Lloyd, D. W., & Tourism, B. (1998). *Pilgrimage and the Commemoration of the Great War in Britain.*

Lochrie, S., Baxter, I. W., Collinson, E., Curran, R., Gannon, M. J., Taheri, B., Thompson, J., & Yalinay, O. (2019). Self-expression and play: Can religious tourism be hedonistic? *Tourism Recreation Research, 44*(1), 2–16. doi:10.1080/02508281.2018.1545825

Lodhi, S. (2017). Terrorism and Its Impact on Economy of Pakistan. *RADS Journal of Social Sciencess & Business Management, 4*(1), 36–50.

Lois-González, R. C., & Santos, X. M. (2015). Tourists and pilgrims on their way to Santiago. Motives, Caminos and Final Destinations. *Journal of Tourism and Cultural Change, 13*(2), 149–164. doi:10.1080/14766825.2014.918985

Louro, A., Padre, S.S.P., Oliveira, R., Frei, O.P., Filipe, J., Ribeiro, A. C. (2015). *A Bíblia Sagrada* (9ª. Edição). Paulus Editora.

Lücking, M. (2014, April). Making'Arab'One's Own: Muslim Pilgrimage Experiences in Central Java, Indonesia. In *Internationales Asien Forum. International Quarterly for Asian Studies* (Vol. 45, No. 1/2, p. 129). Arnold Bergsträsser Institut.

Lücking, M. (2019). Travelling with the Idea of Taking Sides: Indonesian Pilgrimages to Jerusalem. *Bijdragen tot de taal-, land-en volkenkunde/Journal of the Humanities and Social Sciences of Southeast Asia, 175*(2-3), 196-224.

Luz, N. (2020). Pilgrimage and religious tourism in Islam. *Annals of Tourism Research, 82*(April), 102915. doi:10.1016/j.annals.2020.102915

MacCannell, D. (1973). Staged Authenticity: Arrangements of Social Space in Tourist Settings. *American Journal of Sociology, 79*(3), 589–603. Advance online publication. doi:10.1086/225585

MacCannell, D. (1976). *The Tourist: A New Theory of the Leisure Class.* Schocken Books.

MacCannell, D. (1992). *Empty Meeting Grounds.* Routledge.

Macedo, L. (2011). La dimensión territorial del turismo religioso. In Turismo espiritual: Una visión iberoamericana. Centro Universitario de los Altos de la Universidad de Guadalajara.

Mackey, J., & Sisodia, R. (2013). Conscious Capitalism: Liberating the Heroic Spirit of Business. Harvard Business Press.

MacLennan, J., & Moore, R. L. (2011). Conflicts Between Recreation Subworlds: The Case of Appalachian Trail Long-Distance Hikers. *The Cyber Journal of Applied Leisure and Recreation Research, 13*(1), 1–17.

Magazine, S. (2014). *Vedic Dharma and the Principles of Economics.* https://www.sanskritimagazine.com/indian-religions/hinduism/vedic-dharma-principle-economics/

Makashenko, A., & Trenin, D. (2002). *The Time of the South: Russia in Chechnya, Chechnya in Russia.* Carnegie Endowment for International Peace.

Make My Trip. (n.d.). https://www.makemytrip.com/blog/top-10-indian-pilgrimages

Malashenko, A., & Filatova, S. (2009). *Twenty years of religious freedom in Russia*. Book Monograph Moscow Center Kornegi.

Malinowski, B. (2018). *The family among the Australian Aborigines: a sociological study*. BoD–Books on Demand.

Malodia, S., & Singla, H. (2017). Using HOLSAT to evaluate satisfaction of religious tourist at sacred destinations: The case of religious travelers visiting sacred destinations in the Himalayas, India. *International Journal of Culture, Tourism and Hospitality Research, 11*(2), 255–270. doi:10.1108/IJCTHR-10-2015-0120

Manzoor, F., & Wei, L. (2018). *A Review of China-Pakistan Economic Corridor and Tourism Opportunities*. Paper presented at the Third International Conference on Economic and Business Management (FEBM 2018). 10.2991/febm-18.2018.18

Maoz, D., & Bekerman, Z. (2010). Searching for Jewish answers in Indian resorts: The postmodern traveler. *Annals of Tourism Research, 37*(2), 423–439. doi:10.1016/j.annals.2009.10.015

Margetts, H. (2006). Cyber parties. In Handbook of party politics (pp. 528-535). London: Sage.

Maria-Irina, A. (2017). Tourism industry in the new Europe: trends, policies and challenges. *Proceedings of the International Conference on Business Excellence*, 493-503.

Mariani, M. M., Felice, M., & Mura, M. (2016). Facebook as a destination marketing tool: Evidence from Italian regional Destination Management Organizations. *Tourism Management, 54*(1), 321–343. doi:10.1016/j.tourman.2015.12.008

Markedonov, S. (2010). *Radical Islam in the North Caucasus: Evolving Threats, Challenges, and Prospects*. A Report of CSIS and Eurasia Program.

Martin, D. (2011). *The Future of Christianity: Reflections on Violence and Democracy, Religion and Secularization*. Ashgate.

Martins-Rodrigues, M. C. (2018). *Empresas incubadoras em Portugal e o seu efeito na sustentabilidade das incubadas*. Chiado Books.

Marx, K. (1894). [1998]: Capital: A Critique of Political Economy. In *Marx and Engels Collected Works* (Vol. 3, pp. 25–60). Progress Publishers.

Masalacı, M. (2014). *Luminous Horasan*. Accessed from https://www.hurriyet.com.tr/isik-sacan-horasanli-26399524 at 16/03/2020

Maslow, A. H. (1954). *Motivation And Personality*. Harper & Row.

Massetti, E. (2009). Explaining regionalist party positioning in a multi-dimensional ideological space: A framework for analysis. *Regional and Federal Studies,* 501-531.

Mathew, P. V., & Kuriakose, V. (2017). A Study on the Indicator Framework of Responsible Tourism in Tourism Accommodation Sector. *Asian Journal of Management, 8*(4), 989–996. doi:10.5958/2321-5763.2017.00153.6

Matsaberidze, M. (2007). *Inauguration Speeches of Georgian President*. Akhali Azri.

Mazumdar, S., & Mazumdar, S. (1993). Sacred space and place attachment. *Journal of Environmental Psychology, 13*(3), 231–242. doi:10.1016/S0272-4944(05)80175-6

Mazumdar, S., & Mazumdar, S. (2004). Religion and place attachment: A study of sacred places. *Journal of Environmental Psychology, 24*(3), 385–397. doi:10.1016/j.jenvp.2004.08.005

MediaClubSouthAfrica. (2016). *Sacred lake to heritage site*. Retrieved from: https://www.mediaclubsouthafrica.com/culture/

Mehdi, S. (2011). Communication, Culture, and the Essentialized Islam. *Communication Studies*, *62*(1), 113–126. doi:10.1080/10510974.2011.540975

Melanie, S., & Nicola, M. R. M. H. (2010). Key Concepts in Tourist Studies. London: SAGE.

Melzer, G. (2014). A Palaeographic Study of a Buddhist Manuscript from the Gilgit Region. *Manuscript Cultures: Mapping the Field*, *1*, 227.

Meparidze, G., & Chachanidze, N. (2019). The Share of Religious Minorities in Georgian History. Tbilisi: Published by CIPDD and OSGF.

Mertens, D. M. (2005). *Research Methods in education and psychology: Interpreting diversity with quantitative and qualitative approaches* (2nd ed.). Sage.

MFA. (n.d.). https://mfa.gov.il/mfa/aboutisrael/state/pages/menachem%20begin.aspx

Michael, N., James, R., & Michael, I. (2018). Australia's cognitive, affective and conative destination image: An Emirati tourist perspective. *Journal of Islamic Marketing*, *9*(1), 36–59. doi:10.1108/JIMA-06-2016-0056

Miguéns, J., Baggio, R., & Costa, C. (2008). Social media and Tourism Destinations: TripAdvisor Case Study. *Advances in Tourism Research*, *26*(28), 26–28.

Mikos von Rohrscheidt, A. (2013). Religion as a subject of cognitive interest to tourists. Analysis of the phenomenon and products on offer for cultural-religious tourism. *Turystyka Kulturowa*, (5), 5-19.

Millán, C. L., Garzon, D., & Navarro, S. (2016). C2C interactions creating value in the route of Santiago. *Journal of Business Research*, *69*(11), 5448–5455. doi:10.1016/j.jbusres.2016.04.153

Milliyet. (2018). *The Holy City of Faiths Jerusalem.* Accessed from https://www.milliyet.com.tr/tatil/inanclarin-kutsal-kenti-kudus-2668670 at 19/03/2020

Mill, R. C., & Morrison, A. M. (1985). *The Tourism System: An Introduction Text.* Prentice-Hall.

Ministry of Environment, Climate Change, Tourism and Hospitality Industry. (2018). http://www.environment.gov.zw

Ministry of Finance and Economic Development. (2019). *Incentives to Boost Tourism.* https://www.zimtreasury.gov.zw

Mkono, M. (2012). Zimbabwe's Tourism Woes in the Last Decade: Hindsight Lessons for African Tourism Planners and Managers. *Tourism Planning & Development*, *9*(2), 205–210. doi:10.1080/21568316.2011.630749

Mohn, P. (1950, october). Jerusalem and the United Nations. *International Conciliation Pamphlet.*

Molinillo, S., Ximénez-de-Sandoval, J. L., Fernández-Morales, A., & Coca-Stefaniak, A. (2016). Hotel assessment through social media: The case of TripAdvisor. *Tourism & Management Studies*, *12*(1), 15–24. doi:10.18089/tms.2016.12102

Moore, C., & Tumelty, P. (2008). Foreign Fighters and the Case of Chechnya: A Critical Assessment. *Studies in Conflict and Terrorism*, *31*(5), 412–433. doi:10.1080/10576100801993347

Most Affordable. (n.d.). *Damascus City Guide*. Accessed from https://www.enuygun.com/ucak-bileti/sam-sehir-rehberi/at

Moyo, D. (2008). Dead Aid: Destroying the Biggest Global Myth of Our Time. London: Allen Lane.

Muhammad, F., Karim, R., Qureshi, J. A., Razzaq, N., Zahra, M., & Ali, I. (2019). Environmental Degradation, Quality of Institutions and Tourism: New Evidence from Pakistan. *International Journal of Economic and Environmental Geology*, *10*(3), 125–128.

Mukherjee, S., Bhattacharjee, S., & Singha, S. (2020). *Religious to Spiritual Tourism-An Era of Paradigm Shift in India.* Available at SSRN 3546903

Muller, R. (2015). The Zion Christian church and Global Christianity: Negotiating a tightrope between localisation and globalisation. *Religion, 45*(2), 174–190. doi:10.1080/0048721X.2014.992111

Muriuki, L. M., Bururia, D., & Mutegi, J. (2018). Determinants of religious tourism branding tactics in Kenya: A cross sectional study. *Tourism and Hospitality Research, 18*(4), 456–469. doi:10.1177/1467358416670937

Murphy, H. C., Chen, M. M., & Cossutta, M. (2016). An investigation of multiple devices and information sources used in the hotel booking process. *Tourism Management, 52*, 44–51. doi:10.1016/j.tourman.2015.06.004

Murray, M. (2014). The cultural Heritage of Pilgrim Itineraries: The Camino de Santiago. *Journeys, 15*(2), 65–85. doi:10.3167/jys.2014.150204

Murray, M., & Graham, B. (1997). Exploring the dialetics of route-based tourism: The caminho de Santiago. *Tourism Management, 18*(8), 513–524. doi:10.1016/S0261-5177(97)00075-7

Mushava, S. (2015). *Zimbabwe: Mega-churches a boon for faith tourism.* Retrieved from: https://allafrica.com/stories/

Muslim, A., Harun, A., Ismael, D., & Othman, B. (2020). Social media experience, attitude and behavioral intention towards umrah package among generation X and Y. *Management Science Letters, 10*(1), 1–12. doi:10.5267/j.msl.2019.8.020

Mustafa, M. H. (2010). Tourism and globalization in the Arab world. *International Journal of Business and Social Science, 1*(1).

Narayan, D. (2001). Voices of the Poor. In D. Belshaw, R. Calderisi, & C. Sugden (Eds.), *Faith in Development: Partnership between the World Bank and the Churches of Africa* (p. 39). Regnum.

Narbona, J., & Arasa, D. (2016). The Role and Usage of Apps and Instant Messaging in Religious Mass Events The Role and Usage of Apps and Instant Messaging in Religious Mass Events. *International Journal of Religious Tourism and Pilgrimage.* doi:10.21427/D7JT6Q

Narcisse, E. (2012). *A playstation 3 game will be part of a church service this sunday.* Retrieved from https://kotaku.com/5909361/this-sunday-video-games-go-to-church

NationsP. O. (2020). Retrieved from https://palestineun.org/

Natsvlishvili, N. (2019). *Catholic Churches in Georgia: History and Architecture.* Publishing House Ilia State University.

Navruz-Zoda, B., & Navruz-Zoda, Z. (2016). The Destination Marketing Development of Religious Tourism in Uzbekistan. *International Journal of Tourism and Pilgrimage, 4*(7), 9–20.

Nengovhela, M., Tshipala, N., & Nyikana, S. (2017). Domestic tourism in practice: An investigation of Vhembe District Municipality residents' travel motivations. *African Journal of Hospitality, Tourism and Leisure, 6*(4), 1–15.

Neuhofer, B., Buhalis, D., & Ladkin, A. (2012). Conceptualising technology enhanced destination experiences. *Journal of Destination Marketing & Management, 1*(1-2), 36–46. Advance online publication. doi:10.1016/j.jdmm.2012.08.001

Neuhofer, B., Buhalis, D., & Ladkin, A. (2014). A typology of technology-enhanced tourism experiences. *International Journal of Tourism Research, 16*(4), 340–350. doi:10.1002/jtr.1958

Neuhofer, B., Buhalis, D., & Ladkin, A. (2015). Smart technologies for personalized experiences: A case study in the hospitality domain. *Electronic Markets, 25*(3), 243–254. Advance online publication. doi:10.100712525-015-0182-1

New York Times. (1946, Dec. 24). Zionists Condemn Palestine Terror. *The New York Times.*

New York Times. (1947, Aug. 16). Irgun's Hand Seen in Alps Rail Blast. *The New York Times.*

News, P. M. (2015). *Nigeria's loss is South Africa's gain-TB Joshua birthday fallout.* Retrieved from: http://www.pmnews.co.ng/

News, T. G. T. I. (2014). *The importance of the religious tourism market.* Retrieved from https://www.eturbonews.com/50998/importance-religious-tourism-market

Ngobese, D. H. D. (2018). A cultural appraisal of sacred sites of the Eastern Free State: Adjunct to African Traditional Religious and Belief Systems. *China-USA Business Review*, *17*(9), 474–486.

Nickerson, R. C., & Eng, J. (2017). Use of mobile technology and smartphone apps on the Camino de Santiago: A Comparison of American and European Pilgrims. In Atas da Conferencia da Associacao Portuguesa de Sistemas de Informacao. doi:10.18803/capsi.v17.260-269

Nicolaides, A. (2016). Moria Tourism: Eastern Orthodox and Roman Catholic Pilgrimage. *African Journal of Hospitality, Tourism and Leisure*, *5*(4).

Niculina, A. (2019). The economic foundation of the development strategies for the cultural religious tourism. *Sustainable Development.*

Nieminen, K. (2012). *Religious tourism-a Finnish perspective.* Academic Press.

Nilsson, M., & Tesfahuney, M. (2016). Performing the "post-secular" in Santiago de Compostela. *Annals of Tourism Research*, *57*, 18–30. doi:10.1016/j.annals.2015.11.001

Ninikashvili, S. (2017). *A Great Synagogue of Tbilisi – Interesting Facts from History.* The Tbilisi Times.

Nobel Prize. (n.d.). https://www.nobelprize.org/prizes/peace/1978/begin/facts/

Nolan, M. L., & Nolan, S. (1992). Religious sites as tourism attractions in Europe. *Annals of Tourism Research*, *19*(1), 67–78. doi:10.1016/0160-7383(92)90107-Z

Norouzi, M. (2018). Feminism and the modern female prose in Iran (A critical look at Feminism in the writing of Shahrnush Parsipur). *Bulletin of the Moscow State Regional University (Russian philology*), 154-163.

Nourollah, Z. (2012). Self-Perceived Communication Competence in Iranian Culture. *Communication Research Reports*, *29*(4), 292–298. doi:10.1080/08824096.2012.723271

Noy, C. (2004). This trip has really changed me: Backpackers' narratives of self-change. *Annals of Tourism Research*, *31*, 78–102. doi:10.1016/j.annals.2003.08.004

NTV. (2017). *Fury in Kigezi as politician takes over Punishment Island aka Akampene.* Retrieved from: http://www.ntv.co.ug/news/local/18/jan/2018/fury-kigezi-politician-takes-over-punishment-island-aka-akampene-22199#sthash.Y0Tuqo4D.dpbs

Nunes, R. I. de S. (2018). *Turismo Holístico em Portugal: Caracterização da Oferta de Retiros.* Universodade de Lisboa, Faculdade de Letras.

Nursanty, E. (2012). *Halal Tourism The New Product In Islamic Leisure Tourism And Architecture.* Department of Architecture.

Nyahunzvi, D. K. (2015). Negotiating livelihoods among Chivi curio traders in a depressed Zimbabwe trading environment. *Anatolia- An International Journal of Tourism and Hospitality Research, 26*(3), 397-407. 10. 1080/13032917.2014.974065

Nyaupane, G. P., Timothy, D. J., & Poudel, S. (2015). Understanding tourists in religious destinations: A social distance perspective. *Tourism Management*, *48*(1), 343–353. doi:10.1016/j.tourman.2014.12.009

Nyikana, S. (2017). Religious tourism in South Africa: Preliminary analysis of a major festival in Limpopo. *African Journal of Hospitality, Tourism and Leisure*, *6*(1), 1–8.

Nyikana, S. (2017). Religious Tourism in South Africa: Preliminary analysis of a major festival in Limpopo. *African Journal of Hospitality, Tourism and Leisure*, *6*(1), 1–9.

Odum, C.J. (n.d.). *Religious tourism. The Nigerian perspective: A survey of religious tourism in Nigeria*. Academic Press.

Okpoko, P. U., Okonkwo, E. C., & Dimiyi, C. (2008). *Understanding Tourism*. Nsukka University of Nigeria Press.

Okrotsvaridze, A., Vadachkoria, M., & Okrotsvaridze, L. (2011). Tourism and Host Management. Academic Press.

Okumus, M. (2000). *Understanding of the Qur'an and the Interpretation Method of Gazali* (Unpublished Doctoral Dissertation). Ankara: Ankara University Institute of Social Sciences.

Oliveira, C. D. M. (2000). *Viagens a santuários: uma modalidade de turismo religioso ou de religiosidade turística?* Boletim Turístico de Administração Hoteleira, Out.

Oliveira, C. D. M. (2004). *Turismo Religioso*. Aleph.

Oliveira, C. D. M. (2018). Matergrafia e patrimônio: Santuários Marianos como espaço simbólico e vetorial da Latinidade. *Ateliê Geográfico*, *12*(3), 170–194. doi:10.5216/ag.v12i3.47188

Oliver, R. (1980). A cognitive model of the antecedents and consequences of satisfaction decisions. *JMR, Journal of Marketing Research*, *17*(4), 46–49. doi:10.1177/002224378001700405

Oliver, R. (1999). Whence Consumer Loyalty? *Journal of Marketing*, *63*(4_suppl1), 33–44. doi:10.1177/00222429990634s105

Oliver, R. L. (1981). Measurement and evaluation of satisfaction processes in retail settings. *Journal of Retailing*, 25–48.

Olsen, D. (2012). Negotiating Identify at Religious Sites: A Management Perspective. *Journal of Heritage Tourism*, *7*(4), 359–366. doi:10.1080/1743873X.2012.722642

Olsen, D. H., & Timothy, D. J. (1999). Tourism 2000: Selling the millennium. *Tourism Management*, *20*(4), 389–392.

Olsen, D. H., & Timothy, D. J. (2006). Tourism and religious journeys. In D. J. Timothy & D. H. Olsen (Eds.), *Tourism, religion and spiritual journeys* (pp. 1–21). Routledge.

Olsen, D. H., & Wilkinson, G. (2016). Are fast pilgrims true pilgrims? The Shikoku pilgrimage. *Annals of Tourism Research*, *61*, 228–230. doi:10.1016/j.annals.2016.08.004

Olya, H. G. T., Lee, C. K., Lee, Y. K., & Reisinger, Y. (2019). What are the triggers of Asian visitor satisfaction and loyalty in the Korean heritage site? *Journal of Retailing and Consumer Services*, *47*(November), 195–205. doi:10.1016/j.jretconser.2018.11.002

Oppermann, M. (2000). Tourism Destination Loyalty. *Journal of Travel Research*, *39*(1), 78–84. doi:10.1177/004728750003900110

Orange Smile. (2020). *Active Rest and Vacation in Damascus - Places of Interest, Entertainment and Nightlife*. Accessed from http://www.orangesmile.com/gezirehberi/sam/eglence--2541260.htm

Orman, S. (2013). *Ghazali: Biography, Truth Research, Philosophical Criticism*. Human Publications.

Orth, M. (2015). The world's most powerful woman. *National Geographic*, (December), 32–59.

Oslen, D. H. (2006). Management issues for religious heritage attractions. In D. J. Timothy & D. H. Oslen (Eds.), *Tourism, Religion and Spiritual Journey* (pp. 104–118). Routledge.

Ostroff, M. (2009, November 24). Neocons Now Claim West Bank and East Jerusalem was 'Terra Nullius' Giving Zionists Right of Occupation. *Jerusalem Post.*

Ozaydin, A. (1991). Baghdad. In *Religious Foundation of Turkey [TDV] Encyclopedia of Islam.* Accessed from https://islamansiklopedisi.org.tr/bagdat#1-genel-bakis

Ozcan, C. C., Bişkin, F., & Şimşek, Ç. (2018). *Regional Economic Effects and Marketing of Religious Tourism*, (August), 250–274. doi:10.4018/978-1-5225-5730-2.ch014

Ozgudenli, O. G. (2007). Nishabur. In *Religious Foundation of Turkey [TDV] Encyclopedia of Islam.* Accessed from https://islamansiklopedisi.org.tr/nisabur--iran at 16/03/2020

Paiano, A., Crovella, T., & Lagioia, G. (2020). Managing sustainable practices in cruise tourism: the assessment of carbon footprint and waste of water and beverage packaging. *Tourism Management, 77*(August), 104016.

Palestine, S. of. (2015). *Bethlehem declarationon religious tourism as a means of fostering socio-economic development of host communities.* Retrieved from https://www.bethlehem.edu/document.doc?id=3240

Palestinian Violence and TerrorimAgaint Israel, 2018: Data, Nature, and Trends. (2019, February 18). Retrieved from Terrorism-Info.org: https://www.terrorism-info.org.il/en/palestinian-violence-terrorism-israel-2018-data-nature-trends-2/

Papashvili, M. (1995). *Georgian- Rome Relations in VI-XX Centuries.* Publishing House Aghmashenebeli.

Papashvili, M. (1998). *Georgian-Roman Relations.* Publishing House Sazani.

Papathanassis, A. (2011). The Long Tail of Tourism: Holiday Niches, and their Impact in Mainstream Tourism. Gabler Verlang.

Parties and movements of political alternatives in modern Europe N°357. (2019). *Reports of Institute of Europe*, 87-112.

Pavlenishvili, E. (2019). *Catholicism in Georgia (XIX century)* (Dissertation). St. Andreas Georgian University.

Pearce, D. G. (2014). Toward an Integrative Conceptual Framework of Destinations. *Journal of Travel Research, 53*(2), 141–153. Advance online publication. doi:10.1177/0047287513491334

Peterson, K., Hahn, H., Lee, A. J., Madison, C. A., & Atri, A. (2016). In the Information Age, do dementia caregivers get the information they need? Semi-structured interviews to determine informal caregivers' education needs, barriers, and preferences. *BMC Geriatrics, 16*(1), 164. doi:10.118612877-016-0338-7 PMID:27662829

Pew Forum. (2018). https://www.pewforum.org/2018/06/13/how-religious-commitment-varies-by-country-among-people-of-all-ages/

Pforr, C., & Hosie, P. (2009). *Crisis management in the tourism industry: Beating the odds?* Ashgate Publishing, Ltd.

Pinto da Silva, F., Brandão, F., & Sousa, B. (2019). Towards socially sustainable tourism in cities: local community perceptions and development guidelines. *Enlightening Tourism, 9*(2), 168-198.

Pinto, S. O. A. (2011). Turismo Religioso – potencial de desenvolvimento da Vila de Arcozelo, Vila Nova de Gaia (Dissertação de Mestrado). Universidade Lusófona de Humanidades e Tecnologias, Lisboa, Portugal.

Piva, E., Cerutti, S., & Raj, R. (2019). Managing the Sacred: A Governance Perspective for Religious Tourism Destinations. *Managing religious tourism/Piva E. Stefania C., Razaq R*, 10-21.

Poddar, H., Goswami, C., & Shastri, M. A. (1951). *Kalyan: Tirthank*. Gita Press.

Pololikashvili, Z. (2020). *The video conference held between him and the prime minister of Georgia, Giorgi Gakharia*. Retrieved from interpressnews.ge: https://www.interpressnews.ge/ka/article/598753-zurab-pololikashvili-giorgi-gaxarias-sruli-mxardachera-agvutkvi-turizmis-industriashi-koronavirusis-gavlenis-ganeitralebisken-mimartuli-epekturi-reagirebis-koordinaciashi

Poplak, R. (2016). *Planes, blames and autocracies: How TB Joshua will get away with murder*. Retrieved from: http:www.dailymaverick.co.za/article/

Post, P., Pieper, J., & Van Uden, M. (1998). *The modern pilgrim: Multidisciplinary explorations of Christian pilgrimage* (Vol. 8). Peeters Publishers.

Potskhveria, G. (2020). *We need a positive example of Tolerance for the World, The coexistence of Georgian People, and Jewish is more than a Tolerance*. Retrieved from 1tv.ge: https://1tv.ge/news/salome-zurabishvili-chven-gvchirdeba-tolerantobis-pozitiuri-magalitebi-msofliostvis-qartvelebisa-da-ebraelebis-tanackhovreba-ki-tolerantobaze-metia/

Pourtaheri, M., Rahmani, K., & Ahmadi, H. (2012). Impacts of religious and pilgrimage tourism in rural areas: The case of Iran. *Journal of Geography and Geology*, *4*(3), 122. doi:10.5539/jgg.v4n3p122

Ppoescu, R.I., & Corboş, R. A. (2012). The role of Festivals and Cultural Events in the Strategic Development of Cities. Recommendations for Urban Areas in Romania. *Informatica Economică*, *16*(4).

Prayag, G., Chen, N. C., Hosany, S., & Odeh, K. (2016). *The Cognitive-Affective Relationship of Tourists to a Heritage Site: The Case of Petra*. Academic Press.

Prayag, G. (2008). Image, Satisfaction and Loyalty: The Case of Cape Town. *Anatolia: An International Journal of Tourism and Hospitality Research*, *19*(2), 205–224. doi:10.1080/13032917.2008.9687069

Prayag, G., & Ryan, C. (2012). Antecedents of tourists' loyalty to Mauritius: The role and influence of destination image, place attachment, personal involvement, and satisfaction. *Journal of Travel Research*, *5*(3), 342–356. doi:10.1177/0047287511410321

Prazeres, J., & Carvalho, A. (2015). Turismo Religioso: Fátima no Contexto dos Santuários Marianos Europeus. *PASOS Revista de Turismo y Patrimonio Cultural*, *13*(5), 1145–1170. doi:10.25145/j.pasos.2015.13.078

Premium Times. (2015). *South Africa threatens to jail Pastor Chris Oyakhilome*. Retrieved from: https://www.premiumtimesng.com/news/

Press TV. (n.d.). https://www.presstv.com/Detail/2019/09/21/606743/Iran-religious-tourism-top-destinations-Mashhad-Qom

Public opinion in the European Union. (2015). *Standard Eurobarometer*, 298.

Puczko, L., & Smith, M. (2008). *Health and Wellness Tourism*. Routledge.

Punjabi, R. (2018). Kashmiriyat: The Mystique of an Ethnicity. *Himalayan and Central Asian Studies, 22*(4).

Puşcaşu, V. (2015). Religious tourism or pilgrimage. *European Journal of Science and Theology*, *11*(3), 131–142.

Qeshelashvili, A. (2018). *Salome Zourabishvili: I am not a Catholic, I am an Orthodox*. Journal Liberal.

Qurashi, J., & Sharpley, R. A. (2017). *SMART Media Technologies impact on the Spiritual Experience of Hajj Pilgrims*. Academic Press.

Qureshi, F., Humayun, S., & Hummayun, S. F. (2015). Influence of Retail Atmospherics on Consumer Perception in Specialized Department Stores. *Management and Administrative Sciences Review*, *4*(5), 758–765.

Raj, R., & Morpeth, N. D. (2007). *Religious tourism and pilgrimage festivals management. An international perspective*. Academic Press.

Raj, R., & Morpeth, N. D. (2007). *Religious tourism and pilgrimage festivals management: An international perspective*. doi:10.1079/9781845932251.0000

Raj, R., Griffin, K., & Korstanje, M. E. (2018). 1 Risk and Safety Challenges Facing Religious Tourism–An Introduction. *Risk and Safety Challenges for Religious Tourism and Events*, 1.

Raj. (2015). *Religious Tourism and Pilgrimage Management: An International Perspective*. Oxfort: Cabi.

Raj, R., & Griffin, K. (2015). *Religious Tourism and Pilgrimage Management an International Perspective*. CAB International. doi:10.1079/9781780645230.0000

Raj, R., & Griffin, K. (2017). *Conflicts, Religion and Culture in Tourism*. CAB International. doi:10.1079/9781786390646.0000

Raj, R., Griffin, K., & Morpeth, N. D. (Eds.). (2013). *Cultural tourism*. CAB International. doi:10.1079/9781845939236.0000

Ramogale, M., & Galane, S. (1997). *Articles from the 1997 festival of American Folklore Programme Book*. Retrieved from: http://www.folklore.si.edu/resources/festival1977/faithin.htm/

Ramos, C. M. Q. G. A. M. (2019). Analysis of the Contribution of ICT to Cultural and Religious Tourism: In Communicating Religious Heritage to Visitors and Tourists. In Research on Socio-Economic Impacts of Religious Tourism and Pilgrimage (pp. 167–194). IGI Global.

Ramukumba, T. (2016). The role of government in support of tourism businesses: A perspective from guest houses in the Eastern district region. *African Journal of Hospitality, Tourism and Leisure*, *5*(1), 1–11.

Ram, Y., Bjork, P., & Weidenfeld, A. (2016). Authenticity and place attachment of major visitor attractions. *Tourism Management*, *52*, 110–122. doi:10.1016/j.tourman.2015.06.010

Ranjha, A. (2010). *Promoting tourism in Abu Dhabi using social media* (Unpublished Master Thesis). Royal Institute of Technology of Stockholm.

Rasgon, A., Lee, W. H., Leibu, E., Laird, A., Glahn, D., Goodman, W., & Frangou, S. (2017). Neural correlates of affective and non-affective cognition in obsessive compulsive disorder: A meta-analysis of functional imaging studies. *European Psychiatry*, *46*, 25–32. doi:10.1016/j.eurpsy.2017.08.001 PMID:28992533

Rashid, A. G. (2018). Religious tourism–a review of the literature. *Journal of Hospitality and Tourism Insights*.

Rasul, F., Fatima, U., & Sohail, S. (2016). Religion Tourism and Entrepreneurial Development (A Case Study Hazrat Data Ganj Bakhsh Shrine). *South Asian Studies, 31*(1).

Raz-Chaimovich, M. (2019). Jerusalem ranked the world's fastest-growing tourist destination. *Globes, Israel Business News*. Retrieved from https://en.globes.co.il/en/article-jerusalem-ranked-worlds-fastest-growing-tourist-destination-1001310945

ReinsteinD. (2014). Retrieved from: https://www.travelweekly.com/Middle-East-Africa-Travel/Insights/South-Africa-becoming-a-more-popular-stop-for-cruise-ships/

Rejman, K., Maziarz, P., Kwiatkowski, C. A., & Haliniarz, M. (2016). Religious tourism as a tourism product. *World Scientific News*, (57), 562-575.

Rinschede, G. (1992). Forms of religious tourism. *Annals of Tourism Research, 19*(1), 51–67. doi:10.1016/0160-7383(92)90106-Y

Ritchie, J., & Hudson, S. (2009). Understanding and meeting the challenges of customer/tourist experience research. *International Journal of Tourism Research, 11*(2), 111–126. doi:10.1002/jtr.721

Rivera, D. E., Fa, M. C., & Villar, A. S. (2019). Delightful tourism experiences: A cognitive or affective matter? *Tourism Management Perspectives, 32*, 100569. doi:10.1016/j.tmp.2019.100569

Robina Ramírez, R., & Fernández Portillo, A. (2018). What role does touristś educational motivation play in promoting religious tourism among travellers? *Annals of Leisure Research, 5398*. Advance online publication. doi:10.1080/11745398.2018.1561309

Rodrigo, N. N. (2012). *Influence of religious factors on tourism in Maramures*. Retrieved from http://writepass.co.uk/journal/2012/12/influence-of-religious factors-on-tourism-in-maramures-northern-part-of-romania/

Rodríguez, Y. (2014). Podemos preocupa a los empresarios españoles. *Bolsamania*. https://www.bolsamania.com/noticias/pulsos/podemos-preocupa-a-los-empresarios-espanoles--625313.html

Rogerson, C.M., Simango, S. & Rogerson, J.M. (2013). From recreation to niche tourism: The transformation of bird-watching in South Africa. *African Journal of Physical, Health Education, Recreation and Dance*, 122-137.

Rojo, D. M. (2007). *Religious tourism: the way to Santiago. MA European Tourism Management*. Bournemouth University.

Ross, E., & Iso-Ahola, S. (1991). Sightseeing tourists' motivation and satisfaction. *Annals of Tourism Research, 18*(2), 226–237. doi:10.1016/0160-7383(91)90006-W

Russell, P. (1999). Religious travel in the new millennium. *Travel & Tourism Analyst*, (5), 39–68.

Ryan, C. (1995). Learning About Tourists from Conservations: The Over-55s in Majorca. *Tourism Management, 16*(3), 207–216. doi:10.1016/0261-5177(95)00005-9

Saayman, A., Saayman, M., & Gyekye, A. (2014). Perspectives on the regional economic value of a pilgrimage. *International Journal of Tourism Research, 16*(4), 407–414. doi:10.1002/jtr.1936

Saayman, M. (2013). *En-route with tourism: An introductory text* (4th ed.). Juta.

Sabateev, A. (2000). *Islam and politics in the Chechen Republic*. Social Sciences and Modernity.

Sachs, J. (2005). *The End of Poverty: How We Can Make it Happen in Our Lifetime*. Penguin.

Safronova-Urbanaeva, I. (2013). *How and why Tibet became the central country of Buddhism*. Humanitarian Vector.

Saher, S., Ahmed, R., Faizan, M., & Burney, A. I. (n.d.). *Critical Analysis Relation between Indo-Pak: Kartarpur Corridor*. Academic Press.

Śākya, M. (2000). Sacred Art of Nepal: Nepalese Paubha Paintings: Past and Present (N. H. Saṅgha, Ed.). Handicraft Association of Nepal.

Salman, N. (2014). *City in which Seljuk was founded: Nishabur*. Accessed from https://gezimanya.com/GeziNotlari/selcuklunun-kuruldgu-kent-nisabur

San Filippo, M. (2001). The religious niche. *Travel Weekly, 60*(18), 12.

Sánchez, V. L., Fernández, M. D., & Lara, J. Á. S. (2017). Economic impact of a religious and tourist event: A Holy Week celebration. *Tourism Economics*, *23*(6), 1255–1274. doi:10.1177/1354816616675996

Saneblidze, N. (2016). *The State of Cultural Heritage of Religious Minorities in Georgia*. Published by Human Rights Center.

Santos, C. M., Ambrósio, V., Correia, A., & Peres, R. (2013). The importance of religious tourism segmentation for tourism destination management: The case of the Island of S. Miguel, Azores. *World Review of Entrepreneurship, Management and Sustainable Development*, *9*(2), 166–178. doi:10.1504/WREMSD.2013.052357

Santos, X. M. (2002). Pilgrimage and Tourism at Santiago de Compostela. *Tourism Recreation Research*, *27*(2), 41–50. doi:10.1080/02508281.2002.11081219

SATOUR. (2018). *South Africa Tourism*. https://www.statista.com

Savola, C. (2003). EI turismo religioso ya mueve millones por el país y eI mundo. Clarín. *Cuadernos Sociedad, Buenos Aires*, *22*(sep), 3.

Scaffidi Abbate, C., & Di Nuovob, S. (2013). Motivation and personality traits for choosing religious tourism. A research on the case of Medjugorje. *Current Issues in Tourism*, *16*(5), 501–506. doi:10.1080/13683500.2012.749844

Scanlan, C. L. (2020). *Preparing for the Unanticipated: Challenges in Conducting Semi-Structured*. In-Depth Interviews. doi:10.4135/9781529719208

Scannell, L., & Gifford, R. (2010). Defining place attachment: A tripartite organizing framework. *Journal of Environmental Psychology*, *30*(1), 1–10. doi:10.1016/j.jenvp.2009.09.006

Scaperlanda, M., & Scaperlanda, M. R. (2004). *The Journey: A Guide for the Modern Pilgrim*. Layiola Press.

Scheyvens, R., & Biddulph, R. (2018). Inclusive tourism development. *Tourism Geographies*, *20*(4), 589–609. doi:10.1080/14616688.2017.1381985

Schlemmer, L. (2008). *Dormant Capital: The Pentecostal Movement in South Africa and its Potential Social and Economic Role*. Centre for Development and Enterprise.

Schuhmacher, E. F. (1973/1993). *Small is Beautiful: Economics as if People Mattered*. Vintage Books.

Scptt, D. W. (2007). Constructing sacred history: Multi-media narratives and the discourse of "museumness" at Mormon Temple Square. *Journal of Media and Religion*, *6*(3), 201–218. doi:10.1080/15348420701530098

Scripps, T. (2019). *Visiting An African Prophetess' Cave*. Retrieved from: https://thestarlitpath.com//

Selcuklu Municipality. (2017). *Nishabur Castle*. Accessed from http://www.selcuklumirasi.com/architecture-detail/nisabur-kalesi at 16/03/2020

Selinger, L. (2004). 'The Forgotten Factor: The Uneasy Relationship between Religion and Development. *Social Compass*, *51*(4), 523–543. doi:10.1177/0037768604047872

Sen, A. (1999). *Development as Freedom*. Oxford University Press.

Seyer, F. & Muller, D. (2011). Religious tourism: Niche or mainstream? In *The Long Tail of Tourism: Holiday niches and their impact on mainstream tourism*. Gabler Verlag.

Seyer, F., & Müller, D. (2011). Religious tourism. In *The Long Tail of Tourism* (pp. 45–56). Gabler. doi:10.1007/978-3-8349-6231-7_6

Shackley, M. (2006). Empty bottles at sacred sites. Religious retailing at Ireland's national shrine. In D. J. Timothy & D. H. Olsen (Eds.), *Tourism, religion & spiritual journeys* (pp. 94–103). Routledge.

Shafaei, F. (2017). The relationship between involvement with travelling to Islamic destinations and Islamic brand equity: A case of Muslim tourists in Malaysia. *Asia Pacific Journal of Tourism Research*, *22*(3), 255–271. doi:10.1080/10941665.2016.1232741

Shafi, A. T. M. (2013, August). Critical Study of Terrorism and Counter-Terrorism and the "Business of Conflicts and Peace": Whose lose, who is Benefitted? *International Law and Policy Research Journal*, *1*(1), 1–15.

Shah, D. I. (2014). The Development of Hindu Temples in Pakistan: An Historical and Archaeological Overview. Pakistan Historical Society. Journal of the Pakistan Historical Society, 62(4), 47.

Shah, S. I. A. (2016). *Economic corridor business tower Gwadar Balochistan, Pakistan* (Doctoral dissertation). Mehran University of Engineering and Technology, Jamshoro.

Shahzad, S. J. H., Shahbaz, M., Ferrer, R., & Kumar, R. R. (2017). Tourism-led growth hypothesis in the top ten tourist destinations: New evidence using the quantile-on-quantile approach. *Tourism Management*, *60*, 223–232. doi:10.1016/j.tourman.2016.12.006

Shaista, A., & Khan, M. M. (2019). Role of Social Media in Promoting Tourism in Pakistan. *Journal of Social Sciences and Humanities*, *1*(1), 101–113.

Shamasastry, R. (1915). *English Translation of Koutilya's Arthashastra*. The Government Press.

Shao, J. (2001). Alleviating Poverty in Africa. In D. Belshaw, R. Calderisi, & C. Sugden (Eds.), *Faith in Development: Partnership between the World Bank and the Churches of Africa* (pp. 19–30). Regnum.

Sharma, V. (2010). *Pilgrim Satisfaction - A Study of Selected Shrines of Northern India* (Unpublished Ph.D. Thesis). Shri Mata Vaishno Devi University, Katra.

Sharma, A. K., & Swamy, G. A. (2019). Opportunity and threats faced on religious tourism development India. *International Journal of Scientific Research (Ahmedabad, India)*, *8*(10).

Sharpley, R. (2002). Rural tourism and the challenge of tourism diversification: The case of Cyprus. *Tourism Management*, *23*(3), 233–244. doi:10.1016/S0261-5177(01)00078-4

Sharpley, R., & Jepson, D. (2011). Rural Tourism: A spiritual experience? *Annals of Tourism Research*, *38*(1), 52–71. doi:10.1016/j.annals.2010.05.002

Shevardnadze, E. (2003). *Shevardnadzeadze demands the prevention of religious violence*. Retrieved from civil.ge: https://civil.ge/ka/archives/129134

Shinde, K. (2018). Governance and Management of Religious Tourism in India. *International Journal of Religious Tourism and Pilgrimage*, *6*(1), 7. Doi.org/10.21427/D7K42m

Shinde, K. A. (2007b). Visiting sacred sites in India: religious tourism or pilgrimage. In R. Raj & N. D. Morpeth (Eds.), *Religious Tourism and Pilgrimage Festivals Management: An International Perspective* (pp. 186–199). CABI. doi:10.1079/9781845932251.0184

Shinde, K. A. (2010). Entrepreneurship and indigenous enterpreneurs in religious tourism in India. *International Journal of Tourism Research*, *12*(5), 523–535. doi:10.1002/jtr.771

Shinde, K. A. (2015). Religious tourism and religious tolerance: Insights from pilgrimage sites in India. *Tourism Review*, *70*(3), 179–196. doi:10.1108/TR-10-2013-0056

Shinde, K. A. (2017). Planning for urbanization in religious tourism destinations: Insights from Shirdi, India. *Planning Practice and Research*, *32*(2), 132–151. doi:10.1080/02697459.2016.1198197

Show, Me. (2017). *Thathe Vondo-The Holy Forest, Soutpansberg, Limpopo.* Retrieved from: https://showme.co.za/tourism/thathe-vondo/

Shtudiner, Z., Klein, G., Zwilling, M., & Kantor, J. (2019). The value of souvenirs: Endowment effect and religion. *Annals of Tourism Research, 74*(October), 17–32. doi:10.1016/j.annals.2018.10.003

Shuo, Y., Ryan, C., & Liu, G. M. (2009). Taoism, temples and tourists: The case of Mazu pilgrimage tourism. *Tourism Management*, *30*(4), 581–588. doi:10.1016/j.tourman.2008.08.008

SIGA. (2012). *Diverse Beliefs: Tourism of Faith Religious Tourism Gains Ground.* Yes Bank.

Sikharulidze, A., Urushadze, M., Khmaladze, G., & Chichinadze, R. (2016). *Islam in Georgia: Politics and Integration.* Publishing House Caucasian House.

Silva, J. A. (2004). Caminhos de Santiago: Uma Europa peregrina. *Theologica*, *39*(2), 331–357.

Silveira, E. J. S. (2004). Turismo religioso popular? Entre a ambiguidade conceitual e as oportunidades de mercado. *Revista de Antropologia Experimental*, *4*, 1–16.

Simon, D. (2006). *Fifty Key Thinkers on Development.* Routledge. doi:10.4324/9780203098820

Simone-Charteris, M., & Stephen, B. (2010). The Development of Religious Heritage Tourism in Northern Ireland: Opportunities, Benefits and Obstacles. *Turizam: medunarodni znanstveno-strucnicasopis, 58*, 229-257.

Singha, H. (2000). *The Encyclopedia of Sikhism (over 1000 Entries).* Hemkunt Press.

Singh, R. P. (2003). *Where the Buddha walked: A companion to the Buddhist places of India.* Indica Books.

Singh, S. (1998). Probing the product life cycle further. *Tourism Recreation Research*, *23*(2), 61–63. doi:10.1080/02508281.1998.11014839

Singh, S. (2004). Religion, heritage and travel: Case references from the Indian Himalayas. *Current Issues in Tourism*, *7*(1), 44–65. doi:10.1080/13683500408667972

Sitnikov, A., Romanov, M., & Odaev, T. (2019). Religiosity in the Chechen Republic and its Influcence on Social and Political Institutions. *Monitoring of Public Opinion: Economic and Social Changes*, 157-183.

Skavronskaya, L., Scott, N., Moyle, B., Le, D., Hadinejad, A., Zhang, R., Gardiner, S., Coghlan, A., & Shakeela, A. (2017). Cognitive psychology and tourism research: State of the art. *Tourism Review*, *72*(2), 221–237. doi:10.1108/TR-03-2017-0041

Slavin, S. (2003). Walking as spiritual practice: The pilgrimage to Santiago de Compostela. *Body & Society*, *9*(3), 1–18. doi:10.1177/1357034X030093001

Slonim, S. (1988). *Jerusalem in America's Foreign Policy, 1947-1997.* Kluwer Law International.

Smallman, C., & Moore, K. (2010). Process studies of tourists' decision-making. *Annals of Tourism Research*, *37*(2), 397–422. Advance online publication. doi:10.1016/j.annals.2009.10.014

Smith, M., & Puczkó, L. (2009). *Health and wellness tourism.* Elsevier.

Smith, M., & Kelly, C. (2006). Holistic tourism: Journeys of the self? *Tourism Recreation Research*, *31*(1), 15–24. doi:10.1080/02508281.2006.11081243

Smith, M., Macleod, N., & Robertson, M. H. (2010). Religious and spiritual tourism. In *SAGE key Concepts: Key concepts in tourist studies* (pp. 140–145). SAGE Publications Ltd., doi:10.4135/9781446251027

Smith, V. (1992). Forms of Religious Tourism. *Annals of Tourism Research, 19*(12).

Smith, V. L. (1992). Introduction: The quest in guest. *Annals of Tourism Research, 19*(1), 1–17. doi:10.1016/0160-7383(92)90103-V

Smith, V. L. (2012). *Hosts and guests: The anthropology of tourism.* University of Pennsylvania Press.

Solutions, N. (n.d.). *Use Cases*. Retrieved from https://hipla.io/crowd-management-at-religious-places/

Songulashvili, M. (2017). Global Dynamics of Interreligious Dialogue: The Road from Peace Complaint to Ecumenical Charter. *Journal of Tolerance and Diversity Institute.*

Sönmez, S. (1998, April). Tourism, Terroris and Political Instability. *Annals of Tourism Research, 25*(2), 416–456. doi:10.1016/S0160-7383(97)00093-5

Souleimanov, E. (2017). *The North Caucasus Insurgency: Dead or Alive?* Strategic Studies Institute and U.S. Army War College Press.

Sousa, B., & Rodrigues, S. (2019). The role of personal brand on consumer behaviour in tourism contexts: The case of madeira. *Enlightening Tourism. A Pathmaking Journal, 9*(1), 38-62. doi:10.33776/et.v9i1.3597

Sousa, B., & Silva, M. (2019). Creative tourism and destination marketing as a safeguard of the cultural heritage of regions: The case of sabugueiro village. *Revista Brasileira de Gestão e Desenvolvimento Regional, 15*(5), 78-92.

Sousa, B., & Rocha, A. T. (2019). The role of attachment in public management and place marketing contexts: A case study applied to Vila de Montalegre (Portugal). *International Journal of Public Sector Performance Management, 5*(2), 189–205. doi:10.1504/IJPSPM.2019.099094

South Africa Tourism. (2016). *Limpopo's Lake Fundudzi*. Retrieved from: https://www.southafrica.net/za/en/aticles/

Spradley, J. (1979). *The Ethnographic Interview*. Holt, Rinehart and Winston.

Staff, T. (2018, December 29). *Israel saw record-breaking 4 Million Tourist in 2018, Says Tourism Minister*. Retrieved from Timeofisrael: https://www.timesofisrael.com/israel-saw-record-breaking-4-million-tourists-in-2018-says-tourism-ministry/

Stake, S., & Torrance, H. (2005). Case study. In B. Somekh & C. Lewin (Eds.), *Research methods in the social sciences* (pp. 33–40). Sage.

State of the Rivers Report. (2001). *Letaba and Luvuvhu Rivers Systems*. Pretoria: Water Research Council.

Statista. (n.d.). https://www.statista.com/statistics/475509/christmas-budget-europe-by-country/

Statistical Data. (n.d.). Retrieved from https://chechenstat.gks.ru/folder/38713

Stayt, H. A. (1968). *The Bavenda*. Frank Cass.

Steil, C. A., & Carneiro, S. S. (2008). Peregrinação, Turismo e Nova Era: Caminhos de Santiago de Compostela no Brasil. *Religião & Sociedade, 28*(1), 105–124. doi:10.1590/S0100-85872008000100006

Stein, R. A. (2010). *Rolf Stein's Tibetica Antiqua* (A.P. McKeown, Ed.). Boston: Brill NV Leiden.

Stiglitz, J. (2006). *Making Globalization Work: The Next Steps to Global Justice*. Allen Lane.

Stokols, D. (1990). Instrumental and spiritual views of people-environment relations. *The American Psychologist*, *45*(5), 641–646. Advance online publication. doi:10.1037/0003-066X.45.5.641

Sugimura, K., Matsushima, K., Hihara, S., Takahashi, M., & Crocetti, E. (2019). A culturally sensitive approach to the relationships between identity formation and religious beliefs in youth. *Journal of Youth and Adolescence*, *48*(4), 668–679. doi:10.100710964-018-0920-8 PMID:30187258

Sunday World. (2014). *Tourism and housing encroaches on sacred lake*. Retrieved from: https://www.sundayworld.co.za/lifestyle/

Suntikul, W., & Dorji, U. (2016). Local Perspectives on the Impact of Tourism on Religious Festivals in Bhutan. *Asia Pacific Journal of Tourism Research*, *21*(7), 741–762. doi:10.1080/10941665.2015.1069365

Suri, K. (2018). Understanding historical, cultural and religious frameworks of mountain communities and disasters in Nubra valley of Ladakh. *International Journal of Disaster Risk Reduction*, *31*, 504–513. doi:10.1016/j.ijdrr.2018.06.004

Sverigedemokraternas princip program. (2018). Landsdagarna.

Szabaciuk, A. (2014). *The Russian Federation's Policy on the North Caucasus. Islam in the North Caucasus: history and challenges of our time*. Central East European Institute Publishing House.

Szafraniec, K. (2017). The Contemporary Context of Youth Socialization: The Specificity of Post-Communist Countries. *Polish Sociological Review*, *2*(198), 167-187.

Tafveez, M. (2017). Role of social media in tourism: A literature review. *International Journal for Research in Applied Science and Engineering Technology*, *5*(11), 633–635. doi:10.22214/ijraset.2017.11099

Taggart, R. (1998). A touchstone of dissent: Euroscepticism in contemporary Western European party systems. *European Journal of Political Research*, 363-388.

Taheri, B. (2016). Emotional Connection, Materialism, and Religiosity: An Islamic Tourism Experience. *Journal of Travel & Tourism Marketing*, *33*(7), 1011–1027. doi:10.1080/10548408.2015.1078761

Tala Madalina Lavinia, P. M. (2008). Dimensions of religious tourism. *Amfiteatru Economic*, *10*(2), 242–253.

Taleb, R. (2015). Fostering sustainable socio-economic development in host communities. In *International Conference on Religious Tourism*. Bethlehem: World Tourism Organization.

Tan, H. (2020). *China's dollar-denominated exports and imports beat expectations in December, Jan 13, 2020*. https://www.cnbc.com/2020/01/14/china-december-2019-trade-imports-exports.html

Tassiopoulos, D. (2011). *New Tourism Ventures* (2nd ed.). Juta.

Tauoatsoala, P., Henama, U. S., & Monyane, M. (2015). Service quality as a fundamental concernstone for visitor satisfaction: Lessons learned from Pretoria, South Africa. *African Journal of Hospitality, Tourism and Leisure*, *4*(1), 1–15.

Tenay, E. (n.d.). *Islamic Cities: Baghdad*. Accessed from https://www.bidunyahaber.org/islam-sehirleri-bagdat/

Tendeiro, I. L. G. S. (2010). *A igreja de Santo António de Lisboa e o turismo religioso italiano*. Dissertação de Mestrado da Escola Superior de Hotelaria e Turismo de Estoril.

Terzidou, M., Scarles, C., & Saunders, M. N. (2018). The complexities of religious tourism motivations: Sacred places, vows and visions. *Annals of Tourism Research*, *70*, 54–65. doi:10.1016/j.annals.2018.02.011

Terzidou, M., Scarles, C., & Saunders, M. N. K. (2017). Religiousness as tourist performances: A case study of Greek Orthodox pilgrimage. *Annals of Tourism Research*, *66*, 116–129. doi:10.1016/j.annals.2017.06.009

Terzidou, M., Stylidis, D., & Terzidis, K. (2018). The role of visual media in religious tourists' destination image, choice, and on-site experience: The case of Tinos, Greece. *Journal of Travel & Tourism Marketing*, *35*(3), 306–319. doi:10.1080/10548408.2017.1304316

The Al Quran Translated into English by Yusuf Ali. (1946). https://quranyusufali.com/

The Christian Science Monitor. (1991). *Honesty and Economy*. https://www.csmonitor.com/1991/0926/26171.html

The Constitution of Georgia. Article 9. (1995). Tbilisi.

The New Age. (2016). *South African still visit SCOAN despite 2014 incident*. Retrieved from: http://www.thenewage.co.za/south-africans-still-visit-scoan-despite-2014-incident/

The right-wing Finns Party does well in Finland's election. (2019). *The Economist*. https://www.economist.com/europe/2019/04/17/the-right-wing-finns-party-does-well-in-finlands-election

The Sunday Mail. (2017, Feb. 12). Religious tourism booms in Zimbabwe. *The Sunday Mail.*

The Sunday Mail. (2018, Mar. 11). Religious tourism racks in big bucks. *The Sunday Mail.*

Thomas, J. (2013). *Social media effects on tourism sector*. Retrieved from https://socialnetworking.lovetoknow.com/Social_Media_Effects_on_Tourism

Thomas, D. R. (2006). A general inductive approach for analyzing qualitative evaluation data. *The American Journal of Evaluation*, *27*(2), 237–246. doi:10.1177/1098214005283748

Thomin, M. (2019). *Moving towards a harmonized heritage tourism approach: an assessment of small museums in Northwest Florida*. Academic Press.

Thornycroft, P. (2017). *TB Joshua visit a ''blessing'' for Zim's religious tourism*. Retrieved from: https://www.iol.co.za/news/special-features/zimbabwe/

Tibet: Problems, Prospects, and U.S. Policy. Foreign Affairs, Defense, and Trade Division, Asian Affairs. (2007). Washington, DC: Congressional Research Service.

Timothy, D. J. (2011). *Cultural Heritage and Tourism: An Introduction* (Vol. 4). Channel View Publications.

Timothy, D. J., & Olsen, D. H. (2006). *Tourism, religion and spiritual journeys*. Routledge. doi:10.4324/9780203001073

Tirca, A., & Stanciulescu, G. C. (2011). Managing the religious tourism experience in Romanian Christian Orthodoxy. *International Journal of Business and Globalisation*, *7*(1), 40. Advance online publication. doi:10.1504/IJBG.2011.040845

Tkaczynski, A., & Arli, D. (2018). Religious tourism and spiritual leadership development: Christian leadership conferences. *Journal of Hospitality and Tourism Management*, *35*, 75–84. doi:10.1016/j.jhtm.2018.03.004

Tomar, C. (2010). Damascus. In *Religious Foundation of Turkey [TDV] Encyclopedia of Islam*. Accessed from https://islamansiklopedisi.org.tr/sam--suriye at 18/03/2020

Tomić, S., Leković, K., & Tadić, J. (2019). Consumer behaviour: The influence of age and family structure on the choice of activities in a tourist destination. *Economic research-. Ekonomska Istrazivanja*, *32*(1), 755–771. doi:10.1080/1331677X.2019.1579663

Tour My India. (n.d.). https://www.tourmyindia.com/pilgrimage/sikh-pilgrimage-tour.html

Tourism & More. (2011). *Using the social media for tourism*. Academic Press.

Tourism, E. T. S. (n.d.). *Places to Visit in Mecca*. Accessed from https://www.etstur.com/Hac-Umre/Mekke-Ziyaret-Yerleri

Tranfield, D., Denyer, D., & Smart, P. (2003). Towards a methodology for developing evidence-informed management knowledge by means of systematic review. *British Journal of Management*, *14*(3), 207–222. doi:10.1111/1467-8551.00375

Trauer, B., & Ryan, C. (2005). Destination image, romance and place experience - An application of intimacy theory in tourism. *Tourism Management*, *26*(4), 481–491. Advance online publication. doi:10.1016/j.tourman.2004.02.014

Triantafillidou, A., Koritos, C., Chatzipanagiotou, K., & Vassilikopoulou, A. (2010). Pilgrimages: The "promised land" for travel agents? *International Journal of Contemporary Hospitality Management*, *22*(3), 382–398. doi:10.1108/09596111011035963

Tribe, J. (2016). *The economics of recreation, leisure and tourism* (5th ed.). Routledge.

Tripathy, A. K., Tripathy, P. K., Ray, N. K., & Mohanty, S. P. (2018). *ITour: The Future of Smart Tourism: An IoT Framework for the Independent Mobility of Tourists in Smart Cities*. IEEE Consumer Electronics Magazine. doi:10.1109/MCE.2018.2797758

Tshipala, N., & Chauke, J. (2020). Thathe Vondo: Is this a missed opportunity for Adventure Tourism development? *African Journal of Hospitality, Tourism and Leisure*, *9*(1), 1–15.

Tsomu, Y., & Roche, G. (2018). *Tibet's Invisible Languages and China's Language Endangerment*. Academic Press.

Tucci, G. (2001). *The Theory and Practice of the Mandala*. Dover Publications Inc.

Turkish Cooperation and Coordination Agency (TIKA). (n.d.). *Palestine*. Accessed from https://www.tika.gov.tr/upload/2016/Prestij%20Eserler/T%C3%BCrkce%20Filistin%20Rehberi.pdf

Turkish Knowledge. (2020). *Nishabur*. Accessed from https://www.turkcebilgi.com/

Turkish Science. (2020). *Ancient Tus City History and Oghuzes*. Accessed from http://www.turkbilimi.com/antik-Tûs-sehri-tarihi-ve-oguzlar/

Turner, M. (2015). *Trip Tips: Serenity meets dark history on Uganda's Lake Bunyonyi*. Retrieved from: https://uk.reuters.com/article/travel-uganda/trip-tips-serenity-meets-dark-history-on-ugandas-lake-bunyonyi-idUKL6N0UR2YG20150227/

Turner, V., & Turner, E. (2011). *Image and Pilgrimage in Christian Culture*. Columbia University Press.

U.N. 3 CAOR, p. 2. (1949). Remarks by the Israeli Delegation. 230-237.

U.N. Doc. A/565. (1948). Proceedings summarized in Annual Report of the Secretary-General on the Work of the Organization, I July I947-30 June 1948. *3 U.N. GAOR, Supp.I.*

ul Haq, M. Z., & Ullah, R. (2019). Tourism demand function for Pakistan. Journal *of* Social Sciences & Humanities, 3(1 & 2), 1–17.

Ulucam, A. (2010). Damascus. In *Religious Foundation of Turkey [TDV] Encyclopedia of Islam*. Accessed from https://islamansiklopedisi.org.tr/sam--suriye#2

UN GA Res. 181(II). (1947, November 29). U.N. Doc. A/RES/181(II) .

UN SC res 242. (1967, November 22). UNS Res. 478. (1980).

Unal, S. (2001). Kaaba. In *Religious Foundation of Turkey [TDV] Encyclopedia of Islam*. Accessed from. https://islamansiklopedisi.org.tr/kabe at 20/03/2020

UNESCO Constitution, Constitution of the United Nations Educational, Scientific and Cultural Organization. (1945). Retrieved from http://portal.unesco.org/en/ev.php-URL_ID=15244&URL_DO=DO_TOPIC&URL_SECTION=201.html

UNESCO. (2014). *Fifth International Festival of Traditional Cultures and Crafts "ETHNORADUGA" devoted to the World Day of Cultural Diversity for Dialogue and Development*. Kaluga Region.

United National World Tourism Organization. (2011). *Religious Tourism in Asia and the Pacific*. UNWTO. Retrieved from http://pub.unwto.org/epages/Store.sf/?ObjectPath=/Shops/Infoshop/Products/1509/SubProd ucts/1509-1

UNWTO. (2011). Religious Tourism in Asia and the Pacific. In Religious Tourism in Asia and the Pacific. doi:10.18111/9789284413805

UNWTO. (2011). *Religious Tourism in Asia and the Pacific*. Retrieved from https://www.e-unwto.org/doi/pdf/10.18111/9789284413805

UNWTO. (2011). *Religious Tourism in Asia and the Pacific*. World Tourism Organization.

UNWTO. (2011). *Religious Tourism in Asia and the Pacific: United Nations world Tourism Organisation*. UNWTO.

UNWTO. (2017). *Tourism Highlight 2017*. https://planeta.com/unwto-tourism-highlights-2017/

UNWTO. (2019). *International Tourism Highlights*. https://www.unwto.org/publications/international-tourism-highlights-2019-edition

UNWTO. (2019). Retrieved from UNWTO.org: https://www.e-unwto.org/doi/pdf/10.18111/9789284421152

UNWTO. (2020). Compendium of Tourism Statistics. *Data (København)*, 2014–2018.

Uriely, N. (2005). The tourist experience: Conceptual developments. *Annals of Tourism Research*, *32*(1), 199–216. doi:10.1016/j.annals.2004.07.008

Urien-Lefranc, F. (2020). From religious to cultural and back again: Tourism development, heritage revitalization, and religious transnationalizations among the samaritans. *Religions*, *11*(2), 86. doi:10.3390/rel11020086

Uzun, M. I. (2001). Kaaba. In *Religious Foundation of Turkey [TDV] Encyclopedia of Islam*. Accessed from. https://islamansiklopedisi.org.tr/kabe at 20/03/2020

Van Der Merwe, P., & Saayman, M. (2013). Who are the South African hunters and why do they hunt? *Journal of Hospitality Management and Tourism*, *4*(1), 9–18. doi:10.5897/JHMT2013.0085

Van Der Waal, B. C. W. (1997). Fundudzi, a unique, sacred and unknown South African lake. *Southern African Journal of Aquatic Sciences*, *23*(1), 42–55. doi:10.1080/10183469.1997.9631387

VanHunks. (2016). *Bewitched Forests and Waters of the VhaVhenda*. Retrieved from: http://www.vanhanks.com/lwveld1/venda1.html./

Vardosanidze, S. (2017). *Georgian Orthodox Apostolic Church in the XIX-XX centuries"(Catholicos-Patriarch of All Georgia His Holiness and Beatitude Ilia II)*. Publishing House Sezani.

Vatchagaev, M. (2005). *The Role of Sufism in the Chechen Resistance*. Erudition-Russian Electronic Library Report: Islam and Politics in the Chechen Republic.

Vaubel, R. (2013). Secession in the European Union. *Economic Affairs*, *33*(3), 288–302. doi:10.1111/ecaf.12028

Veal, A. J. (2006). *Research methods for leisure and tourism: A practical guide* (3rd ed.). Pearson Education.

Verma, A., & Shukla, V. (2019). *Analyzing the Influence of IoT in Tourism Industry*. SSRN Electronic Journal. doi:10.2139srn.3358168

Verma, M., & Sarangi, P. (2019, August). Modeling attributes of religious tourism: A study of Kumbh Mela, India. *Journal of Convention & Event Tourism*, *20*(4), 296–324. doi:10.1080/15470148.2019.1652124

Vidino, L. (n.d.). *How Chechnya Became a Breeding Ground for Terror*. Middle East Forum. Retrieved 04 02, 2020, from https://www.meforum.org/744/how-chechnya-became-a-breeding-ground-for-terror

Vilas Boas, N. F. S. (2012). *A Pastoral do turismo: da peregrinação ao Santuário. Dissertação de mestrado*. Universidade Católica Portuguesa.

Visser, G., & Baker, C. (2004). A Geography of British Geographic Tourists in South Africa. *Geography (Sheffield, England)*, *89*(3), 226–239. doi:10.1080/00167487.2004.12094100

Vittengle, J. R., & Holt, C. S. (2000). Getting Acquainted: The Relationship of Self-Disclosure and Social Attraction to Positive Affect. *Journal of Social and Personal Relationships*, *17*(1), 53–66. doi:10.1177/0265407500171003

Viveport. (2017). *Terra-Cotta Warriors*. Retrieved from https://www.viveport.com/apps/2220d4be-38bf-4fbc-93b9-e74b90b2dcb9/VR_Terra-Cotta_Warriors/

Vrij, A., Fisher, R. P., & Blank, H. (2017). A cognitive approach to lie detection: A meta-analysis. *Legal and Criminological Psychology*, *22*(1), 1–21. doi:10.1111/lcrp.12088

Vukonik, B. (2002) Religion, tourism and economies: a convenient symbiosis. *Tourism Recreation Research, 27*(2), 59-64.

Waldman, P., & Pope, H. (2011, Sept. 21). 'Crusade' Reference Reinforces Fears War on Terrorism Is Against Muslims. *The Wall Street Journal*.

Waldo, M. (1976). *Travel guide to South Africa: Revised and Updated Edition*. MacMillan Publishing.

Walzer, M. (2002, Fall). The Four Wars of Israel/Palestine. *Dissent*.

Wang, R., & Hao, J. X. (2018, July). Difference on Destination Image and Travel Options: An Exploratory Text-Mining Study. In *2018 15th International Conference on Service Systems and Service Management (ICSSSM)* (pp. 1-5). IEEE.

Wang, D., Li, X., & Li, Y. (2013). China's "smart tourism destination" initiative: A taste of the service-dominant logic. *Journal of Destination Marketing & Management*, *2*(2), 59–61. Advance online publication. doi:10.1016/j.jdmm.2013.05.004

Wang, W., Chen, J. S., & Huang, K. (2016). Religious Tourist Motivation in Buddhist Mountain: The Case from China. *Asia Pacific Journal of Tourism Research*, *21*(1), 57–72. doi:10.1080/10941665.2015.1016443

Weber, M. (1905). The Protestant Ethic and the Spirit of Capitalism. London: George Allen and Unwin.

Web, S. A. C. (2020). *Cult Pakistan: Forgotten mysteries, bygone strangeness & odd folk| Nadeem F.* Paracha–South Asia Citizens Web.

Weintraub, E. R. (2002). *How Economics Became a Mathematical Science?* Duke University Press. doi:10.1215/9780822383802

Werbner, P. (2003). *Pilgrims of love: The anthropology of a global Sufi cult*. Indiana University Press.

Widmer, A. (2012). Ministering to the Pioneers of Prosperity. In M. Schlag & J. A. Mercado (Eds.), *Free Markets and the Culture of the Common Good* (pp. 167–176). Springer. doi:10.1007/978-94-007-2990-2_12

Williams, S. (2016). Iran: The Essential Guide to Customs & Culture. In Tips on Social Etiquette in Iran. Kuperard.

Willson, G. (2011). The search of inner peace: Considering the spiritual movement in Tourism. *The Journal of Tourism and Peace Research*, 16–26.

Wiltshier, P., & Griffiths, M. (2016). Management Practices for the Development of Religious Tourism Sacred Sites: Managing expectations through sacred and secular aims in site development; report, store and access. *International Journal of Religious Tourism and Pilgrimage*, *4*(7), 1–8.

Wiltshier, P., & Griffiths, M. (2019). *Managing Religious Tourism*. CABI.

Wiśniewski, Ł. (2018). Religious tourism in Christian sanctuaries: The implications of mixed interests for the communication of the faith. *Church. Communication and Culture*, *3*(3), 199–220. doi:10.1080/23753234.2018.1537674

Witham, L. (2010). *Marketplace of the Gods: How Economics Explains Religion*. Oxford University Press. doi:10.1093/acprof:oso/9780195394757.001.0001

Wong, C. U. I., McIntosh, A., & Ryan, C. (2013). Buddhism and tourism. Perceptions of the monastic community at pu-tuo-shan, china. *Annals of Tourism Research*, *40*(1), 213–234. doi:10.1016/j.annals.2012.09.004

Wong, C. U. I., McIntosh, A., & Ryan, C. (2014). Visitor Management at a Buddhist Sacred Site. *Journal of Travel Research*, *55*(5), 675–687. doi:10.1177/0047287514563164

Wong, C. U. I., Ryan, C., & McIntosh, A. (2013). The Monasteries of Putuoshan, China: Sites of Secular or Religious Tourism? *Journal of Travel & Tourism Marketing*, *30*(6), 577–594. Advance online publication. doi:10.1080/10548408.2013.810996

Woodward, S. C. (2003). Faith and tourism: accommodating visitor and worshipper in the historic city. *6th US/ICOMOS International Symposium, Managing Conflict and Conservation in Historic Cities.*

Woo, T.-T. (2019). Religion and Politics in the People's Republic of China: An Appraisal of Continuing Mistrust and Misunderstanding. *Religions*, 31.

World, D. (2010). *Religious and pilgrimage tourism. In: the e-newsletter for tourism destination professionals.* Retrieved from http://www.destinationworld.info/newsletter/feature65.html

Wright, O. (2016). Euroscepticism on the rise across Europe as analysis finds increasing opposition to the EU in France, Germany and Spain. *Independent*. https://www.independent.co.uk/news/world/europe/euroscepticism-on-the-rise-across-europe-as-analysis-finds-increasing-opposition-to-the-eu-in-france-a7069766.html

Wright, K. (2008). *The Christian travel planner*. Thomas Nelson.

Writer, S. (2011). *14m Americans scanned QR and bar codes with their mobiles*. Retrieved from https://internetretailing.net/themes/14m-americans-scanned-qr-and-bar-codes-with-their-mobiles-in-june-2011

WTTC. (2018). *Travel and Tourism economic impact 2018 world*. https:www.wttc.org/-/media/files/reports/2018/power-and-performance-rankings-2018.pdf

WTTC. (2020). *Travel & tourism: Economic of all net new jobs created.* Retrieved from https://wttc.org/en-us/Research/Economic-Impact

Wu, M.-Y., & Pearce, P. L. (2014). Chinese recreational vehicle users in Australia: A netnographic study of tourist motivation. *Tourism Management*, *43*, 22–35. doi:10.1016/j.tourman.2014.01.010

Xavier, M. S. (2018). *Sacred spaces and transnational networks in American Sufism: Bawa Muhaiyaddeen and contemporary shrine cultures*. Bloomsbury Publishing.

Xiang, Z., & Fesenmaier, D. R. (2017). Big data analytics, tourism design and smart tourism. In Analytics in Smart Tourism Design (pp. 299–307). Berlin: Springer.

Xiang, Z., & Iis, T. (2015). Special Issue on Smart Destinations: Foundations, Analytics, and Applications. *Journal of Destination Marketing & Management*, *4*(3), 143–201. doi:10.1016/j.jdmm.2015.07.001

Yājñika, A., & Sheth, S. (2005). *The Shaping of Modern Gujarat: Plurality, Hindutva, and Beyond.* Penguin Books India.

YalçınM. F. (n.d.). *Medina.* Accessed from https://www.islamdusunceatlasi.org/detail/place38 at 20/03/2020

Yarlykapov, A. (2003). Islamic education in the North Caucasus in the past and present. *Herald of Eurasia.*

Yarlykapov, A. (2017). *Islam and Tradition in the North Caucasus. Ethnic and Religious Diversity of Russia. Russian Academy of Science.* Institute of Ethnology and Anthropology.

Yasaroglu, M. K. (2001). Kaaba. In *Religious Foundation of Turkey [TDV] Encyclopedia of Islam.* Accessed from. https://islamansiklopedisi.org.tr/kabeat 20/03/2020

Yeoman, I. (2009). *Tomorrow's tourist: scenarios & trends.* Routledge. doi:10.4324/9780080878904

Yesil, M. M. (2013). The Social Media Factor in the Development and Promotion of Tourism Destination. *International Periodical for the Languages. Literature and History of Turkish or Turkic*, *8*(7), 733–744.

Yilmaz, G. (2010). *Central Damascus of Syria.* Accessed from https://gezimanya.com/GeziNotlari/suriyenin-merkezi-sam at 18/03/2020

Yolal, M., Gursoy, D., Uysal, M., Kim, H., & Karacaoglu, S. (2016). Impacts of festivals and events on residents' well-being. *Annals of Tourism Research*, *61*, 1–18. doi:10.1016/j.annals.2016.07.008

Yoon, Y., & Uysal, M. (2005). An examination of the effects of motivation and satisfaction on destination loyalty: A structural model. *Tourism Management*, *26*(1), 45–56. doi:10.1016/j.tourman.2003.08.016

Yorgason, E., & della Dora, V. (2009). Geography, religion, and emerging paradigms: Problematizing the dialogue. *Social & Cultural Geography*, *10*(6), 629–637. doi:10.1080/14649360903068100

Yousaf, S., & Xiucheng, F. (2018). Humanizing stigmatized places: Inter-group contact and attitude change toward Pakistan and Iran in the 'Humans of New York'Facebook space. *Journal of Business Research.*

Yousefi, J. (2018). *Importance-performance Analysis of Religious Tourism Services (The Case of the City of Birjand).* Academic Press.

Yuksel, G. (2020). *My Jerusalem Travel Notes.* Accessed from https://www.gorkemliyollar.com/kudus-gezi-notlarim/

Yuksel, A., Yuksel, F., & Bilim, Y. (2010). Destination attachment: Effects on customer satisfaction, cognitive, affective and conative loyalty. *Tourism Management*, *31*(2), 274–284. doi:10.1016/j.tourman.2009.03.007

Zalewska, J. (2019). Practice Theory Revisited: How Flexible Meta-habit Complements Habitus. *Polish Sociological Review*, *1*(205), 65–84.

Zamani-Farahani, H., & Eid, R. (2016). Muslim world: A study of tourism & pilgrimage among OIC Member States. *Tourism Management Perspectives*, *19*, 144–149. doi:10.1016/j.tmp.2015.12.009

Zamani-Farahani, H., & Henderson, J. C. (2010). Islamic tourism and managing tourism development in islamic societies: The cases of Iran and Saudi Arabia. *International Journal of Tourism Research*, *12*(1), 79–89. doi:10.1002/jtr.741

Zamani-Farahani, H., & Musa, G. (2012). The relationship between Islamic religiosity and residents' perceptions of socio-cultural impacts of tourism in Iran: Case studies of Sare'in and Masooleh. *Tourism Management, 33*(4), 802–814. doi:10.1016/j.tourman.2011.09.003

Zambia Eye. (2018). *Botswana govt shuts down Bushiri's church.* Retrieved from: https://zambianeye.com/botswana-govt-shuts-down-bushiris-church/

Zambia Observer. (2017). Retrieved from: http://www.zambianobserver.com/prophet-uebert-angel-refused-entry-to-zambia/

Zarnowitz, V., & Moore, G. H. (1977). The Recession and Recovery of 1973-1976. Explorations in Economic Research, 4(4).

Zeng, B., & Gerritsen, B. (2014). What do we know about social media in tourism? A review. *Tourism Management Perspectives, 10*, 27–36. doi:10.1016/j.tmp.2014.01.001

Zhang, J. (2018). To brand Gandhara, the Ancient Buddhist Centre: Pakistan's potential to develop Buddhist tourism. *Religious Tourism in Asia: Tradition and Change Through Case Studies and Narratives*, 127.

Zhou, Z. (2016). Implications of China' Approved Tourism Destination Status (ADS) on Zimbabwe's Tourism Environment: Insights from Local and Government Tourism Groups in Zimbabwe. *International Journal of Hospitality and Tourism Systems, 9*(1), 44–52.

Zhou, Z. (2016). Post 2010 evaluation of Zimbabwe as a preferred tourist destination. *African Journal of Hospitality, Tourism and Leisure, 5*(1), 1–15.

Zhou, Z. (2018). The Tourism Sector: A bright light on Zimbabwe's depressed economic environment. *African Journal of Hospitality, Tourism and Leisure, 7*(1), 1–15.

Zhu, W., Zhang, L., & Li, N. (2014). Challenges, Function Changing of Government and Enterprises in Chinese Smart Tourism. *Enter 2014.*

Zimbabwe National Statistical Agency. (2012). *Census 2012 National Report.* www.zimstat.co.zw/sites/default/files/img/publications/Population/NationalReport.pdf

ZTA. (2018). *Zimbabwe Tourism Authority Annual Statistics Report.* ZTA.

Zurher, E. (2013). Buddhism in China. In J. A. Silk (Ed.). *Sinica Leidensia.* Academic Press.

Алиев, М. [Aliev, M.]. (2010). *С благословения Аллаха: основные законы, обряды, традиции и обычаи в Исламе* [With the blessing of Allah: the basic laws, rites, traditions and customs in Islam]. Academic Press.

Вайнштейн, Г. (2018). *Трансформация западноевропейского политического ландшафта и институциализация антисистемной политики* [Transformation of the Western European Political Landscape and the Institutionalization of Antisystem Policy]. World Economy and International Relations.

Гаджиев, К. (2016). Метаморфозы идеологий в современной мировой политике [Metamorphoses of ideologies in modern world politics]. В Party-political systems and political ideologies in Western countries at the beginning of the XXI century: factors of evolution and directions of transformation, 112-124.

Каменева, М. [Kameneva, M.]. (2009). Россия и Иран: история и современность [Russia and Iran: history and modernity]. *East. Afro-Asian societies: history and modernity,* 157-164.

Кошманова, О. [Koshmanova, O.]. (2018). *Краткий очерк иранского коммуникативного поведения* [A brief outline of Iranian communicative behavior]. www.pags.ru/science/conferences/E-Conference/Koshmanova.doc

Кутлаки, С. [Kutlaki, S.]. (2011). Среди Иранцев. Путеводитель по нравам и обычаям самой закрытой страны мира [Among the Iranians. Guide to the mores and customs of the most closed country in the world]. Moscow: United Press.

Оленич, Т. К. (2016). Исламский фактор в миграционной политике Запада [Islamic factor in the migration policy of the West]. *Регулирование миграционных процессов на юге России: политические, юридические и правоохранительные аспекты*, 25-29.

Оленич, Т., & Трима, Е. (2013). Иран: общество традиций (заметки на полях путешественника) [Iran: Society of Traditions (traveler's notes)]. Mariupol, 112.

Хенкин, С. (2018). Перестройка партийной системы Испании [Reorganization of the party system of Spain]. *Actual problems of Europe*, 85-108.

Шарипова, Р. [Sharipova, R.]. (2014). Иранская семья: история и современность [Iranian family: history and modernity]. *Россия и мусульманский мир* [Russia and the Muslim world], *6*(264), 124-142.

Этикет общения в иранской культуре: обмен любезностями [Communication Etiquette in Iranian Culture: Courtesy Sharing]. (2012). *Karavan*, *11*, 60-65.

About the Contributors

Emilia Alaverdov, Ph.D. in Political Science, associate professor, Faculty of Law, and International Relations since 2011. She received her bachelor's degree in Languages and Literature, has been awarded a qualification of Philologist of English Language and Literature and Spanish Language and Literature, Ilia Chavchavadze State University, Tbilisi, Georgia. Has a master's degree in International Relations and Politics from the Diplomatic Academy of Georgia. Member of Editorial board of the following journals: Journal US-China Foreign Language, David Publishing Company, IJOPEC Publication, London-Istanbul; DergiPark Academic; Journal of Education, Culture, and Society, Foundation Pro Scientia Publica; International Journal of Social and Educational Sciences; European Journal of Economics Law and Politics, ELP. She is the editor of 3 books, the author of about 20 scientific papers and co-author of 5 books, participants of 6 Erasmus + Projects and 48 international scientific conferences (majority abroad), and organizer of various international scientific events. Her research interests are religion, migration, religious tourism, and European and Russian studies. She is fluent in the following languages: English, Russian, Georgian, Azerbaijani, and Greek.

* * *

Muhammad Mehtab Alam is MPhil Scholar at Riphah Institute of Public Policy, Riphah University Islamabad Pakistan.

Mekushkin Andrey Alexandrovich is a priest, an assistant professor at Don State Technical University. He is a Candidate of Sciences in Philosophy. He conducts active educational, missionary, educational activities; participates actively in research programs and projects, including international ones. In 1987 Andrey Alexandrovich entered the philological faculty of the Tashkent Order of Friendship of Peoples of the State Pedagogical Institute named after Nizami and graduated with honors with a diploma as a teacher of Russian language and literature in 1993. He worked as a teacher of Russian language and literature, a laureate of the competition "Teacher of the Year". Andrey Alexandrovich entered Moscow Theological Seminary in 1999 graduated from the Moscow Theological Academy (was awarded the qualification of "theologian") in 2011. Andrey Alexandrovich was ordained deacon by Metropolitan Vladimir of Rostov and Novocherkassk in 1995; he has ordained a priest on August 13, 1995; and he was appointed a senior priest of St. Nicholas parish of St. Mankovo-Berezovskaya Milyutinsky district of Rostov region. Andrey Alexandrovich became a senior priest of the Ascension Church of the village of Milyutinskaya, Milyutinsky district of the Rostov region in 1997. He has been a senior priest, Saint Michael of the Archangel parish, since February 7, 2006, in Rostov-on-Don. Andrey Alexandrovich was elevated to the rank of archpriest In 2008. On December 8, 2014, he was appointed a senior priest of the parish of the Church of Midpentecost in Rostov on-Don, as well as the spiritual guardian of the private educational institution "Svyato-Prepolovenskaya Orthodox secondary school". He became the head of the Diocesan Department of Religious Education and Catechesis in 2016. Mekushkin Andrey Alexandrovich defended his thesis and gained a degree as a Candidate of Sciences in Philosophy in 2013. In 2015, he was appointed head of the Department of Orthodox Culture and Theology of the Don State Technical University. He also was the General Director of the ANO Center for Continuing Education Cyrillic from 2014 to 2017.

Lwazi Apleni currently teaches tourism at the University of Zululand, and has published extensively in academic journals and books on carbon tax, sustainable tourism, rural tourism, and e-commerce in tourism. To date, he has contributed more than 13 journal articles in collaboration with some distinguised scholars in the field of tourism. Furthermore, his research work has been cited more than 20 times. He has presented research papers in places like Casablanca, Atlanta, Boston just to mention a few.

Eda Avcı (Vice President of IFITT Turkey Cahepter) is a lecturer in Efes Vocational School at Dokuz Eylul University since 2015. Eda started her tourism education with a vocational high school in 2000 and completed her undergraduate education in Adnan Menderes University, School of Tourism and Hotel Management, Travel Management and Tourism Guidance Program. After completed her master degree at Adnan Menderes University Institute of Social Sciences, Tourism Management Program, she received her Ph.D. at Aydın Adnan Menderes University Institute of Social Sciences, Department of Tourism Management in 2018. Her research interests lie in the area of tourist guidance, package tour management, tourism marketing, cultural tourism, faith tourism and tourism technologies. In addition to academic studies, she has been working as a professional tourist guide since 2009.

Luciana Barbieri da Rosa graduated in Business Administration - Bachelor's Degree from Faculdade de Educação São Luis - FESL (2007) and Full Degree from the Special Teacher Education Graduation Program for Professional Education - PEG / UFSM. Specialist in Environmental Education from the Federal University of Santa Maria / RS (UFSM-2009) and Specialist in Public Management from the Federal University of Santa Maria (UAB / UFSM - 2015). Master in Business Administration from the Postgraduate Program in Business Administration PPGA (UFSM - 2013). PhD in Business Administration from the Postgraduate Program in Business Administration PPGA (UFSM - 2019). Acted as a substitute professor in the Department of Administrative Sciences - UFSM (2013-2015), tutor of the Specialization in Public Management UAB / UFSM (2017-2018), Professor of the Faculty Palotina - FAPAS (2018) and tutor of the Specialization in Health Management - UAB / UNIR (2019). Operates in the areas of: Innovation, Sustainability, Strategy, Marketing and Production. Editorial Advisor to ReA - UFSM Management Magazine.

Muhammad Waseem Bari is an assistant professor at Lyallpur Business School, Government College University Faisalabad, Pakistan. Dr. Waseem earned his Ph.D. degree from the Beijing Institute of Technology, China. He has broad insight into the topics of organizational sciences and has published several papers in high-quality impact factor journals. Dr Waseem is also an editor of two books on tourism, article editor, and reviewer of several impact factor journals. He has vast experience in banking and the FMCG industries. His research areas are knowledge management, psychological contract, innovation, organizational changes, and tourism.

Cristina Castro, Master Student in Tourism Management at Polytechnic Institute of Cávado and Ave, Portugal. Researcher in Tourism and Hospitality.

Manana Darchashvili, doctor of History, professor of Georgian Technical University; researcher at Research Institute for Political Studies. She graduated from the Tbilisi State University, Faculty of History in 1985. Then she received her second higher education in Legal Studies at Tbilisi Business and Marketing Institute in 1999. Later she pursued her advanced study course at Tbilisi State University and became a Doctor of History in 2005. Research Interest: Elections and Political Parties, International Relations, Religious Factor in International Relations, Cultural Politics, Political Parties, International Organizations. She participated in 45 international conferences, is the author of - 36 scientific articles and 2 study courses, editor of 3 monographs, supervisor of 2 Ph.D. theses.

Aytekin Demircioglu was born was 1977 in Turkey. Graduate Education (1999): Gazi University, Gazi Education Faculty, Philosophy and Related Sciences, Ankara, Turkey, Master Education (2007): Gazi University, Institute of Education, Philosophy and Related Sciences, Ankara, Turkey, Doctorate (PhD) (2013): Gazi University, Institute of Education, Philosophy and Related Sciences, Ankara, Turkey, Associate Professor (2015): Istanbul University, Institute of Education, Philosophy and Related Sciences, Istanbul, Turkey.

Dinesh Dhankhar holds the position of Assistant Professor in the Department of Tourism & Hotel Management, Kurukshetra University, Kurukshetra with a doctoral degree in the subject. He has ten years of teaching experience with the number of publications in his credit.

Unathi Sonwabile Henama is a specialist in disruptive technology and teaches marketing in tourism at the Tshwane University of Technology. He has published extensively and presented paper at local and international conferences. He is a sought after public commentator in South Africa and is popularly known as the #AcademicSangoma

Aniqa Ijaz, Ph.D. scholar and Lecturer in Economics at University of Okara She joined University of Okara last year. She earned her Master of Philosophy in Economics degree from the University of the Punjab, Lahore. And Master of Science in Economics from the same department and university. Her areas of research include efficiency measurement and tourism sector.

Gizem Kayar (President of IFITT Turkey Chapter) has received her MSc and PhD degrees from Computer Science department of Freiburg University / Germany, specifically from Computer Graphics group. Later she worked as Post-doc at Massachusetts Institute of Technology for a year. Her research interests include (but not limited to) Computer Graphics, game programming, serious games and Virtual/Augmented/Mixed Reality applications in E-health and especially in e-Tourism. Dr. Kayar worked at Simsoft as a senior Software Engineer, as a faculty member at TED University Computer Engineering department for 4 years, at Yasar University for a season and currently is a faculty member at Mugla Sitki Kocman University Computer Engineering department. She has been leading the Digital Game

Development Certificate Program of TEDU CEC - TOGED since 2016, is a volunteer mentor at Turkey's Engineer Girls (TMK) and is the founding President of IFITT Turkey Chapter.

Qurrah-tul-ain Khan is an assistant professor. Currently, she is working as a faculty member in the Department of Public Administration, University of Kotli, Kotli, Azad Jammu and Kashmir. Her research interests are leadership behaviors, knowledge management, and organizational changes.

Paulo Lopes, Master Student in Tourism Management at Polytechnic Institute of Cávado and Ave, Portugal. Researcher in Tourism and Hospitality.

Maria Emília Luís, Master Student in Tourism Management at Polytechnic Institute of Cávado and Ave, Portugal. Researcher in Tourism and Hospitality.

Maria Carolina Martins-Rodrigues is an International PhD in Management, University of Extremadura, Spain Post-graduate in Knowledge management, University of Belgrano, Buenos Aires, Argentina, and a research fellow of CinTurs – Research Center for Tourism, Sustainability and Well-Being, Portugal- University of Algarve. Her research interests currently are Sustainability, innovation, Knowledge Management, Entrepreneurship, Social Responsibility and Business Management. Has co-authored over 20 articles and book chapters and published in several scientific journals. She has also organized Iberian Conference on Entrepreneurship and international Conferences of Sustainability and Innovation. Founder member and vice-president of the direction of Empreend - Portuguese Association for Entrepreneurship. Was Professor in University Aberta (UAb) and University Moderna. In TAX Administration was Tax Administration Technician Advisor acting as coordinator in the International Relations Services Directorate (DSRI). and was Head of the Tax Service.

Tamara Stanislavovna Olenich is a philosopher, researcher of processes in religion, professor of Don State Technical University. She is a Doctor of Philosophy. She is a specialist on the state of religious confessions within the post-Soviet space; she explores informal youth movements, destructive cults, and sects. She actively participates in research programs and projects, including international ones. She was born in the family of a serviceman in the town of Kizyl-Arvat (Turkmenistan). In 1993 she graduated from the Rostov State Pedagogical Institute. Since 1993 she has been working at Mariupol State University (MGI). She began her career as a laboratory assistant in the department of philological disciplines of Mariupol State University. From 1994 to 1998 she was working as an Assistant and Senior lecturer of the Department of Socio-Political Disciplines, from 1999 to 2001 – she was working as Deputy Dean of the Economics and Law Faculty. In 1996 she successfully defended her thesis and got the degree of Ph.D. in Philosophy in the North-Caucasian Scientific Centre of the Higher School in the city of Rostov-on-Don. In 1999 she gained the nostrification of the Ph.D. thesis at the Institute of Philosophy. Named after G. Skovoroda (Kiev) as well as the re-certification in Ukraine and received a Ph.D. in Philosophy. In 2003 she received the academic title of Associate Professor. June 29, 2005, she defended her thesis to obtain the degree of Doctor of Philosophy (Doctor Habilitatus) in Rostov State University. From 2005 to 2007 Tamara Olenich held the post of pro-rector for scientific work at Mariupol State University. From 2007 to 2012 she was the head of the Department of Philosophy and Sociology. In 2009, Tamara Olenich successfully passed the re-attestation process of her doctoral dissertation at Kyiv National University named after T. Shevchenko and received a diploma of Doctor of

Philosophy. In 2014 Tamara Olenich got the post of professor at the Department of Orthodox Culture and Theology of the Don State Technical University, Rostov-on-Don. In 2005, Tamara Olenich jointly with the Mariupol State University received a grant from the General Secretariat for Greek Affairs, the Greek Consulate General in Mariupol. She is the laureate of the competition for the title "Mariupolets (Mariupolchanka) 2005" (The main inhabitant of Mariupol-city) in the nomination "For a significant contribution to the development of science". In 2006 she was awarded the Certificate of Honour of the Donetsk Regional State Administration for her high professionalism and experience. In 2011, she was awarded the Certificate of Honour of the Ministry of Education and Science, Youth and Sports of Ukraine for long-term conscientious work, significant personal contribution to the training of highly qualified specialists, and fruitful scientific and pedagogical activity.

Saba Pipia works as a teaching assistant at the Institute of International Law of Ivane Jvakhishvili Tbilisi State University, Faculty of Law. He teaches International Humanitarian Law and International Criminal Law. He holds PhD degree and currently pursues his post-doc research at the same University. Previously he carried out research visits at the University of Groningen (The Netherlands), Jerusalem Institute of Justice (Israel), Peace Palace Library (The Netherlands), Aristotle University (Greece), Max Planck Institute of Comparative and Private International Law (Germany) and Max Planck Institute for Comparative Public Law and International Law (Germany). His research interests cover areas of International Humanitarian Law and International Criminal Law. He has published articles in scientific publications in Georgia and abroad.

Salome Pipia pursues her PhD studies in Conflict Resolution at Tbilisi State University. Before she was visiting researcher at the University of Iasi (Romania) under the framework of Erasmus Mundus program. Areas of her research include EU Policies towards Post-Soviet states and the role of int'l organizations in conflict resolution in Eastern Europe. Currently she is invited senior teacher in the department of Politics and International Relations at Georgian Technical University. She teaches courses such as: International Conflict Resolution, Small Countries in Int'l Relations, Art of Negotiation and the Religion in Int'l Relations. Throughout her research, she attended numerous scientific conferences in Georgia and abroad and produced several articles in Georgian.

Natalia Salnikova is a PhD in History, Associated Professor Head of the Humanities Department of the Donetsk Law Institute of the Ministry of Internal Affairs of Ukraine 1997 - 2002 Faculty of History of Mariupol Institute of Humanities of Donetsk National University. Speciality "History", Qualification "Historian, teacher of history". The contributor of more than 70 research thesis in domestic and foreign scientific journals. Research interests: history of Ukraine, regulatory and documentation management, archival institutions of Ukraine, national identity, information space and information technology, gender identity issues, migration policy.

Rizwan Shabbir is Assistant Professor in the Lyallpur Business School, Government College University, Faisalabad. He got his PhD of Marketing Management from Huazhong University of Science and Technology, China. He has earned his MS from Lund University, Sweden. His research interests include value-creation, industrial marketing, branding and cross-cultural difference in decision making.

Olesya Anatolievna Shestopalova–Todarchuk is an educational psychologist, assistant professor of Don State Technical University, Candidate of Sciences in Pedagogy; practicing psychologist, an Orthodox teacher. Olesya Anatolievna Shestopalova is engaged in psychological support for children and adolescents; also a researcher of modern childhood development, an active participant in research programs and projects. Olesya Shestopalova was born in Ed.D in Educational Psychology in Rostov-on Don. Olesya Anatolievna Shestopalova was pursuing a Ph.D. in Pedagogy at Rostov State Pedagogical University In the period from 1999 to 2002. In 2005 she successfully defended the thesis and gained a degree as Candidate of Sciences in Pedagogy in Pedagogy (13.00.08 - theory and methods of vocational education) at Rostov State Pedagogical University in Rostov-on-Don. Olesya Anatolievna Shestopalova was working as an associate professor at the Department of Education Management Southern Federal University from 2003 to 2015. In 2016 Olesya Shestopalova got the post as an assistant professor at the Department of Orthodox Culture and Theology of Don State Technical University, Rostov-on-Don.

Lakhvinder Singh is working as Assistant Professor of Tourism in Government College, Kaithal affiliated to Kurukshetra University, Kurukshetra, Haryana, India. He holds Ph.D. Degree in with UGC-NET qualification in Tourism Administration & Management. He has more than three years of teaching experience and has more than 30 research publications in his credit. He is in the Editorial/Advisory Board of different Journals of International and National repute. His areas of interest include tourist behaviour, tourism marketing, tourism impact, culture & religious tourism, e-tourism and ICT in tourism.

Bruno Sousa is a Professor in Polytechnic Institute of Cavado and Ave (IPCA, Portugal) Head of Master Program - Tourism Management - PhD Marketing and Strategy. He was Market Analist at Sonae Distribuição – Modelo e Continente, S.A. (2006 to 2009) and he was Marketing Assistant - Jornal O Jogo at Controlinveste (2005) - Teaching Award of the School of Economics and Management of the University of Minho 2015/2016 - Best Thesis in Tourism Award - ICIEMC 2015 - Management Graduation, University of Minho Award - Best performance (2006) - Merit Scholarship for Students in Public Higher Education Awards of Merit Scholarship by University of Minho in 2001/02 - 2002/03 - 2003/04 Rresearch centre: CiTUR and Applied Management Research Unit (UNIAG). He is author or co-author of several papers and her research interests include tourism management, marketing and strategy. Editorial board member of several peer reviewed scientific journals and ad-hoc reviewer of several peer-reviewed scientific journals. Member of the scientific committee of several national and international congresses and conferences.

Maria José Sousa (Ph.D. in Management) is a University Professor and a research fellow at ISCTE/ Instituto Universitário de Lisboa. Her research interests currently are public policies, information science, innovation and management issues. She is a best seller author in ICT and People Management and has co-authored over 70 articles and book chapters and published in several scientific journals (e.g. Journal of Business Research, Information Systems Frontiers, European Planning Studies, Systems Research, and Behavioral Science, Computational and Mathematical Organization Theory, Future Generation Computer Systems and others), she has also organized and peer-reviewed international conferences, and is the guest-editor of several Special Issues. She has participated in several European projects of innovation transfer and is also External Expert of COST Association - European Cooperation in Science and Technology and President of the ISO/TC 260 – Human Resources Management, representing Portugal in the International Organization for Standardization.

Kalsoom Sumra is Assistant Professor at Centre for Policy Studies COMSATS University Islamabad Pakistan and having more than 15 years of extensive experience in teaching, research and management with expertise in program development, policy development, planning, monitoring and evaluation of academic projects. A talent for analyzing problems, developing and simplifying procedures and finding innovative solutions. With proven ability to work effectively with persons from other cultures and all walks of life. Having comfortable working stamina in fast paced setting through oriented ability. Well organized and an excellent team player with a proven ability to work proactively in a complex and busy environment.

Kateryna Tryma is a leading researcher of Department of Policy and Governance in Higher Education, Institute of Higher Education, NAESU. Her responsibilities include enhancing the learning environment through instruction, applied research, and service that support the institutional mission. Prior to becoming Leading Researcher she worked as a Senior Lecturer at Mariupol State University in Mariupol. She gained broad theoretical, practical and teaching experience with specializations in the following areas: global civil society and global democracy, NGOs, e-civil organizations and information security. Kateryna got PhD in Political Sciences in 2011, the theme of thesis: "Globalization as a factor for the transformation of Ukrainian civil society (NGOs analysis)". She is the contributor of more than 30 research theses. She has 2 Masters degrees: Company Economics, Donetsk National University of Economics and Trade named after M. I.Tugan-Baranovsky, Kryvyi Rih and International Relations, Mariupol State University (2005). Scientific interests: NGOs, civil society, the European Union.

Waleska Yamakawa Zavatti Campos graduated in Pedagogy from the University of São Paulo-USP - Ribeirão Preto Faculty of Philosophy, Sciences, and Letters. Bachelor of Business Administration from the Federal University of Goiás - Faculty of Administration, Accounting and Economic Sciences. Currently, Master's degree in Administration from the Pontifical Catholic University of Rio de Janeiro/IAG Business School. Postgraduate in External Control and Public Governance (2017). Postgraduate in Public Administration (2014). Master of Business Administration in People Management (2012). Research interests: innovation, leadership, and sustainability.

Index

A

African Spirituality 283, 286-289, 296
Alentejo 44, 46, 51-52, 58
Al-Halil 25-26, 34-35, 39, 41
Anthropomorphism 66, 82
App 2, 6, 73, 76, 81-82, 132

B

Baghdad 25-26, 28, 30-32, 34-35, 39-42
Behavioral approach 149
beliefs 2-3, 12, 24, 43, 46-47, 51, 66, 68, 76, 80, 84, 114, 131, 155, 158, 163, 165, 171, 175, 189, 192, 273, 278, 286-289, 296
Belligerent Occupation 135, 148
Blog 126-127, 132-133, 184, 266-267
Bon 178, 191
Buddhism 2, 49, 52, 54, 114-115, 118-119, 123, 152, 175-181, 183-191, 194, 212, 214, 227, 231, 273, 302, 306
Buddhist Knowledge 49, 186, 191
Buddhist Practice 186, 189, 191

C

Challenges 8, 12, 23, 45, 51-52, 57, 62, 76, 81, 101-102, 111, 115, 135-137, 146-147, 159, 167-171, 194, 201, 206, 216, 227, 232-233, 235, 237-238, 241, 243, 245, 250, 254, 259-260, 262-264, 271, 278, 293, 306
Chat 38, 93, 266
Chechen Republic 11-14, 16-19, 21-24
Christianity 3, 27, 32-33, 39, 43, 114-115, 120-121, 123, 130, 133, 148, 177, 185, 194, 197, 201, 204, 212, 214, 235, 245, 247, 249, 273, 283, 288-289, 300-301, 304
Cognitive 47, 61, 64, 149-150, 152-155, 158-160, 164-166, 168, 170-172, 193, 209, 217, 223, 228, 280, 298
Collaboration 69, 152, 209, 245
Comment 92, 266
Computer Reservation Systems (CRS) 266
Conflict 14-15, 23, 81, 101, 107, 135-142, 146, 148, 156, 165, 185
Consumer behavior 47, 263
Content Marketing 252, 266
Culture 2-3, 14, 17, 19, 21-23, 30-33, 48-49, 51, 54, 64, 66, 68, 73, 77, 81, 83-84, 88-90, 92, 94-97, 99-100, 108-109, 113-114, 122, 128-129, 134, 147, 150-152, 156, 164-165, 168, 170, 175, 177-181, 183-185, 188-189, 191-194, 196, 199-202, 204, 206-207, 209, 222-223, 235-236, 248, 263-264, 268-271, 275, 285, 290, 294, 296, 298, 302-305, 308-309
Curcan 25-26, 28-29, 39, 41

D

Damascus 26, 28, 30-32, 34, 39-42
Destination,Marketing 250
Development 1-2, 5-13, 15, 17-23, 30, 44-45, 48-49, 51, 54, 57-58, 61-62, 67-68, 70, 75-76, 79-80, 84, 87, 90, 94-96, 98-104, 106-110, 112, 114, 119-122, 130, 132-133, 135, 137-138, 145, 149, 153-161, 163-172, 176-177, 179, 181, 184, 186-189, 192-194, 196, 198, 200-208, 211, 220-221, 225, 227-228, 230-235, 237-240, 245-248, 250-251, 254, 257, 262, 264-265, 271-273, 278, 282, 284-285, 291-292, 295, 302, 304, 306, 308

F

Factor 2, 12, 14-15, 18, 22, 60, 66, 84-85, 88, 95, 98-100, 103, 105, 108-110, 112, 133, 145, 178, 189, 203-205, 219, 228, 265
Faith 2-3, 24, 26, 38, 43, 49, 51, 56, 66, 68, 71, 73, 80-82, 107, 113-114, 123, 133, 142, 146, 149,

179, 181, 186, 204, 209, 216-217, 220, 222, 226, 231, 234-236, 249, 269-271, 285, 294, 296-299, 301-305, 308-309
Follower 252, 266
Free State Province 287, 296

G

Gamification 65-66, 69, 73-76, 78, 82
Geo-politics 113-114, 121-122, 130
Georgia 11, 21, 43, 135, 175, 192-194, 196-200, 202-209
Ghazali 25-35, 39-43

H

Hajj 3, 13, 17, 19, 37, 70, 72, 114, 127, 129, 167, 184-185, 214, 227, 229, 246, 279-280, 299-300, 302-304, 309
Historical Monuments 38, 192, 209
Holy Places 3-4, 6, 11-12, 19, 21, 28, 65-66, 68, 135, 139-141, 148, 151, 184, 189, 196, 209, 211-212, 234, 286, 300-305

I

ideologies 106, 110, 113-115, 120-123, 130, 158, 164, 183
Intellectual Learning Excursions 297-300, 302-305, 309
intercultural communication 83-84, 90, 95-96
International Tourists 6, 8, 284, 296
Interreligious Dialogue 194, 204-205, 208-209
IoT 65-67, 69-71, 75-76, 80, 82
Islam 2-3, 5, 11-17, 19, 22-25, 27, 29, 32-33, 36-43, 48, 85-87, 89-90, 96, 109, 112, 114-115, 119-120, 122-123, 148, 152, 177, 194-195, 199, 208, 212, 220, 227, 229, 249, 273, 299-300, 302, 304-307, 309
Islamic Cities 25, 42-43
Islamic Cities,Tus 25
Islamic Education 11-12, 15-17, 19, 24
Islamization 98, 109, 112

J

Jerusalem 2, 10, 25-26, 28, 32-34, 38-41, 43, 48, 66, 127, 129, 135-143, 146-148, 167, 170, 197-198, 204, 225, 273, 304
Jihad 11, 14-15, 24, 122

L

landscape 2, 12, 21, 66, 68, 100, 103, 110, 168, 188, 223, 225-226, 233, 235, 246, 280
Limpopo Province 286-287, 289, 292, 296
Literature Review 27, 57, 99, 137, 168, 193, 210, 212, 235, 264-265, 270, 285, 299

M

Mainline Church 249
Mainstream Tourism Industry 249
Management 9, 11, 43-45, 58-70, 75-82, 105, 110, 137, 147, 149-152, 154-156, 158-172, 174, 178, 187, 191, 193, 207, 210, 214, 224-231, 234, 246-248, 252, 254, 259, 264-266, 278, 280-281, 283, 295, 300, 305, 307-308
Mecca 2-3, 13, 19, 25-26, 28, 34-39, 41-43, 66, 70, 72, 126, 129, 196, 234-235, 237, 301-302, 304
Media 67, 96, 123, 142-143, 183, 210-212, 220-225, 229-230, 241-242, 244, 248, 250, 252-267, 272, 279, 299
Medina 10, 13, 25-26, 28, 36, 38-40, 42-43, 126, 196
Modern Society 84, 99, 132
monasteries 3, 80, 176, 179-180, 182, 185-188, 192, 282
Motivations 44-49, 59, 75-76, 167, 169, 171, 210, 214-217, 220-223, 225-226, 230, 234, 246, 278, 282, 285, 294, 299-300, 305-306, 308
Muslims 1-3, 5, 10, 15-17, 19-20, 24-25, 28, 33, 35-39, 43, 66, 85-87, 95, 101, 109-111, 114-115, 121-124, 129, 131, 134, 136, 138, 140, 145, 150, 152, 184, 192, 194-196, 204, 234, 249, 299-304
mystic experience 65
Mysticism 65-66, 73, 77, 82

N

Niche tourism 44, 67, 82, 233, 236, 285, 289, 295
Nishabur 25-26, 28-30, 39-42
North Caucasus 11-17, 19-24

P

Pakistan 1-9, 126, 149-153, 156-159, 162, 164-167, 169-172, 210-211, 222, 265, 297
Pentecostal or Charismatic Church 249
Phenomenon 10-11, 15, 46, 48-49, 51, 58, 83, 103-104, 107, 137, 143, 145, 150, 158, 170, 175, 182-183, 191, 211, 217, 219, 235, 239, 251, 253, 269, 280
Pilgrim 1-2, 10, 25-26, 28, 35-37, 50-51, 57, 61, 66, 76, 136, 140, 146, 153, 183-184, 190, 214, 216-217,

221, 227-228, 234, 243, 265, 268, 270, 301, 307
Pilgrimage 2-3, 5, 9, 11, 21, 26-27, 32-34, 36-38, 45, 48, 50, 57, 59-63, 66, 68, 70-72, 75-77, 79-82, 114, 127, 134-135, 137-138, 141-142, 147-148, 156, 168-169, 172, 175-177, 182-186, 188-192, 194, 196, 198, 200, 206, 209-212, 214-216, 224-226, 228-232, 234, 237, 244, 246-249, 251, 257, 266, 269, 272, 277-281, 285-287, 289, 292-293, 297-302, 304-309
Place Attachment 48, 57-58, 60-62, 64, 79
Place Dependence 57-58, 64
Place Identity 57-58, 64
Platform 18, 22, 45, 68, 77, 105, 242, 252, 254, 267
Political Instability 135, 137-138, 142-143, 145, 147-148
Political Party 98, 100, 112
Post 13, 21, 122, 129, 146-147, 224, 242, 248, 263, 266-267, 289, 304, 307
Promotion 6, 56, 75, 149, 159-161, 200, 202-203, 205, 216, 222, 250, 252-253, 256-257, 259, 262, 265, 271-272

Q

QR 65-66, 69, 72-73, 75-76, 81-82
QR&Mobile apps 65-66, 75
Quality of Life 20, 67, 263, 296

R

Religion 2-3, 5, 11-15, 17, 19, 22, 24, 27, 31, 43, 51-52, 54-55, 57, 60-64, 66, 68, 73, 76, 78-79, 83, 87, 90, 95, 98-100, 109-110, 112-115, 119, 121-123, 127, 129-130, 132-134, 137-138, 141, 147-149, 152-153, 155, 159, 163, 166, 168, 170, 175, 177-182, 184-185, 188-189, 191-192, 194-195, 197, 200-201, 203-204, 206, 209, 211-212, 214-215, 220, 222, 224, 230, 235-236, 243, 246-247, 249, 263, 265-266, 274-275, 279, 281, 285, 287, 297-298, 300, 302-306, 308-309
Religious Cultusrist Events 309
Religious Destinations 1-3, 5, 7, 45, 65-66, 155-156, 159-161, 166, 224, 229, 251, 256, 263, 269-270
Religious Diversity 24, 193-196, 200, 202-203, 209, 246
Religious Events 19, 57, 72, 155, 182, 236, 239-240, 243-244, 269, 274, 297-298, 300, 309
Religious Festival 234, 249
Religious Journeys 62, 82, 215, 297, 303, 309
Religious Relics 206, 209
Religious Shrine 249
Religious Tourism 1-3, 5-12, 21-22, 25, 27-32, 34-35, 39, 43-46, 48-52, 57-70, 72-73, 75-82, 114, 126-127, 129-131, 135-139, 142, 145-156, 159-160, 164-172, 175-177, 182-185, 187-189, 192-196, 198, 200-201, 203-207, 209-214, 216-217, 220-243, 245-263, 265-266, 268-275, 277-283, 285-287, 292-301, 304-308
Religious Tourism Destination 10, 78, 160, 164, 168, 250, 255, 257, 262, 292
ReligiousTourism 25

S

Sacred 1-2, 4, 8, 22, 25, 32-34, 38-39, 43, 45, 51, 54, 57, 61, 63, 68, 79, 90, 127, 135-136, 138, 140-142, 146, 148-149, 154-156, 159, 169-172, 175-176, 183-185, 188-191, 198, 201, 209, 211, 214-215, 217, 219-221, 223-225, 228, 230-231, 236, 246, 248-249, 251, 253, 263, 265, 268-278, 282, 286-288, 290-291, 294-301, 303-305, 308
Sacred Cities 25, 43
sacred geography 175-176, 188, 263
Segmentation 44, 62, 220
Share 6-7, 13-14, 27, 50, 65, 68, 87, 153, 207, 216, 221-223, 245, 249, 251-254, 256-257, 262, 267, 285
Small Party 98, 103, 112
Smart Destination 65, 67, 77-78, 82
Social Media 67, 183, 210-212, 221, 223-224, 229, 250-267
sociocultural competence 83-84
sociocultural type 175
Socio-Economic Development 1-2, 10-11, 20, 79-80
Soft Image 149, 153, 159, 161-163, 166
South Africa Thathe Vondo Lake Fundudzi Makhetha Mantsopa Cave Church SS Mendi 283
Special Interest Tourism (SIT) 285, 296
Spiritual Pilgrimage 232, 249
Spiritual Tourism 2, 44, 46-47, 49-51, 54, 57-58, 64, 68, 78, 82, 214, 249, 296, 298-299, 306, 308
Spirituality 27, 44-46, 50, 52, 54-55, 58, 60, 64, 118-119, 140, 185, 187, 211, 220, 272, 283, 287-289, 291, 293, 296, 302
stakeholders 65, 68-70, 77, 138, 150, 160-161, 166, 232, 237, 239, 244, 250-251, 253, 260-261
Stress 44, 46-47, 49-50, 117, 159, 210

T

Terrorism 20, 23, 68, 110, 122-123, 133-135, 143-148, 150, 169-170
The United Nations 136, 139-141, 147, 204-205, 208, 234, 269

Thematic Analysis 210
Tibet 175-191
Tolerance 130, 194-195, 200, 202-204, 208-209, 242, 281
Tourism 1-3, 5-13, 17-18, 20-22, 25, 27-32, 34-36, 38-39, 41-52, 54, 56-70, 72-73, 75-83, 98-99, 108-112, 114, 126-127, 129-132, 135-140, 142-143, 145-161, 164-172, 175-177, 182-196, 198-207, 209-214, 216-217, 219-240, 242-243, 245-266, 268-275, 277-287, 289-301, 304-309
Tourism Development 2, 6, 45, 98-99, 108, 110, 112, 156, 161, 168, 171, 176, 187, 192-194, 206-207, 227-228, 231, 233-235, 238-239, 245, 282, 291-292, 295, 306
Tourism religious tourism sanctuary cities mass tourism: Cultural Tourism Tourism of Events Pilgrims: Pilgrim religious: 268
Tourism Stakeholder 249
Tourists 1-2, 6-8, 12, 17, 19-21, 35, 45, 47-48, 51, 60-62, 65-68, 70, 75-78, 80, 82, 96, 108, 110, 112, 127-129, 136, 139-140, 142-143, 145, 149, 151-160, 164-167, 169-170, 175-177, 183-185, 187-188, 199, 201-203, 211, 213-217, 219-224, 226, 228-230, 237, 250-255, 257, 259, 262-264, 268-270, 272, 274-280, 284-285, 289, 291-292, 296-305
tradition 8-9, 16, 24, 52-54, 83-85, 90-92, 95-96, 109, 120-121, 123, 169, 176, 178, 182-187
Traditional 11, 13-14, 16-17, 22-24, 31, 50-51, 53, 56, 67-68, 90, 98, 100, 104-105, 107, 109, 135, 178, 183, 187, 203-204, 208, 234-236, 249, 252-253, 264, 283, 285-291, 294, 296, 302, 305-306
Traditional Indigenous African Church 249
Traditional Society 24
traditions 2, 12, 15-16, 24, 35, 54, 68, 76, 83-85, 87, 90, 95-96, 108, 121, 175, 181, 184-185, 187-189, 193, 195, 197, 199, 202, 204-205, 209, 285, 287, 296

U

Unique Experience 68, 82
United Nations 136, 139-141, 147-148, 177, 204-205, 208, 225, 234, 248, 269
Usage 69-70, 72-74, 77, 79, 88, 222, 250, 252, 256, 258-259, 261-262

V

values 12-13, 26, 47, 54, 68, 84, 87, 94, 102, 104, 106, 109, 113, 118, 120, 146, 159, 163, 165-166, 175, 177, 189, 200-201, 204, 209, 216-217, 220-222, 224-225, 227, 235-236, 254, 280, 287
Viral 242, 259, 267

W

Wahhabism 11, 13-14, 22, 24
War 14-15, 23, 31-32, 34, 38, 46, 100, 107, 115, 118-119, 122, 134-136, 138-142, 146, 148, 150, 288, 307
War,Belligerent Occupation 135
Wellness Tourism 44, 48-49, 56, 62-63, 249

Z

Zimbabwe Tourism Authority 232-233, 249
Zimbabwe tourism sector 232

www.ingramcontent.com/pod-product-compliance
Lightning Source LLC
LaVergne TN
LVHW060409300926
849162LV00006B/1386